Ecology in Agriculture

This is a volume in the

PHYSIOLOGICAL ECOLOGY series
Edited by Harold A. Mooney

A complete list of books in this series appears at the end of the volume.

Ecology in Agriculture

Edited by

Louise E. Jackson
Department of Vegetable Crops
University of California, Davis
Davis, California

Academic Press

San Diego   London   Boston   New York   Sydney   Tokyo   Toronto

Grape vines photo by Rich Knauel.

This book is printed on acid-free paper. ∞

Academic Press
a division of Harcourt Brace & Company
525 B Street, Suite 1900, San Diego, California 92101-4495, USA
http://www.apnet.com

Academic Press Limited
24-28 Oval Road, London NW1 7DX, UK
http://www.hbuk.co.uk/ap/

Library of Congress Cataloging-in-Publication Data

Ecology in agriculture / edited by Louise E. Jackson.
        p   cm. -- (Physiological ecology)
    Includes bibliographical references and index.
    ISBN 0-12-378260-0 (hardcover : alk. paper)
    1. Agricultural ecology.   I. Jackson, Louise E.   II. Series.
    S589.7.E255   1997
    630'.1'577--dc21   97-25875
                                                        CIP

PRINTED IN THE UNITED STATES OF AMERICA
97   98   99   00   01   02   MM   9   8   7   6   5   4   3   2   1

Contents

Part II
Biotic Interactions and Processes

Part III
Ecosystem Processes

Contributors

Arnold J. Bloom (145), Department of Vegetable Crops, University of California, Davis, Davis, California 95616

A. G. Condon (79), CSIRO, Division of Plant Industry and Cooperative Research Centre for Plant Sciences, Canberra, ACT 2601, Australia

David M. Eissenstat (173), Department of Horticulture, The Pennsylvania State University, University Park, Pennsylvania 16802

E. Fereres (117), Department of Agronomy, University of Cordoba, and Institute of Sustainable Agriculture (CSIC), Cordoba 14080, Spain

Maria R. Finckh (203), Institute of Plant Sciences, Swiss Federal Institute of Technology, CH-8092 Zürich, Switzerland

Eric R. Gallandt (291), Department of Applied Ecology and Environmental Sciences, University of Maine, Orono, Maine 04469

C. Gimenez (117), Department of Agronomy, University of Cordoba, Cordoba 14080, Spain

Vincent P. Gutschick (39), Department of Biology, New Mexico State University, Las Cruces, New Mexico 88003

A. E. Hall (79), Department of Botany and Plant Sciences, University of California, Riverside, Riverside, California 92521

Louise E. Jackson (3), Department of Vegetable Crops, University of California, Davis, Davis, California 95616

George W. Koch (3), Department of Biological Sciences, Northern Arizona University, Flagstaff, Arizona 86001

Rik Leemans (415), National Institute of Public Health and the Environment (RIVM), NL-3720 BA Bilthoven, The Netherlands

D. K. Letourneau (239), Department of Environmental Studies, University of California, Santa Cruz, Santa Cruz, California 95064

Matt Liebman (291), Department of Applied Ecology and Environmental Sciences, University of Maine, Orono, Maine 04469

F. Orgaz (117), Department of Agronomy, University of Cordoba, Cordoba 14080, Spain

G. Philip Robertson (347), W. K. Kellogg Biological Station and Department of Crop and Soil Sciences, Michigan State University, Hickory Corners, Michigan 49060

Kate M. Scow (367), Department of Land, Air, and Water Resources, University of California, Davis, Davis, California 95616

Martin S. Wolfe (203), Institute of Plant Sciences, Swiss Federal Institute of Technology, CH-8092 Zürich, Switzerland

Preface

An ecological viewpoint of agriculture focuses on describing the relationship between organisms and their biotic and abiotic environments and on understanding the biological processes that control the distribution, behavior, and fitness of individuals; the interactions between organisms; and the flow of materials and energy through the agricultural ecosystem. The intent of this book is to illustrate how the fundamental principles of ecology operate in agricultural settings and how they can be applied to solve practical problems in crop production and environmental management.

In the past few years, agricultural research has moved toward an increasing emphasis on long-term sustainability of food and fiber production, environmental quality, and ways to promote ecosystem health. These are complex topics. Such new research approaches require a scope broader than the traditional agronomic approach of maximizing yields with efficient use of resource inputs. Ecological principles and methodologies have been helpful in addressing these issues. It has been necessary, however, to bridge a long-standing gap between the two disciplines of ecology and agronomy to find practical solutions to such agricultural problems as minimizing inputs of limited resources, decreasing pesticide application by exploiting biological mechanisms of pest reduction, and mitigating long-term changes in soil and environmental degradation.

Ecological processes in agricultural settings can be examined at several hierarchical levels. Autecology concerns the relationships of organisms to their environment and the processes that regulate water and nutrient use, productivity, and crop yield. Community ecology is the study of interactions between organisms, including processes such as parasitism, predation, and competition involving crop plants and populations of pathogens, insect pests, and weeds. Systems ecology encompasses a broader scope of interactions and considers how biotic and abiotic processes control the flow and availability of carbon and nutrients in the ecosystem as a whole.

Ecophysiological responses of crops to the environment are a result of direct selection by breeders for traits to increase yield and yield reliability, as well as of inadvertent selection for successful adaptation to specific agricultural environments. During crop evolution, most of the differences in ecophysiology can be attributed to changes in the regulation of development and resource partitioning rather than to changes in rates of assimilation and nutrients. There are some inherent constraints on maximizing photosynthesis due to ability to acquire light, CO_2, water, and nutrients that may explain this observation. Optimization of crop growth and yield can best be achieved by considering cost–benefit analyses of resource acquisition and allocation and by studying trade-offs between whole plant integrated responses to stress in genotypes from different agricultural environments.

Community ecology in crop fields is actually very complex in time and space, despite the lack of plant diversity in most agricultural ecosystems. For example, there are many strategies by which herbivores exploit crops and by which crops defend themselves, indicating complicated evolutionary and ecological relationships that must be considered in developing crop protection methods using ecological approaches. Ecologically based management of pathogens, insect pests, and weeds tends to invoke methods that increase spatial, temporal, and community diversity and to require multiple approaches to avoid deleterious effects of other organisms on crop plants.

Ecosystem processes in agriculture require an understanding of small- and large-scale interactions between organisms and their environment and of how the integrated behavior among all biota controls the assimilation, transfer, and losses of carbon and nutrients from the ecosystem. Detailed information of this nature is relatively scarce, especially for community structure and function of soil organisms. This often necessitates lumping organisms into functional groups and recognizing that some elements of spatial and temporal diversity must be overlooked in the effort to find an appropriate scale for expressing flows and transformations of materials. A better grasp of system-level processes promises solutions to such problems as maintaining soil quality over the long term, reducing groundwater pollution due to agricultural practices, and predicting the response of agriculture to atmospheric changes and global warming.

Our goal for this book is to highlight examples of ecological approaches to studying agriculture at these three levels. We deal only with crop-based agriculture and not with animal production. It has not been possible to cover every important topic, but we hope that the book will provide insight on concepts, patterns, and methodology for examining ecology in an array of agricultural settings.

LOUISE E. JACKSON

I

Plant Responses
to the Environment

1

The Ecophysiology of Crops and Their Wild Relatives

Louise E. Jackson and George W. Koch

I. Introduction

Domestication of wild plants for human utilization involved changes in selection pressures and, consequently, in population structure and in evolution of traits associated with resource acquisition, allocation, and reproduction. Selection by crop breeders emphasizes yield, yield stability, and yield per unit resource applied, whereas natural selection operates to continue the survival of species, emphasizing traits that may be at odds with breeders' goals. Criteria for fitness change dramatically under the two selection regimes. In wild populations, seed or propagule production and dissemination are the most important attributes that define the fittest individual (i.e., the individual that leaves the greatest number of descendants). With agricultural selection, production of harvestable yield is the prime consideration for fitness. Even when the seed is the harvestable unit, yield rather than survivorship in the next generation is the most important factor in selection. The two selection processes have fundamentally different implications for plant growth and allocation of resources.

Ecophysiological differences between crops and their wild relatives depend on the taxa compared. In many cases, crops are completely interfertile with their putative progenitors [e.g., cultivated and wild lettuce, *Lactuca sativa* and *L. serriola*, which should be considered conspecific (Kesseli *et al.*, 1991)] despite large morphological and genetic differences. Reproductive isolation is maintained between these plants by human cultural practices, and they are best viewed as populations of a single species subject to diversifying and directional selection. In some cases, selection has advanced so far that the wild progenitor of a crop is not known (e.g., maize until recently), and hybridization with close relatives results in sterile or no offspring. These examples yield less useful information on ecophysiological changes during crop evolution because comparison with immediate progenitors is impossible and the closest extant relative may not share a common environment of origin.

The purpose of this review is to examine the ecological origins of crop plants and to assess the impact of domestication and subsequent agricultural selection on physiological processes and ecologically important traits including rates of photosynthesis and nutrient acquisition, resource allocation patterns, and phenological schedules. The focus is on crops derived from Mediterranean annual species, due to the availability of information for these plants. From an ecological point of view, comparison of crops and wild relatives expands our understanding of evolution and adaptation to the environment when plants of similar genetic backgrounds are subject to differences in selection pressures. From a practical perspective, such information helps to identify successful strategies for purposeful selection of crops growing under various management regimes and indicates possible pitfalls in direct selection for physiological traits in breeding programs.

We first discuss the process of domestication of plants, reviewing what is known about the origin of crops as wild species and the various pathways of genotypic and phenotypic change leading to early cultivars. Next, we review the strategies of selection for crop improvement, emphasizing the trade-offs between selection for maximal yields under high-input conditions and relatively stable yield across a range of environments. Several case studies of the comparative ecophysiology of crops and their wild relatives serve to illustrate the idea that physiological features have been generally conserved during domestication and selection, whereas changes in morphological characteristics have been more amenable to change and have contributed to much of the improvement in yield. We conclude by speculating about the potential for extant wild germplasm to contribute to further improvements in crops in the face of changing demands on modern agricultural ecosystems.

II. Domestication

A. Crop Origins

The progenitors of many of the crops grown in the temperate, northern hemisphere were ruderal species native to frequently disturbed areas in the Mediterranean Basin. Examples include wheat, barley, oat, sunflower, radish, and lettuce (Smartt and Simmonds, 1995). Many ruderal species share attributes such as (1) high rates of resource acquisition and vegetative growth, (2) allocation patterns that respond flexibly to changing resource availability, (3) low levels of defensive chemicals, (4) early onset of reproduction, (5) self-compatibility, (6) high seed production, (7) adaptations for short- and long-distance dispersal, and (8) variable seed dormancy (Baker, 1974; Chapin, 1980). Some of these attributes are obviously advantageous for crops in cultivated environments (termed preadaptations), whereas others, such as variable seed dormancy, are clearly problematic for maximizing harvestable yield. In general, early successional species have a greater breadth of response for such parameters as germination, survivorship, and growth to gradients of resource availability than do species typical of later successional stages (Bazzaz, 1987). Large niche breadth means that these species can establish and survive in unpredictable environments such as those that occur in disturbed or sparsely occupied sites.

B. Pathways of Domestication

A broad definition of plant domestication is given as "human intervention in the reproductive system of the plant, resulting in genetic and/or phenotypic modification" and domestication of wild plants is thought to have occurred through three main types of pathways (Harris and Hillman, 1989, as cited in Frankel *et al.,* 1995): (1) "plastic" phenotypic change from somatic genetic change, involving vegetative propagation; e.g., root and tuber crops such as yam, or cross-pollinated fruit species; (2) gradual genotypic change, involving outbreeding taxa, e.g., beets, carrots, and other vegetables; and (3) rapid genotypic change, involving inbreeding taxa; e.g., cereal and legume domestication in the Near East.

The first pathway of domestication concerns clonal propagation of perennial plants as a rapid means to domesticate native plants with desirable characteristics. Reliance on vegetatively propagated plants, such as yams, sweet potatoes, and bananas, was common in indigenous agricultural systems in the warm, humid tropics possibly due to the higher success of growth and establishment of plants from buds, rhizomes, and stems in this type of environment. In temperate fruit crops that are cross-pollinated, such as olive, figs, and apples, vegetative propagation and grafting were solutions to the problems of wide genetic segregation with sexual reproduction.

A slower process of domestication occurred for cross-pollinated plants that were only capable of reproducing by seed. Populations of plants with an outbreeding reproductive system are composed of many heterozygous genotypes, which may be different from one another (Hamrick and Godt, 1989). Cross-pollination results in a new assembly of genotypes in each generation, but each of these assemblies will be on average similar from one generation to the next. Breeding of cross-pollinated crops can be done by increasing the frequency of those genes in the gene pool that produce a higher frequency of superior genotypes, but this can be a slow process that requires mass selection of bulked seed from several superior plants (Briggs and Knowles, 1967). In outcrossing plants, selection for homozygotes typically leads to loss of vigor. Control of cross-pollination is a recent development to create inbred lines for producing vigorous hybrids.

In taxa subject to the third pathway of domestication, the inbreeding reproductive system would have hastened the selection process for successful domestication. Self-pollinating plants breed true, facilitating perpetuation of superior genotypes and retention of preadaptive traits that were conducive to cultivation (Hawkes, 1983; Zohary and Hopf, 1988; Evans, 1993). Low levels of outcrossing [e.g., typically <1% in Near Eastern wild wheats (Hillman and Davies, 1990)] would have supplied genetic variation for further selection. A trait in the cereals and grain legumes in the Near East (e.g., einkorn and emmer wheat, barley, peas, lentils, chickpeas, and bitter vetch) that could be considered preadaptive is the production of seeds with large, edible storage reserves. It apparently evolved as a mechanism for these annual plants to grow rapidly during the sporadic rains at the end of the long summer drought (Zohary and Hopf, 1988). The predominantly inbreeding mating system of these crops would have facilitated the selection of new, desirable traits, such as more uniform germination, and in the annual cereal crops, a nonbrittle rachis that prevented shattering of the inflorescence and scattering of the seed with ripening. In this case, grains were probably harvested at unripe stages until homozygous recessive mutants with the semitough rachis trait were discovered (Hillman and Davies, 1990). Thereafter, fixation for this trait could have theoretically occurred in 20–30 years, greatly enhancing recovery of grain.

Another aspect of inbreeding that affected domestication is the possibility of local genetic differentiation in response to variation in natural habitats (Heywood, 1991). For example, high site-to-site environmental variability would be expected given the wide span of ecological habitats in which the progenitors of cereal grains are hypothesized to have been distributed (Harlan and Zohary, 1966). In Mediterranean annual species, genetically distinct ecotypes occupy specific niches so that there is a close association between certain genotypes and microhabitats. For example, in *Avena barbata* in Israel, where it is endemic, and in California, where it was introduced 300 years ago, distinct phenotypes and combinations of allozymes occur in "xeric" and "mesic" habitats (Hamrick and Allard, 1972, 1975; Kahler *et al.*, 1980). Research in Israel has shown that wild wheat (*Triticum turgidum* ssp. *dicoccoides*) shows high genetic diversity between and within populations, indicating a capacity for microsite specificity and responsiveness to the high interannual variation in weather (Nevo *et al.*, 1982, 1991; Carver and Nevo,

1990). This situation would have provided variation from which early plant breeders could choose, allowing selection of genotypes suitable for a wide range of agricultural settlements.

In early selections of self-pollinating large-grained annual grasses for cultivation, humans had to deal with growth patterns and genetic population structures that ensured success in a variable climatic regime. For example, human selection increased the uniformity of seed production within single plants and within populations by increasing genetic similarity. There may also have been selection for reduced phenotypic plasticity, which would be expected to be high initially in ruderal species adapted to temporally and spatially heterogeneous environments (Grime, 1994). In fact, phenotypic plasticity in traits related to productivity, such as number of tillers and shoot dry weight, is very high in wild wheat, *T. turgidum* ssp. *dicoccoides* (Anikster *et al.,* 1991). Prolonged flowering and a protracted reproductive phase are likely to increase the variability in allocation to vegetative and reproductive structures among plants in a population, especially if short-term stresses occur intermittently during the life cycle (Chiariello and Gulmon, 1991). In wild annual grasses from the Mediterranean Basin, such as *Bromus mollis,* translocation of carbon into the developing inflorescences and senescence of vegetative tissue occur gradually so that all stored reserves are not immediately invested in seeds during early developmental stages of inflorescences that may or may not make it to maturity (Jackson *et al.,* 1987). In wild barleygrasses (*Hordeum* spp.), new inflorescences are still emerging while seeds are dehiscing from older inflorescences (Chapin *et al.,* 1989). In wild plants, selection for asynchronous development of reproductive structures on a single plant would be more likely than uniform maturation to assure production of offspring under variable conditions, but uniform maturation is a valuable attribute for cultivated production.

Another example of changes in growth patterns with domestication is in the timing of vegetative and reproductive allocation. Annual plants in environments with predictable periods of stress, such as the summer drought in regions with Mediterranean-type climates, typically show a fairly sharp transition from vegetative to reproductive growth (Chiariello and Gulmon, 1991). With the removal of the end of season stress by provision of nutrients and water, we might expect selection to favor genotypes that retain the ability to continue vegetative growth after the onset of reproduction, thereby prolonging the period of resource acquisition and enhancing seed output.

In outbreeding species, archaeological evidence indicates that the domestication process was much slower. For example, the earliest cultivated maize is essentially genetically identical to the oldest accession of wild maize, which occurred nearly 2000 years earlier (Doebley, 1994). Heterozygosity and variability do result from cross-pollination, causing a new assemblage of gene recombinations to occur in each generation, but on the average these assemblages will be similar from generation to generation, as long as genetic material is not lost from or introduced into the population (Briggs and Knowles, 1967). Outcrossing increases genetic heterogeneity in and among populations (Hamrick and Godt, 1989), and most populations of outcrossing plants show relatively low levels of microspatial genetic structuring, which limits the ability for humans to select potentially desirable genotypes from natural plant communities.

III. Environmental Considerations and Agricultural Selection Strategies

For many crops, there has been a gradual change from their use in subsistence farming to intensive large-scale commercial farming operations. This progression has generally, although not always, been accompanied by increased inputs of water, fertilizer, and pesticides to reduce the risk of crop failure. Plant resource availability was initially greater in early agricultural sites than in the ecosystems from which crop progenitors originated because these sites were located in more resource-rich areas such as floodplains and lake margins (Delcourt, 1987). Light availability would have increased due to removal of competitors, although planting densities may have had an opposing influence. As inputs have continued to increase up to the present day, yields have increased, but so has the annual variability in yield; this can be attributed to several factors including greater responsiveness of modern varieties to resource inputs and increased susceptibility and impact of pest outbreaks on more genetically uniform plantings that is caused by widespread adoption of a few varieties (Anderson and Hazell, 1989).

The advent of agriculture also involved changes in the ecological setting of cultivated species because they were planted in stands or had their interspecific competitors removed. In ancient agricultural sites in Europe and North America, the pollen record shows that forest trees declined while ruderal species increased, and that perennial ruderal species such as grasses and plantains decreased after the development of the plow (Delcourt, 1987). Reduced interspecific competition favored the spread of annual crops and more frequent crop rotation. Forest clearing and transition of shrublands to pastures by grazing also permitted the establishment of horticultural trees from beyond their native ranges. For example, the distribution of pistachios and olives extended throughout southern Europe after 3000 BP, away from their native habitats in coastal areas along the Mediterranean Sea (Delcourt, 1987). Thus, human land use promoted changes in range expansion of early horticultural taxa.

Although resource availability to crop plants has generally increased through time, in situations in which low-input, subsistence agriculture has persisted, landraces have been systematically selected for performance in adverse and variable environments. [A landrace is defined as a set of populations (or clones) of a crop species with distinct morphological characteristics that are maintained by farmers, who recognize them as all belonging to the same entity (Harlan, 1975)]. Under low-yielding conditions, yields are often highly variable, and so the main breeding objective is yield stability or reliability to minimize the likelihood of crop failures (Ceccarelli *et al.,* 1991). For example, in Bavaria, an old landrace of barley outyielded a modern cultivar in low-input farming systems because its longer straw provided better competition against weeds, but with present-day farming methods, the yield of the modern cultivar was much higher than the landrace (Fischbeck, 1989). In all farming systems, the Bavarian landrace had a lower coefficient of variation for yield. In another example, a landrace of maize in Malawi was less responsive to environmental conditions (as measured by an environmental index equal to the average yield of all treatments at each location) than a modern variety, but the landrace performed better on the poorer sites (Hildebrand, 1984). Similar results were observed between barley selections made at high-yielding (i.e., expected yields of 3–5 t ha^{-1}, average rainfall of 332 mm) and low-yielding (i.e., expected yields of 1 t ha^{-1}, average rainfall of 201 mm) environments in Syria without supplemental irrigation (Ceccarelli and Grando, 1991). Genotypes developed through repeated selection in low-yielding environments had a higher probability of producing high yields under low-yielding conditions than genotypes selected under high-yielding environments (Fig. 1). Because they produce poorly under high-yielding environments, their performance was more similar across a range of environmental conditions. Landraces of barley from low-yielding environments in Syria tend to have a prostrate habit, cold tolerance, a short grain filling period, low early vigor, and late heading. These attributes promote successful establishment and growth during the cold winter and into the drying spring growing season so that good yields are produced in most years, but would presumably constrain growth in warmer, moister, and more fertile conditions, in which a longer period of reproductive growth would be advantageous (Ceccarelli *et al.,* 1991). These results suggest that one avenue toward yield reliability is a growth cycle that assures a moderate amount of productivity under low-yielding conditions but that inherently limits the ability to produce well under more favorable conditions. Another avenue is that these landraces are genetically heterogeneous and are composed of an assemblage of genotypes, each with a different combination of traits that are likely to be advantageous for specific combinations of environmental stresses that occur in this region of highly variable climatic conditions (Ceccarelli, 1994).

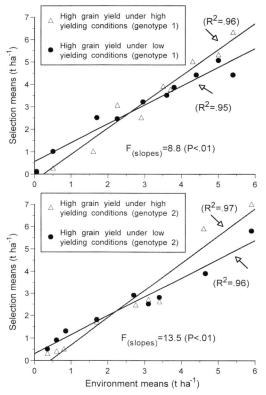

Figure 1 Linear regression of environmental means and yields of two sets of varieties: genotypes selected either for high grain yield in high-yielding conditions or high grain yield under low-yielding conditions. *y* axis values are the means of the selection groups and *x* axis values are an environmental index equal to the mean yield of all treatments at each location and time. Top, 332 genotypes; bottom, 234 genotypes. From Ceccarelli and Grando (1991). Selection environment and environmental sensitivity in barley. *Euphytica* **57**, 157–167, with kind permission from Kluwer Academic Publishers.

Detailed ecophysiological studies have not been made of landraces from subsistence farming systems. It is possible that they may share some characteristics of the "stress resistant syndrome" described in Chapin *et al.* (1993), such as low relative growth rate, low capacities for resource acquisition, and low rates of tissue turnover, while still remaining somewhat sensitive to the variation in resource availability associated with low-intensity cultivation practices (e.g., low, unpredictable fertility and water availability).

Along with a conservative growth strategy, the population of diverse genotypes may serve as another form of insurance against environmental variability, and considerable diversity exists in some landraces (Ceccarelli *et al.*, 1991, 1994). In Syria, a region of highly variable climatic conditions, landraces of barley are genetically distinct and they are composed of assemblages of genotypes, each with a different combination of traits that is likely to be advantageous for a specific combination of environmental stresses. Selection for genetic heterogeneity assures reliability by making sure that at least one genotype is successful in a given time and place. For example, in Rwanda, where beans are the main staple of the diet, households grow varietal mixtures of typically 6–10, but up to 25 components, changing the composition of the mixtures to accommodate different soil conditions, crop systems, and weather (Sperling and Berkowitz, 1994; Sperling, 1995). Use of these mixtures stabilizes productivity. Interestingly, these farmers have unique and accurate ability to evaluate new genotypes and were more competent than breeders in making selections that outperformed existing varieties.

Breeders have selected most modern cultivars under conditions of high fertility and water availability. Modern cultivars obviously perform well under these conditions and are very responsive to added inputs of fertilizer and water (Evans, 1993). Genetic improvements are responsible for a high fraction of the large increases in crop yield during the past century (Fehr, 1984). In wheat, for example, a tremendous increase in production has occurred during this century. During the first 50 years, production doubled due to increased land area, but thereafter wheat production increased 250% in only 40 years, with little increase in land area, and genetic improvement is estimated to contribute about 50% of this increase in grain yield compared with other factors including management practices (Slafer *et al.*, 1994). Although most efforts toward genetic improvement have targeted high-input systems, some modern varieties outperform older varieties even under lower-input conditions (Evans, 1993). Among 12 winter wheat varieties released between 1908 and 1978, the modern, highest yielding dwarf varieties produced 40% higher grain yields under high (104 kg N ha^{-1}) and low nitrogen (38 kg N ha^{-1}) than the oldest variety, which had greater stem weights and later anthesis (Austin *et al.*, 1980). Total aboveground biomass was similar in all varieties so that the increase in grain yield in dwarf varieties was mainly due to higher harvest index. This type of result seems to suggest that "Green Revolution" genotypes are well suited to an enormous range of growing conditions and apparently contradicts the results of comparisons presented previously (Hildebrand, 1984; Ceccarelli *et al.*, 1991) in which genotypes selected under low-yielding conditions produce higher yields under poor growing conditions. It should be noted, however, that the "low" N treatment in experiments such as those of Austin *et al.* (1980) may still provide many times more available nitrogen than occurs under actual farmers' fields in low-productivity environments and in this respect may not be an adequate test of plant response to an infertile soil.

An important distinction must be made between environmental conditions in subsistence and modern commercial farming systems. For example, the yields in the Syrian study (Ceccarelli *et al.,* 1991), even in the "high-yielding" environment, were considerably lower than the 8 or 9 t ha^{-1} yields that occur in North American barley production (Boukerrou and Rasmussen, 1990; Jedel and Helm, 1994). To fully evaluate yield reliability, breeding programs should include very low-yielding environments (Evans, 1993). Information on site-specific environmental conditions (e.g., soil fertility, soil moisture, and crop evapotranspiration) is rare in most breeding trials but would be useful for evaluating the severity of environmental conditions. Moreover, if local adaptation to both high- and low-productivity environments is a goal of a breeding program, then it is probably most advantageous to create two separate selection programs. Apparently, few breeding programs have purposely selected for performance in poor environments (Simmonds, 1991).

IV. Comparison of Crops and Their Wild Relatives

The history of crop domestication and selective breeding has apparently produced much greater change in the regulation of development and partitioning than in rates of primary assimilation (Evans, 1993). Features such as tissue-specific rates of photosynthesis, nutrient uptake, or respiration have undergone relatively little change. Instead, increased allocation of acquired resources to the harvested portions of crops has driven yield improvements. It should be noted, however, that physiological comparisons of crops and their progenitors have focused on only a few species and generally have not assessed physiological performance under a wide range of growth and measurement conditions. Thus, it is possible that selection for specific physiological traits has not occurred and there may be a potential for introducing useful physiological features from the wild relatives of crop species (see Section V).

Three examples illustrate the point that breeding has resulted in large changes in developmental and allocation patterns, but that physiological traits have changed much less and have had a lower impact on the selection of successful crop plants: photosynthesis, growth, and allocation in wheats; N acquisition and utilization in barleys; and water relations in sunflowers. In each example, we start by explaining evolutionary relationships between wild and cultivated taxa, then discuss differences in physiological characteristics between wild and cultivated taxa and between old and modern cultivars, and end with a discussion of differences in yield, allocation patterns, and phenology. These examples purposefully explore the relationships between closely related wild and cultivated plants that in most cases are interfertile. Other situations exist in which distantly related wild species have been used as sources of germplasm for physiological traits in crops, e.g., salinity tolerance in wheat (Dvorak and Ross, 1986) and tomato (Tal and Shannon, 1983). In these cases, the physiology of the wild species is very distinct from the cultivated species, but these differences do not reflect a direct evolutionary pathway and therefore such examples are not included here.

A. Photosynthesis, Growth, and Allocation in Wild and Cultivated Wheats

Wheat is the crop for which differences in photosynthesis and allocation patterns have been most thoroughly compared between wild and cultivated taxa and between old and modern cultivars. The main cultivated wheats are members of a polyploid series of species (see van Slageren, 1994, for a taxonomic review). There are two wild diploid taxa ($2n = 2x = 14$, genomes AA), *Triticum monococcum* ssp. *aegilopoides* and *Triticum urartu*. Einkorn wheat (*T. monococcum* ssp. *monococcum*) is the diploid domesticated crop derived from *T. monococcum* ssp. *aegilopoides*. At the tetraploid level, the wild taxon is *T. turgidum* ssp. *dicoccoides,* wild emmer ($2n = 4x = 28$, genomes AABB). The cultivated taxa at the tetraploid level consist of cultivated emmer (*T. turgidum* ssp. *dicoccon*) and several other subspecies of *T. turgidum* (macaroni or durum wheats and Polish and Persian wheats). The tetraploid taxa are of hybrid origin, involving *T. urartu* as the A genome parent and a still unidentified B genome donor that if not extant, is most genetically similar to the species of the *Sitopsis* section of the genus *Aegilops*. Finally, at the hexaploid level, there are no wild taxa, nor is there any evidence of any in the past. The cultivated taxa comprise the species *T. aestivum* ($2n = 6x = 42$, genomes AABBDD), including the bread or common wheats and spelta, compact, macha, and others. The species is of hybrid origin. The AB genome donor was probably a cultivated emmer and the D genome donor was the wild species *Aegilops tauschii* ($2n = 2x = 14$, genomes DD). Cultivars at all levels, diploid through hexaploid, are under production today. However, production of the primitive cultivars, einkorn, emmer, spelta, macha, etc., is very limited. Plant breeding is very active in many countries for durum wheat and bread wheat.

The wild progenitors of wheat occupy a wide span of habitats in the Mediterranean region that tend to be ruderal areas, open pastures, or open woodlands. Growing seasons are short, variable, and often water limited. Einkorn wheats occur as weedy plants along roadsides and in wheat fields but also grow in dense stands in noncultivated areas (Harlan and Zohary, 1966). *Triticum urartu,* in contrast, is found in localized areas of basaltic soils in the Near East (Waines *et al.,* 1993). Wheat was probably domesticated at the same time as barley and, along with barley, served as the major cereal for the first permanently occupied villages in the Near East. Large genetic variation in *T. turgidum* ssp. *dicoccoides* still occurs within populations and between geographic areas in Israel (Nevo *et al.,* 1982). Landraces of wheat are still in production in low-resource environments in the subtropics and tropics. In Turkey, for example, modern, improved cultivars are prevalent in irrigated or valley bottom areas, but many farmers still grow both landraces and modern cultivars in marginal areas (Brush, 1995). In temperate regions, production of modern wheat cultivars generally occurs in rainfed conditions, and planting often occurs in the fall or winter in summer–dry environments so that the crop experiences suboptimal environmental conditions sometime during the growing season (e.g., light or temperature).

Net photosynthetic rates per unit leaf area are lower for bread and durum wheats than for their diploid ancestors and relatives. Evans and Dunstone (1970) grew populations of eight wild and cultivated wheat taxa (*T. monococcum* ssp. *aegilopoides, T. monococcum* ssp. *monococcum, Aegilops speltoides, Ae. tauschii, T. turgidum* ssp. *dicoccoides, T. turgidum* ssp. *dicoccon, T. turgidum* ssp. *durum, T. aestivum* ssp. *spelta,* and *T. aestivum* ssp. *aestivum* corresponding, respectively, to the following species names given by Evans and Dunstone as *Triticum boeoticum, T. monococcum, Ae. speltoides, Ae. tauschii, T. dicoccoides, T. dicoccum, T. durum, T. spelta,* and *T. aestivum*) in the greenhouse and measured net photosynthetic rates of the flag and pentultimate leaves by gas exchange and carbon partitioning with $^{14}CO_2$. The diploid species had the highest rates (Fig. 2), which were approximately twice as high as those

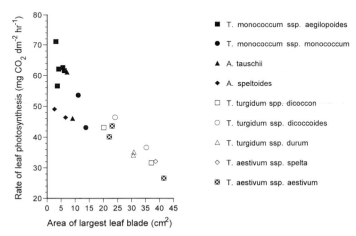

Figure 2   Leaf area and net photosynthetic rate per unit leaf area of the flag leaves of wild and cultivated wheats. From Evans and Dunstone (1970).

of the cultivated taxa when grown at high light (approximately 1500 μmol photons m^{-2} sec^{-1}). The D genome in the hexaploid wheats was associated with the lowest net photosynthetic rates per unit leaf area. In another study, differences in net photosynthetic rates among *Triticum* genotypes were less pronounced at tillering, but hexaploid wheats did have significantly lower rates than diploid and tetraploid wheats (Johnson *et al.*, 1987). Photosynthetic characteristics in the flag leaf appear to be more divergent among wheat taxa than in lower leaves, and this could be related to differences in canopy structure and reproductive sink demand in wild and cultivated taxa. Diploid wheats (Evans and Dunstone, 1970) produce more tillers, have slower rates of grain growth, and produce lower dry weight per ear and lower grain size than tetraploid and hexaploid wheats.

Flag leaves of cultivated wheat are larger than in wild taxa and for this reason net photosynthesis per leaf blade is higher in cultivars despite their lower area-specific rates (Evans and Dunstone, 1970). Flag leaves of cultivars also remain active well into the period of grain development, whereas in wild taxa they begin to senesce prior to completion of grain fill. It is unclear whether this difference in flag leaf senescence is genetically based or dependent on resource availability, but in either case it has a large impact on seed production because seed yield in wheat is strongly related to the duration of flag leaf display during grain filling and to the integrated photosynthesis over the life of the flag leaf (Stoy, 1965; Rawson *et al.*, 1983). The importance of flag leaf assimilation is clearly seen in studies in which it has been labeled with $^{14}CO_2$; labeling 2 weeks after anthesis shows a greater quantity and proportion of fixed carbon translocated to the ear in cultivated compared to wild wheats (Evans and Dunstone, 1970).

Apparently, linked structural and physiological changes have occurred during the development of modern wheat cultivars. From early diploid and tetraploid wild progenitors to modern cultivated hexaploids, there has been an increase in mesophyll cell size and leaf area, and decreased leaf thickness, with the result that total mesophyll cell surface area has decreased (Dunstone and Evans, 1974). The same authors suggested that the decreased surface to volume ratio of larger mesophyll cells increases mesophyll resistance to CO_2 and may partially explain the lower rates of photosynthesis per unit leaf area in the hexaploid cultivated taxa. Recent studies indicate that, in fact, the cultivated hexaploid taxa have lower N content per unit leaf area, in conjunction with lower leaf mass per unit area (g m^{-2}) and that this, not increased mesophyll resistance, explains the reduced photosynthesis of cultivated taxa (Singh and Tsunoda, 1978; Evans, 1993). LeCain *et al.* (1989) and Morgan *et al.* (1990) compared leaf anatomy and gas exchange in nearly isogenic semidwarf and tall winter wheat lines. Consistent with the aforementioned studies, these authors concluded that the higher rates of net photosynthesis in the semidwarf line were due to greater mesophyll photosynthetic capacity rather than reduced mesophyll resistance. Thus, it appears that between progenitors and modern cultivars, and within cultivars, most variation in photosynthetic capacity is a function of the amount of photosynthetic machinery per unit area or mass.

Consistent with a reduction in the amount of photosynthetic machinery per unit leaf area, light saturation is reached at lower light intensities in the cultivated taxa (Evans and Dunstone, 1970). A trade-off exists between maximizing photosynthetic capacity of individual leaves, which requires high concentrations of resources in the leaf, and maximizing light capture by the canopy, which can be achieved by reducing leaf mass (and thus photosynthetic machinery) per unit area and increasing total leaf area. Maximizing light capture through increased leaf area generally appears to advantageous in terms of wheat yield, although it may make the crop more vulnerable to drought and high temperature stress associated with stomatal closure (see Chapter 2, this volume).

In the higher light conditions associated with lower cover in natural ecosystems, the high net photosynthetic rates per unit leaf area and small leaves of the wild taxa would be expected to confer rapid carbon gain. Smaller leaves are also advantageous for maintaining low leaf temperatures and for minimizing N and C loss when environmental conditions promote early senescence. Leaves of the wild species *T. monoccocum* ssp. *aegilopoides* are pubescent, have higher stomatal density, and have more deeply recessed stomata on the upper leaf surface than a wheat cultivar; these features maintain a favorable water balance under limited soil moisture and high VPD. Also, the wild species has a higher rate of photosynthesis across a range of soil moisture conditions than cultivated taxa (Singh and Tsunoda, 1978). High photosynthetic rates per unit leaf area would allow rapid C gain during the short periods when optimal environmental conditions do occur and may be especially important given the annual variability in moisture and temperature regimes. In fact, native populations of *T. turgidum* ssp. *dicoccoides,* wild emmer wheat, show substantial variation in photosynthetic characteristics among and within geographic regions in Israel, indicating ecological adaptation to specific environmental regimes in the region; in controlled environment studies, the highest net photosynthetic rates per unit leaf area and per mole chlorophyll were found in accessions from the margins of the most xeric areas rather than more humid or mesic areas (Carver and Nevo, 1990; Nevo *et al.,* 1991).

For many years, researchers in this field have been faced with the apparent paradox that harvestable yield in cultivated taxa has typically not been correlated with intraspecific, heritable differences in net maximum photosynthetic rate per unit leaf area (P_{max}) (Evans, 1993). The reasons for the lack of correlation are not well understood. Austin (1989) has speculated that the basis for this apparent paradox may be that (1) microclimate may preclude the benefits from high P_{max} because of low resource availability, e.g., in rainfed systems, low light levels early in the season and water stress later prevent the utilization of high photosynthetic capacity; (2) pleiotropic effects may mean that high P_{max} is associated with smaller or fewer leaves, which would reduce total light interception, as would higher leaf mass to area ratio for the same total leaf mass; (3) similarly, there may be close linkages between genes for high P_{max} and other undesirable genes; and (4) sink limitations could also confound the potential benefits of high P_{max}—CO_2 fixation would be inhibited unless sufficient demand for photosynthate exists elsewhere in the plant and organic phosphate is returned to the chloroplast for continued synthesis of ATP.

Gutschick (1984a,b, 1987) suggested that the lower leaf mass per unit leaf area and thicker leaves in cultivated wheats, as observed by Singh and Tsunoda (1978) and Dunstone and Evans (1974), lead to lower photosynthetic rate per unit area but higher photosynthetic rate per unit leaf mass. Data on rates of photosynthesis per unit mass for the various wheat taxa are not available, but comparisons among taxa may depend on the basis of expression (mass or area). Photosynthetic rate per unit leaf mass is likely a more useful basis for assessing costs and benefits of resource partitioning to leaves.

Constraints on the relationship between seed size and growth-related traits were the basis of a hypothesis to explain the tendency for lower photosynthetic rates per unit leaf area in cultivated plants (Chapin *et al.*, 1989, 1993). A comparison of wild and cultivated barleys suggested that agricultural selection for large seed size has led to large seedlings with large leaves and a high ratio of photosynthetic leaf area to mass of meristematic sinks (i.e., a high source to sink ratio). This relative sink limitation during early growth may downregulate photosynthesis, both phenotypically and through genetic selection. Selection for lower photosynthetic rates is also consistent with greater self-shading in cultivated taxa (see above). These apparent causal relationships between seed size, photosynthetic source to sink ratio, and canopy architecture are thought to have resulted in lower relative growth rates in cultivated barley (Chapin *et al.*, 1989, 1993).

Some recent studies have shown positive associations between net photosynthetic rates and increased yield, contrary to the general lack of correlation of these traits that has been described previously. In the high-radiation environment of Israel, a high-yielding modern cultivar of spring bread wheat does have a higher photosynthetic capacity than older cultivars, indicating that genetic improvement has occurred in both net photosynthesis per unit leaf area and grain yield (Blum, 1990). A survey of Australian bread wheats showed higher net photosynthetic rate, leaf N, and chlorophyll in more recent cultivars, changes that are associated with the introduction of the *Rht* dwarfing genes into these semidwarf wheats, allowing higher fertilizer application without risk of lodging (Watanabe *et al.*, 1994).

Respiration accounts for a large fraction of the carbon budget of wheat. Amthor (1989) cites a number of studies in which nearly as much carbon was lost in respiration as was accumulated in biomass over the growing season. In wheat after anthesis, respiratory losses equal 40% of daytime net photosynthesis (Denmead, 1976). Total shoot respiration during grain filling may amount for as much as 75% of final grain yield (Pearman *et al.*, 1981). Increased efficiency of respiration could significantly increase crop productivity, but little is known of taxonomic variation in respiration or its components among wheat cultivars and their wild relatives. Evans (1993) suggested that the efficiency of maintenance processes is more variable than that of primary biosynthetic pathways, which may have little room for efficiency gains. Among winter rye, triticale, and wheat, there was significant variation in specific dark respiration rates of leaves (rates for triticale and rye were less than those for wheat in two temperature regimes) and a strong negative correlation between leaf respiration rate and crop biomass accumulation (Winzeler *et al.*, 1989). The authors concluded that low respiratory losses from maintenance processes might make available more soluble carbohydrate for delivery to growing organs.

Maintenance processes that might be subject to selection include rates of protein turnover, regulation of concentration gradients, and adjustment of membrane composition. Because these processes may be involved in acclimation to changing environmental conditions, selection for reduced maintenance respiration may result in decreased capacity to adjust to stress (Evans, 1993). A widely touted apparent inefficiency of respiration is the so-called alternative pathway of the mitochondrial electron transport chain, the activity of which represents a loss of energy conservation (ATP production) and so might lead to a reduction of plant growth (Amthor, 1989). In wheat, alternative pathway activity has been found to vary with tissue type, age, nutrition and water stress, and time during the day/night cycle (summarized in Amthor, 1989). Crop production improvements based on selection for reduced alternative pathway engagement seem sensible, but much remains to be learned about the regulation, adaptive significance, and proper measurement of the alternative pathway.

Numerous studies have shown that total aboveground biomass has not changed during the course of wheat breeding (Slafer *et al.*, 1994). Also, most studies indicate that canopy radiation interception is similar between old and modern cultivars and that leaf area has also not changed between old and modern cultivars (Feil and Geisler, 1989; Slafer *et al.*, 1994). The genetic improvement in yield potential is largely due to increased reproductive partitioning and to an increase in the number of grains produced per land area, but not to the weight of individual grains (Slafer *et al.*, 1994).

The increase in harvest index has occurred through stem shortening and reduced production of non-spike-bearing tillers (Evans, 1993). Spike dry weight and spike:stem ratio begin to diverge between old and modern cultivars at preanthesis, and modern cultivars show earlier anthesis and faster growth rates (Austin *et al.*, 1993; Slafer and Andrade, 1993; Slafer *et al.*, 1994). In modern cultivars, there is a shorter period from sowing to ear emergence and a longer interval from ear emergence to maturity and, therefore, for grain growth (Evans, 1993). Lower grain N concentration is associated with greater harvest index in modern cultivars, and together these changes result in little difference in total N uptake between old and modern cultivars (Evans, 1993; Austin *et al.*, 1993; Slafer *et al.*, 1994). Another factor related to increased reproductive partitioning is lower root allocation in modern cultivars (Siddique *et al.*, 1990).

Architectural attributes of leaves do show differences between old and modern cultivars and may be a potential source of improvement in biomass production and yield. There has been a trend toward more erect leaves, a curved leaf position, and large flag leaves (Feil, 1992). Modern cultivars in Australia with more erect leaves were more efficient in the conversion of photosynthetically active radiation to biomass than older cultivars, and this more than compensated for lower ground cover (Siddique *et al.*, 1989). In another study, yields were higher in wheat cultivars with erect upper leaves compared to more horizontal leaf placement (Innes and Blackwell, 1983). It has been suggested that the optimal situation might be erect upper leaves before anthesis, then reorientation to a more horizontal position when lower leaves start to senesce; apparently, genetic variation for such a trait does exist (Borojevic, 1986).

Morphological adaptations and canopy architecture can also improve plant water use with respect to the timing of water availability (Ludlow and Muchow, 1990). In Mediterranean-type climates, in which water availability is high in the early, cool part of the season but is low during seed filling, early production of leaf area to achieve rapid canopy closure surface results in higher transpiration compared to ground surface evaporation and, consequently, more efficient water use (Passioura, 1994; Ludlow and Muchow, 1990). However, early vigor can be deleterious in very dry environments in which stored water is the main source of availability, and the water supply can be depleted before the seed-filling stage (Passioura, 1994).

In summary, comparison of wild and cultivated wheats indicates that some physiological differences do exist in photosynthetic characteristics, but as in many other crops (Evans, 1993), improvement of yield in wheat has largely depended on changes in allocation patterns, phenology, and developmental patterns rather than on increased photosynthesis and above-ground biomass production (Fig. 3). An important missing link, however, is research on *T. urartu*, the wild species that contributed the A genome

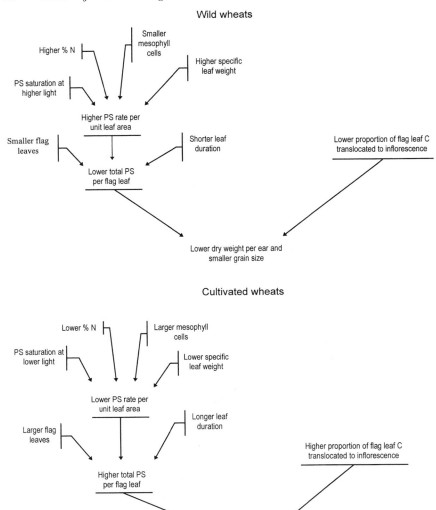

Figure 3   Relationships between physiological and morphological factors contributing to reproductive allocation and grain size in wild and cultivated wheats. Adapted from Chapin *et al.* (1989).

to the tetraploid and hexaploid wheats. Preliminary studies indicate that it has fairly high levels of carbon isotope discrimination, suggesting lower water use efficiency than other wild *Aegilops* and *Triticum* species, but accessions were highly variable for this and other growth traits (Waines *et al.,* 1993). Further work is also needed to understand the lack of correspondence between photosynthetic rate per unit leaf area and harvestable yield. In terms of practical implications for crop breeding, however, a worthwhile approach seems to be to focus on selecting plants for architecture and morphology of leaves and canopies to maximize radiation interception during optimal growth periods.

B. Nitrogen Acquisition and Utilization in Wild and Cultivated Barleys

Nitrogen is the mineral element required by plants in greatest quantity and its availability limits productivity in many natural and agricultural ecosystems. Cultivated soils typically have greater concentrations of inorganic N than their undisturbed counterparts with more of that N present as nitrate compared to ammonium (Smith and Young, 1975; Woods, 1989). Nitrogen availability varies temporally in cultivated soils due to fertilization schedules and in both cultivated and native soils as a result of variation in environmental factors (e.g., temperature and moisture) that affect microbial activity and N uptake by plants and microbes. These differences in the soil N environment suggest that changes in characteristics related to the acquisition and utilization of N may have occurred during the development of crop genotypes from their wild ancestors.

Hordeum vulgare ssp. *spontaneum* is the wild progenitor of cultivated barley, *H. vulgare* ssp. *vulgare,* with which it is interfertile (see von Bothmer *et al.,* 1995, for a taxonomic review). Wild barley grows in steppes and deserts of the Near East but now is mainly found as a weed in cultivated fields (Harlan and Zohary, 1966; Evans, 1993). *Hordeum jubatum,* however, is a distantly related perennial species that grows in diverse habitats in North America. *Hordeum jubatum* and *H. vulgare* can be crossed but the hybrids are sterile. Major taxonomic differences between the wild and cultivated taxa include brittle vs tough rachis structure and loose vs clasping glumes, respectively. Barley has experienced a long period of domestication, beginning at about 8000 BC (Harlan, 1995), typically in low-yielding environments, and landraces are still prevalent in some areas (see above).

Nitrogen acquisition requires that an ammonium ion, nitrate ion, or amino acid encounters a root surface, that the root have a transport mechanism that permits net influx of the N compound, and that the plant assimilate the N compound into organic substances. The relative importance of root growth characteristics, uptake kinetics, and N metabolism depends on the nature of the root environment and the spatial and temporal availability of N (Clarkson, 1985; Robinson *et al.,* 1991). In many situations, N uptake may depend more on root architecture and distribution than on kinetic parameters (Bloom, 1994). This may be so because spatial variability in soil N and depletion zones around roots prevent a continuous N supply in the immediate rhizosphere of the entire root system (Robinson *et al.,* 1991). For barley, physiological and agronomic aspects of N uptake and utilization have been studied in wild and cultivated genotypes from a range of high- and low-yielding conditions.

Several lines of evidence indicate that the physiology of N uptake and assimilation is similar in wild and cultivated taxa of barley and in older and newer cultivars. First, based on studies of ion uptake from dilute solution cultures, wild and cultivated barley seedlings have similar uptake kinetics for ammonium and nitrate (Bloom, 1985). Mean values of the apparent K_m (20 mmol m^{-3}) and V_{max} (1.43 μmol g$_{root}$$^{-1}$ min^{-1}) for net ammonium uptake were similar for *H. vulgare* ssp. *spontaneum, H. vulgare* ssp. *vulgare,* and *H. jubatum.* Net nitrate influx was also similar among the taxa with mean K_m and V_{max} values of 65 mmol m^{-3} and 1.28 μmol g$_{root}$$^{-1}$ min^{-1}, respectively. The lower uptake rates of nitrate uptake in wild and cultivated genotypes from colder soils were attributed to the relative lack of importance of nitrate as a N source in such soils due to the temperature sensitivity of nitrification (Bloom and Chapin, 1981).

These results indicate that there has been relatively little selection for physiological differences between wild and cultivated barleys in kinetic parameters for N absorption. There also is no evidence for selection for differences in ammonium and nitrate preferences, except in cold soils (see above), despite higher nitrate availability in cultivated soils than in undisturbed soils. In fact, there is evidence for a physiological preference for mixed ammonium and nitrate nutrition because it facilitates pH and ionic balance and diminishes organic acid synthesis (Cox and Reisenauer, 1973; Bloom and Sukrapanna, 1990; see Chapter 5, this volume).

In studies of primitive and modern cultivars of barley grown in solution culture at constant relative addition rates of N, nitrate uptake characteristics were similar but N accumulation and partitioning differed (Mattson *et al.,* 1991, 1992a,b, 1993). The modern cultivars performed well under high and low N levels, whereas the primitive cultivar did poorly at low N levels. All cultivars had essentially similar V_{max} for nitrate uptake for a given nutritional status, but the primitive cultivar, Laevigatum, had a tendency for higher V_{max} in comparison to its relative growth rate and during the grain-filling stage. This could be related to lower levels of stored nitrate in its shoots and roots and to a larger root system. It is interesting to speculate that the slightly higher V_{max} and larger root system in the primitive cultivar may allow it to exploit temporal variation in N, which should be less important under the relatively constant and high availability of N in modern agricultural situations.

A number of studies have examined the partitioning of biomass and N among plant parts in relation to the timing of phenological events in wild and cultivated barleys. Corke *et al.* (1988) found that the wild progenitor of cultivated barley, *H. vulgare* ssp. *spontaneum,* recovered a higher percentage of total plant N in the grain than the cultivated *H. vulgare* ssp. *vulgare,* while also having higher N content in the vegetative structures and grain and more N per plant at the onset of anthesis. The wild subspecies has also been shown to produce more leaves and leaf biomass, have a higher root:shoot ratio, reach anthesis later, and initiate spikelets over a longer time span than the cultivar (Corke *et al.,* 1989; Chapin *et al.,* 1989; Kernich *et al.,* 1995). Wild barleys show a great deal of variation in growth rate and N accumulation, even under nonlimiting growth conditions. Rates of total N uptake and total biomass production in some accessions of *H. vulgare* ssp. *spontaneum* can be very similar to cultivated barley. However, greater changes appear to have occurred in whole plant N partitioning than in the total accumulation of N per plant during the course of evolution of cultivated barley from its wild progenitors.

The series of studies by Mattson *et al.* (1991, 1992a,b, 1993) also compared biomass and N allocation patterns in primitive and modern barley cultivars. The primitive cultivar had lower ear weight, N content of the ears, and fraction of total N allocated to the ears and sterile flowers than the modern cultivars, especially under low N growth conditions (Table I). These differences were attributed to less total translocation of N to reproductive structures in the primitive cultivar, a feature that may guard against N allocation to weak sinks, such as sterile flowers. This may ensure a continuous N supply to successful reproductive structures by gradually translocating N from vegetative organs instead of initially investing in many tillers, all of which may not produce viable seeds. Selection for shorter stems, and a reduced propensity for lodging under high N inputs, has apparently been associated with greater N translocation to reproductive structures (Evans, 1993).

The development of modern varieties of barley has resulted in higher grain yield, lower grain protein concentration, higher harvest indices (e.g., Corke and Atsmon, 1990; Jedel and Helm, 1994), and lower root growth at depth in the soil profile (Brown *et al.,* 1987). Many studies have shown that high N availability promotes earlier growth and increased growth rates in modern cultivars (see Evans, 1993). With high soil N supply, reproductive growth can rely on current N uptake rather than on remobilization of stored N. For many taxa, higher N availability in modern agricultural systems has made possible the selection of a longer reproductive growth phase, longer leaf life span, and, consequently, longer duration of photosynthetic activity. Modern improvements that have reduced lodging under high N

Table I Effects of High and Low Nitrogen Treatments on Nitrate Uptake and Partitioning in Primitive and Modern Cultivars of Barley[a]

Cultivar	Description	Treatment	V_{max} nitrate uptake (N uptake per unit N day^{-1})	Ear weight (g plant^{-1})	Ear N content (mg ear^{-1})	Ear N partitioning (% of total N)
Golf	High-yielding modern cultivar with low grain protein concentration	High nitrogen	0.7	16.1	227.0	74.5
		Low nitrogen	0.5	4.8	52.0	58.8
Mette	High-yielding modern cultivar with high grain protein concentration	High nitrogen	0.6	10.0	146.0	72.4
		Low nitrogen	0.4	5.6	81.2	70.4
Laevigatum	Low-yielding primitive land race with high grain protein concentration	High nitrogen	1.0	7.8	106.0	56.4
		Low nitrogen	0.7	1.4	15.8	23.6

[a] In the high nitrogen treatment, relative addition rates of nitrate were added to sustain stable relative growth rates at high levels during the first 7 weeks of growth, then were decreased gradually for the next 13 weeks. In the low nitrogen treatment, N additions were 45% lower than in the high nitrogen treatment. Data on V_{max} from Mattson et al. (1992b) and on biomass and N partitioning from Mattson et al. (1992a).

supply are increased culm diameter and shorter stature (Jedel and Helm, 1994).

An intriguing hypothesis is that a relatively simple genetic change may activate a suite of phenotypic alterations resulting in plant types suited for either high-resource (high rates of resource acquisition and high relative growth rate) or low-resource (low rates of resource acquisition and low relative growth rate) environments (Chapin *et al.*, 1993). Patterns of selection for characteristics related to soil fertility may vary in different agricultural systems: With traditional modes of low-input cultivation, selection may be for genotypes that produce well in low fertility soils, whereas modern industrial agriculture favors high productivity in highly fertile soils. With this in mind, differences in N uptake among wild and cultivated barleys can be best attributed to selection for changes in biomass and N partitioning between vegetative and reproductive structures and to phenological adaptations to specific environmental regimes rather than to the kinetic parameters of ammonium and nitrate uptake. Root system size and distribution and plasticity in response to N availability may be key factors determining the actual rates of N uptake, but little information currently exists on the relevant architectural characteristics and the efficiency of root N uptake in soil among wild and cultivated taxa.

C. Water Relations in Wild and Cultivated Sunflowers

Comparison of water relations in sunflowers provides an opportunity to examine differences in patterns of response to water deficits that may have arisen during selection of crop plants from drought-adapted wild ancestors. The cultivated sunflower, *Helianthus annuus,* is native to the western part of North America in summer–dry environments with occasional, brief rainstorms (Heiser, 1976). Wild populations of this species exist in open, dry locations and especially in disturbed habitats and waste areas. This section of the genus *Helianthus* consists mainly of taprooted, annual, diploid species. Archaeological evidence suggests that Native Americans used wild sunflower in the western states, but that actual domestication occurred in the central states in the first millenium BC and that domesticated sunflower subsequently was carried throughout central and eastern North America from Canada to Mexico (Heiser, 1976). Hybrid sunflowers now dominate current production and are the result of crosses of *H. annuus* with *H. petiolaris,* an annual species from the south and southwest of North America that provided cytoplasmic male sterility. Sunflower is typically produced as a rainfed, nonirrigated crop but is grown in environments that differ greatly in water availability during the growing season.

Wild and cultivated sunflowers show surprisingly few differences in tissue water relations in response to soil water deficits. In a field study with irrigated and nonirrigated treatments of two cultivars of *H. annuus* (a modern hybrid and an Native American variety), the wild annual *H. petiolaris,* and the wild perennial *H. nuttalli* from locally moist sites, relative water content at zero turgor, osmotic potential at zero turgor, and the volumetric modulus of elasticity were similar among all taxa. The only variation was in the osmotic adjustment under water stress, a response that was seen in the cultivars but not in the wild species (Sobrado and Turner, 1983). In another study on water deficits, maximum photosynthetic rates were similar for cultivated *H. annuus* and wild *H. petiolaris,* as were declines in photosynthesis with decreasing leaf water potentials (Sobrado and Turner, 1986).

Despite similarities in physiology, dry matter accumulation and seed yield are much lower in wild sunflowers, apparently due to greater leaf area development in the cultivated sunflowers. Maximum leaf area in *H. petiolaris* was about one-third that of *H. annuus* (Sobrado and Turner, 1986). In both taxa, however, leaf expansion was highly responsive to water availability and was greatly reduced under nonirrigated conditions.

As in most plants, leaf expansion in cultivated sunflower is more sensitive to water deficits than photosynthesis; net photosynthetic rates are fairly constant across a range of moderate water stress, whereas leaf expansion responds rapidly to slight changes in water status (English *et al.,* 1979; Rawson *et al.,* 1980; Takami *et al.,* 1982, Rawson and Turner, 1982; Gimenez and Fereres, 1986). In a comparison of four cultivars of sunflower, relationships between leaf expansion and leaf water potential and between leaf expansion and turgor potential were very similar, indicating limited genotypic variation in these features (Takami *et al.,* 1982). Interestingly, in one cultivar, leaves that were water stressed grew larger than leaves that had never been stressed.

Adjustment of leaf area display is an important means of regulating transpiration in sunflower under drought conditions. If water stress occurs during the preanthesis period, sunflowers reduce leaf expansion, thereby decreasing interception of radiation by reducing leaf area index (Connor and Sadras, 1992). After anthesis, wilting and leaf senescence become more important means for reducing leaf area during drought. Although stomatal conductance is less sensitive to water deficits in sunflower than in other crops, sunflower is not a profligate water user because it regulates leaf area to decrease transpiration when water deficits occur.

Cultivated sunflowers have deep explorative root systems with low specific root length, low root length density, and high conductivity (Connor and Sadras, 1992). Genotypic variation in root distribution and its response to water deficits has been observed. For example, short-season genotypes have shallower root distributions, lower leaf area index, and produce less biomass and grain yield under drought than do long-season genotypes (Gimenez and Fereres, 1986). Under drought conditions, the short-season genotypes have only slightly lower leaf water potential, slightly higher osmotic potential, and slightly lower stomatal conductance for a given leaf water potential than do the long-season genotypes. Long-season genotypes maintain constant, low leaf water potential during the drought period, and apparently regulate water balance by accelerated loss of leaf area toward the end of the growing season. These differences among sunflower genotypes indicate that patterns of growth, allocation, and development contribute more strongly to genotypic responses to water deficits than do differences in tissue water relations and osmotic adjustment.

No data currently exist to compare wild *H. annuus* with cultivated *H. annuus,* so it is not possible to determine changes in response to water deficits that may have resulted from selection of cultivated sunflowers from wild germplasm. Based on genotypic differences among cultivated sunflowers and between cultivated *H. annuus* and other *Helianthus* species, it appears that tissue water relations are not widely divergent. Rather, with respect to water stress, genotypes differ mainly in patterns of leaf area development and display, as well as in root growth and distribution in the soil profile.

V. Potential Contribution of Wild Germplasm to Crop Ecophysiology

Examination of the ecological origins of crop progenitors and relatives provides valuable insights into the types of genetic variability in wild populations available to breeding programs for yield improvement and for more efficient use of resources. The progenitors of Mediterranean annual crops tend to have considerable niche breadth and strong capacity for phenotypic plasticity, and they exhibit high genetic variation within and among populations. A broad niche does not preclude high genetic variation among individuals. In fact, a combination of genetic polymorphism and high phenotypic plasticity can be considered to be an effective strategy for some ruderal plants (Meerts, 1995). This suggests that extensive genetic variability probably exists for yield-determining characteristics that could potentially be used in breeding programs, but also that wild genotypes have a huge number of permutations in the specific combinations of physiological, morphological, and developmental traits. Consequently, it is important to understand how suites of characteristics are linked evolutionarily and operate together to create high productivity and stress resistance (Chapin *et al.,* 1993), especially when the goal is to produce crop genotypes with greater resource use efficiency.

Although there are clear differences in agricultural selection strategies for high- and low-productivity environments, this and other literature reviews (e.g., Evans, 1993) indicate that when growth and yield of interfertile wild genotypes, old cultivars, and modern cultivars of the same taxonomic species are compared, most of the variation is attributable to morphological, allocational, and developmental traits rather than to tissue-specific rates of photosynthesis, nutrient assimilation, or other key metabolic processes. It should be recognized, however, that there are, in fact, very few examples for which such comparisons have been extensively explored. Moreover, it is difficult to create an experimental framework to examine a wide enough range of environmental conditions to allow full expression of phenotypic potential and to measure adequately a broad scope of physiological, morphological, and developmental characteristics.

It is important to note some examples in which possible physiological differences in yield-determining characteristics may occur among crop cultivars or between closely related genotypes of crops and progenitors. One example is the genotypic variation within crop species with respect to leaf gas-exchange efficiency (CO_2 uptake in photosynthesis/H_2O loss in photosynthesis) and water use efficiency (ratio of C gained to water lost at the whole plant level). Although photosynthetic capacity and stomatal conductance can contribute to leaf gas-exchange and water use efficiency, and thus affect yields under water deficits, such changes do not occur independently but are typically associated with specific morphological and developmental characteristics, such as root distribution, canopy size, ground cover in relation to seasonal water supply, and flowering times. This topic is discussed in detail by Condon and Hall (see Chapter 3, this volume). Another example is the concomitant increase in photosynthetic rate and wheat yield in high-radiation environments in Australia (Watanabe *et al.*, 1994) that was described previously. Although increased photosynthetic rate is a component of the gain in yield potential in modern Australian wheats, morphological characteristics have also been changed to increase the efficiency of conversion of photosynthetically active radiation (Siddique *et al.*, 1989). A third example is that of decreased yield in response to water deficits in cultivated lettuce (*L. sativa*), which may be related to the absence of adjustment in leaf osmotic potential; in *L. serriola*, the probable progenitor of cultivated lettuce (Kesseli *et al.*, 1991), osmotic changes do occur and yield is not decreased at similar water deficits (Gallardo *et al.*, 1996). Associated with this physiological difference between the taxa, however, are large morphological differences in root architecture and distribution that affect water uptake and consequently also play a role in response of yield to water deficits (Jackson, 1995; Gallardo *et al.*, 1996). A fourth example is the increase in stomatal conductance, photosynthetic capacity, and yield in modern cultivars of Pima cotton that are bred for hot, intensively irrigated areas (Cornish *et al.*, 1991; Lu *et al.*, 1994; Radin *et al.*, 1994). Inadvertent selection has occurred for greater transpiration and, thus, lower leaf temperatures. Along with these changes in physiological attributes, there has also been reduction in leaf area.

These examples point to the potential for physiologically based characteristics of improved yield, but they also indicate that simultaneous assessment of the contribution of related morphological, allocational, and developmental characteristics to yield will ultimately provide a better understanding of the bases for yield improvement. Throughout crop evolution, along with the intentional selection for yield and yield reliability, there has always been inadvertent selection for other traits that explain the complex adaptation of crops to their environments, as well as their potential ability to produce well in new environments. Recognition of costs vs benefits of these complex adaptations is a central focus of the ecological "adaptive strategy" of plants and is inherent in the idea that suites of traits contribute to a stress-resistant syndrome (Grime, 1977; Chapin *et al.*, 1993). Breeding for a stress-resistant syndrome that encompasses linkages between physiological, allocational, morphological, and developmental suites of characteristics is undoubtedly a daunting task but may become simpler with techniques using genetic molecular markers. Chapin *et al.* (1993), however, hypothesize that changes in a few major regulatory genes, such as for regulatory hormones, could occur in response to strong directional selection as might be imposed by a novel type of environmental stress, simultaneously affecting an array of related plant characteristics. Polygenic adaptation involving many genes, each with a small effect, is hypothesized to be more likely to occur when the new stress regime is fairly close to the original environment. The role of these two evolutionary possibilities in crop adaptation to stress is unknown, but at least this hypothesis considers the integration between various kinds of plant characteristics in determining productivity and yield.

Genetic improvement in crops for complexly inherited traits such as stress tolerance has depended in the past on being able to hybridize a cultivar with a genotype possessing the desired trait. In many cases, wild species are not closely enough related to crops to exchange genes easily and this has precluded their use as gene resources for crop improvement, especially if the trait is under complex genetic control. The successful transfers by means of wide crosses have almost all been single genes conferring disease or pest resistance (e.g., for wheat see Dvorak and McGuire, 1990; for tomato see Rick and Chetelat, 1995; Rick, 1995). There have been several advances, however, in molecular genetics and breeding that promise to facilitate the transfer or exchange of gene complexes, including molecular markers linked to very specific chromosome regions, strategies for marker-assisted selection of single genes and quantitative traits, development of transformation protocols for most major crops, and the discovery of conserved regions of genomes across wide taxonomic boundaries, e.g., in the grass family. Concomitant progress in understanding the ecophysiology and autecology of related wild species would further enhance their potential as gene resources for crop improvement.

Salt tolerance in cereals is a good example of what has been involved in utilizing species that are not in the direct lineage of the crop for crop improvement. The major obstacles to using wild germplasm for salinity tolerance are that (1) it is a genetically complex trait, (2) identification and isolation of component genes are difficult, (3) transferring genes from the wild species to crops is difficult, and (4) there is an incomplete understanding of the physiological mechanisms of salt tolerance. In the grass tribe Triticeae, which contains the crops wheat, barley, and rye, there are many wild species that are apparent sources of salt tolerance (e.g., McGuire and Dvorak, 1981). Two diploid species in particular, because of their lower chromosome number, are more obvious targets. Both are perennial wheat grasses. *Lophopyrum elongatum* is native to saline marsh areas along the eastern Mediterranean coast and *Thinopyrum bessarabicum* is native to coastal habitats along the Black Sea. By studying wheat with chromosome additions and substitutions from these species, it has been possible to identify chromosomal regions that govern salt tolerance and to ascertain that it is a polygenic trait in the Triticeae (Gorham and Wyn Jones, 1990; Zhong and Dvorak, 1995). In particular, molecular cytogenetic study of the chromosomes from these species in wheat has identified regions and genes responsible for K/Na discrimination and tolerance to sudden salt stress. The progress that has been made with *L. elongatum* has been with a single genotype collected by chance decades ago. Imagine what progress could have been made had the species been well studied ecogeographically and had accessions been evaluated ecophysiologically for salt tolerance before choosing genotypes for breeding purposes.

To summarize, exotic germplasm has not generally been exploited for traits to improve stress resistance and resource acquisition in crops; instead, breeders have tended to rely on other closely related cultivars for sources of germplasm. For wild taxa that can be crossed with their crop relatives, one of the molecular methods that should facilitate the identification and transfer of useful genes is the ability to dissect and tag with linked markers major quantitative trait loci (QTLs). For example, markers for the QTLs regulating stress responses have been found by examining the probability of coincidence between molecular markers and specific physiological and agronomic traits (see Lee, 1995). This is accomplished by crossing widely different inbred parents, such as wild and cultivated members of an interfertile pair of taxa, and advancing by single seed descent to later generations, e.g., F_3 to F_7 populations. A molecular map is made from DNA extracted from one of these later generations, and physiological and morphological measurements are also made at this stage. Various statistical approaches can be used to determine the minimum number and location of genes regulating these traits. Several studies have shown the potential for QTL analysis to aid in selection of yield-determining characteristics and responses to environmental stresses (e.g., Martin *et al.*, 1989; Breto *et al.*, 1994; Causse *et al.*, 1995; Lebreton *et al.*, 1995). Also, associations between traits can be examined by comparing the coincidence between the locations on the genetic map for a set of measured plant characteristics, allowing the identification of suites of characteristics that are under the same genetic control.

VI. Concluding Remarks

In summary, this chapter has shown that many of the major changes in crop evolution from closely related wild plants have involved morphological and developmental traits rather than assimilatory and metabolic processes, and that the ecophysiological trade-offs between different patterns of resource acquisition and allocation can have a large impact on crop yield. As Sinclair (1994) points out, physiological limits to crop yields probably have been reached under experimental conditions, and our current challenge is to focus on how to achieve reliable yields in a sustainable manner. It will surely become easier to transfer advantageous traits from wild relatives to crops, but success will depend on the integration of the new trait into the complex biology of the whole plant, and this will require a better understanding of costs and benefits of various aspects of form and function in crop plants than is currently available.

This review raises many questions about crop evolution and adaptation of annual Mediterranean crop species. Do wild plants from resource-poor environments necessarily have higher resource use efficiencies? How does selection for plasticity vs genetic uniformity operate in high- and low-resource environments? Why are physiological traits largely conserved among related genotypes? What is the potential for further crop improvement via changes in physiological vs morphological, allocational, and developmental characteristics? Answers to these questions will surely depend on integrated analysis of many types of plant characteristics, on better understanding of evolutionary and genetic relationships between taxa, and on scaling up from physiological attributes to functioning of the whole plant in relation to environmental conditions.

Acknowledgments

We greatly appreciate the help of Patrick McGuire with updating taxonomic relationships in the Triticeae and for providing critical reviews of the manuscript. Arnold Bloom and Robert Pearcy made many useful suggestions to improve the text.

References

Amthor, J. S. (1989). "Respiration and Crop Productivity." Springer-Verlag, New York.

Anderson, J. R., and Hazell, P. B. R. (1989). Synthesis and needs in agricultural research and policy. *In* "Variability in Grain Yields: Implications for Agricultural Research and Policy in Developing Countries" (J. R. Anderson and P. B. R. Hazell, eds.), pp. 339–356. Johns Hopkins Univ. Press, Baltimore, MD.

Anikster, Y., Eshel, A., Ezrati, S., and Horovitz, A. (1991). Patterns of phenotypic variation in wild tetraploid wheat at Ammiad. *Isr. J. Bot.* **40,** 397–418.

Austin, R. B. (1989). Genetic variation in photosynthesis. *J. Agric. Sci. Camb.* **112,** 287–294.

Austin, R. B., Bingham, J., Blackwell, R. D., Evans, L. T., Ford, M. A., Morgan, C. L., and Taylor, M. (1980). Genetic improvements in winter wheat yields since 1900 and associated physiological changes. *J. Agric. Sci. Camb.* **94,** 675–689.

Austin, R. B., Ford, M. A., Morgan, C. L., and Yeoman, D. (1993). Old and modern wheat cultivars compared on the Broadbalk wheat experiment. *Eur. J. Agron.* **2,** 141–147.

Baker, H. F. (1974). The evolution of weeds. *Annu. Rev. Ecol. Syst.* **5,** 1–24.

Bazzaz, F. A. (1987). Experimental studies on the evolution of niche in successional plant populations. *In* "Colonization, Succession and Stability" (A. J. Gray, M. J. Crawley, and P. J. Edwards, eds.), pp. 245–251. Blackwell, Oxford.

Bloom, A. J. (1985). Wild and cultivated barleys show similar affinities for mineral nitrogen. *Oecologia* **65,** 555–557.

Bloom, A. J. (1994). Crop acquisition of ammonium and nitrate. *In* "Physiology and Determination of Crop Yield" (K. J. Boote, J. M. Bennett, T. R. Sinclair, and G. M. Paulsen, eds.), pp. 303–309. American Society of Agronomy, Madison, WI.

Bloom, A. J. (1997). Nitrogen as a limiting factor: Crop acquisition of ammonium and nitrate. *In* "Ecology in Agriculture" (L. E. Jackson, ed.). Academic Press, San Diego.

Bloom, A. J., and Chapin, F. S., III (1981). Differences in steady-state net ammonium and nitrate influx by cold- and warm-adapted barley varieties. *Plant Physiol.* **68,** 1064–1067.

Bloom, A. J., and Sukrapanna, S. (1990). Effects of exposure to ammonium and transplant shock upon the induction of nitrate absorption. *Plant. Physiol.* **94,** 85–90.

Blum, A. (1990). Variation among wheat cultivars in the response of leaf gas exchange to light. *J. Agric. Sci. Camb.* **115,** 305–311.

Borojevic, S. (1986). Genetic changes in morpho-physiologic characters in relation to breeding for increased wheat yield. *In* "Genetic Improvement in Yield of Wheat," pp. 71–85. Crop Sci. Soc. America Spec. Publ. No. 13.

Boukerrou, L., and Rasmussen, D. D. (1990). Breeding for high biomass yield in spring barley. *Crop Sci.* **30,** 31–35.

Breto, M. P., Asins, M. J., and Carbonell, E. A. (1994). Salt tolerance in *Lycopersicon* species. III. Detection of quantitative trait loci by means of molecular markers. *Theor. Appl. Genet.* **88,** 395–401.

Briggs, F. N., and Knowles, P. F. (1967). "Introduction to Plant Breeding." Reinhold, New York.

Brown, S. C., Keatinge, J. D. H., Gregory, P. J., and Cooper, P. J. M. (1987). Effects of fertilizer, variety and location on barley production under rainfed conditions in Northern Syria. 1. Root and shoot growth. *Field Crops Res.* **16,** 53–66.

Brush, S. B. (1995). In situ conservation of landraces in centers of crop diversity. *Crop Sci.* **35,** 346–354.

Carver, B. F., and Nevo, E. (1990). Genetic diversity of photosynthetic characters in native populations of *Triticum dicoccoides*. *Photosynth. Res.* **25,** 119–128.

Causse, M., Rocher, J. P., Henry, A. M., Charcosset, A., Prioul, J. L., and de Vienne, D. (1995). Genetic dissection of the relationship between carbon metabolism and early growth in maize, with emphasis on key-enzyme loci. *Mol. Breeding* **1,** 258–272.

Ceccarelli, S. (1994). Specific adaptation and breeding for marginal conditions. *Euphytica* **77,** 205–2219.

Ceccarelli, S., and Grando, S. (1991). Selection environment and environmental sensitivity in barley. *Euphytica* **57,** 157–167.

Ceccarelli, S., Acevedo, E., and Grando, S. (1991). Breeding for yield stability in unpredictable environments: single traits, interaction between traits, and architecture of genotypes. *Euphytica* **56,** 169–185.

Chapin, F. C., III (1980). The mineral nutrition of wild plants. *Annu. Rev. Ecol. Syst.* **11,** 233–260.

Chapin, F. S., III, Groves, R. H., and Evans, L. T. (1989). Physiological determinants of growth rate in response to phosphorus supply in wild and cultivated *Hordeum* species. *Oecologia* **79,** 96–105.

Chapin, F. S., III, Autumn, K., and Pugnaire, F. (1993). Evolution of suites of traits in response to environmental stress. *Am. Nat.* **142,** S78–S92.

Chiariello, N. R., and Gulmon, S. L. (1991). Stress effects on plant reproduction. *In* "Responses of Plants to Multiple Stresses" (H. A. Mooney, W. E. Winner, E. J. Pell, and E. Chu, eds.), pp. 161–188. Academic Press, San Diego.

Clarkson, D. T. (1985). Factors affecting mineral nutrient acquisition by plants. *Annu. Rev. Plant Physiol.* **36,** 77–115.

Condon, A. G., and Hall, A. E. (1997). Adaptation to diverse agricultural environments: Genotypic variation in water use efficiency within crop species. *In* "Ecology in Agriculture" (L. E. Jackson, ed.). Academic Press, San Diego.

Connor, D. J., and Sadras, V. O. (1992). Physiology of yield expression in sunflower. *Field Crops Res.* **30,** 333–389.

Corke, H., and Atsmon, D. (1990). Wild barley (*Hordeum spontaneum* Koch) and its potential utilization in barley protein improvement. *Isr. J. Bot.* **39,** 271–286.

Corke, H., Nevo, E., and Atsmon, D. (1988). Variation in vegetative parameters related to the nitrogen economy of wild barley, *Hordeum spontaneum,* in Israel. *Euphytica* **39,** 227–232.

Corke, H., Avivi, N., and Atsmon, D. (1989). Pre- and post-anthesis accumulation of dry matter and nitrogen in wild barley (*Hordeum spontaneum*) and in barley cultivars (*H. vulgare*) differing in final grain size and protein content. *Euphytica* **40,** 127–134.

Cornish, K., Radin, J. W., Turcotte, E. L., Lu, Z., and Zeiger, E. (1991). Enhanced photosynthesis and stomatal conductance of pima cotton (*Gossypium barbadense* L.) bred for increased yield. *Plant Physiol.* **97,** 484–489.

Cox, W. J., and Reisenauer, H. M. (1973). Growth and ion uptake by wheat supplied nitrogen as nitrate or ammonium or both. *Plant Soil* **38,** 363–380.

Delcourt, H. R. (1987). The impact of prehistoric agriculture and land occupation on natural vegetation. *Trends Ecol. Evol.* **2,** 39–44.

Denmead, O. T. (1976). Temperate cereals. *In* "Vegetation and the Atmosphere" (J. L. Monteith, ed.), Vol. 2, pp. 1–31. Academic Press, London.

Doebley, J. (1994). Genetics and the morphological evolution of maize. *In* "The Maize Handbook" (M. Freeling and V. Walbot, eds.), pp. 66–77. Springer-Verlag, New York.

Dunstone, R. L., and Evans, L. T. (1974). Role of changes in cell size in the evolution of wheat. *Aust. J. Plant Physiol.* **1,** 157–165.

Dvorak, J., and McGuire, P. E. (1990). Triticeae, the gene pool for wheat breeding. *In* "Genome Mapping of Wheat and Related Species: Proceedings of a Public Workshop" (P. E. McGuire, H. Corke, and C. O. Qualset, eds.), Report No. 7. Univ. of California Genetic Resources Conservation Program, Davis, CA.

Dvorak, J., and Ross, K. (1986). Expression of tolerance of Na, K, Mg, Cl, and SO_4 ions and sea water in the amphiploids of *Triticum aestivum* and *Elytrigia elongata. Crop Sci.* **26,** 658–660.

English, S. D., McWilliam, J. R., Smith, R. C. G., and Davidson, J. L. (1979). Photosynthesis and partitioning of dry matter in sunflower. *Aust. J. Plant Physiol.* **6,** 149–164.

Evans, L. T. (1993). "Crop Evolution, Adaptation and Yield." Cambridge Univ. Press, Cambridge, UK.

Evans, L. T., and Dunstone, R. L. (1970). Some physiological aspects of evolution in wheat. *Aust. J. Biol. Sci.* **23**, 725–41.

Fehr, W. R. (1984). "Genetic Contributions to Yield Gains of Five Major Crop Plants." Agronomy Society of America, Madison, WI.

Feil, B. (1992). Breeding progress in small grain cereals—a comparison of old and modern cultivars. *Plant Breeding* **108**, 1–11.

Feil, B., and Geisler, G. (1989). Uptake and distribution of nitrogen in old and new German wheats. *J. Agron. Crop. Sci.* **162**, 49–56.

Fischbeck, G. (1989). Variability in winter wheat and spring barley yields in Bavaria. *In* "Variability in Grain Yields: Implications for Agricultural Research and Policy in Developing Countries" (J. R. Anderson and P. B. R. Hazell, eds.), pp. 118–126. Johns Hopkins Univ. Press, Baltimore, MD.

Frankel, O. H., Brown, A. H. D., and Burdon, J. J. (1995). "The Conservation of Plant Biodiversity." Cambridge Univ. Press, Cambridge, UK.

Gallardo, M., Jackson, L. E., and Thompson, R. B. (1996). Shoot and root physiological responses to localized zones of soil moisture in cultivated and wild lettuce (*Lactuca* spp.). *Plant Cell Environ.* **19**, 1169–1178.

Gimenez, C., and Fereres, E. (1986). Genetic variability in sunflower cultivars under drought. II. Growth and water relations. *Aust. J. Agric. Res.* **37**, 583–597.

Gorham, J., and Wyn Jones, R. G. (1990). Utilization of Triticeae for improving salt tolerance in wheat. *In* "Wheat Genetic Resources: Meeting Diverse Needs" (J. P. Srivastava and A. B. Damania, eds.), pp. 269–277. Wiley, Sussex, UK.

Grime, J. P. (1977). Evidence for the existence of three primary strategies in plants and its relevance to ecological and evolutionary theory. *Am. Nat.* **111**, 1169–1194.

Grime, J. P. (1994). The role of plasticity in exploiting environmental heterogeneity. *In* "Exploitation of Environmental Heterogeneity by Plants" (M. M. Caldwell and R. W. Pearcy, eds.), pp. 1–19. Academic Press, San Diego.

Gutschick, V. P. (1984a). Photosynthetic model for C_3 leaves incorporating CO_2 transport, propagation of radiation, and biochemistry. 1. Kinetics and their parametrization. *Photosynthetica* **18**, 549–568.

Gutschick, V. P. (1984b). Photosynthetic model for C_3 leaves incorporating CO_2 transport, propagation of radiation, and biochemistry. 2. Ecological and agricultural utility. *Photosynthetica* **18**, 569–595.

Gutschick, V. P. (1987). "A Functional Biology of Crop Plants." Timber Press, Portland, OR.

Hamrick, J. L., and Allard, R. W. (1972). Microgeographical variation in allozyme frequencies in *Avena barbata. Proc. Natl. Acad. Sci. USA* **69**, 2100–2104.

Hamrick, J. L., and Allard, R. W. (1975). Correlations between quantitative characters and enzyme genotypes in *Avena barbata. Evolution* **29**, 438–442.

Hamrick, J. L., and Godt, M. J .W. (1989). Allozyme diversity in plant species. *In* "Plant Population Genetics, Breeding and Genetic Resources" (A. H. D. Brown, M. T. Clegg, A. L. Kahler, and B. S. Weir, eds.), pp. 43–63. Sinauer, Sunderland, MA.

Harlan, J. R. (1975). Our vanishing genetic resources. *Science* **188**, 618–621.

Harlan, J. R. (1995). "The Living Fields: Our Agricultural Heritage." Cambridge Univ. Press, Cambridge, UK.

Harlan, J. R., and Zohary, D. (1966). Distribution of wild wheats and barley. *Science* **153**, 1074–1080.

Harris, D. R., and Hillman, G. C. (1989). Introduction. *In* "Foraging and Farming: The Evolution of Plant Exploitation" (D. R. Harris and G. C. Hillman, eds.), pp. 1–8. Unwin Hyman, London.

Hawkes, J. G. (1983). "The Diversity of Crop Plants." Harvard Univ. Press, Cambridge, MA.

Heiser, C. B. (1976). Sunflowers. *In* "Evolution of Crop Plants" (N. W. Simmonds, ed.), pp. 36–38. Longman, London, UK.

Heywood, J. S. (1991). Spatial analysis of genetic variation in plant populations. *Annu. Rev. Ecol. Syst.* **22,** 335–355.

Hildebrand, P. E. (1984). Modified stability analysis of farmer managed, on-farm trials. *Agronomy J.* **76,** 271–274.

Hillman, G. C., and Davies, M. S. (1990). Domestication rates in wild-type wheats and barley under primitive cultivation. *Biol. J. Linnean Soc.* **39,** 39–78.

Innes, P., and Blackwell, R. D. (1983). Some effects of leaf posture on the yield and water economy of winter wheat. *J. Agric. Sci. Camb.* **101,** 367–376.

Jackson, L. E. (1995). Root architecture in cultivated and wild lettuce (*Lactuca* spp.). *Plant Cell Environ.* **18,** 885–894.

Jackson, L. E., Houpis, J. L. J., and Diemer, M. W. (1987). The role of leaf position in the ecophysiology of an annual grass during reproductive growth. *Am. Midl. Natl.* **117,** 56–62.

Jedel, P. E., and Helm, J. H. (1994). Assessment of western Canadian barleys of historical interest: I. Yield and agronomic traits. *Crop Sci.* **34,** 922–927.

Johnson, R. C., Kebedie, H., Mornhinweg, D. W., Carver, B. F., Rayburn, A. L., and Nguyen, H. T. (1987). Photosynthetic differences among *Triticum* accessions at tillering. *Crop. Sci.* **27,** 1046–1050.

Kahler, A. L., Allard, R. W., Krzakowa, M., Wehrhahn, C. F., and Nevo, E. (1980). Associations between isozyme phenotypes and environment in the slender wild oat (*Avena barbata*) in Israel. *Theor. Appl. Genet.* **56,** 31–47.

Kernich, G. E., Halloran, G. M., and Flood, R. G. (1995). Variation in developmental patterns of wild barley (*Hordeum spontaneum* L.) and cultivated barley (*H. vulgare* L.). *Euphytica* **82,** 105–115.

Kesseli, R. V., Ochoa, O., and Michelmore, R. W. (1991). Origin of *Lactuca sativa* (lettuce). *Genome* **34,** 430–436.

Lebreton, C., Lazic-Jancic, V., Steed, S., Pekic, S., and Quarrie, S. A. (1995). Identification of QTL for drought responses in maize and their use in testing causal relationships between traits. *J. Exp. J. Exp. Bot.* **46,** 853–865.

LeCain, D. R., Morgan, J. A., and Zerbi, G. (1989). Leaf anatomy and gas exchange in nearly isogenic semidwarf and tall winter wheat. *Crop Sci.* **29,** 1246–1251.

Lee, M. (1995). DNA markers and plant breeding programs. *Adv. Agron.* **55,** 265–344.

Lu, Z-M., Radin, J. W., Turcotte, E. L., Percy, R., and Zeiger, E. (1994). High yields in advanced lines of pima cotton are associated with higher stomatal conductance, reduced leaf area and lower leaf temperature. *Physiol. Plant.* **92,** 266–272.

Ludlow, M. M., and Muchow, R. C. (1990). A critical evaluation of traits for improving crop yields in water-limited environments. *Adv. Agron.* **49,** 107–153.

Martin, B., Nienhuis, J., King, G., and Schaefer, A. (1989). Restriction fragment length polymorphisms associated with water use efficiency in tomato. *Science* **243,** 1725–1728.

Mattson, M., Johansson, E., Lundborg, T., Larsson, M., and Larsson, C.-M. (1991). Nitrogen utilization in N-limited barley during vegetative and generative growth. I. Growth and nitrate uptake in vegetative cultures grown at different relative addition rates of nitrogen. *J. Exp. Bot.* **42,** 197–205.

Mattson, M., Lundborg, T., Larsson, M., and Larsson, C.-M. (1992a). Nitrogen utilization in N-limited barley during vegetative and generative growth. II. Method for monitoring generative growth and development in solution culture. *J. Exp. Bot.* **43,** 15–23.

Mattson, M., Lundborg, T., Larsson, M., and Larsson, C.-M. (1992b). Nitrogen utilization in N-limited barley during vegetative and generative growth. III. Post-anthesis kinetics of net nitrate uptake and the role of the relative root size in determining the capacity for nitrate acquisition. *J. Exp. Botany* **43,** 25–30.

Mattson, M., Lundborg, T., and Larsson, C.-M. (1993). Nitrogen utilization in N-limited barley during vegetative and generative growth. IV. Translocation and remobilization of nitrogen. *J. Exp. Botany* **44,** 537–546.

McGuire, P. E., and Dvorak, J. (1981). High salt tolerance potential in wheatgrasses. *Crop Sci.* **21,** 702–705.

Meerts, P. (1995). Phenotypic plasticity in the annual weed, *Polygonum aviculare. Bot. Acta* **108,** 414–424.

Morgan, J. A., LeCain, D. R., and Wells, R. (1990). Semidwarfing genes concentrate photosynthetic machinery and affect leaf gas exchange of wheat. *Crop Sci.* **30,** 602–608.

Nevo, E., Golenberg, E., and Beiles, A. (1982). Genetic diversity and environmental associations of wild wheat, *Triticum dicoccoides* in Israel. *Theor. Appl. Genet.* **62,** 241–254.

Nevo, E., Carver, B. F., and Beiles, A. (1991). Photosynthetic performance in wild emmer wheat, *Triticum dicoccoides*: Ecological and genetic predictability. *Theor. Appl. Genet.* **81,** 445–460.

Passioura, J. B. (1994). The yield of crops in relation to drought. *In* "Physiology and Determination of Crop Yield" (K. J. Boote, J. M. Bennett, T. R. Sinclair, and G. M. Paulsen, eds.), pp. 343–359. American Society of Agronomy, Madison, WI.

Pearman, I., Thomas, S. M., and Thorne, G. N. (1981). Dark respiration of several varieties of winter wheat given different amounts of nitrogen fertilizer. *Ann. Bot.* **47,** 535–546.

Radin, J. W., Lu, Z., Percy, R., and Zeiger, E. (1994). Genetic variability for stomatal conductance in Pima cotton and its relation to improvements of heat adaptation. *Proc. Natl. Acad. Sci. USA* **91,** 7217–7221.

Rawson, H. M., and Turner, N. C. (1982). Recovery from water stress in five sunflower (*Helianthus annuus* L.) cultivars. II. The development of leaf area. *Aust. J. Plant Physiol.* **9,** 449–460.

Rawson, H. M., Constable, G. A., and Howe, G. N. (1980). Carbon production of sunflower cultivars in field and controlled environments. II. Leaf growth. *Aust. J. Plant Physiol.* **7,** 575–86.

Rawson, H. M., Hindmarsh, J. H., Fischer, R. A., and Stockman, Y. M. (1983). Changes in leaf photosynthesis with plant ontogeny and relationships with yield per ear in wheat cultivars and 120 progeny. *Aust. J. Plant Physiol.* **10,** 503–514.

Rick, C. M. (1995). Tomato, *Lycopersicon esculentum. In* "Evolution of Crop Plants" (J. Smartt and N. W. Simmonds, eds.), pp. 452–457. Longman, New York.

Rick, C. M., and Chetelat, R. T. (1995). Utilization of related wild species for tomato improvement. *Acta Hort.* **412,** 21–38.

Robinson, D., Linehan, D. J., and Caul, S. (1991). What limits nitrate uptake from soil? *Plant Cell Environ.* **14,** 77–85.

Siddique, K. H. M., Belford, R. K., Perry, M. W., and Tennant, D. (1989). Growth, development and light interception of old and modern wheat cultivars in a Mediterranean-type environment. *Aust. J. Agric. Res.* **40,** 473–487.

Siddique, K. H. M., Belford, R. K., and Tennant, D. (1990). Root:shoot ratios of old and modern, tall and semi-dwarf wheats in a mediterranean environment. *Plant Soil* **121,** 89–98.

Simmonds, N. W. (1991). Selection for local adaptation in a plant breeding programme. *Theor. Appl. Genet.* **82,** 363–367.

Sinclair, T. R. (1994). Limits to crop yield? *In* "Physiology and Determination of Crop Yield" (K. J. Boote, J. M. Bennett, T. R. Sinclair, and G. M. Paulsen, eds.), pp. 509–532. American Society of Agronomy, Madison, WI.

Singh, M. K., and Tsunoda, S. (1978). Photosynthetic and transpirational response of a cultivated and a wild species of *Triticum* to soil moisture and air humidity. *Photosynthetica* **12,** 280–283.

Slafer, G. A., and Andrade, F. H. (1993). Physiological attributes related to the generation of grain yield in bread wheat cultivars released at different eras. *Field Crops Res.* **31,** 351–367.

Slafer, G. A., Satorre, E. H., and Andrade, F. H. (1994). Increases in grain yield in bread wheat from breeding and associated physiological changes. *In* "Genetic Improvements of Field Crops" (G. A. Slafer, ed.), pp. 1–68. Dekker, New York.

Smartt, J., and Simmonds, N. W. (1995). "Evolution of Crop Plants." Longman, Essex, UK.

Smith , S. J., and Young, L. B. (1975). Distribution of nitrogen forms in virgin and cultivated soils. *Soil Sci.* **120,** 354–360.

Sobrado, M. A., and Turner, N. C. (1983). Influence of water deficits on the water relations characteristics and productivity of wild and cultivated sunflowers. *Aust. J. Plant Physiol.* **10,** 195–203.

Sobrado, M. A., and Turner, N. C. (1986). Photosynthesis, dry matter accumulation and distribution in the wild sunflower *Helianthus petiolaris* and the cultivated sunflower *Helianthus annuus* as influenced by water deficits. *Aust. J. Plant Physiol.* **10,** 195–203.

Sperling, L. (1995). Results, methods and institutional issues in participatory selection: The case of beans in Rwanda. *In* "Participatory Plant Breeding" (P. Eyzaguirre and M. Iwanaga, eds.), pp. 44–56. International Plant Genetics Resources Institute, Rome.

Sperling, L., and Berkowitz, P. (1994). "Partners in Selection: Bean Breeders and Women Bean Experts in Rwanda." Consultative Group on International Agricultural Research, Washington, DC.

Stoy, V. (1965). Photosynthesis, respiration, and carbohydrate accumulation in spring wheat in relation to yield. *Physiol. Plant.* Suppl. IV.

Takami, S., Rawson, H. A., and Turner, N. C. (1982). Leaf expansion of four sunflower (*Helianthus annuus* L.) cultivars in relation to water deficits. III. Diurnal patterns during stress and recovery. *Plant Cell Environ.* **5,** 279–286.

Tal, M., and Shannon, M. C. (1983). Salt tolerance in the wild relatives of the cultivated tomato: Responses of *Lycopersicon esculentum, L. cheesemanii, L. peruvianum, Solanum pennellii* and F1 hybrids to high salinity. *Aust. J. Plant Physiol.* **10,** 109–117.

van Slageren, M. W. (1994). "Wild Wheats: A Monograph of *Aegilops* L. and *Amblyopyrum* (Jaub. & Spach) Eig (Poaceae)." Veenman Drukkers, Wageningen, The Netherlands.

von Bothmer, R., Jacobsen, N., Baden, C., Jørgensen, R.B., and Linde-Laursen, I. (1995). "An Ecogeographical Study of the Genus *Hordeum*." 2nd Ed., Systematic and Ecogeographic Studies on Crop Genepools 7. International Plant Genetic Resources Institute, Rome.

Waines, J. G., Rafi, M. M., and Ehdaie, B. (1993). Yield components and transpiration efficiency in wild wheats. *In* "Biodiversity and Wheat Improvement" (A. B. Damania, ed.), pp. 173–186. Wiley, Chichester, UK.

Watanabe, N., Evans, J. R., and Chow, W. S. (1994). Changes in the photosynthetic properties of Australian wheat cultivars over the last century. *Aust. J. Plant Physiol.* **21,** 169–183.

Winzeler, M., McCullough, D. E., and Hunt, L. A. (1989). Leaf gas exchange and plant growth of winter rye, triticale, and wheat under contrasting temperature regimes. *Crop Sci.* **29,** 1256–1260.

Woods, L. E. (1989). Active organic matter distribution in the surface 15 cm of undisturbed and cultivated soil. *Biol. Fertil. Soil* **8,** 272–278.

Zhong, G-Y. and Dvorak, J. (1995). Evidence for common genetic mechanisms controlling the tolerance of sudden salt stress in the tribe Triticeae. *Plant Breeding* **114,** 297–302.

Zohary, D., and Hopf, M. (1988). "Domestication of Plants in the Old World." Clarendon, Oxford.

2

Photosynthesis, Growth Rate, and Biomass Allocation

Vincent P. Gutschick

I. Introduction

Photosynthesis is the ultimate source of all growth and, as such, it has been subjected to intense natural selection in wild plants and in crop ancestors. Consequently, its components of light interception, photochemical reactions, and patterns of photosynthate use (respiration, biomass allocation, etc.) have been nearly optimized. Thus, breeding for leaf photosynthesis has often been counterproductive when directed toward higher rates, or beneficial when breeding has inadvertently and indirectly achieved lower leaf rates but higher whole-plant rates (Gifford and Evans, 1981; Khan and Tsunoda, 1970; but see Radin, 1991). Overall, breeding for photosynthesis has enabled few discrete gains in yield other than those achieved by leaf erectness (reviewed by Trenbath and Angus, 1975), which improves light distribution, or by dwarfing, which allows higher ratios of leaf to support structures in addition to the nonphotosynthetic benefit of improved harvest index. These two routes of improvement in yield trade off some gains in potential growth rate against increased vulnerability to competition for light. They suit the crop for a new ecological status in agriculture. Similarly, changes in management have implicitly recognized the new ecological regime in taking plants from natural to agricultural environments. Higher planting densities have enabled earlier canopy closure, which increases the season-total light interception per ground area (Thornley, 1983) and also reduces soil evaporation (Richards, 1991) that otherwise helps limit water available for transpiration that supports photosynthesis.

What other opportunities for photosynthetic gains—and for avoidance of pitfalls—exist in an ecological view of crop breeding and management? Consider, for example, that plants as a whole exhibit wide variations in relative growth rate that must be explained (Poorter, 1991), and perhaps exploited further, in our slower-growing crops such as tree crops. Consider, too, the now well-developed appreciation of the trade-offs in resource-use efficiency. Plants vary, even within a species, in the physiologically (genetically) determined "setpoint" of leaf internal CO_2 concentration (C_i) (measured either as moles per volume or as partial pressure). A higher C_i confers greater photosynthesis per unit mass (Condon *et al.*, 1987) but lower water-use efficiency (WUE). Exploitation of the ability to breed for C_i is advanced (Ehleringer *et al.*, 1993) but further progress awaits. For example, it is not fully appreciated that the range that can be achieved in WUE is strongly limited by physiological and biophysical feedbacks (Jarvis and McNaughton, 1986; Gutschick and Cunningham, 1989). Furthermore, the rational economic balancing of yield against WUE is not yet widely implemented (often because water laws are stunningly irrational; Quirk, 1978) nor based on process-level understanding of breeding and management. Only a process-based understanding is readily and cheaply transferred to new species, new sites, and changing climates. In a more comprehensive ecological context, photosynthesis exhibits wide plasticity to respond to variations in resource availability and what we may term modulators: environmental components such as temperature or salinity that are not physical resources but that scale up or down the utility of light, nutrients, and water. One prediction I am most willing to make is that most of the future gains in photosynthesis will be based on ideotype breeding that considers the whole plant in an ecological context and on management founded on process-level understanding, such as in the GOSSYM model for cotton (McKinion *et al.*, 1989). Quick fixes based on introducing novel genes, however ingeniously chosen, are unlikely to be effective.

II. Basic Concepts

A. Ways of Expressing Photosynthetic Performance

Photosynthesis per se is readily defined biochemically, as well as operationally, by the process of measurement, as by gas exchange. Some attention to how we may express the rate(s) of photosynthesis is rewarding.

Gross photosynthesis (denoted as P_g), which is initial CO_2 fixation by Rubisco enzyme, should be distinguished from net photosynthesis or net assimilation (P_n). The latter debits the carbon gain for respiration (R_d), which is expressed as $P_n = P_g - R_d$. Respiration and photosynthesis follow different time courses so that they cannot be untangled by short-term measurements, and, indeed, measurement of respiration is intrinsically difficult. However, in the longer term, respiration can be related to growth and maintenance (Penning de Vries, 1975; Thornley, 1977). For a given plant biochemical composition, the portion of R_d supporting growth is virtually a constant fraction of P_g. The portion of R_d expended for maintenance is an exponential function of temperature (though this acclimates) and a fairly linear function of nitrogen content, which reflects growth rate.

Note: The Appendix gives a brief concordance between my notation and that of Condon and Hall (see Chapter 3, this volume).

Photosynthesis involves the balance of photosynthetic carbon reduction (PCR) and photosynthetic carbon oxidation (PCO) (Farquhar *et al.*, 1980). The latter is unfortunately and misleadingly called photorespiration. However, the PCR–PCO balance is determined almost uniquely by the Rubisco enzyme kinetics and by temperature; it does not respond to growth or maintenance demands. (The balance is, of course, heavily favored by the C_4 pathway that pumps up CO_2 levels inside the bundle sheath cells, virtually eliminating the PCO cycle, at the cost of a modest ATP demand to run the CO_2 pump and *perhaps* a sensitivity to low temperatures.) The ratio of carbon oxidation to carbon reduction is the product of (i) a collection of Rubisco kinetic constants, which push the PCO/PCR ratio upward as temperature rises because the oxygenation capacity of Rubisco activates more strongly than does the carboxylation capacity (Farquhar *et al.*, 1980); and (ii) the ratio of O_2 to CO_2 partial pressure at the chloroplast, which is under stomatal control. It is almost certain that PCO cannot be significantly reduced by developing variant Rubisco, contrary to earlier high hopes (Morell *et al.*, 1992).

A most important distinction is between photosynthesis per unit leaf mass ($P_{L,m}$) and the much more commonly reported photosynthesis per unit leaf area ($P_{L,a}$). They are related by the mass per leaf area, $m_{L,a}$, which is sometimes misleadingly called specific leaf weight (I have been guilty of this). Mass per leaf area is the inverse of area per mass, which is called specific leaf area (SLA) or leaf area ratio (LAR). We have $P_{L,m} = P_{L,a}/m_{L,a} = P_{L,a} \cdot \text{SLA}$. Photosynthesis per unit mass is much more directly related to relative growth rate of the plant (RGR). Consider first the absolute growth rate of the whole plant, which equals the whole plant photosynthetic rate times the conversion efficiency (β) from raw photosynthate to final plant biomass. The value of β varies little unless plant biochemical composition changes notably (Penning de Vries *et al.*, 1974). RGR is the absolute growth rate divided by whole-plant mass, so it is simply the near-constant β multiplied by the whole-plant photosynthesis rate and divided by total plant mass. Thus, it equals β times the whole-plant photosynthetic rate per unit mass ($P_{p,m}$). Leaves do the greatest share of photosynthesis, so we may express $P_{p,m}$ as the leaf photosynthetic rate per mass times the fraction of whole-plant mass represented by leaves. This mass fraction is sometimes called the leaf weight ratio (LWR). There are some important developmental and ecological variations in LWR that will be explored. An important point to note about $P_{L,m}$ itself is that it typically decreases with increases in leaf investment in photosynthetic machinery per area, that is, as $m_{L,a}$ and $P_{L,a}$ increase (Gutschick, 1987).

One should also distinguish photosynthetic capacity from photosynthetic rate. Capacity is properly measured as the light-saturated and CO_2-saturated rate. It is set by the biochemical investment, particularly in Rubisco enzyme, in individual leaves. It is only realized under conditions of light saturation and minimal stress. Water stress, in particular, induces stomatal closure that lowers the internal CO_2 concentration [see Eq. (1) for a quantification of this effect]. Long–term water stress leads to photoinhibition, the decrease in ability of light absorption to drive the generation of reductant for CO_2 fixation. Plants broadly regulate capacity to realizable rates, as discussed in Section II,C, but changes in capacity are neither rapid nor efficient in using resources; it costs metabolic energy to break down capacity in one leaf and move photosynthetic capacity elsewhere.

Commonly, the goal is to estimate whole-plant photosynthetic rate (P_p) or whole-canopy photosynthetic rate per unit ground area (P_{can}). The whole-plant rate can be composed from individual leaf rates, for example, as the sum of $P_{L,a}$ for each leaf times the area of each leaf. The canopy rate might be composed as an integral over physical depth $[P_c = \int dz \overline{P}(z)]$, of the average photosynthetic rate of all leaves at that depth $[\overline{P}(z)]$. The canopy rate might also be obtained as an integral over depth expressed as cumulative leaf area index (L), that is, $P_c = \int dL \overline{P}(L)$. Whatever formalism is chosen, the challenge is to scale from individual leaves—often, one measures the top leaves only—to the rates for leaves elsewhere in the canopy. The scaling might be achieved by comprehensive sampling, which is tedious. It is also not informative of how leaf-to-leaf scaling will occur, for example, at other times of day, with different lighting conditions, etc. Thus, it is often desirable to have a process-level understanding of how photosynthetic rates respond to environment and of how the environment varies with canopy position. This requires verifiable models of light and wind penetration as well as of heat and gas transport in canopies that determine the distribution of leaf and air temperature and of water vapor and CO_2 concentrations. Models of an accuracy and attendant complexity, as appropriate to the questions being asked, are available (see Norman, 1993; Baldocchi, 1993).

The environment varies diurnally so that photosynthesis varies diurnally. The environment varies on shorter time scales, too, as when clouds pass over or wind moves the leaves. Time-integrated photosynthesis is commonly predicted rather well if one assumes that the leaves reach a series of steady states in each short-term environment. There are exceptions for rapid changes in light intensity that are discussed in detail near the end of Section III,A. On longer time scales (several days to years), the canopy develops, with changes in leaf area and leaf physiology. Commonly, we must compose diurnal and seasonal averages of photosynthetic rates (per leaf, per plant, and per ground area) from short-term or "instantaneous" rates. Again, data and resulting process models are helpful.

B. Outline of Responses of Photosynthesis to Environment

Although we are aware that scaling from leaf to canopy involves some unique phenomena, we may first consider how the environment and physiology determine leaf photosynthetic performance.

In the short term, a leaf's physiology is fairly fixed. This is the scale of up to a few days, barring extreme events such as freezing that can reset the physiology quickly. We then consider photosynthesis to be determined by the aerial environment—specifically, light (as incident irradiance; I_L), temperature (T), CO_2 concentration, humidity, and wind speed. The roles of each factor and their interactions have been reviewed and reformulated a number of times (see Massman and Kaufmann, 1991), and $P_{L,a}$ has correspondingly been expressed as products and sums of empirical functions of each variable. These formulations are, to some degree, exercises in curve fitting. They do not provide formulas that readily transfer to other species and conditions. Neither do they incorporate some important, close-to-universal physiological regularities, such as the pacing of stomatal conductance (g_s) to $P_{L,a}$ that keeps C_i nearly constant as light, temperature, etc. are varied (Wong *et al.*, 1985). A more fully process-based formulation is available (Collatz *et al.*, 1991; V. Gutschick, manuscript in preparation) that I now summarize.

Light-saturated photosynthesis depends only on Rubisco enzyme activity, thus, on the amount of enzyme per leaf area (a physiological setpoint), the leaf temperature (T_L), and the concentration or partial pressure of its CO_2 substrate (C_i). (The concentration of oxygen is considered constant, which is very accurate for land plants.) There are concise and accurate enzyme-kinetic models for both C_3 plants (Farquhar *et al.*, 1980) and C_4 plants (Collatz *et al.*, 1992). A simple exposition of the C_3 model sets the light-saturated photosynthetic rate as

$$P_{L,a}^{sat} = A_{max}\frac{C_i - \Gamma}{C_i + K_{CO}}, \tag{1}$$

where A_{max} is the rate of photosynthesis when Rubisco is operating at CO_2 saturation, Γ is the CO_2 compensation point, a function only of leaf temperature, as is the effective Michaelis constant for CO_2 binding, $K_{CO} = K_C(1 + O/K_O)$.

Light-limited photosynthesis occurs with a rate proportional to I_L,

$$P_{L,a}^{LL} = Q_0 I_L = \left[Q_{00} \frac{1 - \phi/2}{1 + \phi} \right] I_L. \tag{2}$$

The initial quantum yield, Q_0, has been written as a function of its limiting value at high CO_2 (Q_{00}) and of ϕ, the ratio of oxygenation to carboxylation rate. The latter is a function only of C_i and the temperature, as environmental variables: $\phi = (V_{o,max}/V_{c,max}) \cdot K_C O/(K_O C_i)$. Here, $V_{o,max}$ and $V_{c,max}$ are the maximal rates of oxygenation and of carboxylation, respectively, for Rubisco enzyme.

The transition between light-limited and light-saturated rate depends on a physiological coupling parameter, or convexity parameter (γ), as used in the empirical Johnson–Thornley equation (1984):

$$P_{L,a} = \frac{1}{2\gamma} \left[P_{L,a}^{sat} + Q_0 I_L - \left\{ \left(P_{L,a}^{sat} + Q_0 I_L \right)^2 - 4\gamma P_{L,a}^{sat} Q_0 I_L \right\}^{1/2} \right]. \tag{3}$$

Convexities are typically high, near 0.9, for plants, giving a sharp transition from light-limited to saturated rates.

Thus, leaf photosynthesis is formulated in terms of physiology ($P_{L,a}^{sat}$, Q_{00}, and "invariants" such as K_{CO}) and light, leaf temperature, and CO_2 as C_i. However, the latter two are not raw environmental variables: They are partly determined by leaf activity, specifically, the stomatal conductance and $P_{L,a}$ itself. C_i equals the free-air CO_2 concentration (C_a) drawn down by the amount needed to keep CO_2 flowing through the combined stomatal and boundary-layer resistance for CO_2 movement ($1/g_{bs}'$:$C_i = C_a - P_{L,a} \cdot P/g_{bs}'$). Here, P is the total atmospheric pressure, needed in this equation if conductance is in molar units. Clearly, $P_{L,a}$ is determined by a feedback loop, and a model must be solved iteratively or otherwise cleverly; this is readily done, and the feedbacks are real (Mott, 1988).

What sets the stomatal conductance (g_s)? It responds to C_i (Mott, 1988) and the transpiration rate (Mott and Parkhurst, 1991; Monteith, 1995), as well as to the temperature. Stomatal constriction in response to high transpiration rates is less pronounced at higher temperatures probably because resupply of water to guard cells is stronger, in part because water's viscosity decreases. (Mechanistic studies have not yet been carried out on this explanation.) Given this balance between transpiration response and temperature response, g_s is close to proportional to leaf-surface relative humidity, as in the empirical equation of Ball *et al.* (1987):

$$g_s = m \frac{P_{L,a} h_s}{C_s} + b. \tag{4}$$

This "Ball–Berry" equation closes the loop for photosynthesis and g_s. One can then solve it (or more accurate, more mechanistic equations, e.g., Dewar, 1995) along with the earlier equations accurately and stably over all environmental conditions. This is challenging mathematically (Collatz *et al.,* 1991; V. Gutschick, manuscript in preparation). The Ball–Berry equation compactly expresses how g_s and photosynthesis are paced to each other. It also expresses the near constancy of C_i. If the total conductance is limited by the stomatal conductance (as in common), and if the residual (cuticular) conductance is small, then at light saturation we have

$$\frac{C_i}{C_a} \approx 1 - \frac{1}{mh_s}. \tag{5}$$

This further assumes that Γ is small relative to C_i; if not, a more complicated formula results, still showing how only m and h_s really control C_i. Nearly constant C_i is important in balancing light use with water use, as discussed below. For C_3 plants, m is about 10, so that C_i/C_a is about $1-1/5 = 0.8$ at a relative humidity of $0.5 = 50\%$ (the true ratio is about $0.7–0.75$, obtained with a more accurate approximation). For C_4 plants, m is about 2 or 3 and C_i/C_a (evaluated with a more accurate approximation) is about 0.25.

What determines the leaf temperature (T_L)? The principles of energy balance are well established (Gates, 1980; Nobel, 1991). At steady state (quite close to true for almost all leaves except thick ones, such as cactus phyllodes; Nobel, 1978), heat gain balances heat loss. Heat gain is radiative, from the absorbed irradiance in the photosynthetically active region (PAR) as well as in the near infrared and thermal infrared. Heat loss is by (i) transpirational cooling, at a rate $\lambda E = \lambda g_{bs}(e_L - e_a)/P$, where g_{bs} is the conductance for water vapor through stomata and the leaf boundary layer, λ is the heat of vaporization of water, and the last factor is the gradient in mole fraction of water vapor from leaf to air (using partial pressures). Ambient e_a is a raw environmental variable, and e_L depends only on leaf temperature so that the solution for T_L is feedback looped, although mathematically well defined; (ii) radiative cooling, at a modified blackbody rate dependent virtually only on T_L; and (iii) convective–conductive cooling, dependent on boundary-layer conductance and the leaf-to-air temperature difference. Thus, leaf temperature is determined by air temperature (T_a), air humidity (e_a), windspeed, the radiative fluxes, plus the leaf behavior as g_s and the leaf boundary layer, set by windspeed and leaf dimensions.

With the previously discussed formulations, leaf photosynthesis, transpiration, stomatal conductance, and temperature are set by the interaction of leaf physiology with six major environmental variables, all of which are measurable and able to be modeled for a leaf anywhere in the canopy: T_a, C_a, e_a, windspeed, PAR irradiance I_L, and the NIR and thermal infrared fluxes. The net results are generally familiar. For example, photosynthetic rate has a pronounced temperature curve, with an optimum near-maximal air temperature in a region. Also, $P_{L,a}$ scales downward at low humidity [see Eq. (4)], which may commonly cause the midday depression in stomatal conductance (*e.g.,* Meinzer *et al.,* 1993).

This presentation has not accounted for several other effects that are important in some regimes: (i) stress, especially water stress (see, for example, Tardieu *et al.,* 1993) and photoinhibition; (ii) response of the leaf to transients, which are notable in the light response (Krall and Pearcy, 1993; Pons and Pearcy, 1992); and (iii) the possibility of product inhibition of photosynthesis on itself, which can be demonstrated with some strong forcing (Goldschmidt and Huber, 1992; Socias *et al.,* 1993) and some normal diurnal buildup (Chatterton, 1973). The third effect appears to be rare, inducible by high CO_2, prolonged day length, etc. These topics are taken up in more detail in the following section.

C. Physiological Acclimation and the Integration of Other Resource Needs

Photosynthesis acclimates to the environment in the longer term. Responses occur both at the leaf level, such as by changes in chlorophyll and enzyme contents, and at the canopy level, by changes in development of leaf area, placement of leaves, branching pattern, etc. The responses will be discussed in some detail in Section III. Some responses are to the immediate driving variables of photosynthesis discussed previously. For example, photosynthesis acclimates to the recent history of light and temperature (Bunce, 1983). Plants typically show a near-optimal response, allocating carbon and nutrient resources to match photosynthetic capacity (as $P_{L,a}^{sat}$) to potential photosynthetic gain leaf-by-leaf throughout the plant or canopy (see deJong and Doyle, 1985; Hirose and Werger, 1987; Gutschick and Wiegel, 1988). Admittedly, the optimum in allocation pattern is quite broad, so it is hard to incur more than several percentage penalties in canopy photosynthesis. Nonetheless, selection pressure to stay in the broad optimal range is strong. Leaves represent one of the largest growth investments of a plant; they must pay back several times their metabolic costs of construction and operation (Mooney and Gulmon, 1982). As a result, plants control investment so that leaves operate near or somewhat above light saturation at the diurnal peak. Leaves respond to temperature with a greater investment in capacity at low T to keep photosynthesis rates high at light saturation (*e.g.,* Mooney *et al.,* 1978). Leaves respond to CO_2, and the clear prospects of climate change have made a signficant industry of elucidating these responses. Of course, plants have had experience with CO_2 variations only over many generations. Their responses, both physiological and genetic, on the time scale up to decades and centuries are unlikely to be strongly adaptive, as with responses to other rare phenomena with low time-averaged selection pressure (see Robinson, 1994). Note that genetic changes require many generations (tens of thousands) with modest selection pressure or fewer generations with intense selection, which are rarely seen (Wallace, 1981). In any event, we have no measures of genetic change in photosynthetic responses over the past century; we do not even know which genes to track. By examining museum specimens, Peñuelas and Matamala (1990) have noted dramatic developmental and physiological changes in stomatal density and leaf nitrogen content in the past three centuries. These changes may be adaptive but are hard to evaluate. Even with living specimens, exposure to high CO_2 gives divergent patterns of photosynthetic and growth responses among species (Poorter, 1993), and among the plants with the highest gains in growth it is difficult to assess if they are responding close to optimally, not only in growth but in tolerance of stresses—water, nutrients, heat, pests, etc.

Plants also respond in photosynthetic physiology (and leaf morphology and phenology) to other environmental variables than we have so far considered. Plants respond to the availability of resources that support photosynthesis on a time-integrated basis—nutrients and water. They must trade off the acquisition and usage efficiencies of resources: light, water, nutrients, and carbon. There are similar compromises in root function, such that the separate resources may be viewed as having votes on the root:shoot ratio (Davidson, 1969). These compromises are discussed in some detail below. Plants also respond in their photosynthetic physiology to biotic factors such as symbionts, herbivores, and diseases. These biotic factors may "simply" alter resource availability (light and nutrients) or they may have specific metabolic and developmental effects that are more complex to describe (e.g., curly top virus, which alters both leaf physiology and light interception). Biotic factors, and some extremes in the abiotic environment (frost, etc.), occur rather stochastically, being described more as intermittent risks than as deterministic variables. They demand a special structure in plant adaptive responses (Gutschick, 1987).

On the larger spatial scales, plants help set their own environment and their ability to respond to the environment. Large regions of vegetation (e.g., hundreds of meters distance or more) generate a canopy boundary layer in the atmosphere that presents a resistance (a low conductance) to the transport of CO_2 and water vapor (Jarvis and McNaughton, 1986). With a sufficient combination of short plant stature, long distance or "fetch," and low windspeed, this resistance becomes so large that stomatal control becomes much less effective (Jarvis and McNaughton, 1986) and the environment alone controls transpiration. Control of photosynthesis remains partly physiological because there is a biochemical "resistance" in it that is lacking in transpiration. In any event, vegetation upwind helps set the temperature and humidity of air over a canopy. The canopy itself, even of more modest fetch, sets some of its temperature and humidity (Raupach, 1989).

Overall, we may view the ecological performance of a plant as following a genetic program to maximize its Darwinian fitness. This view, that plants may have some objective function (a combination of fecundity, etc.) to maximize, can help us synthesize their adaptive responses. Consequently, we can understand that there are trade-offs that limit raw photosynthesis and yield. For example, the pressure to repress competitors for light and water can lead to complexes of adaptive responses ("strategies") that forgo some canopy photosynthesis for height development (Givnish, 1982) or that reduce water-use efficiency and thus limit season-total photosynthesis, in order to get a greater share of water competitively (e.g., Lajtha and Barnes, 1991). I have given (Gutschick, 1987, pp. 17–20) a four-component view of the adaptive response program: (i) Plants tend to minimize the energy used in acquiring a unit of any resource (light, nutrient, or water); (ii) plants often have to trade off short-term benefits for greater long-term benefits, such as by developing a taproot at the expense of short-term water uptake that would be better served with fine roots; (iii) resource uses are compromised among each other: for example, high nitrogen-use efficiency might demand low leaf nitrogen content and attendant low g_s and low $P_{L,a}$, but these may depress light-use efficiency: low g_s leads to low C_i and low efficiency of using both light at low levels and Rubisco investment (carbon) at high light; (iv) plants reduce risk at some costs in resource use, such as by terminating vegetative growth (giving up some yield) well before the frost hazard is imminent. There are many risks, and they may not be obvious, especially in the response programs that plants have evolved. Cowan (1987) gives an intriguing discussion of risk amelioration in drought stress.

Furthermore, we can realize that the genetic program of a crop plant evolved in a different environment and has a suboptimal fit to the agricultural environment. This genetic program is a misfit not just to the abiotic environment, such as temperature optima or inappropriate responses in nitrogen nutrition (Gutschick, 1981) or photoperiod (Garner and Allard, 1920). I have given a summary (Gutschick, 1987, pp. 3–4) of the mismatch in objective function between wild growth and agriculture: (i) wild plants tend to maximize fitness, of which the chain of photosynthesis → growth → yield is only one component. For example, some plants must time reproduction to achieve seed predator satiation (see Ashton *et al.*, 1988), possibly fruiting before or after the time that maximizes yield. More certainly, plants do not respond to costs of inputs (nutrients, water, etc.) with the same relative value/weighting as does the farmer. (This presentation combines my first and fifth points in the Gutschick, 1987.); (ii) wild progenitors of crops were often ruderals, in the three-strategy classification of Grime (1979); they grew rapidly to garner maximal shares of resources. In agriculture, in contrast, we ask for good cooperative growth, without self-thinning, with sharing of light and nutrients, etc. Dwarfing helps optimize agricultural yield but is contrary to protection from overtopping. Consequently, it is contrary to the genetic programs of most plants and perhaps in more ways than we have yet elucidated. In general, we have substituted herbicides and cultural practices for competitive ability of plants to ensure that they get resources abundantly; (iii) risk management in wild plant responses is often much different from what is desirable in agriculture. Seed shatter and staggered germination to give a seed bank are desirable in wild growth, whereas they are anathema in most of agriculture. Low leaf N content may achieve a near-optimal compromise between photosynthetic rate and risk of herbivory in wild plants, but in agriculture the latter risk may be reduced with pesticides; a high N content is demanded to balance costs and benefits in agriculture; and (iv) the agricultural environment is not the same as the original environment. We might breed in proper cold tolerance, photoperiodic behavior, etc.—perhaps fixing almost any environmental mismatch we recognize. How many mismatches do we not recognize? To these points, I add two: (i) The plant is not fully optimized for its original environment. There is always a lag in coevolution of competitors or predators and prey (Stenseth and Maynard Smith, 1984); (ii) much of agriculture is based on hybrid production from inbred lines. The mentality that uniform heterozygotes are superior is widespread. In wild plants, in contrast, both inbreeding and outbreeding extremes are suboptimal. There is an optimal degree of outbreeding for individual traits related to photosynthesis as well as for adaptive complexes of genes (Thornhill, 1993). Part of this legacy is that genetic variability is both beneficial and selected. Another part is that we have ended up with a tiny subset of the genetic complexes in our crops. Many alleles work best with specific combinations (adaptive complexes; Hedrick *et al.,* 1978) of the alleles at other loci. We do not have all the best combinations. This realization should help us breed and manage plants.

Next, we examine some of the ecological responses of photosynthesis in more detail.

III. Some Detailed Adaptive Responses of Photosynthesis

A. Light Use

Effective use of light requires both good light interception and efficiency in use of intercepted light. Let us attend to interception first. Both the formal and the practical descriptions of how plants intercept light are rather complete (see, *e.g.,* Ross, 1981; Asrar, 1991; Myneni and Ross, 1994).

An ecological perspective is also well developed (*e.g.,* Caldwell and Pearcy, 1994).

Light interception by leaves is the first stage in light use. One measure, quite incomplete by itself, is light penetration to various depths or, in row-structured or otherwise nonuniform canopies, to various positions. The simplest case is that of a canopy that is laterally uniform at least in a statistical sense, over some averaging "window," perhaps 1 m². Any reference on light interception (see above) formulates the penetration fraction for the direct beam (P_{pen}) as exp(-KL). Here, L is the cumulative leaf area index to the chosen depth and K is the extinction or attenuation coefficient. The value of K depends on the statistical distribution of leaf orientation angles and on the view angle. Simply, K equals the average fraction of total leaf area presented along a view direction (that is, the average cosine of the angle between leaf normals and the view direction) divided by the cosine of the view angle. The latter factor corrects from slanted optical path to nominal vertical path.

Some effects are immediately apparent in the formula. First, the total interception by the canopy reaches saturation as [1-exp(-KL)]. More leaves mean more interception, though at decreasing average irradiances per unit leaf area. This dimunition calls for leaf area to be deployed economically, with leaves at the top having more photosynthetic capacity and lower leaves having less. Aging and senescence naturally achieve this scaling with depth. The dimunition of capacity in deep leaves can be delayed or reversed by reexposure to high light (Johnston *et al.,* 1969). It is natural to inquire what the optimal distribution is for total leaf mass or for leaf nitrogen content (deJong and Doyle, 1985; Hirose and Werger, 1987; Gutschick and Wiegel, 1988). It is typically found that plants are near the optimal distribution. It is also predicted that canopy total photosynthesis is fairly insensitive to changes in the distribution of leaf mass or N if it is anywhere near the optimum. There is little reward in looking to reoptimize this aspect of canopy structure.

Penetration and total interception are only part of the story. Also important is the distribution of irradiances on leaves. For a given leaf angle distribution (LAD) and direction of illumination, one can calculate the probability distribution of irradiances [$p(I_L)$] either analytically or numerically (Gutschick, 1984). Briefly, the more the leaf blade aligns with illumination direction (vertical leaves for overhead sun, for example), the lower will be the average I_L and the greater will be the total area of leaf illuminated for a fixed degree of canopy total light interception. Such considerations underlie the model-inspired breeding for greater leaf erectness (Trenbath and Angus, 1975). Consider the highly simplified case in which leaf photosynthesis saturates at one-third of full sunlight. Replacing horizontal leaves with leaves at 60° off vertical will double the illuminated leaf area while keeping $P_{L,a}$ saturated. Of course, there are co-occurring increases in costs of leaf construction and maintenance. Considering these costs, as well as diffuse light and second interceptions, etc., in real canopies gains in photosynthetic are modest (on the order of 10%) but significant economically. The erect-leaf hypothesis is not yet exploited in a number of crops.

The simple formulae for light penetration overlook the statistical clumping of leaves within a neighborhood. That is, real leaves are often placed with either more or less overlap in shadowing than in the Poisson distribution of leaf centers assumed in the formulae. Clumped leaves intercept less light than expected based on LAI and LAD, and overdispersed leaves (leaf mosaics) intercept more. Trees typically have clumped leaves (Baldocchi, 1989; Oker-Blom *et al.,* 1991), and crops are often overdisperse. Overdisperse geometry would appear to be more efficient in leaf display, if leaves meet the corresponding criteria for high photosynthetic capacity to exploit the high irradiances of first interception. Overdispersion may bear costs of structures such as longer petioles (R. Pearcy, manuscript in preparation). (Trees' photosynthetic strategies, in leaf display and in carboxylation capacity, appear far from optimal and are puzzling, certainly to the author.)

Crops are often nonuniform laterally, being arranged in rows. The description of light interception is more complicated but quite tractable (Norman and Welles, 1983; Wang and Jarvis, 1990). Greater plant density from smaller row spacing is desirable to get earlier canopy closure, for both light use and water use (soil evaporation is repressed; Richards (1991). Smaller spacing within rows has been chosen, accounting for some significant gains in yield (for a brief review and model, see Thornley, 1983). Between-row spacing is constrained by requirements for mechanical harvesting and weed control. The complexities of light interception in orchards and grapevine trellis systems have yielded to some empirical insights. For example, pruning of vines and trees is needed for access to fruit. Although it clearly decreases canopy light interception and canopy total photosynthesis, the effect is transient. Remaining leaves acclimate quickly and leaf area regrows well before the time of reproduction (Downton and Grant, 1992). In thinning of developing fruit, fruit quality is kept high by enforcing patterns that keep fruit nearest the leaves that have highest light interception and thus are the strongest carbohydrate sources (Corelli-Grappadelli and Coston, 1991; other direct effects of light on fruit skin are cited).

Crops have been bred for uniformity in stature as well as in many other attributes. Uniform stature gives the most efficient light interception per unit leaf area, but the effect of nonuniform height on canopy photosynthesis per LAI is very modest (Gutschick, 1991). An extension of nonuniform height is nonuniform species composition, as in intercropping. A number of studies (*e.g.,* Harris *et al.,* 1987) show gains in overall light interception efficiency with intercropping, most of which are the result of better temporal continuity of cover when crops have differing phenologies. Differences in species or stature *per se* would not be expected to alter light interception efficiency, at least not for the better.

Dwarfing improves light interception per mass of shoot. Although interception of light per leaf area is only mildly improved (by lessened shading of leaves by stems), the plant can deploy a greater fraction of its mass as leaves and less as stems. Dwarf plants are more readily overtopped by competing weeds, of course, and are a tenable choice only in very intensive agriculture.

Leaves of some plants, especially legumes, track the sun (Forseth and Ehleringer, 1983; Reed and Travis, 1987). That is, the LAD changes dynamically over the day. Tracking typically acts to maximize light interception per leaf area for unstressed plants, whereas water-stressed plants may switch to paraheliotropism, avoiding the transpirational load (Reed and Travis, 1987). Interesting trade-offs occur in solar tracking. There is a higher average irradiance and higher interception per unit leaf area. This will translate to higher photosynthesis per unit leaf mass only in certain conditions: (i) LAI is low to moderate. If LAI is high, light interception is virtually complete with or without tracking. Tracking only serves to concentrate irradiance on top leaves, which are light saturated; photosynthesis per unit mass of the whole canopy then declines; and (ii) leaf temperatures do not rise unduly. If the leaves are operating at or near their optimal T_L, then diaheliotropic orientation may only increase T_L without increasing $P_{L,a}$. It will also decrease water-use efficiency, in general.

Thus far, we have not considered diffuse light, either as diffuse skylight or as light scattered from other leaves. Diffuse sunlight may be considered as a set of collimated beams from different sky angles. One emergent property is that it penetrates a canopy almost deterministically, rather than probabilistically (Gutschick, 1984). If the diffuse irradiance onto a horizontal surface is D_0 at the top of the canopy, it is approximately $D_0 e^{-K_d^L}$ on every leaf at depth L, where K_d is an effective extinction coefficient for diffuse light (it varies somewhat with depth, as glancing rays are removed). Diffuse light from skylight and light scattered from other leaves and soil accounts for a significant fraction of canopy photosynthesis. Its contribution is very high in cloudy climates (Ort and Baker, 1988). There is a modest opportunity to exploit it by using low-chlorophyll mutants of plants (Gutschick, 1984; tested by Pettigrew *et al.*, 1989). In these mutants, light-saturated photosynthesis is little affected or slightly increased, whereas light sharing with the deeper canopy is increased. There is a penalty in early growth at low LAI, where light absorption is reduced. Yield increases up to about 8% might be achieved. However, mutants that bred true (vs the more common heterozygotes) are not readily available.

Interception of the direct solar beam is not a yes–no proposition that gives either full sun or full shade. The sun has a finite angular size as viewed from the earth so that shadows are graded in the penumbra (Miller and Norman, 1971; Oker-Blom, 1985). In broad-leaved plants, passage of light to nearby canopy layers is little altered by penumbral effects. However, penumbra changes the irradiance distribution $[P(I_L)]$ significantly on grasses and conifers. The change is favorable. Simply, the irradiance distribution is pushed toward intermediate values. Photosynthetic rate at half of full sunlight exceeds the rate averaged between full and no sunlight because the light response of $P_{L,a}$ is convex. Narrower leaves might be beneficial (Wells *et al.*, 1986). Although penumbral effects would favor canopy photosynthesis, narrower leaves have higher boundary-layer conductances and maintain leaf temperatures close to air temperatures. Broad-leaved plants with high $P_{L,a}$ have attendant high conductance and transpirational cooling. Their leaf temperature may be 8°C below air temperature, which allows higher water-use efficiency.

Overall, the nuances of light distribution among leaves give the whole canopy a different efficiency of using light than that of single leaves. The strongest effect is that canopy photosynthesis increases almost linearly with photon flux density (PPFD) out to PPFD values that are about twice as large as those that begin to saturate single-leaf photosynthesis. This extended linearity has led to the concept that light-use efficiency, as biomass produced per mole or per Joule of photons intercepted, is a near constant with different light levels. Some authors even argue that various species show similar efficiencies, but (i) the variation in slope is clear, (ii) it is informative about the crop physiology, and (iii) saturation at highest light intensities is also clear (Monteith, 1994). The saturating PPFD is also lower when a greater proportion of sunlight is a high-PPFD direct solar beam (Norman and Arkebauer, 1991), which is fully consistent with the concepts of light interception previously discussed.

One more aspect of light interception is the effect of interception by nonphotosynthetic tissue-support structures and senescent leaves. In crops, these tissues are generally modest, intercepting less than 5% of light until the whole crop matures. In wild grasses and some woody species, much light is intercepted. This limits the productivity of unburned grasslands, for example (Collins and Wallace, 1990). Nonetheless, nonphotosynthetic light interception may be important in limiting yield in some crops, as in rice, in which leaves are shaded by panicles. The flag leaves that provide most of the photosynthate for grain filling are most affected. Breeding for panicle height is promising (Setter *et al.*, 1995).

After light is intercepted, it is used to drive the light-dependent biochemical reactions of photosynthesis. The quantum yield (Q) between absorbed light and electron transport is important. (I refer to $Q = P_{L,a}/I_L$; its initial value at low irradiance, Q_0, is only one measure of this efficiency). The value of Q is affected by the distribution of leaf irradiances and hence by LAD and other aspects of light interception discussed previously. It is also affected by leaf temperature (Ehleringer and Björkman, 1977) and ambient CO_2 concentration [see Eq. (2) for Q_0, and the discussion]. For C_3 plants, Q_0 matches that of C_4 plants (about 0.05 mol CO_2 per mole photons) near 25°C, but it is a declining function of increasing temperature. In contrast, Q evaluated for light-saturated photosynthesis rises until the optimal T is reached, declining precipitously thereafter. For C_4 plants, Q_0 is almost unaffected by temperature. Their light-saturation degree is less and their temperature optima are typically higher (Pearcy and Ehleringer, 1984) so that Q at high light is also less T-dependent. A more complete discussion of light-use efficiency is given by Sinclair and Horie (1989). These contrasts between C_3 and C_4 plants have many ramifications, as in weed control and in the desirability of introgressing the C_4 pathway into C_3 crops (not a likely route to improving crops). The ramifications extend to crop management, such as choice of crop to suit mean temperatures.

The discussion to date has treated photosynthesis as if leaf irradiance (I_L) were constant; that is, as if photosynthesis were in steady state. Fluctuations in I_L cause three phenomena. The biochemical steps can average out rapid variations in I_L (on a timescale faster than pools of intermediate metabolites react). Consider alternation between two irradiances, for example, 300 and 900 μmol m^{-2}sec^{-1}. Let the light-response curve be approximated by a Blackman curve: $P_{L,a} = 0.05\ I_L$, up to $I_L = 600$, and $P_{L,a} = 30$ μmol m^{-2}sec^{-1} after that. A leaf slowly alternating between the irradiances would have an average photosynthetic rate of $0.5*(15 + 30) = 22.5\ \mu$mol m^{-2}sec^{-1}. Rapidly alternating I_L would give the rate appropriate to the average $I_L = 600$ in these units or $P_{L,a} = 30\ \mu$mol m^{-2}sec^{-1}, a superior rate. Rapid fluctuations are common in grass canopies in the wind and may occur at frequencies exceeding 10 Hz (Norman and Tanner, 1969). Fluctuations in irradiance are accompanied by lags in adjustment of stomatal conductance. When I_L declines, stomatal conductance stays higher than the steady-state value appropriate to the new I_L; photosynthesis is modestly elevated by the higher C_i allowed (and transpiration is much elevated). Conversely, photosynthesis is much limited during rises in I_L, or sunflecks. R. W. Pearcy and co-workers have made extensive studies of responses of both crop plants (*e.g.*, Pons and Pearcy, 1992) and forest understory plants (Pearcy and Pfitsch, 1994) to sunflecks of varied durations, from seconds to minutes. Both stomatal and biochemical responses were quantified. The net effect of stomatal lags can readily be inferred as a decrease in average photosynthetic rate, and this may be observed (Denmead, 1968). Fluctuations in the short term (seconds and minutes) and medium term (hours) irradiance and in other driving variables of photosynthesis induced regulatory responses in the enzymes of carbon fixation. Leaves show marked metabolic flexibility (Geiger and Servaites, 1994), especially in controlling the degree of activation of Rubisco enzyme to match needed and potential carbon fluxes. Remaining lags in adjustment of intermediate metabolite pools cause small (on the average) reductions in photosynthesis under varying light for overstory plants (cf. Woodrow and Berry, 1988; Gross *et al.,* 1991).

Plants acclimate to the irradiance regime, approximately to the total daily irradiance, as well as to temperature (Bunce, 1983). Essentially, leaves shift their light-saturated rate to keep the utilization of available irradiance rather high. Leaves undergoing a change to higher irradiance increase their investment in Rubisco enzyme. The investment in electron-transport capacity keeps pace so that it is not limiting (Terashima and Evans, 1988); this capacity is metabolically inexpensive relative to Rubisco. Pigment concentrations per volume decrease, which is adaptive because light capture becomes less limiting and because this allows light penetration to use the full depth of the leaf mesophyll (Cui *et al.,* 1991). Given that light scattering in the leaf remains high, leaf total absorptivity in the PAR region drops slightly, for example, from 0.90 for shade leaves to 0.85 for sun leaves; this is a secondary effect of acclimation for light penetration through the leaf. The saturating irradiance level is, of course, a function of temperature. As temperature changes, Rubisco becomes more or less limiting. Acclimatory changes in Rubisco content per leaf area occur. No fully quantitative studies have been made of whether the degree or rate of change of photosynthetic capacity with changes in I_L or T_L are optimal. Certainly, the leaf cannot adjust its capacity very much on a diurnal course to follow diurnal changes in light and temperature. The cost of catabolizing and then building up capacity is too high, and the kinetics are also limiting. On the seasonal scale, there are both fast and slow changes in carboxylation capacity.

Light is not always a resource, as my colleague Marilyn Ball phrases it—just as the same is true for water, nutrients, and CO_2. Because the ability to use high irradiance for PCR and PCO is limited, high irradiance will lead to photoinhibition (PI) of the efficiency of electron transport (Baker and Bowyer, 1994) and, later, of Rubisco activity. Low PCR capacity can occur because the temperature is low or because the stomata are closed from water stress. PI induced by low T is distinct from low-T damage per se in that it depends on high light exposure in order to develop (Lundmark and Hällgren, 1987). Similarly, low-g_s induction of PI is distinct from water-stress effects *per se.* Photoinhibition commonly occurs in recurrent episodes, such as a run of very cold, clear mornings or a run of drought with stomatal closure early in the photoperiod. The initial manifestation of PI as a reversible downregulation is adaptive in that it prevents oxidative damage to the photosystems. Downregulatory PI is reversible in fractions of an hour and bears a modest cost in lost time to fix CO_2 after stress relief. Its insufficiency can lead to PI construed as damage, which is reversible only with the (greater?) metabolic cost of repair and with more persistent lost opportunity costs. The blend of both types of PI has been estimated to reduce diurnal photosynthesis on the order of 10% in willow leaves exposed to low temperature and high light (Ögren and Sjöström, 1990) or in a variety of field crops (Baker *et al.*, 1994). In most crops, the significance of PI is unassessed (especially for drought-induced PI) but probably similar. The consequences of low-*T*-induced PI for eucalyptus seedlings regenerating in clear-cuts (Ball *et al.*, 1991) are large. The tree seedlings can even be put into negative carbon balance in severe circumstances, leading to death. Acclimation to photoinhibition is probably modest and limited and, again, not well assessed. It is limited because any process that overlies PI to reduce excess electron transport looks like PI itself, with the exception of strategies of leaf-angle shifts (rolling, drooping, and paraheliotropism).

B. Use of Carbon Dioxide

Over a day or a season, the concentration of CO_2 in ambient air (C_a) typically varies little. The canopy is rather well ventilated so that CO_2 transport is high relative to the photosynthetic sink and respiratory sources of CO_2. Exceptions occur in dense crops wherein windspeed is low so that CO_2 may be depleted (Raupach, 1989) or in forests with high litter decomposition rates and soil respiration that elevate the concentration of CO_2 (Broadmeadow *et al.*, 1992).

Plants with the C_3 pathway may be considered CO_2-limited in that short-term increases in C_a will increase photosynthetic rates on both area and mass bases. This increase is a manifestation of the monotonic increase of P_{La} with C_i inside the leaf. Given the CO_2 limitations, why do C_3 plants generally restrict stomatal conductance to give C_i near 70% of C_a (Bell, 1982) rather than a higher value? The simple answer is that high C_i decreases water-use efficiency very dramatically as C_i/C_a approaches unity. The trade-off of photosynthesis vs WUE is rather finely poised, as discussed in the following section. Plants with the C_4 pathway respond little to CO_2 in their photosynthetic rate but decrease their stomatal conductance and improve their WUE (*e.g.*, Owensby *et al.*, 1993).

Long-term responses of photosynthesis and related physiology (photosynthate transport, etc.) to changes in C_a show a mixed picture (Bowes, 1993; see Körner *et al.*, 1995, for an interpretation). Commonly, plants at high C_a exhibit notable downregulation of P_{La} and of underlying N and Rubisco contents per leaf area. These responses complicate the prediction of crop performance under future climates. They may be related to improving nitrogen-use efficiency if N is limiting or to a nonadaptive nature of responses to a rare phenomenon or to other causes (see Robinson, 1994).

C. Water Use

This topic is also treated in Chapter 3 in this book, so the discussion here will focus more on adaptive responses.

Transpiration of water to support photosynthetic carbon gain is the only significant season-long use of water in the plant. Water for cellular growth is less than 1% of transpiration. Transpiration purely for leaf cooling at high temperature may occur, but the effect is hard to assign quantitatively and is limited to extreme conditions that account for a small part of season total water use. We might enlarge the accounting of water use by transpiration to include a significant part of water that is "intercepted," having fallen on leaves and remained there to evaporate directly over time. The evaporating water suppresses the water-vapor gradient, allowing photosynthesis to occur with low transpirational water loss. Wronski (1984) calculates that in typical conditions, intercepted water is about 60% as effective as transpired water in supporting photosynthesis. (The water beads on leaves also depress photosynthesis approximately in proportion to fractional area occupied; they depress transpiration much more.)

The most concise measure of water's role in photosynthesis is the water-use efficiency, WUE $= P/E$. Here, both photosynthesis and (evapo)transpiration must be expressed on a common basis—per leaf area or per canopy area; instantaneously, as a diurnal average, or as a seasonal average; etc.—as appropriate to the question being considered (Jarvis and McNaughton, 1986; Gutschick, 1987; Gutschick and Cunningham, 1989).

The central trade-off to consider is of photosynthesis against WUE, as set primarily by the value of C_i. For short-term gas exchange,

$$\text{WUE} = \frac{g_{bs}'(C_a - C_i)}{g_{bs}(e_L - e_a)} \approx \frac{0.62(C_a - C_i)}{(e_L - e_a)}. \tag{6}$$

A remarkable body of work has centered on viewing C_i as subject to optimization, by either natural or artificial selection (Cowan and Farquhar, 1977; Givnish, 1987). Equation (6) is often viewed in the approximation that the vapor-pressure deficit (VPD) in the denominator is independent of C_i. However, C_i can be altered, for example, decreased, and in a variety of ways. At one extreme, g_s might be reduced while holding a constant photosynthetic capacity (as $P_{L,a}^{sat}$). At the other extreme, g_s might be held constant while increasing $P_{L,a}^{sat}$. There is a continuum of gradations in between. The first option, of lower g_s, leads to hotter leaves and a greater VPD. This is one major compromise in WUE set by biophysical feedbacks. Another is slower canopy closure with lower g_s and lower $P_{L,a}$, leading to greater time-integrated soil evaporation. There are other plant strategies to alter WUE, such as by variations in leaf angle distribution, including heliotropism. Overall (Gutschick and Cunningham, 1989), one expects that WUE cannot be altered by more than about 15% upward from common crop values. The setpoint of C_i at about 0.75 of C_a is thus an appropriate one in C_3 plants for mesic conditions. Lower C_i values are appropriate to arid zones, where water has a high cost of acquisition by roots or there is a high expectation value of drought-induced damage. [Note, however, that ecotypes from more arid zones appear to have higher C_i than mesic ecotypes when grown in common gardens. See Read and Farquhar (1991). My interpretation is that C_i would be pushed too low for (nutrient-)efficient photosynthesis by the humidity response of stomata in arid environments, were it not for a higher basic setpoint.] The C_i values of C_4 plants are dramatically lower (about 0.2 C_a). The advantages for water-use efficiency, among other performance measures, are renowned (Pearcy and Ehleringer, 1984). The trade-offs involved in the C_4 setpoint are not well explored.

Not only leaf function in photosynthesis and transpiration but also canopy structure and light interception respond to water-use constraints. Leaf area development is rapidly curtailed by low soil water availability (Davies and Zhang, 1991). This is clearly adaptive in that transpiration insupportable with current leaf area becomes more problematic rapidly with increasing leaf area. Recent work (Blackman and Davies, 1985; Masle and Passioura, 1987; Tardieu *et al.*, 1993) emphasizes that the primary signal is in response to soil water content (SWC) and not leaf water potential. This befits SWC as the best indicator of longer-term prospects of water availability. One may ask if canopy LAI development is curtailed at the appropriate level for a given time trajectory of water stress. That is, if LAI development were less conservatively curtailed, might the crop be better able to use water made available later, even with greater WUE? The question considers the balance between probabilistic gains of water use against probabilistic losses of function and even losses of leaf area (carbon investment) by leaf death if the plant overshoots in transpiration. The question has not been addressed in research, although Jones (1983) has addressed the complementary question about appropriate curtailment of g_s; he phrases the problem as the contrast of optimistic and pessimistic stomata. While we consider the topic of drought, it is also worth noting that WUE is only one component of drought tolerance, with other contributions by osmotic adjustment, root:shoot investment, leaf reflectance, etc. (*e.g.*, Turner *et al.*, 1986). In fact, unstressed WUE has been observed repeatedly to correlate negatively with plausible measures of drought tolerance (Grieu *et al.*, 1988; Thomas, 1986). The optimal balance of WUE, drought tolerance, and photosynthetic performance remains elusive.

Photosynthetic capacity can be decreased reversibly by water stress (Chaves, 1991; Kaiser, 1987). As stress develops, the first change is typically stomatal closure without significant change in mesophyll photosynthetic capacity. This is adaptive in WUE and protective of mesophyll capacity. Further stress can cause, first, reversible capacity changes, then damage.

Damage debits the whole-plant carbon gain in tissue loss, repair cost, and lost opportunity cost of forgone photosynthesis after stress relief. Reversible and damaging decreases in photosynthetic capacity are not correlated well with leaf water potential but are correlated with relative water content. Leaf water potential changes can be ameliorated by osmotic adjustment, acclimation of cell extensibility, etc. (Turner *et al.,* 1986). Only when a plant has lost control does RWC decline significantly. RWC decline and damage are rather rare in wild plants, whereas they are more common and formative of performance in crops. Tolerance of developed low RWC is much less important in crop yield than is prevention through adaptive responses in water use (stomatal control, leaf area development, etc.).

D. Nutrient Use

The agricultural ecology of mineral nutrition is covered in several other chapters in this book (Chapter 5, this volume), so only considerations most specific to photosynthesis will be raised here.

Photosynthesis is supported by both water use in transpiration and nutrient use in the biochemical apparatus. In analogy with WUE, one may define a nutrient-use efficiency (nUE) (NUE will be taken to mean nitrogen-use efficiency, specifically). There are some conceptual problems in the definition (Gutschick, 1993), but it is reasonable to take nUE = (total C gained during the residence time of a nutrient)/(total nutrient content responsible for the C gain). We have argued (Gutschick, 1993; Gutschick and Kay, 1995) that nUE, effectively taken as $1/f_n = 1/$(fractional nutrient content in tissue), is set passively and is not actively regulated. Nutrient uptake rate achievable in soil conditions tends to drive f_n upward. Photosynthetic utilization of the nutrient tends to drive carbon gain that dilutes f_n. Other studies of nitrogen allocation are consistent with this interpretation, *e.g.,* Hirose and Werger (1994). In any event, it has been found, and well interpreted, that NUE is negatively correlated with WUE (Field *et al.,* 1983). Basically, leaves with higher N content have high $P_{L,a}$. If stomatal conductance does not keep exact parallelism with $P_{L,a}$, then higher N leaves will have lower C_i than reference leaves. Consequently, they will have WUE. On the other hand, $P_{L,a}^{sat}$ shows saturation as N content per area increases (Field and Mooney, 1983), so that NUE = $P_{L,a}^{sat}/N_a$ declines as N_a increases.

E. Temperature Modulation of Photosynthesis

Leaf temperature is determined partly by light interception, which has been discussed previously. This section will consider leaf temperature as already determined and will examine the consequences and the (adaptive) responses.

In the normal operating range of leaves, temperature effects are reversible: upon restoration of any temperature chosen as a reference, the photosynthetic rate will be restored, as will the patterns of light response, etc. There is no cumulative gain or loss in capacity from time spent at any temperature. The light-saturated rate $P_{L,a}^{sat}$ shows a strong thermal activation up to a peak where a separate deactivation process sets in very strongly. Before the peak, the T dependence of photosynthesis is that of Rubisco enzyme (Kirschbaum and Farquhar, 1984) so that it is similar for all plants. The deactivation process that sets the peak location and the rapid decline thereafter is the inhibition of photosystem II. Both light-dependent (photo-inhibition) and -independent deactivation occur (Havaux, 1994). Deactivation sensitivity differs among plant species and ecotypes and can acclimate in any given genotype from changes in electron transport functions and membrane composition. Light-limited photosynthesis, along the line of the initial quantum yield, shows simpler T trends. In C_4 plants, there is no change. In C_3 plants, there is a uniform, virtually linear decline of rate with temperature (Ehleringer and Björkman, 1977). In Eq. (2), the partition factor ϕ drops linearly with T.

Persistent or hysteretic effects of temperature are of two types: damage (often acute) and developmental or acclimatory responses. The acclimatory responses to both light and temperature have been discussed under Light Use. Damage occurs from either too low or too high temperatures. Low T damage may be purely thermal or else requires high light exposure (photoinhibition, which was discussed earlier). Thermal damage appears to always involve membrane functional changes (Li and Christersson, 1993). Low-temperature tolerance is very important economically; its extension by 1 or 2°C could notably increase crop production in temperate zones. Low T damage can occur from chilling as well as freezing. In freezing, damage occurs from membrane changes, from mechanical stress, and from dehydration of cytoplasm as water moves to the apoplastic ice (Guy, 1990). Plants can harden to chilling and freezing stresses. Hardening is partly a photosynthetic phenomenon but major contributions come from acclimation in transport systems, buds, etc.

Damage to photosynthesis at high T may be purely thermal. It may also be indirect, a result of water stress and perhaps compounded photoinhibitory damage, which was discussed earlier. High-temperature thermal damage has been studied primarily for molecular biological responses rather than for kinetics and costs in the whole plant. The conditions to cause it are rare but of high impact. I believe that high T damage is strongly formative of desert vegetation, but, as is the case for crops, information is inadequate to assess its effects. It is plausible that induction of stomatal opening by damagingly high temperatures is not simply a loss of control; rather, it is initially adaptive, serving to limit thermal damage.

F. Herbivory, Pests, and Diseases Affecting Photosynthesis

Herbivory by insects and large animals during the plant's vegetative growth stage typically has its major effect in removing photosynthetic tissue rather than in damaging critical transport pathways or removing growing points to limit recovery (Richards and Caldwell, 1985). The immediate effect of herbivory is a resetting of leaf area index and exposure of leaf area to higher solar radiation than previously encountered. Given that canopy photosynthesis (P_c) saturates as LAI increases, roughly as [1-exp(-K*LAI)], a reduction of LAI at high LAI only reduces P_c marginally. If $K = 0.7$ and LAI $= 4$, then a 10% loss of LAI reduces P_c by only 2%. This loss estimate applies well to distributed herbivory, such as insects. Large-animal herbivory selectively removes top leaves, which leaves behind lower N leaves less acclimated to high light. Some plant species and ecotypes are well adapted to such herbivory. They rapidly reallocate total stored carbohydrate to support regrowth (Chapin and McNaughton, 1989), and they shift the distribution of nitrogen to near-optimal patterns. Tolerant ecotypes and species must have sufficient meristems to use the reallocated resources for fast regrowth; not all do (Richards and Caldwell, 1985). Some ecotypes even have higher total productivity (though not standing biomass) with herbivory than without. However, they are almost always lower in productivity than ungrazed plants with low herbivory tolerance that optimize canopy structure for total photosynthesis (Polley and Detling, 1988). There is significant confusion over definitions of ''compensatory growth'' in the literature. Season-integrated productivity is one good measure, as is reproductive biomass produced. Many studies leave the definition vague or implicit, and few studies involve measurement of canopy photosynthesis, either per ground area or per mass.

Continued low-level herbivory need not affect reproductive yield significantly if moderately high LAI can still be attained *and* if water is not limiting. Herbivory-tolerant ecotypes can even increase their fractional allocation to reproduction, especially by tillering in grasses. If water limits total biomass accumulation, then loss of photosynthetic tissue directly subtracts from total yield possible. Furthermore, plants whose aboveground biomass has been reduced by herbivory often have their competitiveness for water reduced relative to ungrazed neighbors; leaf water potentials often decline (Senock *et al.*, 1991).

Herbivory tolerance in photosynthetic performance need not translate into dominance or even persistence in mixed-species communities. Herbivores can show sufficiently strong preferences for species so that the tolerant species becomes rarer (Brown and Stuth, 1993).

The risk of herbivory is related to photosynthetic performance in that high N leaves are more attractive to many herbivores (Williams *et al.*, 1989, and references therein). Thus, there is an evolved trade-off, which we have reset in agriculture by limiting herbivory. (The final percentage loss of productivity is still about one-third, however; Pimentel *et al.*, 1978.) Following herbivory, carbon gain is supported by remaining leaves and, after a lag, by regrowth of leaf area. Newly exposed ''shade'' leaves have lower photosynthetic capacity (Hodgkinson *et al.,* 1971) and may even lose function for awhile as a result of photoinhibition in the suddenly high irradiance. If the crop canopy had been of high LAI, the rate gain of leaf mass is accelerated after herbivory, but this compensatory growth (Chapin and McNaughton, 1989) does not proceed on a fundamentally different curve of LAI vs time than that in the original crop growth.

Some pests, diseases, and parasites selectively impact foliage development or photosynthetic function. Others, such as mistletoe (Davidson *et al.,* 1989), alter plant water relations and indirectly have their greatest impact by reducing cumulative photosynthesis as estimated by (water available)*WUE. A third category includes stressors that impact nonphotosynthetic functions such as transport (e.g., yellow virus of sugarbeets) and have very indirect effects on photosynthesis. In the first category of direct impact, an example is curly-top virus that both reduces foliage expansion and causes foliar chlorosis with loss of photosynthetic rate. Many leaf-spot diseases—fungal, bacterial, and viral—selectively kill areas of leaf tissue with little other effect, acting much like herbivory. Still other diseases target photosynthetic functions more intricately (Goodman *et al.,* 1986). The phaseola toxins attack membranes, and the mosaic viruses clear chloroplasts. In any event, the course of disease and its impact on photosynthesis and productivity is determined by the disease organism and the plant disease-resistance physiology, not by photosynthetic acclimation responses.

IV. Use of Photosynthate and Adaptive Patterns Therein

A. Four Basic Uses and the Partitioning of Respiration

Photosynthate has four major fates or uses: growth, storage, expenditure for tissue maintenance [mostly protein turnover (Penning de Vries *et al.,* 1974) and ion balance against leakage], and expenditure for operation (root ion uptake, especially). In this scheme, respiration is not a distinct use but rather a part of each of the four uses. For example, in growth, the energy use in synthesizing plant matter is in rather well-defined proportion to the final dry matter produced. One gram of photosynthate (plus appropriate mineral nutrients) will yield about 0.7 g carbohydrate with 0.3 g respiratory loss, 0.55 g protein with 0.45 g respiratory loss, or 0.4 g lipid with 0.6 g respiratory loss (Penning de Vries *et al.,* 1974). The proportions are virtually invariant with growth rate, temperature, or other conditions. The total growth respiration to construct a leaf depends on the relative proportions of carbohydrate, protein, and lipid. An average retention for herbaceous annuals is about 0.6 g tissue from 1 g photosynthate. Defensive compounds may be added as another category within growth. They represent a small fraction of biomass in annual crops but large fractions in some woody perennials, especially if we count lignins in a separate category. The direct metabolic costs of synthesizing the compounds can be quantified, with some care (*e.g.,* Gershenzon, 1994). Total costs are more difficult to assess (see Fineblum and Rausher, 1995). They may include interference of defensive compounds with other metabolism.

Storage, in the scheme just given, means long-term storage in special tissues, as distinct from short-term storage of starches and sugars in leaves before export. In storage, the retention fraction is quite high (about 0.9) because typical storage compounds are little modified from raw photosynthate. In maintenance and operation, photosynthate is fully respired for energy. The rate of respiration for maintenance, in particular, is very dependent on environment, in contrast to growth respiration. It depends on temperature, tissue composition, and activity or stress level in the tissue. For example, the rate of turnover of photosynthetic systems, and the attendant respiratory rate, depends on light level and N content. High-light leaves have high N and high respiratory rates (van der Werf *et al.,* 1993). There is no avoidance of this cost by any acclimation.

Respiration also occurs ahead of all four fates, as energy is used to drive phloem transport to get photosynthate and nutrients to their site of use. Transport costs may be moderately significant to allow for at least one active transport episode in moving between sites. Overall, it is sometimes practical to lump respiration into just two categories: (1) growth respiration (including storage as delayed growth, taking the respiration to drive storage as "growth respiration"), and (ii) "maintenance" respiration to drive the sum of maintenance, operation, and transport.

Total respiration may be measured, though with considerable difficulty both in technique and in concept. Note that respiration of a leaf or shoot may be measured, but part of that respiration is attributable to other tissues and organs. For example, respiration drives nitrate reduction and the synthesis of amino acids in leaves, but these are subsequently exported to support growth throughout the plant. To attribute respiratory costs accurately, one must measure respiration rates and the materials exchange rates.

Over a whole growth season of a grain crop, respiration may consume two-thirds of the original photosynthate, much of it in maintenance and operational costs (Akita *et al.,* 1990). The proportion of initial photosynthate used for lifetime maintenance does not differ as markedly among plants as do respiratory rates. Slow-growing plants have lower protein turnover rates, as a fraction of protein per unit time, especially in leaves. Their respiratory rates per unit mass are correspondingly low relative to fast-growing plants. The product of mean respiratory rate and lifetime (of plant or of leaves) is moderately well conserved between slow and fast-growing plants. Still, there are cases in which maintenance costs are such a large fraction of total photosynthate as to tip the balance in plant viability. In trees, woody biomass far exceeds leaf biomass. Bole respiration for maintenance and transport is large enough (Ryan *et al.*, 1995) that it may be critical in temperature limits for tree species.

B. Patterns of Photosynthate Use, Environmental Responses, and Optimization

Growth may produce both vegetative and reproductive structures. In the vegetative stage of growth, photosynthate is largely reinvested in more leaves and roots to acquire yet more resources. Because the leaf is not the (reproductive) end product, leaves are a temporary investment that must pay back much more than their cost of construction and operation—about 5- to 10-fold commonly (Mooney and Gulmon, 1982). The large excess over construction costs allows for support of maintenance costs, herbivory losses, etc. Of course, part of the construction cost can be paid back by scavenging of leaf C and nutrient contents at senescence, but only about one-fourth of construction costs can be recovered. Only about half of the leaf mass or of specific (phloem-mobile) nutrients can be recovered (Vitousek, 1982), largely because enough of the metabolic and transport system must remain intact to the end. The metabolic cost of catabolism and transport cuts the final biomass to about one-fourth the initial mass. Recovery of some nutrients such as N, however, is nearer to one-half. The cost–benefit balance in senescence of leaves or roots is a moderately shallow function of organ age: Senescence bears a modest cost in lost function of the old leaf, while bearing a benefit of nutrient and energy scavenging that, too, is modest because scavenging is so imperfect. Thus, plants must maintain moderately strong control over initiating senescence. In any case, senescence of nonreproductive parts is necessary for good yield to recover both their biomass and specific nutrients, which generally decline in availability late in the growing season (Cregan, 1983). Nonsenescing soybean mutants have significantly lower yields (Noodén *et al.*, 1979).

Leaves can meet the high payback requirement fairly easily if they are not deployed or incremented during stress (low water, low light, etc.). Indeed, plants have evolved sharp signals to curtail leaf development during water stress, as noted earlier. Responses to nutrient stress are far less marked, given that (i) nutrient availability declines far more gradually with time than does water availability and (ii) nutrients can be scavenged from old leaves for reuse, but this is dramatically not true for water; water storage in plants is a very small fraction of water throughput in transpiration. Still, the payback ratio taken as photosynthesis over leaf lifetime divided by construction costs does depend on nutrient content. A low nutrient content leads to low photosynthetic rate—even per unit of construction cost. Correspondingly, it demands longer leaf lifetime. This pattern is observed in contrasts between species or nutrient treatments (Williams *et al.*, 1989) among perennials. The question is more difficult to answer for the same contrast among annuals. Many annual crops keep a large fraction of their leaves to maturity, even if at reduced function. Furthermore, leaf senescence in annuals is driven more by overtopping by new leaves than by programmed senescence, photodamage, or other aging phenomena: An early leaf supplied with high irradiance maintains high photosynthetic capacity (Johnston *et al.*, 1969); more general control is discussed by Thomas and Stoddardt (1980). Loss of hydraulic connection to the root, caused by xylem cavitation, is, however, important for leaf death in some species and conditions (Tyree and Sperry, 1988). Salt stress is a strong driver of senescence. If senescence exceeds the rate of new leaf expansion, a plant rapidly loses productivity and dies. Better salt tolerance may arise either from decreased salt transport to leaves or from the ability to withstand a higher salt concentration before senescence is induced, such as might be achieved with better ion compartmentation (Munns, 1993).

For the plant as a whole, allocation patterns between leaf and stem and between root and shoot are important for productivity. In herbaceous crops, whether annual or perennial, the fraction of shoot mass as leaves, or L:s ratio, is rather uniformly large (near one-half). Modest increases in L:s are possible, as by dwarfing; see also Sims and Pearcy (1994) for a remarkable study in wild plants. In wheat, a limiting L:s slightly greater than 0.5 is projected (Austin, 1988). The gain in yield can be of the same order as the relative increase in L:s. First, the harvest index is increased. A given reproductive mass (m_R) is reached with lower total mass (including leaves, stems, and roots). Thus, if nutrients or water limit the total mass, then a greater m_R is possible with high L:s. Second, a given leaf mass (hence, potential seed-filling rate) is reached earlier in the season. As noted earlier, the relative growth rate is proportional to photosynthetic rate per mass of whole plant ($P_{p,m}$). In turn, $P_{p,m}$ equals the rate per mass of leaves ($P_{L,m}$) times the fraction of plant as leaves (L:s)$r/(1 + r)$, where r is the root:shoot mass ratio. thus, RGR rises in direct proportion to the L:s ratio. Faster growth is often valued in itself for marketing, and it also reduces the time that soil is exposed, hence, the evaporative loss of water from soil; more plant mass can be made from the same initial soil water content.

The root:shoot ratio (r) has a moderately significant effect on RGR (Gutschick and Kay, 1995). It acclimates rather strikingly to water and nutrient availability (Davidson, 1969; Gutschick, 1993, and references therein). When r adjusts to balance the nutrient (or water) gain rate of roots (expressed as carbon gain these can support) with the carbon gain rate of shoots, then RGR is maximized. Together with changes in uptake capacity per root mass, RGR may be doubled relative to that of unacclimated plants (Gutschick and Kay, 1995). In this contribution to RGR, r can be shown to be more important and long-lasting in effect than increases in uptake capacity. The root:shoot ratio is conditioned by stress episodes to stay larger than immediate requirements for functional balance. If, in contrast, r is kept to its optimum value for functional balance, harvest index is also improved modestly, compared to r values inflated by plant responses to reduce risk. For example, a change in r from 0.25 to 0.20 would improve the harvest index by the fraction $\delta r/(1 + r)$, or about $0.05/1.2 = 4\%$. One often-overlooked but important fact about root:shoot allocation (as well as other plant performance measures) is that benefits to the individual plant may not accrue in the stand. A high r may increase water acquisition in plants grown individually under water limitation. However, allocation does not increase water availability per ground area; it merely redistributes it among plants. Selection for allocation patterns must be evaluated in stands, not pot-grown individuals.

Storage of carbohydrate may occur in stems, tubers, and other organs, with later remobilization for vegetative or reproductive growth. (Lipids, in contrast, are used almost solely for seeds' growth reserves.) Wheat stores carbohydrate in stems to support later grain filling (Bell and Incoll, 1990; Kiniry, 1993). Alfalfa stores carbohydrate in its root crown for regrowth after cutting or to start a new season (Hendershot and Volenec, 1989). Grasses also store root carbohydrates that are important for regrowth (Richards and Caldwell, 1985). How is storage valuable for growth *or* risk amelioration? There is a respiratory cost in storing carbohydrate and in its remobilization so that photosynthate run into and out of storage has modestly lower energy value than current photosynthate. However, it is available at high rates when current photosynthesis may be low, as in regrowth. Storage thus helps ameliorate risks from competition in early growth or from low repair capacity during regrowth from damage. The benefits in field situations will require quantitative modeling, calibrated with experimental data. Estimating the potential to optimize storage schedules and degrees awaits such modeling.

Operational costs are energy expenditures for processing raw resources, such as reducing nitrate or N_2. They might be lumped into growth costs because the resources are directly used for growth, but the operations do occur in distinct tissues and organs, while supporting growth in general. Operational costs are not significant in photosynthesis itself, though costs of phloem transport of photosynthate may be significant. Phloem loading may be driven by active transport (in the apoplastic loading alternative) or it may be essentially passive (symplastic alternative; see van Bel *et al.*, 1994); costs of the former may be quantified. Phloem flow is sustained by phloem unloading, which occurs principally by simple metabolic use of the sugars (Sung *et al.*, 1994); this bears no extra cost.

Operational costs are significant in mineral nutrition and water acquisition that support photosynthesis. See, for example, Bloom *et al.* (1992) and Gutschick (1981) regarding operational costs in N nutrition. In water acquisition, there may be significant costs for redistributing root solutes moved by water flow (Steudle *et al.,* 1993; Rygol *et al.,* 1993). There do not appear to be acclimation processes in photosynthesis that would ameliorate these indirect operational costs significantly.

C. Reproduction: New Patterns of Growth, Operation, Maintenance, and Storage

In most crop plants, the agronomically important yield is reproductive. In turn, in most plants, reproduction is a discrete growth stage in which the plant converts vegetative buds to floral buds. One may ask how this conversion should be scheduled. Cohen (1976) and Paltridge *et al.* (1984) derived compelling conclusions from two basic ideas: (i) The vegetative-to-reproductive ($V \rightarrow R$) shift should occur late enough that the canopy is large and has a high canopy total photosynthetic rate P_c, and (ii) the shift should occur early enough to allow time for seed fill. Cohen showed from control theory that the optimal behavior is "bang-bang" style, with total conversion from V to R growth at one time. [Why, then, are any crops indeterminate? This question is addressed after discussing resource availability. Nonetheless, Gay *et al.* (1980) have ascribed the superiority of one soybean cultivar over another as resulting from more complete or more determinate $V \rightarrow R$ switching in the first cultivar.]

The time at which the $V \rightarrow R$ switch should occur is set in part by the time when the growing season ends. This is not predictable from any information that can be sensed by plants or humans. The $V \rightarrow R$ switch is thus risky. Plants use conservative estimates, especially photoperiod, to indicate the season's progress independently of immediate environmental conditions. The useful season length is also a function of some resource availability. Drought and nutrient stresses can accelerate the $V \rightarrow R$ switch (Angus and Moncur, 1977), as is appropriate if water or nutrient availability is terminal. However, with some stress levels and some varieties of wheat, maturation is delayed under water stress (Angus and Moncur, 1977) which is appropriate if water is expected later in the season. Earlier maturation at low N availability is also observed: So is the converse (later maturation) or even reversal of the $V \rightarrow R$ shift if N is made available copiously late in the season (Loomis *et al.,* 1976). Resource availability schedules may underlie the occurrence of indeterminate reproduction (King and Roughgarden, 1982), though the question has not been adequately explored. Several other environmental conditions might favor indeterminacy, such as schedules of seed predation (Ashton *et al.,* 1988) or of episodes of unfavorable temperatures—indeterminacy can help spread the risk of choosing times for reproduction. A further discussion of reproductive timing is given in Gut-schick (1987).

The extent of biomass allocation to reproduction, or harvest index, is high in annual plants, cropped or wild. Increasing the harvest index (HI) has been one of plant breeding's major contributions to yield increases (Austin, 1988). Some grains are near theoretical limits (Austin, 1988) esti-mated from the need for support structures and limits on scavenging the vegetative biomass. What prevents maximal HI in various crops? Tallness does. To have minimal support structure that allows high HI, plants must be short, which is affordable only when height competition for light inter-ception is naturally low or is artifically curtailed, as noted earlier. Allocation to roots also is a diversion from HI. Reduced allocation is possible under stable conditions of high water and nutrient availability (cf. Davidson, 1969). However, current favorability is not strongly indicative of the future. Plants probably allocate excess root biomass as a hedge against risk, though this hypothesis would require careful testing.

V. Synthesis and Conclusions

Photosynthetic performance has been improved at the whole-plant or whole-stand level in a variety of crops by several routes. There have been some discrete breeding advances (erect leaves, and dwarfing), some inad-vertent gains in breeding (lower photosynthetic rate per leaf area and higher rate per mass), and some improvements in management (higher planting density). The discrete breeding gains and management gains came from viewing plant and stand performance as a system, in contrast to focusing on one or a few traits. In a systems view, one readily sees trade-offs in function, such as between photosynthetic rate per leaf area and leaf area index, between WUE and yield, or between WUE and nitrogen-use efficiency. These trade-offs achieved fairly close balance during the evolu-tion of crop ancestors so that the whole plant (or population) attained nearly optimal performance in wild growth. In the systems view, however, one also sees, with some insight, conditions in which the trade-off can be shifted to achieve net gains in agriculture because the functions traded off have very different values in the ecology of agriculture than in the ecology of wild growth. A good example is leaf erectness, in which gains in photosyn-thetic performance are shared among plants; this is desirable in agriculture but selected against in competitive wild growth. In complementary fashion, a systems view helps avoid futile efforts, such as (i) seeking to reduce photosynthetic carbon oxidation or photorespiration, which is built into the Rubisco enzyme (Morell *et al.,* 1992); (ii) trying to optimize the distribu-tion of photosynthetic capacity in the canopy, which is already on a very flat optimum; or (iii) pursuing gains in WUE near the limits set by biophysical feedbacks, in which trade-offs (such as against yield) become severe. There are some pleasant surprises, too, such as the realization that rainfall that stays on the leaves, to evaporate therefrom, is almost as valuable as rain reaching the soil and being transpired.

We should assume that new insights will come from such systems viewpoints on agriculture or agricultural ecology. In breeding, the use of ideotypes (the whole-stand view) will be profitable (*e.g.,* Tsunoda, 1983). In crop management, the use of rather comprehensive models such as GOSSYM will likewise be rewarding. Whether in breeding or management, the greatest promise lies in the following:

1. Using process-based models of plant and stand performance in contrast to statistical "black-box" models calibrated to local conditions. Only process-based understanding can be transferred to new species or varieties, new locations, and new environments. Only such models make us aware of the trade-offs we face in different functions. Such models can be put together from very robust, well-verified models of individual processes, *e.g.,* the Farquhar–von Caemmerer–Berry (1980) model of C_3 photosynthesis, models of any desired resolution in light interception, energy balance models for leaves and canopies, models of isotopic discrimination that indicate WUE, models of growth and maintenance respiration, etc. The final model appropriate to an investigation need not be huge with many parameters and attendant mathematical complexity. See, for example, the history of the erect-leaf hypothesis (Trenbath and Angus, 1975).

Of course, what is mathematically complex to one person may appear straightforward to another (who may see biological complexity where the former sees simplicity). It is important to distinguish mathematical complexity in processes (many equations, parameters, and feedbacks) from complexity in mathematical solution. A problem that is simple to state mathematically may require extensive computation, but that is not much of a concern: Computing time is cheap and getting cheaper. One's efforts should always go to keeping the model conceptually tractable. To balance the effort, recall Einstein's dictum, "One should simplify as much as possible, but no further." To do so, breeders, agronomists, and physiologists must pay more attention than has been customary to biophysical processes. Some fruits of this attention are seen in the work of Farquhar, Cowan, Richards, Condon, Givnish, Norman, and others who should not take offense if I keep the list short.

Process models and a process understanding by their nature generate improved guides to progress. For example, it is more profitable to track photosynthetic rate per unit mass than rate per area.

2. Constructing hypotheses, not explorations or "fishing expeditions." The focus afforded by looking for evidence of a specific hypothesis helps

one see other patterns. Also, agronomic systems are sufficiently complex that it takes considerable discipline to see even powerful patterns. As the geologists are reputed to say, "I wouldn't have seen it if I hadn't believed it." (This is not an endorsement to seek only proof, not falsification, if the latter is merited.)

3. Attending to plant responses that reduce whole-season risk. Some risks to yield or quality are now managed both quantitatively and well: We have effective breeding to adjust photoperiodic triggering of the vegetative-to-reproductive switch. Other opportunities await, as in balancing root development (in excess of immediate need for functional balance) against risk of drought damage, or stomatal control of transpiration against the risk of high-temperature damage.

4. Evaluating crop performance in the stand, not the individual, whether one is making models or testing them on real crops. Only in the stand does one see some major feedbacks and trade-offs in function, such as in light interception and water balance.

Agriculture faces a potentially problematic "brave new world" with the developing changes in climate—rising atmospheric CO_2 content, global and regional shifts in mean temperatures, changes in temperature extremes, and highly regionalized changes in precipitation means and in storm intensities (Houghton *et al.*, 1990). We should expect plant responses to not be near the optimum in agriculture or in wild growth. It will be up to all of us—breeders, agronomists, physiologists, climatologists, policymakers, and more—to define and attain the new optima using the full insight of the systems viewpoint.

VI. Appendix

Correspondence of Variables with Those of Condon and Hall (Chapter 3, This Volume)

Quantity	My notation	Condon and Hall
Leaf photosynthetic rate per area	$P_{L,a}$	A
Light-saturated photosynthetic rate per area	$P_{L,a}^{sat}$	A_{cap}
Leaf transpiration rate per area	E	T
Leaf-internal CO_2 concentration, especially as partial pressure	C_i	c_i
Ambient CO_2 partial pressure	C_a	c_a
Total atmospheric pressure	P	P_a
Water-use efficiency	WUE	W_{ET}, W_T, and W^a
As transpiration ratio of leaf		A/T
Leaf mass per area	$m_{L,a}$	$1/SLA$

[a] As defined for different types of water-use efficiency.

Acknowledgments

This synthesis of ideas was made possible by research supported by Grant NA16RC0435 from the National Atmospheric and Oceanic Administration, by the Department of Energy's National Institutes of Global Environmental Change, and by Grant DEB 94-11971 from the National Science Foundation's Long-Term Ecological Research Program. I thank two anonymous reviewers for insightful and accurate comments that led me to understand several topics more deeply.

References

Akita, S., Parao, F. T., Laza, R. C., Blanco, L. C., and Coronel, V. P. (1990). Physiological basis of rice yield potential improvement in the tropics. *In* "Proceedings of the International Congress of Plant Physiology" (S. K. Sinha, P. V. Sane, S. C. Bhargava, and P. K. Agrawal, eds.), Vol. 1, pp. 60–74. Society for Plant Physiology and Biochemistry, New Delhi.

Angus, J. F., and Moncur, M. W. (1977). Water stress and phenology in wheat. *Aust. J. Agric. Res.* **28**, 177–181.

Ashton, P. S., Givnish, T. J., and Appanah, S. (1988). Staggered flowering in the Dipterocarpaceae: New insights into floral induction and the evolution of mast fruiting in the aseasonal tropics. *Am. Nat.* **132**, 44–66.

Asrar, G. (ed.) (1991). "Theory and Applications of Optical Remote Sensing." Wiley-Interscience, New York.

Austin, R. B. (1988). New opportunities in breeding. *HortScience* **23**, 41–45.

Baker, N. R., and Bowyer, J. R. (eds.) (1994). "Photoinhibition of Photosynthesis: From Molecular Mechanisms to the Field." Bios, Oxford.

Baker, N. R., Farage, P. K., Stirling, C. M., and Long, S. P. (1994). Photoinhibition of crop photosynthesis in the field at low temperatures. *In* "Photoinhibition of Photosynthesis: From Molecular Mechanisms to the Field" (N. R. Baker and J. R. Bowyer, eds.), pp. 349–363. Bios, Oxford.

Baldocchi, D. D. (1989). Turbulent transfer in a deciduous forest. *Tree Physiol.* **5**, 357–377.

Baldocchi, D. D. (1993). Scaling water vapor and carbon dioxide exchange from leaves to a canopy: Rules and tools. *In* "Scaling Physiological Processes: Leaf to Globe" (J. R. Ehleringer and C. B. Field, eds.), pp. 77–114. Academic, San Diego.

Ball, J. T., Woodrow, I. E., and Berry, J. A. (1987). A model predicting stomatal conductance and its contribution to the control of photosynthesis under different environmental conditions. *In* "Progress in Photosynthesis Research" (J. Biggins, ed.), Vol. 4, pp. 221–224. Nijhoff, Dordrecht.

Ball, M. C., Hodges, V. S., and Laughlin, G. P. (1991). Cold-induced photoinhibition limits regeneration of snow gum at tree line. *Funct. Ecol.* **5**, 663–668.

Bell, C. J. (1982). A model of stomatal control. *Photosynthetica* **16**, 486–495.

Bell, C. J., and Incoll, L. D. (1990). The redistribution of assimilate in field-grown winter wheat. *J. Exp. Bot.* **41**, 949–960.

Blackman, P. G., and Davies, W. J. (1985). Root to shoot communication in maize plants of the effects of soil drying. *J. Exp. Bot.* **36**, 39–48.

Bloom, A. J., Sukrapanna, S. S., and Warner, R. L. (1992). Root respiration associated with ammonium and nitrate absorption and assimilation by barley. *Plant Physiol.* **99**, 1294–1301.

Bowes, G. (1993). Facing the inevitable: Plants and increasing atmospheric CO_2. *Annu. Rev. Plant Physiol. Mol. Biol.* **44**, 309–332.

Broadmeadow, M. S. J., Griffiths, H., Maxwell, C., and Borland, A. M. (1992). The carbon isotope ratio of plant organic material reflects temporal and spatial variations in CO_2 within tropical forest formations in Trinidad. *Oecologia* **89**, 435–441.

Brown, J. R., and Stuth, J. W. (1993). How herbivory affects grazing tolerant and sensitive grasses in a central Texas grassland: Integrating plant response across hierarchical levels. *Oikos* **67**, 291–298.

Bunce, J. A. (1983). Photosynthetic characteristics of leaves developed at different irradiances and temperatures: An extension of the current hypothesis. *Photosynth. Res.* **4**, 87–97.

Caldwell, M. M., and Pearcy, R. W. (1994). "Exploitation of Environmental Heterogeneity by Plants." Academic Press, San Diego.

Chapin, F. S., III, and McNaughton, S. J. (1989). Lack of compensatory growth under phosphorus deficiency in grazing-adapted grasses from the Serengeti Plains. *Oecologia* **79**, 551–557.

Chatterton, N. J. (1973). Product inhibition of photosynthesis in alfalfa leaves as related to specific leaf weight. *Crop Sci.* **13**, 284–285.

Chaves, M. M. (1991). Effects of water deficits on carbon assimilation. *J. Exp. Bot.* **42**, 1–16.

Cohen, D. (1976). The optimal timing of reproduction. *Am. Nat.* **110**, 801–807.

Collatz, G. J., Ball, J. T., Grivet, C., and Berry, J. A. (1991). Physiological and environmental regulation of stomatal conductance, photosynthesis and transpiration: A model that includes a laminar boundary layer. *Agric. Forest Meteorol.* **54**, 107–136.

Collatz, G. J., Ribascarbo, M., and Berry, J. A. (1992). Coupled photosynthesis–stomatal conductance model for leaves of C_4 plants. *Aust. J. Plant Physiol.* **19**, 519–538.

Collins, S. L., and Wallace, L. L. (1990). "Fire in North American Tallgrass Prairies." Univ. of Oklahoma Press, Norman.

Condon, A. G., Richards, R. A., and Farquhar, G. D. (1987). Carbon isotope discrimination is positively correlated with grain yield and dry matter production in field-grown wheat. *Crop Sci.* **27**, 996–1001.

Corelli-Grappadelli, L., and Coston, D. C. (1991). Thinning patterns and light environment in peach tree canopies influence fruit quality. *HortScience* **26**, 1464–1466.

Cowan, I. R. (1987). Economics of carbon fixation in higher plants. *In* "On the Economy of Plant Form and Function" (T. J. Givnish, ed.), pp. 133–170. Cambridge Univ. Press, Cambridge, UK.

Cowan, I. R., and Farquhar, G. D. (1977). Stomatal diffusion in relation to leaf metabolism and environment. *Symp. Soc. Exp. Biol.* **31**, 471–505.

Cregan, P. B. (1983). Genetic control of nitrogen metabolism in plant reproduction. *In* "Strategies of Plant Reproduction" (W. J. Meudt, ed.), pp. 243–262. Allanheld, Osmum, Granada.

Cui, M., Smith, W. K., and Vogelmann, T. C. (1991). Chlorophyll and light gradients in sun and shade leaves of *Spinacia oleracea*. *Plant Cell Environ.* **14**, 493–500.

Davidson, N. J., True, K. C., and Pate, J. S. (1989). Water relations of the parasite : host relationship between the mistletoe *Amyema linophyllum* (Fenzl) Tieghem and *Casuarina obesa* Miq. *Oecologia* **80**, 321–330.

Davidson, R. L. (1969). Effects of soil nutrients and moisture on root/shoot ratios in *Lolium perenne* L. and *Trifolium repens* L. *Ann. Bot.* **33**, 571–577.

Davies, W. J., and Zhang, J. (1991). Root signals and the regulation of growth and development of plants in drying soil. *Annu. Rev. Plant Physiol. Mol. Biol.* **42**, 55–76.

deJong, T. M., and Doyle, F. (1985). Seasonal relationships between leaf nitrogen content (photosynthetic capacity) and leaf canopy light exposure in peach (*Prunus persica*). *Plant Cell Environ.* **8**, 701–706.

Denmead, O. T. (1968). Carbon dioxide exchange in the field: Its measurement and interpretation. *In* "Agricultural Meteorology, Proceedings of the WMO Seminar," pp. 445–482. Bureau of Meterorology, Melbourne.

Dewar, R. C. (1995). Interpretation of an empirical model for stomatal conductance in terms of guard cell function. *Plant Cell Environ.* **18**, 365–372.

Downton, W. J. S., and Grant, W. J. R. (1992). Photosynthetic physiology of spur pruned and minimal pruned grapevines. *Aust. J. Plant Physiol.* **19**, 309–316.

Ehleringer, J., and Björkman, O. (1977). Variation in quantum yield for CO_2 uptake in C_3 and C_4 plants. *Plant Physiol.* **59**, 86–90.

Ehleringer, J., Hall, A. E., and Farquhar, G. D. (eds.) (1993). "Stable Isotopes and Plant Carbon/Water Relations." Academic Press, San Diego.

Farquhar, G. D., von Caemmerer, S., and Berry, J. A. (1980). A biochemical model of photosynthetic CO_2 assimilation in leaves of C_3 species. *Planta* **149**, 78–90.

Field, C. B., and Mooney, H. A. (1983). Leaf age and seasonal effects on light, water, and nitrogen use efficiency in a California shrub. *Oecologia* **56**, 348–355.

Field, C. B., Merino, J., and Mooney, H. A. (1983). Compromises between water-use efficiency and nitrogen-use efficiency in five species of California evergreens. *Oecologia* **60**, 384–389.

Fineblum, W. L., and Rausher, M. D. (1995). Tradeoff between resistance and tolerance to herbivore damage in a morning glory. *Nature* **377**, 517–520.

Forseth, I., and Ehleringer, J. R. (1983). Ecophysiology of two solar tracking desert winter annuals. III. Gas exchange responses to light, CO_2 and VPD in relation to long-term drought. *Oecologia* **57**, 344–351.

Garner, W. W., and Allard, H. A. (1920). Effect of the relative length of day and night and other factors of the environment on growth and reproduction in plants. *J. Agric. Res.* **18**, 553–606.

Gates, D. M. (1980). "Biophysical Ecology." Springer-Verlag, New York.

Gay, S., Egli, D. B., and Reicosky, D. A. (1980). Physiological aspects of yield improvement in soybeans. *Agron. J.* **72**, 387–391.

Geiger, D. R., and Servaites, J. C. (1994). Dynamics of self-regulation of photosynthetic carbon metabolism. *Plant Physiol. Biochem.* **32**, 173–183.

Gershenzon, J. (1994). Metabolic costs of terpenoid accumulation in higher plants. *J. Chem. Ecol.* **20**, 1281–1328.

Gifford, R. M., and Evans, L. T. (1981). Photosynthesis, carbon partitioning, and yield. *Annu. Rev. Plant Physiol.* **32**, 485–509.

Givnish, T. J. (1982). On the adaptive significance of leaf height in forest herbs. *Am. Nat.* **120**, 353–381.

Givnish, T. J. (1987). Optimal stomatal conductance, allocation of energy between leaves and roots, and the marginal cost of transpiration. *In* "On the Economy of Plant Form and Function" (T. J. Givnish, ed.), pp. 171–213. Cambridge Univ. Press, Cambridge, UK.

Goldschmidt, E. E., and Huber, S. C. (1992). Regulation of photosynthesis by end-product accumulation in leaves of plants storing starch, sucrose, and hexose sugars. *Plant Physiol.* **99**, 1443–1448.

Goodman, R. N., Király, Z., and Wood, K. R. (1986). "The Biochemistry and Physiology of Plant Disease." Univ. of Missouri, Columbia.

Grieu, R., Guehl, J. M., and Aussenac, G. (1988). The effects of soil and atmospheric drought on photosynthesis and stomatal control of gas exchange in three coniferous species. *Physiol. Plant.* **73**, 97–104.

Grime, J. P. (1979). "Plant Strategies and Vegetation Processes." Wiley, Chichester, UK.

Gross, L. J., Kirschbaum, M. U. F., and Pearcy, R. W. (1991). A dynamic model of photosynthesis in varying light taking account of stomatal conductance, C_3 cycle intermediates, photorespiration and Rubisco activation. *Plant Cell Environ.* **14**, 881–893.

Gutschick, V. P. (1981). Evolved strategies of nitrogen acquisition by plants. *Am. Nat.* **118**, 607–637.

Gutschick, V. P. (1984). Statistical penetration of diffuse light into vegetative canopies: Effect on photosynthetic rate and utility for canopy measurement. *Agric. Meteorol.* **30**, 327–341.

Gutschick, V. P. (1987). "A Functional Ecology of Crop Plants." Croom Helm, London/ Timber Press, Beaverton, OR.

Gutschick, V. P. (1991). Joining leaf photosynthesis models and canopy photon-transport models. *In* "Photon–Vegetation Interaction: Applications in Optical Remote Sensing and Plant Ecology" (R. B. Myneni and J. Ross, eds.), pp. 501–535. Springer-Verlag, Berlin.

Gutschick, V. P. (1993). Nutrient-limited growth rates: Roles of nutrient-use efficiency and of adaptations to increase nutrient uptake. *J. Exp. Bot.* **44,** 41–51.

Gutschick, V. P., and Cunningham, G. L. (1989). A physiological route to increased water-use efficiency in alfalfa. Report 239, New Mexico Water Resource Research Institute, Las Cruces, NM.

Gutschick, V. P., and Kay, L. E. (1995). Nutrient-limited growth rates: Quantitative benefits of stress responses and some aspects of regulation. *J. Exp. Bot.* **46,** 995–1009.

Gutschick, V. P., and Wiegel, F. W. (1988). Optimizing the canopy photosynthetic rate by patterns of investment in specific leaf mass. *Am. Nat.* **132,** 67–86.

Guy, C. L. (1990). Cold acclimation and freezing stress tolerance: Role of protein metabolism. *Annu. Rev. Plant Physiol. Plant Mol. Biol.* **41,** 187–223.

Harris, D., Natarajan, M., and Willey, R. W. (1987). Physiological basis for yield advantage in a sorghum/groundnut intercrop exposed to drought. I. Dry-matter production, yield, and light interception. *Field Crops Res.* **17,** 259–272.

Havaux, M. (1994). Temperature-dependent modulation of the photoinhibition-sensitivity of photosystem II in *Solanum tuberosum* leaves. *Plant Cell Physiol.* **35,** 757–766.

Hedrick, P. W., Jain, S., and Holden, L. (1978). Multilocus systems in evolution. *Evol. Biol.* **11,** 101–184.

Hendershot, K. L., and Volenec, J. J. (1989). Shoot growth, dark respiration, and nonstructural carbohydrates of contrasting alfalfa genotypes. *Crop Sci.* **29,** 1271–1275.

Hirose, T., and Werger, M. J. A. (1987). Maximizing daily canopy photosynthesis with respect to the leaf nitrogen allocation pattern in the canopy. *Oecologia* **72,** 520–526.

Hirose, T., and Werger, M. J. A. (1994). Photosynthetic capacity and nitrogen partitioning among species in the canopy of a herbaceous plant community. *Oecologia,* **100,** 203–212.

Hodgkinson, K. C., Smith, N. G., and Miles, G. E. (1971). The photosynthetic capacity of stubble leaves and their contribution to the growth of the lucerne plant after high level cutting. *Aust. J. Agric. Res.* **23,** 225–238.

Houghton, J. T., Jenkins, G. J., and Ephraims, J. J. (1990). "Climate Change: The IPCC Scientific Assessment." Cambridge Univ. Press, Cambridge, UK.

Jarvis, P. G., and McNaughton, K. G. (1986). Stomatal control of transpiration: Scaling up from leaf to region. *Adv. Ecol. Res.* **15,** 1–47.

Johnson, I. R., and Thornley, J. H. M. (1984). A model of instantaneous and daily canopy photosynthesis. *J. Theor. Biol.* **107,** 531–545.

Johnston, T. J., Pendleton, J. W., Peters, D. B., and Hicks, D. R. (1969). Influence of supplemental light on apparent photosynthesis, yield, and yield components of soybeans (*Glycine max* L.). *Crop Sci.* **9,** 577–581.

Jones, H. G. (1983). "Plants and Microclimate." Cambridge Univ. Press, Cambridge, UK.

Kaiser, W. M. (1987). Effect of water deficit on photosynthetic capacity. *Physiol. Plant.* **71,** 142–149.

Khan, M. A., and Tsunoda, S. (1970). Evolutionary trends in leaf photosynthesis and related leaf characters among cultivated wheat species and its wild relatives. *Jpn. J. Breeding* **20,** 133–140.

King, D., and Roughgarden, J. (1982). Graded allocation between vegetative and reproductive growth for annual plants in growing seasons of random length. *Theor. Pop. Biol.* **22,** 1–16.

Kiniry, J. R. (1993). Nonstructural carbohydrate utilization by wheat shaded during grain growth. *Agron. J.* **85,** 844–849.

Kirschbaum, M. U. F., and Farquhar, G. D. (1984). Temperature dependence of whole-leaf photosynthesis in *Eucalyptus pauciflora* Sieb. ex Spreng. *Aust. J. Plant Physiol.* **11,** 519–538.

Körner, Ch., Pelaez-Riedl, S., and van Bel, A. J. E. (1995). CO_2 responsiveness of plants: A possible link to phloem loading. *Plant Cell Environ.* **18,** 595–600.

Krall, J. P., and Pearcy, R. W. (1993). Concurrent measurements of oxygen and carbon dioxide exchange during lightflecks in maize (*Zea mays* L.). *Plant Physiol.* **103,** 823–828.

Lajtha, K., and Barnes, F. J. (1991). Carbon gain and water use in pinyon pine-juniper woodlands of northern New Mexico: Field versus phytotron chamber measurements. *Tree Physiol.* **9,** 59–67.

Li, P. H., and Christersson, L. (eds.) (1993). "Advances in Plant Cold Hardiness." CRC Press, Boca Raton, FL.

Loomis, R. S., Ng, E., and Hunt, W. F. (1976). Dynamics of development in crop production systems. *In* "CO_2 Metabolism and Plant Productivity" (R. H. Burris and C. C. Black, eds.), pp. 269–286. University Park Press, Baltimore, MD.

Lundmark, T., and Hällgren, J.-E. (1987). Effects of frost on shaded and exposed spruce and pine seedlings planted in the field. *Can. J. For. Res.* **17,** 1197–1201.

Masle, J., and Passioura, J. B. (1987). The effect of soil strength on the growth of young wheat plants. *Aust. J. Plant Physiol.* **14,** 643–656.

Massman, W. J., and Kaufmann, M. R. (1991). Stomatal response to certain environmental factors: A comparison of models for subalpine trees in the Rocky Mountains. *Agric. For. Meteorol.* **54,** 155–167.

McKinion, J. M., Baker, D. N., Whisler, F. D., and Lambert, J. R. (1989). Application of the GOSSYM/COMAX system to cotton crop management. *Agric. Syst.* **31,** 55–65.

Meinzer, F. C., Goldstein, G., Holbrook, N. M., Jackson, P., and Cavelier, J. (1993). Stomatal and environmental control of transpiration in a lowland tropical forest tree. *Plant Cell Environ.* **16,** 429–436.

Miller, E. E., and Norman, J. M. (1971). A sunfleck theory for plant canopies. II. Penumbra effect: Intensity distributions along sunfleck segments. *Agron. J.* **63,** 739–742.

Monteith, J. L. (1994). Validity of the correlation between intercepted radiation and biomass. *Agric. For. Meteorol.* **68,** 213–220.

Monteith, J. L. (1995). A reinterpretation of stomatal responses to humidity. *Plant Cell Environ.* **18,** 357–364.

Mooney, H. A., and Gulmon, S. L. (1982). Constraints on leaf structure and function in reference to herbivory. *BioScience* **32,** 198–206.

Mooney, H. A., Björkman, O., and Collatz, G. J. (1978). Photosynthetic acclimation to temperature in the desert shrub, *Larrea divaricata*. I. Carbon dioxide exchange characteristics of intact leaves. *Plant Physiol.* **61,** 406–410.

Morell, M. K., Paul, K., Kane, H. J., and Andrews, T. J. (1992). Rubisco: Maladapted or misunderstood? *Aust. J. Bot.* **40,** 431–441.

Mott, K. A., and Parkhurst, D. F. (1991). Stomatal responses to humidity in air and helox. *Plant Cell Environ.* **14,** 509–515.

Mott, K. J. (1988). Do stomata respond to CO_2 concentrations other than intercellular? *Plant Physiol.* **86,** 200–203.

Munns, R. (1993). Physiological processes limiting plant growth in saline soils: Some dogmas and hypotheses. *Plant Cell Environ.* **16,** 15–24.

Myneni, R. B., and Ross, J. (eds.) (1994). "Photon–Vegetation Interactions." Springer-Verlag, Berlin.

Nobel, P. S. (1978). Surface termperatures of cacti—Influences of environmental and morphological factors. *Ecology* **59,** 986–996.

Nobel, P. S. (1991). "Physicochemical and Environmental Plant Physiology." Academic Press, San Diego.

Noodén, L. D., Kahanak, G. M., and Okatan, Y. (1979). Prevention of monocarpic senescence in soybeans with auxin and cytokinin: An antidote for self-destruction. *Science* **206,** 841–843.

Norman, J. M. (1993). Scaling processes between leaf and canopy levels. *In* "Scaling Physiological Processes: Leaf to Globe" (J. R. Ehleringer and C. B. Field, eds.), pp. 39–76. Academic Press, San Diego.

Norman, J. M., and Arkebauer, T. J. (1991). Predicting canopy photosynthesis and light-use efficiency from leaf characteristics. *In* "Modelling Crop Photosynthesis—From Biochemistry to Canopy" (K. J. Boote and R. S. Loomis, eds.), pp. 75–94. Crop Sci. Soc. Am., Madison, WI.

Norman, J. M., and Tanner, C. B. (1969). Transient light measurements in plant canopies. *Agron. J.* **61,** 847–849.

Norman, J. M., and Welles, J. M. (1983). Radiative transfer in an array of canopies. *Agron. J.* **75,** 481–488.

Ögren, E., and Sjöström, M. (1990). Estimation of the effect of photoinhibition on the carbon gain in leaves of a willow canopy. *Planta* **181,** 560–567.

Oker-Blom, P. (1985). The influence of penumbra on the distribution of direct solar radiation in a canopy of Scots pine. *Photosynthetica* **19,** 312–317.

Oker-Blom, P., Lappi, J., and Smolander, H. (1991). Radiation regime and photosynthesis of coniferous stands. *In* "Photon–Vegetation Interactions" (R. B. Myneni and J. Ross, eds.), pp. 469–499. Springer-Verlag, Berlin.

Ort, D. R., and Baker, N. R. (1988). Consideration of photosynthetic efficiency at low light as a major determinant of crop photosynthetic performance. *Plant Physiol. Biochem.* **26,** 555–565.

Owensby, C. E., Coyne, P. I., Ham, J. M., Auen, L. M., and Knapp, A. K. (1993). Biomass production in a tallgrass prairie ecosystem exposed to ambient and elevated CO_2. *Ecol. Appl.* **3,** 644–653.

Paltridge, G. W., Denholm, J. V., and Connor, D. J. (1984). Determinism, senescence and the yield of plants. *J. Theor. Biol.* **110,** 383–398.

Pearcy, R. W., and Ehleringer, J. (1984). Comparative ecophysiology of C_3 and C_4 plants. *Plant Cell Environ.* **7,** 1–13.

Pearcy, R. W., and Pfitsch, W. A. (1994). The consequences of sunflecks for photosynthesis and growth of forest understory plants. *Ecol. Stud.* **100,** 343–359.

Penning de Vries, F. W. T. (1975). The cost of maintenance processes in plant cells. *Ann. Bot.* **39,** 77–92.

Penning de Vries, F. W. T., Brunsting, A. H. M., and van Laar, H. H. (1974). Products, requirements and efficiency of biosynthesis: A quantitative approach. *J. Theor. Biol.* **45,** 339–377.

Peñuelas, J., and Matamala, R. (1990). Changes in N and S leaf content, stomatal density and specific leaf area of 14 plant species during the last three centuries of CO_2 increase. *J. Exp. Bot.* **41,** 1119–1124.

Pettigrew, W. T., Hesketh, J. D., Peters, D. B., and Woolley, J. T. (1989). Characterization of canopy photosynthesis of chlorophyll-deficient soybean isolines. *Crop Sci.* **29,** 1025–1029.

Pimentel, D., Krummel, J., Gallahan, D., Hough, J., Merrill, A., Schreiner, I., Vittum, P., Koziol, F., Back, E., Yen, D., and Fiance, S. (1978). Benefits and costs of pesticide use in U. S. food production. *BioScience* **28,** 772–784.

Polley, H. W., and Detling, J. K. (1988). Herbivory tolerance of *Agropyron smithii* populations with different grazing histories. *Oecologia,* **77,** 261–267.

Pons, T. L., and Pearcy, R. W. (1992). Photosynthesis in flashing light in soybean leaves grown in different conditions. II. Lightfleck utilization efficiency. *Plant Cell Environ.* **15,** 577–584.

Poorter, H. (1991). "Interspecific Variation in the Relative Growth Rate of Plants: The Underlying Mechanisms." Proefschrift, Utrecht, The Netherlands.

Poorter, H. (1993). Interspecific variation in the growth response of plants to an elevated ambient CO_2 concentration. *Vegetatio* **104/105,** 77–97.

Quirk, J. P. (1978). The simple economics of water. *Eng. Sci.* **Sept./Oct.,** 22–26.

Radin, J. W. (1991). Enhanced photosynthesis and stomatal conductance of Pima cotton (*Gossypium barbadense* L.) bred for increased yield. *Plant Physiol.* **97,** 484–489.

Raupach, M. R. (1989). Turbulent transfer in plant canopies. *In* "Plant Canopies: Their Growth, Form and Function" (G. Russell, B. Marshall, and P. G. Jarvis, eds.), pp. 41–61. Cambridge Univ. Press, Cambridge, UK.

Read, J., and Farquhar, G. (1991). Comparative studies in Nothofagus (Fagaceae). I. Leaf carbon isotope discrimination. *Funct. Ecol.* **5,** 684–695.

Reed, R., and Travis, R. L. (1987). Paraheliotropic leaf movements in mature alfalfa canopies. *Crop Sci.* **27,** 301–304.

Richards, J. H., and Caldwell, M. M. (1985). Soluble carbohydrates, concurrent photosynthesis and efficiency in regrowth following defoliation: A field study with *Agropyron* species. *J. Appl. Ecol.* **22,** 907–920.

Richards, R. (1991). Crop improvement for temperate Australia: Future opportunities. *Field Crops Res.* **26,** 141–169.

Robinson, J. M. (1994). Speculations on carbon dioxide starvation, Late Tertiary evolution of stomatal regulation and floristic modernization. *Plant Cell Environ.* **17,** 343–354.

Ross, J. (1981). "The Radiation Regime and Architecture of Plant Stands." Junk, The Hague.

Ryan, M. G., Gower, S. T., Hubbard, R. M., Waring, R. H., Gholz, H. L., Cropper, W. P., Jr., and Running, S. W. (1995). Woody tissue maintenance respiration of four conifers in contrasting climates. *Oecologia* **101,** 133–140.

Rygol, J., Pritchard, J., Zhu, J. J., Tomos, A. D., and Zimmermann, U. (1993). Transpiration induces radial turgor pressure gradients in wheat and maize roots. *Plant Physiol.* **103,** 493–500.

Senock, R. S., Sisson, W. B., and Donart, G. B. (1991). Compensatory photosynthesis of *Sporobolus flexuosus* (Thurb.) following simulated herbivory in the northern Chihuahuan desert. *Bot. Gaz.* **152,** 275–281.

Setter, T. L., Conocono, E. A., Egdane, J. A., and Kropff, M. J. (1995). Possibility of increasing yield potential of rice by reducing panicle height in the canopy. I. Effects of panicles on light interception and canopy photosynthesis. *Aust. J. Plant Physiol.* **22,** 441–451.

Sims, D. A., and Pearcy, R. W. (1994). Scaling sun and shade photosynthetic acclimation of *Alocasia macrorrhiza* to whole-plant performance—I. Carbon balance and allocation at different daily photon flux densities. *Plant Cell Environ.* **17,** 881–887.

Sinclair, T. R., and Horie, T. (1989). Leaf nitrogen, photosynthesis, and crop radiation use efficiency: A review. *Crop Sci.* **29,** 90–98.

Socias, F. X., Medrano, H., and Sharkey, T. D. (1993). Feedback limitation of photosynthesis of *Phaseolus vulgaris* L. grown in elevated CO_2. *Plant Cell Environ.* **16,** 81–86.

Stenseth, N. C., and Maynard Smith, J. (1984). Coevolution in ecosystems: Red Queen evolution or stasis? *Evolution* **38,** 870–880.

Steudle, E., Murrmann, M., and Peterson, C. A. (1993). Transport of water and solutes across maize roots modified by puncturing the endodermis. *Plant Physiol.* **103,** 335–349.

Sung, S. S., Shieh, W. J., Geiger, D. R., and Black, C. C. (1994). Growth, sucrose synthase, and invertase activities of developing *Phaseolus vulgaris* L. fruits. *Plant Cell Environ.* **17,** 419–426.

Tardieu, F., Zhang, J., and Gowing, D. J. G. (1993). Stomatal control by both [ABA] in the xylem sap and leaf water status: A test of a model for droughted or ABA-fed field-grown maize. *Plant Cell Environ.* **16,** 413–420.

Terashima, I., and Evans, J. R. (1988). Effects of light and nitrogen nutrition on the organization of the photosynthetic apparatus in spinach. *Plant Cell Physiol.* **29,** 143–155.

Thomas, H. (1986). Water use characteristics of *Dactylis glomerata* L., *Lolium perenne* L., and *L. multiflorum* Lam. plants. *Ann. Bot.* **57,** 211–223.

Thomas, H., and Stoddardt, J. L. (1980). Leaf senescence. *Annu. Rev. Plant Physiol.* **31,** 83–111.

Thornhill, N. W. (ed.) (1993). "The Natural History of Inbreeding and Outbreeding." Univ. of Chicago Press, Chicago.

Thornley, J. H. M. (1977). Growth, maintenance and respiration: A re-interpretation. *Ann. Bot.* **41,** 1191–1203.

Thornley, J. H. M. (1983). Crop yield and planting density. *Ann. Bot.* **52,** 257–259.

Trenbath, B. R., and Angus, J. F. (1975). Leaf inclination and crop production. *Field Crop Abstr.* **28,** 231–244.

Tsunoda, S. (1983). Photosynthetic strategy in rice and wheat. *In* "Proceedings of the XV International Congress of Genetics," pp. 255–266. Oxford/IBH, New Delhi.

Turner, N. C., O'Toole, J. C., Cruz, R. T., Yambao, E. B., Ahmad, S., Namuco, O. S., and Dingkuhn, M. (1986). Responses of seven diverse rice cultivars to water deficits. II. Osmotic adjustment, leaf elasticity, leaf extension, leaf death, stomatal conductance and photosynthesis. *Field Crops Res.* **13,** 273–286.

Tyree, M. T., and Sperry, J. S. (1988). Do woody plants operate near the point of catastrophic xylem dysfunction caused by dynamic water stress? *Plant Physiol.* **88,** 574–580.

van Bel, A. J. E., Ammerlaan, A., and van Dijk, A. A. (1994). A three-step screening procedure to identify the mode of phloem loading in intact leaves. *Planta,* **192,** 31–39.

van der Werf, A., van Nuenen, M., Visser, A. J., and Lambers, H. (1993). Effects of N-supply on the rates of photosynthesis and shoot and root respiration of inherently fast-growing and slow-growing monocotyledonous species. *Physiol. Plant.* **89,** 563–569.

Vitousek, P. (1982). Nutrient cycling and nutrient use efficiency. *Am. Nat.* **119,** 553–572.

Wallace, B. (1981). "Basic Population Genetics." Columbia Univ. Press, New York.

Wang, Y. P., and Jarvis, P. G. (1990). Influence of crown structural properties on PAR absorption, photosynthesis, and transpiration in Sitka spruce: Application of a model (MAESTRO). *Tree Physiol.* **7,** 297–316.

Wells, R., Meredith, W. R., Jr., and Williford, J. R. (1986). Canopy photosynthesis and its relationship to plant productivity in near-isogenic cotton lines differing in leaf morphology. *Plant Physiol.* **82,** 635–640.

Williams, K., Field, C. B., and Mooney, H. A. (1989). Relationships among leaf construction cost, leaf longevity, and light environment in rain-forest plants of the genus *Piper. Am. Nat.* **133,** 198–211.

Wong, S.-C., Cowan, I. R., and Farquhar, G. D. (1985). Leaf conductance in relation to rate of CO_2 assimilation. 1. Influence of nitrogen nutrition, phosphorus nutrition, photon flux density, and ambient partial pressure of CO_2 during ontogeny. *Plant Physiol.* **78,** 821–825.

Woodrow, I. E., and Berry, J. A. (1988). Enzymatic regulation of photosynthetic CO_2 fixation in C_3 plants. *Annu. Rev. Plant Physiol.* **39,** 533–594.

Wronski, E. (1984). A model of canopy drying. *Agric. Water Management* **8,** 243–262.

3

Adaptation to Diverse Environments: Variation in Water-Use Efficiency within Crop Species

A. G. Condon and A. E. Hall

I. Introduction

It is axiomatic that, for plants to assimilate carbon dioxide and grow, they must use water because carbon dioxide diffuses into the leaves through the same stomatal pores from which water is transpired. This has led to the notion that where water supply is a limitation to crop growth, increasing the efficiency of this exchange of carbon dioxide for water should lead to greater crop growth and yield. Cowan and Farquhar (1977) provided a stimulus for research in this area in a classic paper that described a theoretical basis for optimal stomatal function that would enhance the efficiency of gas exchange. The extent to which stomata function optimally is not known but experimental studies (Hall and Schulze, 1980) indicated that they, at least approximately, follow the model of Cowan and Farquhar (1977). The theoretical analysis of Cowan and Farquhar (1977) also indicated that the stomatal responses that are optimal would be different in different environments. This suggests that, by seeking out and exploiting genotypic variation in the efficiency of exchange of CO_2 for water, it should be possible to improve the adaptation of crop species to the environments in which they are grown. For many years this path toward varietal improvement has remained essentially unexplored, largely because identifying genotypic variation in water-use efficiency within crop species has proved too tedious. The proposition that carbon isotope analysis may provide an efficient means to assess genotypic variation in the water-use efficiency of leaf gas exchange (Farquhar *et al.,* 1982) has stimulated renewed interest in the prospect of gaining yield increases through enhancing water-use efficiency.

There are many processes and influences that can confound any simple relationship between the water-use efficiency of leaf gas exchange and the final yield obtained from the water consumed in crop growth. In discussing the significance of these processes and influences, we first need to define "water-use efficiency" in terms appropriate to the scale on which it is being measured. Such definition is doubly necessary because the term water-use efficiency has come to mean different things to different people, depending on the scale of measurement (e.g., from stomata to irrigation schemes) and/or the units of exchange being considered (e.g., from molecules of CO_2 to tons of biomass or grain).

In this chapter we will describe the instantaneous ratio of the rates of CO_2 assimilation (A) and transpiration (T) at the leaf level, $A/T,$ as gas-exchange efficiency. The term "transpiration efficiency" (W) will be used to describe the ratio of carbon gained to water transpired at the whole plant level, where the carbon gain has typically been measured as plant dry matter. At the crop level, the term water-use efficiency (W_{ET}) will be used to describe the amount of dry matter production per unit of total crop water use (ET), i.e., both transpiration by the crop (T) and evaporation from the soil surface (E). "Crop transpiration efficiency" (W_T) will be used where the evaporation component has been accounted for (and subtracted from the total crop water use) or is negligible [for correspondence between this notation and that of Gutschick (Chapter 2, this volume) see Section VI, Appendix, in Chapter 2]. Grain yield will not be considered explicitly in terms of water-use efficiency, i.e., as a ratio with some measure of water use. Rather, it will be seen as the outcome of factors influencing the amount of dry matter produced and its partitioning.

The impact on crop performance of genotypic variation in A/T depends on how the ratio of A/T at the leaf level influences the functioning of crop canopies. These influences can be evaluated by asking three questions: (i) What is the impact of variation in A/T on crop growth, (ii) what is the impact of variation in A/T on crop water use, and (iii) how do growth and water use interact over the crop's duration to produce the final outcome— grain yield? These questions need to be asked in the context of both the crop species being grown and the environment in which the crop is growing. Water availability is the dominant source of diversity among the agricultural environments we will consider. The first two questions will be addressed in Section II, which will deal with mechanisms of genotypic variation in water-use efficiency and "scaling up" from the leaf to the crop canopy level. The third question will be addressed on a theoretical level in Section III and then on a practical level in Section IV, under which we will review results from field experiments that have used stable carbon isotope analysis to explore the association between crop performance and water-use efficiency.

II. Mechanisms of Genotypic Variation in Water-Use Efficiency

A. Leaf Level

At the leaf level, gas-exchange efficiency is defined as the ratio of the rates of net CO_2 uptake (A) and transpiration (T). Both CO_2 uptake in photosynthesis and water loss through transpiration occur almost exclusively through the stomatal pores in the surface of the leaf and can be described as diffusion processes. Accordingly, the rates of these processes are determined by the conductances to diffusion of CO_2 and water vapor through the stomata and the concentration differences in CO_2 and water vapor between the ambient atmosphere outside the leaf boundary layer and inside the substomatal cavities.

$$A = g_c(c_a - c_i)/P_a, \tag{1}$$

where g_c is the conductance to diffusion of CO_2, c_a and c_i are the atmospheric and intercellular partial pressures of CO_2, respectively, and P_a is atmospheric pressure, and

$$T = g_w(e_i - e_a)/P_a, \tag{2}$$

where g_w is the conductance to diffusion of water vapor and e_i and e_a are, respectively, the water vapor pressures inside the leaf (which is close to the saturation value except with very negative leaf water potentials) and in the atmosphere. The instantaneous ratio of the rates of these two processes, i.e., the leaf gas-exchange efficiency, can be summarized as follows:

$$A/T = c_a(1 - c_i/c_a)/1.6\nu, \tag{3}$$

where 1.6 is the ratio of the diffusivities of water vapor and CO_2 in air and ν is the vapor pressure difference $(e_i - e_a)$. Equation (3) indicates that there are two possible sources of genotypic variation in A/T, i.e., c_i/c_a (to which A/T is negatively related) and ν (to which A/T is inversely related).

Genotypic variation in the value of the ratio c_i/c_a may reflect variation among genotypes in either leaf photosynthetic capacity (A_{cap}), i.e., the amount and activity of photosynthetic machinery per unit leaf area, or genotypic variation in stomatal conductance as it affects g_w and A, or both A_{cap} and g_w. The interrelationships among A, g_w, and c_i for three "genotypes" of a C_3 crop species are depicted schematically in Fig. 1. On a leaf area basis, an increase in A_{cap} at constant g_w (e.g., from 1 to 2 in Fig. 1) will

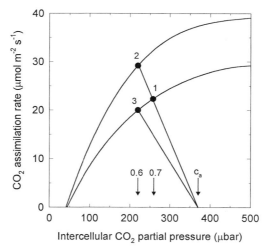

Figure 1   A schematic representation of genotypic variation in gas-exchange characteristics of a C_3 species [after Farquhar and Sharkey (1982)]. Two sets of lines are plotted on the figure. The two curved lines rising from near the origin represent the dependence of A on c_i. This dependence is typically determined for a leaf by measuring A in a gas-exchange cuvette and altering the CO_2 partial pressure inside the cuvette so as to span a range from near zero to well above ambient. Variation in the initial slopes of these curves reflects variation in the photosynthetic "demand" (A_{cap}). The curved lines are intersected by two straight lines originating at the ambient CO_2 partial pressure (c_a). The slope of these lines is the stomatal conductance to CO_2 under ambient conditions, which governs the "supply" of CO_2 to the photosynthetic machinery. The intersections of the straight and curved lines determine the operating values of A and c_i for three genotypes of a C_3 species under ambient conditions. For genotype 1, a "standard genotype," the intersection point is set such that the value of c_i/c_a equals 0.7, a typical value for C_3 species. For genotypes 2 and 3 the operating set points each give a value of c_i/c_a of 0.6. According to Eq. (3), a change from 0.7 to 0.6 in the value of p_i/p_a should result in a 33% increase in instantaneous gas-exchange efficiency. For genotype 2, this increase in A/T is achieved through a large increase in A_{cap}. There is also a large increase in A at the operating point, relative to genotype 1. For genotype 3, the increase in A/T is achieved through a decrease in g_w (indicated by the smaller slope of the line from c_a to the operating point of genotype 3) but at the expense of a small decrease in A relative to genotype 1.

result in an increase in A but no change in T. Thus, A/T will increase with increased A_{cap}. Because A is greater but there has been no change in g_w, there will be a greater drawdown of CO_2 inside the leaf, i.e., the value of c_i/c_a will be lower (e.g., from 0.7 to 0.6 in Fig. 1). With constant A_{cap}, c_i/c_a will also vary with changes in g_w (e.g., from 1 to 3 in Fig. 1). In this case, to achieve an increase in A/T, g_w must decrease. Decreases in g_w will reduce T more than they will reduce A, in most cases, because of the additional factors affecting photosynthesis. The net result will be an increase in A/T.

Nevertheless, an increase in A/T achieved via reduced g_w will be at the expense of a reduction in the rate of photosynthesis. It should also be recognized that there can be concurrent variation in both A_{cap} and g_w. Often it has been observed that both change in the same direction and in proportion such that there is little net change in c_i/c_a or A/T, but this is not always the case (Schulze and Hall, 1982).

Apart from a decrease in A, there is likely to be another penalty associated with a decrease in g_w. Unless the leaf boundary layer conductance is very large (which it often is inside a leaf gas-exchange cuvette but which it often is not outside such a device), leaf temperature will increase as g_w decreases. An increase in leaf temperature will cause an increase in the gradient driving transpiration, i.e., v, and therefore an increase in T per unit g_w. The net result is that the increase in A/T as g_w decreases will not be as great as predicted from Eq. (3) with the assumption that v is an independent variable (which it is not). In addition, increases in leaf temperature could influence A by either decreasing A if the environment is hotter than the optimal range of temperatures or increasing A if the environment is very cool.

Researchers have evaluated the possibility of increasing leaf gas-exchange efficiency by screening genotypes for stomatal characteristics, but they have had limited success. Jones (1987) has reviewed this work and suggests that major reasons for the lack of success are the inherent variability and the strong phenotypic component of stomatal characters, be they aspects of stomatal morphology, such as density or size, or stomatal conductance or resistance. Some examples of successful attempts at increasing leaf gas-exchange efficiency through screening for stomatal characters are the work on barley (*Hordeum vulgare* L.) by Miskin *et al.* (1972), who screened for variation in stomatal density (one component of g_w), and the work on *Vicia faba* by Nerkar *et al.* (1981), who screened on the basis of stomatal resistance (the inverse of g_w). Leaf cooling can also be achieved through greater g_w, but under most conditions this would be expected to result in lower gas-exchange efficiency. Under very hot atmospheric conditions, however, stomatal cooling may keep leaf temperatures closer to those that are optimal for A and also minimize damage to the photosynthetic apparatus, thereby maintaining A_{cap} (Radin *et al.*, 1994).

With respect to A_{cap}, there has been a long history of interest in attempting to exploit genetic variation in the rate of photosynthesis on a leaf area basis. This research has mainly involved attempts to increase productivity with less emphasis on changing gas-exchange efficiency. A recent example is the comparison of okra leaf and normal leaf cotton (*Gossypium hirsutum* L.) genotypes (Pettigrew *et al.*, 1993). Photosynthetic capacity per unit leaf area was found to be greater in okra leaf cotton as was gas-exchange efficiency but total leaf area per plant was smaller, resulting in productivity being no greater for okra leaf cotton. In contrast, the increases in yield achieved in Pima cotton (*Gossypium barbadense* L.) through several decades of breeding and selection for yield in hot environments appear to be associated with increases in both A_{cap} and g_w (Cornish *et al.*, 1991; Lu and Zeiger, 1994), but the increases in g_w appear to have been more important (Lu *et al.*, 1996).

The vapor pressure difference between the leaf and the air (ν) is subject to genetic manipulation at the leaf level by various means. The vapor pressure inside the leaf is strongly affected by leaf temperature. Characters that increase the reflectivity of leaves, such as pubescence (hairiness) and glaucousness (waxiness), reduce the energy load on the leaf and, thereby, reduce leaf temperature and ν compared with leaves with less reflectivity.

Selection for increased glaucousness has been associated with increased yield and water-use efficiency in several crop species including wheat (*Triticum aestivum* L. and *T. turgidum* L.) (Johnson *et al.*, 1983; Richards *et al.*, 1986), barley (Baenziger *et al.*, 1983), and sorghum (*Sorghum bicolor* L.) (Jordan *et al.*, 1983). Clawson *et al.* (1986) compared soybean (*Glycine max* L.) lines that were isogenic except for different levels of pubescence. In one of two genetic backgrounds used, water-use efficiency (measured as aboveground biomass per unit ET) was greater in the line with dense pubescence. There was no difference between isogenic lines with the other genetic background.

Leaves that are more tangential to the sun's direct beam have reduced radiant energy loads. This leaf orientation could result in increased gas-exchange efficiency when radiation levels are greater than necessary to saturate photosynthesis. Innes and Blackwell (1983) demonstrated that selecting for more erect leaves in winter wheat resulted in greater water-use efficiency for biomass production. In several dicotyledonous species, notably many legumes, light-avoiding leaf movements (paraheliotropism) occurring under drought (Shackel and Hall, 1979) probably enhance gas-exchange efficiency.

B. Whole Plant Level

For whole plants, transpiration efficiency (W) is the ratio of total dry weight accumulated to total plant water use. This ratio is not merely arrived at by a simple integration of the instantaneous rates of A and T. Additional factors must be accounted for. There are respiratory losses of carbon that are not coupled to transpiration. These include respiration by nonleaf tissues during the day and by all parts of the plant during the night. There is also water loss that is not directly coupled to carbon gain. This would include water loss through the cuticle during the day and through the cuticle and incompletely closed stomata at night. Equation (4) extends the mathematical description of leaf gas-exchange efficiency to account for these additional factors affecting plant W:

$$W = [c_a(1 - c_i/c_a)(1 - \phi_c)]/[1.6\nu(1 + \phi_w)]. \qquad (4)$$

In this equation the parameters ϕ_c and ϕ_w account for the losses of carbon and water, respectively, that do not take place during photosynthetic gas exchange. If aboveground dry matter production only is being considered, then another factor, $1 - r$, should be placed in the denominator, with r representing the proportion of plant carbon in the roots.

A substantial proportion of the carbon fixed by plants is used in respiration (ϕ_c). Estimates of this proportion vary from about 0.3 to 0.5 depending on the species and stage of plant development. Genotypic variation in the ratio of dark respiration to net photosynthesis has been observed in the leaves of some species (Heichel, 1971; Jones and Nelson, 1979; Winzeler *et al.*, 1989) and could contribute to enhanced W. For example, Wilson (1975) identified differences in mature-leaf dark respiration rate among *Lolium perenne* genotypes and selected plants with low leaf respiration rates and achieved increases in productivity (Wilson and Jones, 1982). Gifford *et al.* (1984) noted, however, that in this pasture species, much of the productivity advantage from low rates of respiration appeared to be associated with faster regrowth after defoliation, a situation unlikely to be encountered by many crop species. Genotypic variation in carbon allocation among different plant organs may be a more important source of variation in whole plant respiration for crop species. For example, a greater carbon allocation to stems, rather than to roots or leaves, may reduce ϕ_c because respiratory costs associated with both growth and maintenance are lower for stems (Stahl and McCree, 1988). Breeding to reduce carbon allocation to roots should result in plants producing greater aboveground biomass provided adequate water and nutrient extraction are sustained.

Genotypic variation in water loss not related to photosynthesis (ϕ_w) also appears to be substantial within some crop species (Jordan *et al.*, 1984; Clarke *et al.*, 1989) but its significance in relation to plant transpiration efficiency (W) remains unclear. For wheat, grain yield and low ϕ_w were related only in very dry environments (Clarke *et al.*, 1989), presumably because low ϕ_w was associated with maintenance of greater plant hydration. There was no relationship found in wet field environments. In a glasshouse study, genotypic variation in ϕ_w failed to account for any of the substantial variation in W observed among a set of 15 wheat genotypes (Clarke *et al.*, 1991). Variation in W among these genotypes was strongly correlated with variation in c_i/c_a estimated using carbon isotope analysis (Condon *et al.*, 1990).

C. Canopy Level

At the field scale there are several factors that may influence the extent to which genotypic variation in crop water-use efficiency or yield reflects genotypic variation in leaf gas-exchange efficiency. Some of these factors relate to scaling leaf gas exchange to canopy gas exchange, i.e., factors associated with scaling in space. Other factors relate to the interaction of the crop with its environment over the many weeks or months of the growing season, i.e., factors associated with scaling over time.

A major potential problem in scaling genotypic variation in leaf gas-exchange efficiency arises if the source of variation in c_i/c_a is in g_w. It has already been noted at the single leaf level that if the boundary layer conductances to the diffusion of water vapor and sensible heat are not very large, leaf temperature and ν (and therefore transpiration per unit g_w) will tend to be greater for genotypes with low g_w. The effect of lower g_w on leaf-to-air vapor pressure difference will be even greater for a crop canopy that extends over a large area. Not only will leaf temperature be higher but also, because of the presence of a substantial canopy boundary layer, the temperature of the air around the canopy will also be higher and the air will be drier. All these "canopy" effects will increase the effective driving force for transpiration (ν), reducing the impact of decreasing g_w on canopy transpiration. As stomatal control over canopy transpiration decreases so too will the impact of changes in g_w on crop transpiration efficiency (W_T) (Jones, 1976; Jarvis and McNaughton, 1986). The size of this canopy effect will depend on both leaf and canopy characteristics. Briefly, the response of W_T to a change in g_w will decrease as (i) the leaf boundary layer resistance increases; (ii) canopy aerodynamic resistance increases; (iii) the absolute temperature increases; and (iv) the resistance to CO_2 uptake by the canopy decreases, such as for a crop under optimal conditions with a large leaf area index and open stomata. Several mathematical models of canopy gas exchange predict that effects of changes in leaf-level g_w on canopy-level W_T can be small (Jones, 1976; Cowan, 1988) and it seems that this may be so in cropping situations that approximate conditions ii and iv. However, for many rainfed crops such conditions may either not apply or only apply for a relatively short period of time.

In environments in which crop growth relies strongly on rainfall during the growing season, an important factor influencing the relationship between leaf gas exchange and crop performance is the variation among genotypes in the soil evaporation component of total crop water use. In such environments soil evaporation may account for between 30 and 60% of total seasonal crop ET (Cooper *et al.*, 1987). Variation among genotypes in soil evaporation is associated with genotypic differences in the rate at which the crop canopy grows and shades the frequently rewetted soil surface from direct sunlight. By reducing evaporative losses through soil shading, genotypes with a fast rate of leaf area growth have much more water available near the soil surface for use in transpiration. In large-scale experiments on wheat (Condon and Richards, 1993) two cultivars differed by ca. 15% in W_T, due to differences in g_w, but there was no difference found in W_{ET}

because soil evaporation was a much greater proportion of total ET for the genotype with lower stomatal conductance. The greater soil evaporation was the result of the relatively slow rate of leaf area development for the genotype with lower conductance. In cereals, high A/T also can result from high A_{cap} (Condon *et al.,* 1990; Morgan and LeCain, 1991). Where this is the case, canopy effects that cause differences in soil evaporation can influence the relationship between A/T and W_{ET}. Among cereal genotypes, high A_{cap} is often associated with the concentration of N into smaller leaves that have higher levels of the photosynthetic enzyme ribulose-1,5-bisphosphate carboxylase oxygenase (EC 4.1.1.39) and lower specific leaf area (SLA, leaf area per unit leaf dry weight). For such genotypes, the rate of leaf area growth is usually relatively slow because of the larger leaf mass per unit leaf area, as was found for the okra leaf cotton genotypes described earlier, and can result in less shading and more soil evaporation. When barley and wheat are grown together in environments with rainfall occurring during periods of rapid crop growth (current rainfall) barley has a substantial yield advantage. A large proportion of the yield advantage of barley is due to its much faster leaf area growth (Lopez-Castaneda and Richards, 1994). Some of this difference is a result of the higher SLA of barley. In peanut (*Arachis hypogaea* L.), genotypic variation in gas-exchange efficiency is dominated by variation in A_{cap} and in this species there seems to be no penalty in terms of early leaf area growth. Rather, genotypes of peanut with high A_{cap} have a higher rate of leaf area development (Wright *et al.,* 1993). Genotypic differences for sunflower (*Helianthus* sps.) reported by Virgona and Farquhar (1996) are similar to those described for peanut.

Another advantage of a fast rate of leaf area development in "current rainfall" environments is that it promotes greater crop growth during that part of the growing season when the evaporative demand and ν are small and water-use efficiency is large. In a comprehensive review of studies from the level of container-grown plants through to crop canopies, Tanner and Sinclair (1983) concluded that the greatest amount of variation in water-use efficiency found for different crops could be accounted for by differences in the vapor pressure deficit of the air (similar to ν) during growth. Furthermore, they suggested that large increases in water-use efficiency could be achieved by growing crops in conditions in which the vapor pressure deficit is small. The smallest vapor pressure deficits occur during the late fall, winter, and early spring and there are several plant characteristics that could be manipulated, either genetically or agronomically, that would increase plant growth during these cooler seasons.

Genotypic variation in phenology is another trait that may be manipulated to increase crop water-use efficiency because it can be used to make possible a more favorable sowing date (Woodruff and Tonks, 1983; Gimeno *et al.,* 1989). Genes that alter sensitivity to photoperiod or vernalization can be manipulated in many species to permit fall sowing of crops that would otherwise flower too early (Murfet, 1977). Often, disease can seriously limit plant growth when crops are grown under cool, moist conditions. Increased disease resistance can be important if it allows sowing time to be changed so as to maximize water-use efficiency. For example, chickpea (*Cicer arietinum* L.) is normally sown in late spring in Syria but increase in resistance to foliar disease permitted the crop to be sown in the fall. The outcome was a doubling of grain yield (Keatinge and Cooper, 1983).

Genotypic variation in partitioning of dry matter influences the relationship between leaf gas-exchange efficiency and crop yield. In the latter half of this century substantial gains in grain yield of wheat have been achieved by increasing the partitioning of dry matter to the grain with little perceptible change in total biomass production or the water-use efficiency of biomass production. These yield gains have been achieved by exploiting genotypic variation in plant height so as to decrease lodging, and they also reduced competition for assimilate between reproductive and vegetative growth (Gifford *et al.*, 1984). For rainfed crops there is also competition for water between the vegetative and reproductive stages of growth. The extent to which available soil water is used during vegetative growth, which often occurs under conditions in which water supply is being frequently replenished, influences the water available during reproductive growth, which frequently takes place under conditions of rapidly diminishing soil water reserves (Fischer, 1981). This competition for available water usually results in harvest index being lower under rainfed conditions than when crops of the same genotype are fully irrigated. For rainfed crops, phenology may be manipulated to reduce the risk of flowering occurring under conditions of severe water deficit. For example, earlier flowering may be adaptive for environments with end-of-season drought, whereas for crops grown in environments with mid-season drought, later flowering times that occur after the dry period may be desirable.

III. Theoretical Associations between Leaf Gas-Exchange Efficiency and Adaptation

The preceding section established that there are numerous factors that may influence the association between crop performance and instantaneous gas-exchange efficiency of leaves. The effects of these factors can vary with the cropping environment and the species being grown. In this section we will attempt to establish the likely impact of genotypic variation in leaf gas-exchange efficiency on crop performance in different types of cropping environments.

A. Well-Watered Environments

In well-watered environments effects of variation in leaf gas-exchange efficiency on crop growth rate will be of primary importance. It is assumed that in these environments there is no limitation on crop growth or yield associated with water supply. If the physiological basis for genotypic variation in leaf-level A/T is g_w, then in well-watered environments greater productivity should result from the high A associated with high g_w. The effect will be greater in hotter environments in which there are additional benefits to the plants from the lower leaf temperatures resulting from the greater evaporative cooling with higher g_w (Lu and Zeiger, 1994). If the physiological basis for genotypic variation in leaf-level A/T is A_{cap}, the situation may not be so straightforward because the relationship between A_{cap} and crop growth rate may vary depending on the species being grown. Crop growth rate is more likely to reflect variation in A per unit ground area than A per unit leaf area. High A_{cap} should result in high A per unit leaf area but not necessarily high A per unit ground area. This is because in many crop species high A_{cap} is achieved by the concentration of leaf N into smaller leaves (Bhagsari and Brown, 1986) resulting in canopies that can intercept less radiation. In this case early crop growth may be faster in genotypes with low A_{cap} and low A/T but more leaf area. However, once canopy closure has been achieved and the crop is intercepting all the available radiation, crop growth is likely to be faster in genotypes with high A_{cap} and high A/T. The outcome over the duration of the cropping season may be a positive association, a negative association, or no association between A_{cap} and yield, depending on the balance between these two phases of crop growth and also on the timing of processes that influence grain number. The relationship between A_{cap} and grain yield should be simpler for those species in which high A_{cap} is associated with faster crop growth throughout the season, i.e., where there is no trade-off with leaf area development. In this case, high A_{cap} should result in greater crop yield in well-watered environments, as would be the case with high g_w. In both of these cases it is likely that canopy water-use efficiency would be increased, due to the increases in biomass production, because under well-watered conditions canopy water use would not change much, even for the genotypes with higher g_w as a consequence of the feedback effects discussed earlier.

B. Water-Limited Environments with Moisture Stored in the Soil

In some water-limited environments crop growth is totally reliant, from shortly after crop establishment, on a finite supply of moisture stored within the soil profile. In these environments transpiration is the major component of crop water use because soil evaporation is small, and higher A/T would be expected to enhance crop biomass production providing that all the available soil water is used by the crop. If the physiological basis for high

A/T is high A_{cap} and there is no trade-off involving reduced leaf area development, then there should be a positive association between grain production and A/T provided that any extra leaf area growth does not lead to excessive early soil water depletion and extreme water deficits during periods that are critical for establishing grain number or seed development. If there is a trade-off involving an association between higher A_{cap} and slower leaf area development then this may be an advantage if it is associated with conservation of water during the vegetative stage making more soil water available to the plant during reproductive growth. Alternatively, there may be a limitation on biomass production and grain yield if the growth rates of the shoot and roots are too slow. The magnitude of this limitation will depend on the size of the soil water store and how quickly it is consumed relative to the timing of processes that determine grain number and size and how completely the available soil water is used. The effect on crop performance of high A/T resulting from low g_w should be similar to that of high A_{cap} coupled with slower leaf area growth. Low g_w should result initially in a slower crop growth rate but at high W_T. Scale effects associated with canopy coupling are unlikely to be important in this kind of cropping environment because large, smooth canopies are unlikely to develop and stomata are unlikely to be wide open. It should be noted that seasonal and diurnal patterns of stomatal opening may be important and that some genotypes might achieve higher seasonal transpiration efficiency by having higher g_w, compared with other genotypes, during times of the season and day when evaporative demands are low, providing that they have lower g_w when evaporative demands are high. This diurnal response appears to be consistent with the concept of optimal stomatal function proposed by Cowan and Farquhar (1977).

C. Variable Environments with Rain during the Cropping Season

In some semiarid environments a large proportion of the crop water supply comes from rainfall during the cropping season and, in most years, the amount of water available is insufficient and limits crop growth, although in some years rainfall is adequate. In environments such as these there is a risk, in some years, associated with a pattern of crop growth and water use that is too profligate such that severe soil water deficits occur during the critical periods when grain number and size are determined. In other years there is a risk associated with a pattern of growth that is too conservative and does not take full advantage of the favorable conditions. Genotypes of those species in which high A_{cap} is associated with fast leaf area growth should be able to use the available water efficiently under either scenario. If most of the rainfall occurs during the early part of the growing season then losses due to soil evaporation should be minimized because of the fast rate of leaf area development of these genotypes. If there are extended periods without rainfall, stored soil moisture should be used effectively for crop growth by these genotypes because of their high W_T. In favorable seasons the fast crop growth rate of these genotypes should result in greater crop yield. Crop genotypes with high A/T due to high A_{cap}, but for which the rate of leaf area development is slow, may not use rainfall early in the season as effectively as genotypes with low A_{cap} but that develop their leaf area faster. Losses to soil evaporation are likely to be greater when the rate of leaf area development is slower, so crop W_{ET} may not be any greater for these high A/T genotypes. If the growing season is short and canopy closure is not achieved, low A_{cap} genotypes that develop leaf area quickly are likely to have a yield advantage because they will probably have achieved greater biomass by flowering or pod set with minimal extra water use. For similar reasons there are likely to be few years, in this type of semiarid environment, in which genotypes with high A/T due to low g_w achieve a yield advantage over low A/T genotypes due to high g_w. In favorable seasons when early season rainfall persists to near flowering or pod set, the faster growth of genotypes with high g_w is likely to be reflected in greater crop yield unless the soil water store is small. If rainfall stops early and there is an extended period without rain during flowering, genotypes with low g_w are likely to have an advantage if there is an appreciable soil water store because the rate of water use should be slower by these genotypes and crop W_T will be greater. If the soil water store is small then low g_w is unlikely to be an advantage. Genotypes that have high g_w during well-watered periods but rapidly develop low g_w with the onset of soil water deficits may have an advantage in this type of variable environment. Consequently, it is useful to consider both the set point for leaf gas-exchange efficiency and the extent that it can be modified by drought-induced changes in g_w (Ehleringer, 1993; White, 1993).

IV. Empirical Observations

In this section we will review the empirical observations from the now numerous field studies in which genotypic variation in leaf gas-exchange efficiency has been evaluated in relation to crop performance. Most of these studies have been facilitated by the realization that, for C_3 species, carbon isotope discrimination (Δ) may provide a relatively simple, indirect assay for genotypic variation in leaf gas-exchange efficiency (Farquhar *et al.*, 1982). The basis for the relationship between Δ and A/T for C_3 species is described briefly. For a more complete description the reader is referred to Farquhar *et al.* (1982, 1989).

A. Theoretical Relationship between Carbon Isotope Discrimination and Leaf Gas-Exchange Efficiency

For C_3 plant species, the ratio of the stable isotopes of carbon ($^{13}C/^{12}C$) in plant tissue is lower than that of the air from which the plant carbon is drawn. This is because there is a net discrimination against $^{13}CO_2$ during photosynthesis and the assimilation of ^{13}C into plant dry matter. This discrimination arises from isotope effects during diffusion of CO_2 into the substomatal cavities, carboxylation by RubP carboxylase, respiration, and other processes. An approximate mathematical description of the impact of these isotope effects on net discrimination is given by

$$\Delta = a(c_a - c_i)/c_a + b(c_i/c_a) - d, \tag{5}$$

which can be rearranged to give

$$\Delta = a + (b - a)(c_i/c_a) - d, \tag{6}$$

where a, b, and d are parameters for isotope effects associated with diffusion of CO_2 in air (a), carboxylation (b), and respiration and other processes (d), and c_a and c_i are the atmospheric and intercellular partial pressures of CO_2, respectively (Farquhar *et al.*, 1982; Hubick *et al.*, 1986). The values of a, b, and d are close to 4, 29, and 3‰, respectively (Farquhar and Lloyd, 1993). Comparison of Eq. (6) with Eq. (3) reveals that, whereas Δ is a positive function of c_i/c_a, instantaneous gas-exchange efficiency is a negative function of c_i/c_a, and where ν is constant, A/T and Δ should be negatively related.

A negative relationship between instantaneous gas-exchange efficiency and carbon isotope discrimination has been demonstrated for wheat (Evans *et al.*, 1986), peanut (Hubick *et al.*, 1988), and bean (*Phaseolus vulgaris* L.) (Ehleringer *et al.*, 1991) using on-line gas-exchange techniques. In these studies, the C-isotope composition of air sampled before and after passing over a leaf in a gas-exchange cuvette was measured and the corresponding gas-exchange parameters were determined. Also, a negative relationship between whole plant transpiration efficiency and Δ has been demonstrated in pot studies for a number of C_3 crop species including wheat (Farquhar and Richards, 1984; Condon *et al.*, 1990; Ehdaie *et al.*, 1991), barley (Hubick and Farquhar, 1989), peanut (Hubick *et al.*, 1986), common bean (Ehleringer *et al.*, 1991), cowpea (*Vigna unguiculata* L. Walp.) (Ismail and Hall, 1992), and sunflower (*Helianthus anuus* L.) (Virgona *et al.*, 1990). The theory proposed by Farquhar *et al.* (1982) relating Δ to A/T is therefore well established at both the leaf and whole plant levels.

B. Value and Limitations of Carbon Isotope Discrimination as a Measure of Crop Water-Use Efficiency

Carbon isotope discrimination appears to be a potentially valuable tool for detecting genotypic variation in gas-exchange efficiency. Genotypic variation in Δ may also provide useful estimates of genotypic variation in W, W_T, or W_{ET}, provided the numerous complexities discussed in Section II are recognized that occur when scaling-up instantaneous A/T to the levels of plant or crop transpiration efficiency or water-use efficiency. The Δ of plant dry matter provides an estimate of instantaneous c_i [and thus, from Eq. (1), an estimate of instantaneous A/g_w] integrated over the time that the carbon in the dry matter was accumulated, which is particularly useful. The estimate will be assimilation weighted because periods when A is high will have a larger influence on Δ of the carbon pool than periods when A is low, and this can be advantageous because periods when A is high will have the greatest influence on crop growth.

Certainly, Δ is much easier to measure than plant or crop transpiration efficiency or water-use efficiency, especially for large numbers of genotypes. Isotope analysis is usually carried out on dried, finely ground samples of plant tissue using isotope ratio mass spectrometry. Preston (1992) provides a comprehensive description of isotope analysis procedures. The procedures can be automated and several hundred samples can be analyzed within a few days, although the process may be expensive.

However, there are clearly several limitations in using Δ to search for genotypic variation in crop water-use efficiency. One major limitation is that genotypic variation in Δ only provides an estimate of genotypic variation in the ratio A/g_w and not necessarily A/T, because ν is neither constant nor an independent variable. Also, it gives no indication of the variation among genotypes in the rates of either A or g_w (or T, for the cases in which g_w and T are related) nor of the mechanisms of the variation in A/g_w among genotypes, i.e., various stomatal or photosynthetic characteristics. As shown in Fig. 1, when A/g_w is relatively high (low c_i), for example, the rate of A (and potentially of plant or crop growth rate) may be either high or low depending on whether A_{cap} or g_w is the source of variation in A/g_w. Another major limitation is that scaling-up leaf-level A/g_w to the whole plant or crop canopy scales is often complex, as discussed in Section II. In particular, in some circumstances, genotypic variation in g_w may be only weakly reflected in genotypic variation in the rate of crop transpiration because of interactions between the crop canopy and the atmosphere. The relationship between g_w and variation in the rate of total crop water use may be weaker still because of additional variation among genotypes in the ratio of soil evaporation to crop transpiration.

A third limitation in the use of Δ to assess genotypic variation in water-use efficiency is that the measured Δ may be only a small component of the total variation in Δ over the course of the growing season or in the plant canopy. For example, leaves in the lower part of peanut canopies had higher Δ than leaves in the upper part of the canopies (Nageswara Rao *et al.*, 1995). Also, both for convenience and to reduce sampling error, Δ often is measured on leaves of specific ages that contain carbon assimilated over a relatively short time. For a given plant tissue, variation in Δ within a set of crop genotypes or breeding lines is typically of the order of

2‰, but environmental and physiological effects on Δ may generate variation three or four times greater than this between plant parts that have grown at different times. These environmental effects are most often associated with stomatal responses to, for instance, seasonal changes in soil water supply and/or vapor pressure deficit of the air (Condon *et al.*, 1992; Condon and Richards, 1993; Hall *et al.*, 1992; Nageswara Rao and Wright, 1994). As a result of this large environmental effect on Δ, genotype ranking for Δ or the range of variation in Δ may change, depending on the plant part sampled. In a drying soil, this could occur if the extent of soil drying is not uniform among genotypes or if genotypes differ in their stomatal response to soil drying and/or vapor pressure deficit. As a result, the sampled variation in Δ measured late in the season as the soil dries may not correspond to earlier variation in Δ when the soil was wet, which is more closely associated with the processes leading to differences in early crop growth and water use (Condon and Richards, 1993; Condon *et al.*, 1993). This problem may be compounded if the tissue sampled for Δ was not laid down at the same time for all genotypes, such as where seeds are sampled for genotypes that began flowering on different dates.

These limitations in using genotypic variation in Δ as an assay for genotypic variation in crop water-use efficiency will be drawn into focus in the next section, in which we will present some case studies on the association between leaf gas-exchange efficiency (estimated using variation in Δ) and crop performance in different environments.

C. Associations between Crop Performance and Carbon Isotope Discrimination: Some Case Studies with Small-Grain Cereals and Grain Legumes

1. Small-Grain Cereals   Several field studies on the association between crop performance and Δ have been conducted with small-grain cereals. In the majority of the studies, a range of wheat or barley genotypes differing in Δ but also in other characteristics, such as flowering time and height, were grown under naturally occurring rainfed conditions. In some studies water supply was manipulated using either supplementary irrigation or rainout shelters. Many of these studies reported associations between Δ and shoot biomass production and/or grain yield. Most of these associations were either positive or "neutral." There have been few reports of negative associations from these field studies. This indicates that, for cereals, factors contributing to or associated with low Δ (i.e., potentially high A/T) may often also contribute to or be associated with low productivity, even in dry environments.

a. Associations between Biomass Production and Carbon Isotope Discrimination   Associations between biomass production and Δ will be considered first because the relation is more direct than that with grain yield. Associations between biomass production and Δ are important in water-limited environments because, if it is assumed that all genotypes use all or nearly all the available water, biomass production should be positively related to crop W_T. Associations between biomass production and Δ in wet environments are also important for at least two reasons. First, they can indicate the extent of the trade-off between potential productivity in favorable environments and any putative productivity advantage that might accrue through high A/T in less favorable environments. Second, they are also important because in many "water-limited" environments some proportion of crop growth (often a large proportion of early crop growth) takes place under well-watered conditions.

In the first published study of the relationship between crop performance and Δ (Condon *et al.,* 1987), positive associations were found between shoot biomass and Δ for wheat grown at two locations in southeastern Australia. Both environments were rainfed, but seasonal rainfall was high for the region and there was little soil water depletion until well into grain filling. In a recent study with a different set of wheat cultivars grown in a larger range of environments, Condon and Richards (1993) found that the direction of the association between shoot biomass and Δ changed from being positive in wet sites and seasons to being negative at one low rainfall, stored-moisture site. For intermediate situations shoot biomass and Δ were not correlated. Ehdaie *et al.* (1991), with a small set of wheat genotypes grown at a field site in California, mainly found positive associations between shoot biomass and Δ when plots either were maintained well-irrigated through to maturity or where drought was imposed after a common date approaching heading. The study by Ehdaie *et al.* (1991) is somewhat difficult to interpret because, as well as variation in Δ, there was considerable variation in both time to ear emergence and height among the genotypes grown, with strong negative associations between both of these traits and Δ. Biomass production and height are often positively associated under well-watered conditions and time to ear emergence can have a major influence on growth under water-limited conditions. It is not clear whether it was height, time to ear emergence, or Δ that was having the greatest influence on the relationships observed in these environments.

In the absence of variation in other characteristics influencing growth, a positive association between biomass production and Δ would be expected under well-watered conditions if the major mechansim for variation in Δ is stomatal conductance (Jones 1994) because c_i/c_a and A should be positively related in this circumstance (Fig. 1). This assumes, of course, that biomass production and A are positively related. For cereals, this should be the case if the mechanism of variation in A is g_w. However, the relationship between A and crop growth rate in cereals may not be positive if the mechanism of the variation in A (and Δ) is A_{cap}. In wheat and barley an increase in A_{cap} is often achieved by the concentration of leaf nitrogen into smaller leaves which therefore have lower SLA. The result, at least during early crop growth, is a reduction in rates of leaf area expansion, which results in less light interception and less canopy photosynthesis and dry matter production (Condon *et al.*, 1993; Lopez-Casteneda *et al.*, 1995). If the duration of the crop is short and only limited time is spent at full light interception the outcome may be a negative association between A_{cap} and biomass and a positive association between biomass and Δ, low A_{cap} being associated with higher Δ (Fig. 1).

Irrespective of whether it is achieved through high g_w or high SLA, a faster rate of crop growth should be accompanied by greater crop transpiration and faster soil water depletion. Condon *et al.* (1993) investigated the relationships between Δ, biomass production, and water use for wheat grown in a rainfed environment in which there was minimal water stress in the period before anthesis but increasing soil water deficit during the postanthesis phase. There was a positive association between biomass at anthesis and Δ but genotypes with high Δ values had also used more water by anthesis, with the result that there was a convergence in total dry matter production among genotypes by the end of the season. In environments in which this and other similar experiments have been conducted (Lopez-Casteneda and Richards, 1994; A. G. Condon, 1996, unpublished data), early crop growth is sustained by regular rainfall events and the "penalty," in terms of soil water depletion, associated with high dry matter production by anthesis is not as great as might be expected. Transpiration does tend to be greater for high Δ genotypes, but a substantial proportion of the extra water consumed in the production of greater biomass by these genotypes is not conserved by low Δ genotypes but is "wasted" due to greater evaporation from the frequently rewetted soil surface under their slower-developing canopies. Perhaps more important, much of the extra water transpired by high Δ genotypes is consumed early in the season during the cool winter and early spring period when the vapor pressure deficit is low and transpiration efficiency is high due to the low evaporative demand. This suggests that in current rainfall environments, in which the relationship between final biomass production and Δ is "neutral" or even positive, it is likely that the relationship between c_i/c_a and Δ still holds but that other factors are equally or even more important in influencing crop water-use efficiency. These factors include the proportion of total crop ET that is actually transpired, rather than evaporated, and the interaction of crop transpiration with the seasonal course of vapor pressure deficit.

A high A/T and low rate of transpiration are more likely to be advantageous in situations in which the crop grows largely on stored soil moisture. In the one environment in which Condon and Richards (1993) found a negative association between biomass production and Δ, growing-season rainfall stopped well before anthesis but there was a substantial amount of subsoil moisture available that had been stored from summer rainfall, i.e., transpiration was very much the dominant component of total crop ET. Under these conditions genotypes with higher A/T (low Δ) produced more total shoot biomass.

   b. *Associations between Grain Yield and Carbon Isotope Discrimination*   Grain yield is the ultimate measure of crop performance and is clearly of greatest interest to plant breeders and agronomists. Most of the studies on the association between grain yield and Δ in small-grain cereals have been conducted in "current rainfall" or Mediterranean environments in which, as discussed previously, biomass production and Δ mainly have been positively associated. Where biomass and grain yield have both been measured, associations between grain yield and Δ have tended to be even more positive than those between biomass and Δ, i.e., there is a tendency for harvest index and Δ to also be positively related. For example, in the original study on wheat by Condon *et al.* (1987), the relative increase in biomass per mil Δ was 24% and for grain yield it was 35%. In the multisite study on biomass and Δ reported by Condon and Richards (1993), relationships between grain yield and Δ were all positive except for the single stored-soil moisture site, where no association between grain yield and Δ was found (Condon and Richards, 1996, unpublished data). There was a negative association between biomass and Δ at the stored-soil moisture site. The field data presented by Ehdaie *et al.* (1991) and by Morgan *et al.* (1993) also indicate positive associations between harvest index and Δ.

   There are several factors that could account for the positive associations between Δ and harvest index often observed with cereals. One major factor is phenology. In environments characterized by end-of-season drought, early flowering is critical for maximizing seed number, harvest index, and grain yield of determinate species such as wheat and barley (Loss and Siddique, 1994). In several of the studies conducted in this type of environment, strong negative associations have been found between Δ and days to heading or anthesis with low Δ genotypes flowering many days later than high Δ genotypes (Ehdaie *et al.*, 1991; Craufurd *et al.*, 1991; Acevedo, 1993; Sayre *et al.*, 1995). In all these studies grain yield has varied several-fold among genotypes, with very low yields in late flowering, low Δ lines for the driest sites and treatments. The huge range in grain yields indicates that in these studies the large variation in flowering date was having by far the greatest influence on yield. However, in studies in which there has been little variation in flowering date (Condon *et al.*, 1987; Morgan *et al.*, 1993; Acevedo, 1993) or in which variation in flowering date has been accounted for statistically (Sayre *et al.*, 1995) grain yield and Δ were positively associated.

Interestingly, phenology was probably again the dominant factor in a rare experiment with cereals that exhibited a negative association between grain yield and Δ (Craufurd *et al.*, 1991). Under conditions of continuous irrigation and low evaporative demand at Cambridge, U.K., late-flowering low Δ barley genotypes outyielded early flowering high Δ genotypes. It seems that the late-flowering lines were in a better position to take advantage of the very favorable growing conditions because of their longer duration. When the same genotypes were grown under rainfed conditions in Syria, a positive association was observed between grain yield and grain Δ at several sites. Early flowering, high Δ genotypes had a very large yield advantage at very dry sites and a small yield advantage at a site with moderate rainfall.

Another factor likely to contribute to the positive association between harvest index and Δ in cereals is the ability of cereals to use stored assimilate to sustain grain growth. This assimilate is laid down in the stems and other organs before and shortly after flowering. Under irrigated conditions grain growth can be sustained almost totally on current photosynthesis. However, as the degree of postanthesis water stress increases so does the dependence on stored assimilate. Depending on conditions during grain filling, preanthesis assimilate may contribute as much as 80% of final grain carbon (Palta and Fillery, 1995). It seems probable that among cereal genotypes growing in a terminal drought environment the contribution of stored assimilate to grain growth will be greater for high Δ genotypes that have produced more dry matter and used more of the soil water store at anthesis. This is demonstrated in Fig. 2, which shows the dependence of grain yield on dry matter production up to anthesis and after anthesis for a set of wheat genotypes of similar duration grown in a relatively favorable Mediterranean climatic zone in southeastern Australia. There was a positive relationship between grain yield and shoot biomass production at anthesis (Fig. 2a) but the relationship between grain yield and postanthesis dry matter gain tended to be negative (Fig. 2b), i.e., there was a tendency for genotypes that had produced more biomass at anthesis to be more dependent on stored assimilate for grain filling. Both harvest index and the ratio of grain yield to postanthesis growth were positively correlated with Δ. Soil water use was not measured, but there was a strong positive association between shoot biomass production at anthesis and Δ. In this study, grain yield increased 16% per mil Δ compared to a 6% increase for shoot biomass.

c. Productivity in Wheat Breeding Lines Selected Using Δ   Much of the available information on the association between productivity and Δ in cereals has been collected using cultivars that differ not only in Δ but also in flowering date, height, and other characteristics that can strongly influence growth and yield. In Fig. 3, data are presented from an ongoing study on wheat by A. G. Condon and R. A. Richards (unpublished results) in which

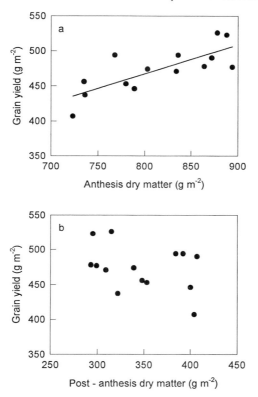

Figure 2   Relationships between final grain yield and aboveground dry matter production for the periods from sowing to anthesis (a) and anthesis to physiological maturity (b) for 14 wheat genotypes grown under rainfed conditions at Moombooldool, southeastern Australia, in 1986. Values are means of individual genotypes averaged over four plots in randomized blocks. a, $r = 0.78$ ($P < 0.01$); b, $r = -0.35$ (n.s.).

the relationship between productivity and Δ is being studied using the progeny of crosses specifically made to generate variation in Δ. The parents used in the crosses had substantial differences in Δ and are cultivars currently grown in Australia. F_2-derived lines were screened at F_3 and F_4 to give a subset of lines that were uniform for flowering time and height. These were tested for variation in Δ at F_5 using dry matter sampled early in the season when there was no water stress. Thirty lines per cross spanning a range of F_5 Δ values were grown in replicated yield trials using F_6 and F_7 families. Results presented here are for two crosses grown at one site over two seasons but this covers the range of relationships observed to date. For both crosses the major source of variation in Δ was g_w.

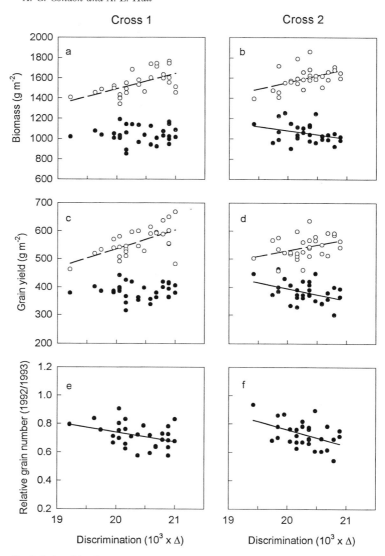

Figure 3   Relationships between measures of crop productivity and carbon isotope discrimination (Δ) among wheat genotypes grown under rainfed conditions at Condobolin, eastern Australia, in 1992 and 1993. Genotypes grown were F_6 (1992) and F_7 (1993) progeny of two crosses between parents with high and low Δ values. Left panels are for cross 1 and right panels are for cross 2. Values are plotted for total aboveground biomass production (a and b), grain yield (c and d), and grain number per square meter given as a ratio of values (e and f). Productivity values shown are means of individual genotypes averaged over three plots in randomized blocks. The values of Δ plotted are genotype means measured on F_5 lines using leaf material sampled early in the 1991 season when there was no water stress. In panels a–d, solid symbols are for 1992 and open symbols are for 1993. Least-squares linear fits are

The 1993 season produced record wheat yields for this site due to regular rainfall events from sowing well into grain filling (May to October rainfall, 292 mm). For both crosses the relationships between productivity (shoot biomass and grain yield) and Δ were strongly positive in 1993 (Figs. 3a–3d). In 1992, May to October rainfall (200 mm) was a little below average. For the lines from cross 1, the relationships between productivity and Δ were both "neutral" (Figs. 3a and 3c). For the lines from cross 2, which on average flowered 1 week later than the lines from cross 1, the relationships between productivity and Δ were negative in 1992 (Figs. 3b and 3d). Of the components of grain yield, the major impact of the drier 1992 season was on grain number, which was lower for all lines in 1992 but much more so for high Δ lines (Figs. 3e and 3f). Because harvest index was not associated with Δ in either cross in either season, the reduction in grain number in 1992 is unlikely to have been the result of a "catastrophic" effect of water stress on seed set but may have been the result of reduced ear growth (and biomass growth) associated with soil water depletion in the period shortly before anthesis when potential grain number is established (Fischer, 1981). The results strongly indicate greater soil water depletion by high Δ lines approaching anthesis in 1992 because there was a larger reduction in grain number for high Δ lines than low Δ lines, and this effect was greatest for high Δ lines from the later-flowering cross.

From the results of this selection study and those studies reviewed earlier relating yield and Δ in diverse genotypes, a reasonable conclusion would be that high A/T (low Δ) is most likely to be associated with greater productivity in cereals only when it is associated with greater seed number per unit area. Due to the large capacity of cereals to translocate stored assimilate to fill grains, variation in kernel weight is unlikely to play a major role. Opportunities for greater seed number associated with high A/T in cereals may be limited to only a few environments either (i) water-limited stored-soil moisture environments in which transpiration makes up the bulk of crop water use or (ii) relatively favorable current rainfall or Mediterranean climates in which there is reasonable soil water storage from excess winter/spring rainfall but in which substantial transpirational depletion of this

---

plotted where correlation coefficients were statistically significant ($P < 0.05$, $n = 30$). Among the lines within each cross there was only small variation for either flowering date or height but lines from cross 1 flowered, on average, 1 week earlier than the lines from cross 2. Condobolin has an average annual rainfall of 450 mm with a relatively even distribution of between 30 and 45 mm per month. Ideally at this location, wheat is sown in late May, flowers in early October, and is harvested in late November. There is often some soil moisture stored at sowing from summer rains. Flowering occurs just after September, which is the driest month (30 mm on average), but in some years rainfall during September and October can be exceptionally high. September rainfall was 24 mm in 1992 and in 1993 it was 76 mm.

stored moisture occurs approaching anthesis. There may also be a few cropping environments in which the seasonal course of vapor pressure deficit favors conservation of soil water for use later in the season.

In well-irrigated or rainfed environments in which soil water depletion before anthesis is minimal, productivity will often be associated with high Δ but this may depend on the major source of genotypic variation in Δ. Among cereal genotypes Δ may vary due to variation in either g_w or A_{cap}. If g_w is the major source of genotypic variation in Δ, a positive association between productivity and Δ is likely under favorable conditions because of the positive relationship between A and g_w. Even though crop water use may be greater, there should be no penalty for growth or yield if irrigation or frequent rainfall is sufficient to maintain the soil water store up to anthesis. Positive relationships between yield and Δ were observed for two of three sets of advanced lines grown by Sayre *et al.* (1995) under irrigation in Mexico and for progeny from two crosses grown at Condobolin in the favorable 1993 season (Fig. 3).

High Δ may not be associated with greater productivity under favorable conditions if a large component of genotypic variation in Δ is due to variation in A_{cap}. In this case, high Δ is likely to be associated with low A_{cap}. In cereals low A_{cap} may be associated with faster early growth because rate of leaf area development will often be faster, but once canopy closure has been achieved growth rate is likely to be faster for genotypes with high A_{cap} (low Δ). If canopy closure is maintained for a long enough period, through irrigation, for instance, the overall effect may be that there is no relationship between productivity and Δ. Sayre *et al.* (1995) found no association between grain yield and Δ for a third set of advanced lines grown under irrigation in Mexico. These were lines from CIMMYT known to vary considerably in A_{cap} (K. D. Sayre, 1995, personal communication). At Condobolin, Condon and Richards also grew progeny from a third cross for which the parents differed mainly in A_{cap} rather than g_w. In both 1992 and 1993 there was no relationship found between productivity and Δ for these genotypes.

In Mediterranean or current rainfall climates in which either rainfall is low or soil water storage capacity is low, grain number and yield will probably be greater in high Δ genotypes irrespective of whether the source of high Δ is high g_w or low A_{cap}. This is because, for either case, early crop growth rate is likely to be faster. Transpiration may be greater for high Δ genotypes but seasonal W_T may also be high because the extra transpiration takes place early in the growing season when the vapor pressure deficit is low. In addition, genotypes with high Δ during the vegetative phase and more rapid growth rates should shade the soil surface more and have less soil evaporation. These genotypes should also produce more biomass by anthesis. This should result in greater grain number and a greater store of assimilates to sustain growth of these grains through translocation. For high Δ cereal genotypes growing in these environments, faster crop growth rate should result in greater productivity with minimal difference in seasonal W_{ET}.

2. Grain Legumes The association between crop performance and Δ has been examined in several grain legume crop species including common bean, cowpea, and peanut. In most studies a range of genotypes of a given species has been grown together in the field with different irrigation treatments, in some cases, and genotypic variation in Δ, shoot biomass production, and grain yield have been measured. In addition, for cowpea, a selection experiment has been conducted that provides information on the association between Δ and grain yield.

a. Common Bean A few studies have been conducted with common bean on the association between productivity and Δ. Under water-limited rainfed conditions on a deep, fertile soil, White *et al.* (1990) observed positive associations between Δ and both biomass production and grain yield for 9 of 10 common bean cultivars that were tested. This result is consistent with the observation by Ehleringer (1990) of a positive association between stomatal conductance and Δ in common bean cultivars. However, when the same genotypes were grown under water-limited rainfed conditions at a site with a shallow root zone, there was no association observed between biomass production or grain yield and Δ (White *et al.*, 1990). Also, Δ was positively correlated with root length density, which varied greatly among these bean genotypes. Consequently, an additional explanation for the genotypic differences in Δ and grain yield is that at the deep-soil site, genotypes with greater root length densities could have had greater access to water and therefore could have suffered smaller water deficits, which could result in greater biomass production, grain yield, g_w, and Δ values. At the second rainfed site at which acid subsoil conditions restricted root penetration, Δ and root length density were again positively associated but the highest grain yields were associated with intermediate values of Δ. In this case it may be argued that genotypes with very low root length density and Δ had very low g_w and grew very slowly such that they had low yield, whereas genotypes with very high root length density and Δ had very high g_w and exhibited rapid initial growth and water use such that they exhausted the moisture in the shallow soil, became severely stressed during reproductive development, and consequently also had low yields (White, 1993). Genotypic differences in root growth are an important component of adaptation to drought in common bean (White, 1993) and severely complicate the relationships among Δ, edaphic conditions, and yield.

b. Cowpea When associations between productivity and Δ have been examined for cowpea, they have tended to be positive under well-irrigated and water-limited rainfed and stored-soil moisture conditions. A broad range of cultivars have been studied in two water-limited summer rainfall tropical environments in northern Senegal. Local cultivars that are well adapted to these environments in Senegal, which had the largest grain yields, tended to have high leaf Δ values (Hall *et al.*, 1994a). When the same set of cowpea cultivars were grown under complete irrigation in a subtropical environment in California, cultivars selected for high grain yield in California had high Δ (Hall *et al.*, 1993, 1994a). The specific cultivars that had high yields and high Δ in California were different from those that performed well and had high Δ in Senegal. Drought consistently causes cowpea to have much lower leaf Δ, which may be related to the sensitive responses of g_w to soil drought in this species that occur before changes in leaf water potential are detected (Bates and Hall, 1981). However, genotypic rankings for Δ were very similar under either well-irrigated conditions or water-limited stored-soil moisture conditions when plants were grown in the same climatic zone (Hall *et al.*, 1990) but not when comparing genotypes across radically different climatic zones (Hall *et al.*, 1994a).

Selection studies have been conducted in California to examine the association between productivity and Δ among progeny of two types of crosses, one involving only low Δ parents and the other only high Δ parents (Hall *et al.*, 1993). Progeny from the low Δ crosses were selected for earliness, high grain yield, and high biomass production under stored-moisture conditions. Progeny from the high Δ cross were selected under irrigated conditions for heat tolerance, high harvest index, and high grain yield. The impact of selection on the association between yield and Δ was determined by growing nine lines selected from the low Δ crosses, six lines selected from the high Δ cross, and three high-yielding local California lines under both well-irrigated and dry stored-soil moisture conditions in California. All the lines began flowering at about the same time. In this experiment, grain yield and Δ were found to be positively correlated under both irrigated and dry stored-soil moisture conditions across all lines. The strongest positive correlation was observed between average grain yield of individual lines across the wet and dry conditions with average leaf Δ (Fig. 4). The data would have been more convincing if they had been obtained with progeny from crosses between low Δ and high Δ parents (e.g., Fig. 3 for wheat) but the data for the progeny from the low $\times$ low Δ cross indicate that low $\times$ high Δ crosses with cowpea could have produced results similar to those presented in Fig. 4.

The studies described in the previous paragraph indicate, for cowpea, that adaptation and grain yield are positively associated with high Δ, and presumably low W (Ismail and Hall, 1992, 1993a; Ismail *et al.*, 1994), and low A/T (Ismail and Hall, 1993b) in both well-watered and water-limited conditions. For well-watered conditions, these results may be partially ex-

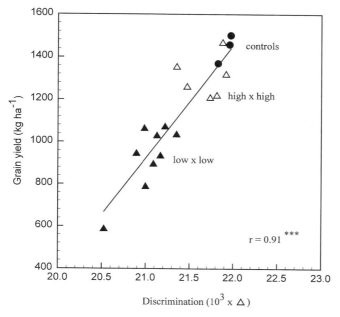

Figure 4   Relationships between average grain yield of cowpea lines and average leaf Δ. The lines were grown in both a well-irrigated environment and a strongly water-limited environment at Riverside, California, in 1992. The water-limited environment involved growth mainly on water stored in the soil. The controls are three advanced breeding lines that are well adapted to the well-irrigated environment, and there were six F_5 lines that had been selected from a cross with high Δ parents for high yield under well-irrigated conditions and nine F_6 lines that had been selected from crosses with low Δ parents for high yield under stored-soil moisture water-limited conditions. All lines began flowering at about the same time and leaves were sampled for Δ at the end of the vegetative stage.

plained by genotypic variation in g_w with higher g_w resulting in higher A, greater biomass production, lower A/T, and higher Δ. However, genotypic variations in Δ in cowpea have been associated with variation in both g_w and A_{cap} (Ismail and Hall, 1993b). Interestingly, cowpea hybrids that have strong early vegetative vigor also have high Δ but only under well-watered conditions (Ismail and Hall, 1993a; Ismail *et al.*, 1994). As for wheat, the positive association is greater for Δ and grain yield than it is for Δ and shoot biomass (Hall *et al.*, 1997). In the selection experiment, the genotypic variation in average grain yield was mainly due to variation in harvest index (Hall *et al.*, 1997). Studies have shown that for cowpea there can be a positive genetic correlation between Δ and harvest index (Menéndez and Hall, 1996), and this has been observed in several other species (Hall *et al.*, 1994b). Another factor is that cowpea lines selected for California have early flowering, and studies have shown that cowpea can also exhibit negative genetic correlations between Δ and days to flowering (Menéndez and Hall, 1995). Negative correlations of this type have been observed in several other species (Hall *et al.*, 1994b).

For water-limited conditions, the positive correlations between Δ and grain yield are more difficult to explain. In the selection experiment, Δ was positively correlated with both grain yield and shoot biomass for the water-limited conditions and the lines began flowering at about the same time and leaves were sampled for Δ at the beginning of flowering. A possible explanation for these results is that the genotypes with higher Δ had greater root growth, which enabled them to access more soil water. This would provide enhanced plant water status and higher g_w, explaining both the higher Δ and the greater biomass production. The progeny from the low Δ crosses were subjected to selection for large total shoot biomass under stored-soil moisture conditions that could have resulted in indirect selection for deeper, more effective root systems. Also, the genotypes with higher Δ may have had more open stomata due to the plants being less sensitive to soil water deficits in the root zone. In addition, there was some selection for earliness and higher harvest index that genetic studies indicate could also indirectly result in some selection for higher Δ.

These results exhibit some differences from those obtained in selection experiments with wheat reported previously. In the wheat experiments, yield and Δ were positively correlated under well-watered conditions but the relationship between yield and Δ became neutral or negative, depending on average flowering time, with increasing dependence on stored soil moisture. The change in the slope of the relationship between yield and Δ in wheat was mainly due to a greater drought-induced decrease in seed number for high Δ lines than low Δ lines. It may be that in the experiment with cowpea, selection for earliness meant that high Δ lines were able to express greater pod set under both watering regimes. Soil water use may have been faster for high Δ genotypes under the stored-moisture conditions, but soil water deficits may not have developed sufficiently rapidly to reduce pod set to below that of the low Δ selections. Having set more pods, high Δ selections may then have also been able to translocate sufficient stored assimilate to retain a yield advantage. This is consistent with the greater harvest index found for high Δ cowpea selections and is similar to the situation found for wheat under rainfed conditions (Fig. 2), in which grain yield and harvest index were positively correlated with Δ.

This scenario is supported by the strong stomatal sensitivity of cowpea to diurnal changes in vapor pressure deficit (Hall and Schulze, 1980). Cowpea is grown as a summer crop in tropical and subtropical regions of the world and in these environments there can be large diurnal variation in the atmospheric vapor pressure deficit. Even under well-watered conditions, it is typical for the stomata of cowpea to partially close by late morning as the vapor pressure deficit rises and to remain so for the remainder of the daylight hours. It appears that in cowpea this diurnal stomatal variation is having a dominant influence on Δ measured in dry matter. For cowpea, studies that have examined the relationship between instantaneous leaf gas-exchange efficiency and Δ measured in dry matter have proved less conclusive than in some other C_3 species. Hall *et al.* (1992) found a negative correlation between A/g_w and Δ when these were compared across watering regimes but there was no association among genotypes within watering regimes. The gas-exchange data were collected exclusively in the afternoon. In a subsequent study, Ismail and Hall (1993a) again found no association between A/g_w and Δ among genotypes measured in the afternoon but did find strong negative associations between A/g_w and Δ for gas-exchange measurements taken in the early morning. These results indicate that for cowpea the isotopic signal of plant carbon may be dominated by carbon fixed during the first few hours of daylight when A and g_w are both highest. Even though stomata are more open at this time, this is by far the most efficient time of day for carbon fixation because the atmospheric vapor pressure deficit is low. For cowpea it may be that genotypes with the highest values of Δ are those with the highest conductance during the early morning period. Genotypes with high g_w and A at this time may be able to maintain relatively high daily transpiration efficiency due to the low vapor pressure deficit during this period, i.e., there may only be a small penalty in terms of crop water use associated with high Δ and faster growth. Thus, whereas for common bean, genotypic variation in the ability to access soil water appears to be complicating the relationship between productivity and Δ, for cowpea, interactions with the aerial environment may be having a substantial influence on this relationship.

   c. Peanut   The association between biomass production and Δ appears to be quite different for peanut than for the small grain cereals, common bean, or cowpea. For peanut the relationship between biomass production and Δ almost invariably has been found to be negative, i.e., there is a positive association between leaf gas-exchange efficiency and crop growth rate (Wright *et al.*, 1993). This association has been observed under both well-watered and water-limited conditions.

   There are two factors contributing to the negative association between biomass production and Δ in peanut. First, the bulk of genotypic variation in Δ for peanut arises from variation in A_{cap}, not g_w (Nageswara Rao *et al.*, 1995). The variation in Δ has been associated with variation in amount of the photosynthetic enzyme ribulose-1,5-bisphosphate caboxylase oxygenase (EC 4.1.1.39) per unit leaf area (Nageswara Rao *et al.*, 1995). Therefore, among genotypes of this species, the tendency is for A to increase (on a leaf area basis) as Δ and c_i/c_a decrease (Fig. 1). The second important factor is that for peanut there does not appear to be a trade-off between genotypic variation in A_{cap} and the rate of leaf area growth, such as was found for many other species, e.g., the cereals and cotton mentioned previously. In peanut, high A_{cap} is associated with a faster rate of leaf area growth. The reason why peanut does not exhibit this trade-off is not clear but it may reflect the fact that peanut is a legume with the ability, in symbiosis with rhizobia, to fix atmospheric nitrogen. In most if not all plant species, genotypic variation in A_{cap} largely reflects variation in nitrogen content per unit leaf area and peanut is no exception (Nageswara Rao and Wright, 1994; Nageswara Rao et al., 1995). In nonlegume species it appears that genotypically achieved increases in leaf N per unit area are associated with reductions in leaf size. This may reflect the fact that the nitrogen enters the plant from finite soil sources. The ability of peanut to maintain an additional supply of nitrogen through symbiotic fixation may be responsible for its ability to exhibit both high A_{cap} and rapid leaf area growth. However, negative correlations between A_{cap} and leaf area have been observed in other legumes that fix atmospheric nitrogen (Bhagsari and Brown, 1986).

Although the physiological basis for the phenomenon is unresolved, the positive association between rate of leaf area growth and A_{cap} clearly has important consequences for the association between biomass production and Δ in peanut. Whether determined using a range of diverse genotypes or within breeding populations, biomass production has almost invariably been found to be positively associated with indicators of leaf gas-exchange efficiency. In the case of peanut, this indicator has not always been Δ. Because variation in A_{cap} is often reflected in variation in SLA, this relatively easily measured characteristic is also used as an indicator of genotypic variation in Δ and c_i/c_a in peanut, especially when large numbers of breeding lines are screened. There is a strong positive association between Δ and SLA in peanut (Nageswara Rao and Wright, 1994).

Because A_{cap} is the major source of variation in A/T in peanut and high A/T is not associated with slow leaf area development, many of the problems associated with scaling genotypic variation in leaf gas exchange to the level of the canopy are not apparent. Crop W_T and Δ have been shown to be negatively correlated in small-plot experiments under well-watered and water-limited conditions (Wright et al. 1988, 1994) and this relationship has been reflected in genotypic variation in biomass production.

A practical problem peanut breeders have encountered in attempting to use selection for leaf gas-exchange efficiency traits to enhance crop performance is the occurrence of a positive association between harvest index and Δ (or SLA). This association has been found in studies with diverse cultivars and also in both early and subsequent generations of

crosses involving parents with high and low Δ values (Wright *et al.*, 1993). As a consequence of this association, there was no increase in pod yield with selection for lower Δ, despite large increases in biomass production. It appears that the association between small partitioning of dry matter to pods and low values of Δ or SLA is due to genetic linkage. Recent selections appear to have generated progeny with both high harvest index and low SLA (G. C. Wright, 1997, personal communication). This indicates the linkage has been broken, and that it should be possible to exploit genotypic variation in A/T to improve crop performance in many peanut production environments, including those in which grain production is the major goal. If the linkage proves intractable, the high biomass production associated with high A/T should still be advantageous in production systems in which both fodder yield for animal production and pod yield for human consumption need to be maximized (Wright *et al.*, 1993).

V. Critical Research Needs and Conclusions

Carbon isotope discrimination has provided physiologists, agronomists, and plant breeders with a measure of physiological activity that can be applied relatively easily to large numbers of crop genotypes. Its use over the past decade has established that there are genotypic differences in leaf gas-exchange efficiency within C_3 crop species that may contribute to crop performance in different environments. There are several critical areas of research that need to be addressed if the potential of Δ as an indicator of this variation is to be exploited effectively.

If Δ is to be used reliably as a selectable trait in plant breeding, then its genetics and inheritance need to be understood more thoroughly. Genetic studies of Δ would be most effective if they also consider the physiological basis of variation in Δ, because this will determine the types of relationships present between crop performance and Δ. Obviously, Δ is much easier to measure than either g_w or A_{cap}. This is one of its main advantages. Nevertheless, if genetic studies of Δ included judicious estimates or measurements of variation in g_w or A_{cap}, then the worth of these studies would be greatly increased. Such studies need not involve direct measurements of g_w or A_{cap} but could involve indirect techniques, such as measurement of canopy temperature as an indicator of differences in g_w (Rees *et al.*, 1993) or measurements of SLA as an inverse indicator of differences in A_{cap} (Nageswara Rao *et al.*, 1995).

More selection studies relating crop performance to Δ also need to be performed using the progeny of crosses that are made specifically for the purpose. These studies should extend the results of the genetic studies reported in Figs. 3 and 4, and evaluations should be performed in different field environments because selection using Δ may be more effective in some environments than in others. The interaction with environment may vary depending on the species. If the primary determinant of adaptation to an environment is flowering time, which is often the case for rainfed environments, then it is important that variation in this trait be minimized so that the impact of variation in gas-exchange efficiency can be adequately assessed. In many of the published studies that have examined the association between yield and Δ in diverse genotypes, there have been large variations in flowering time. Often, late flowering has been associated with low Δ. This association may be due to genetic linkage, it may be the result of selection pressure for earliness combined with fast growth, or it may be an artifactual result of the protocol used for sampling variation in Δ. For example, using grain samples for determining Δ is likely to be ineffective in determining relationships between gas-exchange efficiency and productivity when the genotypes exhibit substantial variation in time to first flower, especially in environments with late drought. This is because the genotypes produce the carbon sampled for Δ at different times and the aerial environment and soil water supply when the carbon is assimilated are likely to differ substantially among the genotypes. Also, the relative contributions to grain carbon of current and stored assimilates may vary widely among genotypes. In addition, the nature of the association between earliness and Δ and other physiological or genetic associations with Δ need to be better understood because they may limit the application of Δ unless they can be overcome.

The positive associations between Δ and biomass production (and grain yield), which have been observed with many crop species and environments, including dry environments in some cases, warrants discussion concerning possible mechanistic and evolutionary causes. In many cases the association is linked to a positive association between Δ and g_w. A recent study with Pima cotton (Lu *et al.*, 1996) provides additional evidence for this phenomenon. Why have past increases in biomass productivity of these annual crops through plant breeding been associated with increases in g_w (e.g., Rees *et al.*, 1993 for wheat)? We speculate that the evolution of these annual crop plants resulted in conservative performance with respect to g_w, that is, a tendency for stomata to be at least partially closed on many occasions. This could have happened if plant performance during very dry years, when conservative stomatal performance may be adaptive, had disproportionate influences on seed production and long-term evolutionary success over many years due to soil "seed banks" being much less effective after 1 year. This leads us to comment on the influence on water use of the less conservative stomatal performance of newer cultivars. In well-watered environments there may be little influence of more open stomata on water use due to the scaling effects discussed earlier. Also, in rainfed well-watered environments there is a high likelihood that the soil water will be quickly replenished. In some dry environments there may be little influence of occasional more open stomata on seasonal water use if it involves more open stomata during times in the season or day when evaporative demands are low, i.e., more optimal stomatal function, as defined by Cowan and Farquhar (1977), or if cultivar differences in transpiration rate are offset by counteracting differences in evaporation from the soil. Consequently, there are circumstances in which genotypes with more open stomata may have similar water use but greater productivity and thus greater water-use efficiency than genotypes with more closed stomata. A final unresolved question concerns the even stronger positive associations that have been observed between Δ and grain yield due to the positive associations between Δ and harvest index. The physiological or genetic mechanisms of this association are largely unknown and warrant further study. An overall conclusion is that there are some crops (e.g., wheat, barley, cowpea, and cotton) in which selection for higher Δ may result in increased yield in many well-watered and possibly even in some water-limited environments. This opportunity for plant breeding may only be present, however, where elements of conservative stomatal function are still present in advanced cultivars and, thus, available for exploitation by their removal.

For water-limited environments, obtaining greater crop yield from the available water supply remains a critical objective of agriculture, and manipulating water-use efficiency at the leaf level, by breeding and selection, may aid in achieving this objective. Crops such as peanut, in which variation in gas-exchange efficiency is due to variation in photosynthetic capacity and which is positively correlated with crop growth rate, appear to offer a major opportunity for achieving this objective. The capacity to exploit carbon isotope discrimination to estimate genotypic variation in leaf-level water-use efficiency provides an opportunity with several crop species for making progress in increasing productivity, either directly through breeding or indirectly by providing a means to understand the important processes limiting crop yield. There are also many other means for increasing crop water-use efficiency and grain yield that are amenable to manipulation through breeding or management or a combination of these two (Richards *et al.*, 1993) that provide additional ways for developing improved cropping systems for dry environments.

References

Acevedo, E. (1993). Potential of carbon isotope discrimination as a selection criterion in barley breeding. *In* "Stable Isotopes and Plant Carbon–Water Relations" (J. R. Ehleringer, A. E. Hall, and G. D. Farquhar, eds.), pp. 399–417. Academic Press, San Diego.

Baenziger, P. S., Wesenberg, D. M., and Sicher, R. C. (1983). The effects of genes controlling barley leaf and sheath waxes on agronomic performance in irrigated and dryland environments. *Crop Sci.* **23,** 116–120.

Bates, L. M., and Hall, A. E. (1981). Stomatal closure with soil water depletion not associated with changes in bulk leaf water status. *Oecologia* **50,** 62–65.

Bhagsari, A. S., and Brown. R. H. (1986). Leaf photosynthesis and its correlation with leaf area. *Crop Sci.* **26,** 127–132.

Clarke, J. M., Romagosa, I., Jana, S., Srivastava, J. P., and McCaig, T. N. (1989). Relationship of excised-leaf water loss rate and yield of durum wheat in diverse environments. *Can. J. Plant Sci.* **69,** 1075–1081.

Clarke, J. M., Richards, R. A., and Condon, A. G. (1991). Effect of drought stress on residual transpiration and its relationship with water use of wheat. *Can. J. Plant Sci.* **71,** 695–702.

Clawson, K. L., Specht, J. E., Blad, B. L., and Garay, A. F. (1986). Water use efficiency in soybean pubescence density isolines—A calculation procedure for estimating daily values. *Agron. J.* **78,** 483–487.

Condon, A. G., and Richards, R. A. (1993). Exploiting genetic variation in transpiration efficiency in wheat: An agronomic view. *In* "Stable Isotopes and Plant Carbon–Water Relations" (J. R. Ehleringer, A. E. Hall, and G. D. Farquhar, eds.), pp. 435-450. Academic Press, San Diego.

Condon, A. G., Richards, R. A., and Farquhar, G. D. (1987). Carbon isotope discrimination is positively correlated with grain yield and dry matter production in field-grown wheat. *Crop Sci.* **27,** 996–1001.

Condon, A. G., Farquhar, G. D., and Richards, R. A. (1990). Genotypic variation in carbon isotope discrimination and transpiration efficiency in wheat. Leaf gas exchange and whole plant studies. *Aust. J. Plant Physiol.* **17,** 9–22.

Condon, A. G., Richards, R. A., and Farquhar, G. D. (1992). The effect of variation in soil water availability, vapor pressure deficit and nitrogen nutrition on carbon isotope discrimination in wheat. *Aust. J. Agric. Res.* **43,** 935–947.

Condon, A. G., Richards, R. A., and Farquhar, G. D. (1993). Relationships between carbon isotope discrimination, water use efficiency and transpiration efficiency for dryland wheat. *Aust. J. Agric. Res.* **44,** 1693–1711.

Cooper, P. J. M., Gregory, P. J., Keatinge, J. D. H., and Brown, S. C. (1987). Effects of fertilizer, variety and location on barley production under rainfed conditions in northern Syria. 2. Soil water dynamics and crop water use. *Field Crops Res.* **16,** 67–84.

Cornish, K., Radin, J. W., Turcotte, E. L., Lu, Z., and Zeiger, E. (1991). Enhanced photosynthesis and stomatal conductance of Pima cotton (*Gossypium barbadense* L.) bred for increased yield. *Plant Physiol.* **97,** 484–489.

Cowan, I. R. (1988). Stomatal physiology and gas exchange in the field. *In* "Flow and Transport in the Natural Environment: Advances and Applications" (W. L. Steffen and O. T. Denmead, eds.), pp. 160–172. Springer-Verlag, Berlin.

Cowan, I. R., and Farquhar, G. D. (1977). Stomatal function in relation to leaf metabolism and environment. *In* "Society for Experimental Biology Symposium, Integration of Activity in the Higher Plant" (D. H. Jennings, ed.), pp. 471–505. Society for Experimental Biology, Cambridge.

Craufurd, P. Q., Austin, R. B., Acevedo, E., and Hall, M. A. (1991). Carbon isotope discrimination and grain yield in barley. *Field Crops Res.* **27,** 301–313.

Ehdaie, B., Hall, A. E., Farquhar, G. D., Nguyen, H. T., and Waines, J. G. (1991). Water-use efficiency and carbon isotope discrimination in wheat. *Crop Sci.* **31,** 1282–1288.

Ehleringer, J. R. (1990). Correlations between carbon isotope discrimination and leaf conductance to water vapor in common beans. *Plant Physiol.* **93,** 1422–1425.

Ehleringer, J. R. (1993). Carbon and water relations in desert plants: An isotopic perspective. *In* "Stable Isotopes and Plant Carbon–Water Relations" (J. R. Ehleringer, A. E. Hall, and G. D. Farquhar, eds.), pp. 155–172. Academic Press, San Diego.

Ehleringer, J. R., Klassen, S., Clayton, C., Sherrill, D., Fuller-Holbrook, M., Fu, Q. A., and Cooper, T. A. (1991). Carbon isotope discrimination and transpiration efficiency in common bean. *Crop Sci.* **31**, 1611–1615.

Evans, J. R., Sharkey, T. D., Berry, J. A., and Farquhar, G. D. (1986). Carbon isotope discrimination measured concurrently with gas exchange to investigate CO_2 diffusion in leaves of higher plants. *Aust. J. Plant Physiol.* **13**, 281–292,

Farquhar, G. D., and Lloyd, J. (1993). Carbon and oxygen isotope effects in the exchange of carbon dioxide between plants and the atmosphere. *In* "Stable Isotopes and Plant Carbon–Water Relations" (J. R. Ehleringer, A. E. Hall, and G. D. Farquhar, eds.) pp. 47–70. Academic Press, San Diego.

Farquhar, G. D., and Richards, R. A. (1984). Isotopic composition of plant carbon correlates with water-use efficiency of wheat genotypes. *Aust. J. Plant Physiol.* **11**, 539–552.

Farquhar, G. D., and Sharkey, T. D. (1982). Stomatal conductance and photosynthesis. *Annu. Rev. Plant Physiol.* **33**, 317–345.

Farquhar, G. D., O'Leary, M. H., and Berry, J. A. (1982). On the relationship between carbon isotope discrimination and the intercellular carbon dioxide concentration in leaves. *Aust. J. Plant Physiol.* **9**, 121–137.

Farquhar, G. D., Ehleringer, J. R., and Hubick, K. T. (1989). Carbon isotope discrimination and photosynthesis. *Annu. Rev. Plant Physiol. Plant Mol. Biol.* **40**, 503–537.

Fischer, R. A. (1981). Optimizing the use of water and nitrogen through breeding of crops. *Plant Soil* **58**, 249–278.

Gifford, R. M., Thorne, J. H., Hitz, W. D., and Giaquinta, R. T. (1984). Crop productivity and photoassimilate partitioning. *Science* **225**, 801–808.

Gimeno, V., Fernandez-Martinez, J. M., and Fereres, E. (1989). Winter planting as a means of drought escape in sunflower. *Field Crops Res.* **22**, 307–316.

Hall, A. E., and Schulze, E.-D. (1980). Stomatal response to environment and a possible interrelation between stomatal effects on transpiration and CO_2 assimilation. *Plant Cell Environ.* **3**, 467–474.

Hall, A. E., Mutters, R. G., Hubick, K. T., and Farquhar, G. D. (1990). Genotypic differences in carbon isotope discrimination by cowpea under wet and dry field conditions. *Crop Sci.* **30**, 300–305.

Hall, A. E., Mutters, R. G., and Farquhar, G. D. (1992). Genotypic and drought-induced differences in carbon isotope discrimination and gas exchange of cowpea. *Crop Sci.* **32**, 1–6.

Hall, A. E., Ismail, A. M., and Menéndez, C. M. (1993). Implications for plant breeding of genotypic and drought-induced differences in water-use efficiency, carbon isotope discrimination, and gas exchange. *In* "Stable Isotopes and Plant Carbon–Water Relations" (J. R. Ehleringer, A. E. Hall, and G. D. Farquhar, eds.), pp. 349-369. Academic Press, San Diego.

Hall, A. E., Thiaw, S., and Krieg, D. R. (1994a). Consistency of genotypic ranking for carbon isotope discrimination by cowpea grown in tropical and subtropical zones. *Field Crops Res.* **36**, 125–131.

Hall, A. E., Richards, R. A., Condon, A. G., Wright, G. C., and Farquhar, G. D. (1994b). Carbon isotope discrimination and plant breeding. *In* "Plant Breeding Reviews, Vol. 12." (J. Janick, ed.), pp. 81–113. Wiley, New York.

Hall, A. E., Thiaw, S., Ismail, A. M., and Ehlers, J. D. (1997). Water-use efficiency and drought adaptation of cowpea. *In* "Advances in Cowpea Research," (B. B. Singh, Raj Mohan, K. Dershiell, and L. E. N. Jackai, eds.), IITA/JIRCAS, IITA, Ibadan, Nigeria. In press.

Heichel, G. H. (1971). Confirming measurements of respiration and photosynthesis with dry matter accumulation. *Photosynthetica* **5**, 93–98.

Hubick, K. T., and Farquhar, G. D. (1989). Genetic variation in carbon isotope discrimination and the ratio of carbon gained to water lost in barley. *Plant Cell Environ.* **12**, 795–804.

Hubick, K. T., Farquhar, G. D., and Shorter, R. (1986). Correlation between water-use efficiency and carbon isotope discrimination in diverse peanut (*Arachis*) germplasm. *Aust. J. Plant Physiol.* **13**, 803–816.

Hubick, K. T., Shorter, R., and Farquhar, G. D. (1988). Heritability and genotype x environment interactions of carbon isotope discrimination and transpiration efficiency in peanut. *Aust. J. Plant Physiol.* **15**, 799–813.

Innes, P., and Blackwell, R. D. (1983). Some effects of leaf posture on the yield and water economy of winter wheat. *J. Agric. Sci.* **101**, 367–375.

Ismail, A. M., and Hall, A. E. (1992). Correlation between water-use efficiency and carbon isotope discrimination in diverse cowpea genotypes and isogenic lines. *Crop Sci.* **32**, 7–12.

Ismail, A. M., and Hall, A. E. (1993a). Inheritance of carbon isotope discrimination and water-use efficiency in cowpea. *Crop Sci.* **33**, 498–503.

Ismail, A. M., and Hall, A. E. (1993b). Carbon isotope discrimination and gas exchange of cowpea accessions and hybrids. *Crop Sci.* **33**, 788–793.

Ismail, A. M., Hall, A. E., and Bray, E. A. (1994). Drought and pot size effects on transpiration efficiency and carbon isotope discrimination of cowpea accessions and hybrids. *Aust. J. Plant Physiol.* **21**, 23–35.

Jarvis, P. G., and McNaughton, K. G. (1986). Stomatal control of transpiration: Scaling up from leaf to region. *Adv. Ecol. Res.* **15**, 1–49.

Johnson, D. A., Richards, R. A., and Turner, N. C. (1983). Yield, water relations, gas exchange, and surface reflectance of near-isogenic wheat lines differing in glaucousness. *Crop Sci.* **23**, 318–325.

Jones, H. G. (1976). Crop characteristics and the ratio between assimilation and transpiration. *J. Appl. Ecol.* **13**, 605–622.

Jones, H. G. (1987). Breeding for stomatal characters. *In* "Stomatal Function" (E. F. Zeiger, G. D. Farquhar, and I. R. Cowan, eds.), pp. 431–443. Stanford Univ. Press, Stanford, CA.

Jones, H. G. (1994). Drought tolerance and water-use efficiency. *In* "Water Deficits: Plant Responses from Cell to Community" (J. A. C. Smith and H. Griffiths, eds.), pp. 193–204. BIOS, Oxford.

Jones, R. J., and Nelson, C. J. (1979). Respiration and concentration of water soluble carbohydrates in plant parts of contrasting tall fescue genotypes. *Crop Sci.* **19**, 367–372.

Jordan, W. R., Monk, R. L., Miller, F. R., Rosenow, D. T., Clark, L. E., and Shouse, P. J. (1983). Environmental physiology of sorghum. I. Environmental and genetic control of epicuticular wax load. *Crop Sci.* **23**, 552–558.

Jordan, W. R., Shouse, P. J., Blum, A., Miller, F. R., and Monk, R. L. (1984). Environmental physiology of sorghum. II. Epicuticular wax load and cuticular transpiration. *Crop Sci.* **24**, 1168–1173.

Keatinge, J. D. H., and Cooper, P. J. M. (1983). Kabuli chickpea as a winter-sown crop in northern Syria: Moisture relations and crop productivity. *J. Agric. Sci.* **100**, 667–680.

Lopez-Castaneda, C., and Richards, R. A. (1994). Variation in temperate cereals in rainfed environments. III. Water use and water-use efficiency. *Field Crops Res.* **39**, 85–98.

Lopez-Castaneda, C., Richards, R. A., and Farquhar, G. D. (1995). Variation in early vigor between wheat and barley. *Crop Sci.* **35**, 472–479.

Loss, S. P., and Siddique, K. H. M. (1994). Morphological and physiological traits associated with wheat yield increases in mediterranean environments. *Adv. Agron.* **52**, 229–276.

Lu, Z., Chen, J., Percy, R. G., Sharifi, M. R., Rundel, P. W., and Zeiger, E. (1996). Genetic variation in carbon isotope discrimination and its relation to stomatal conductance in Pima cotton (*Gossypium barbadense* L.). *Aust. J. Plant Physiol.* **23**, 127–132.

Lu, Z.-M., and Zeiger, E. (1994). Selection for higher yields and heat resistance in Pima cotton has caused genetically determined changes in stomatal conductance. *Physiol. Planta.* **92**, 273–278.

Menéndez, C. M., and Hall, A. E. (1995). Heritability of carbon isotope discrimination and correlations with earliness in cowpea. *Crop Sci.* **35**, 673–678.

Menéndez, C. M., and Hall, A. E. (1996). Heritability of carbon isotope discrimination and correlations with harvest index in cowpea. *Crop Sci.* **36**, 233–238.

Miskin, K. E., Rasmussen, D. C., and Moss, D. N. (1972). Inheritance and physiological effects of stomatal frequency in barley. *Crop Sci.* **12**, 780–783.

Morgan, J. A., and LeCain, D. R. (1991). Leaf gas exchange and related leaf traits among 15 winter wheat genotypes. *Crop Sci.* **31**, 443–448.

Morgan, J. A., LeCain, D. R., McCaig, T. N., and Quick, J. S. (1993). Gas exchange, carbon isotope discrimination, and productivity in winter wheat. *Crop Sci.* **33**, 178–186.

Murfet, I. C. (1977). Environmental interaction and the genetics of flowering. *Annu. Rev. Plant Physiol.* **28**, 253–278.

Nageswara Rao, R. C., and Wright, G. C. (1994). Stability of the relationship between specific leaf area and carbon isotope discrimination across environments in peanut. *Crop Sci.* **34**, 98–103.

Nageswara Rao, R. C., Udaykumar, M., Farquhar, G. D., Talwar, H. S., and Prasad, T. G. (1995). Variation in carbon isotope discrimination and its relationship to specific leaf area and ribulose-1,5-bisphosphate carboxylase content in groundnut genotypes. *Aust. J. Plant Physiol.* **22**, 545–551.

Nerkar, Y. S., Wilson, D., and Lawes, D. A. (1981). Genetic variation in stomatal characteristics and behaviour, water use and growth of five *Vicia faba* L. genotypes under contrasting soil moisture regimes. *Euphytica* **30**, 335–345.

Palta, J. A., and Fillery, I. R. P. (1995). N application increases pre-anthesis contribution of dry matter to grain yield in wheat grown on a duplex soil. *Aust. J. Agric. Res.* **46**, 507–518.

Pettigrew, W. T., Heitholt, J. J., and Vaughn, K. C. (1993). Gas exchange differences and comparative anatomy among cotton leaf-type isolines. *Crop Sci.* **33**, 1295–1299.

Preston, T. (1992). The measurement of stable isotope natural abundance variations. *Plant Cell Environ.* **15**, 1091–1097.

Radin, J. W., Lu, Z. M., Percy, R. G., and Zeiger, E. (1994). Genetic variability for stomatal conductance in Pima cotton and its relation to improvements of heat adaptation. *Proc. Natl. Acad. Sci. USA* **91**, 7217–7221.

Rees, D., Sayre, K., Acevedo, E., Sanchez, T. N., Lu, Z., and Zeiger, E. (1993). Canopy temperature of wheat: Relationship with yield and potential as a technique for early generation selection. Wheat Special Report No. 10, p. 30. CIMMYT, Mexico.

Richards, R. A., Rawson, H. M., and Johnson, D. A. (1986). Glaucousness in wheat: Its development and effect on water-use efficiency, gas-exchange and photosynthetic tissue temperature. *Aust. J. Plant Physiol.* **13**, 465–473.

Richards, R. A., Lopez-Castaneda, C., Gomez-Macpherson, H., and Condon, A. G. (1993). Improving the efficiency of water use by plant breeding and molecular biology. *Irrig. Sci.* **14**, 93–104.

Sayre, K. D., Acevedo, E., and Austin, R. B. (1995). Carbon isotope discrimination and grain yield for three bread wheat germplasm groups grown at different levels of water stress. *Field Crops Res.* **41**, 45–54.

Schulze, E.-D., and Hall, A. E. (1982). Stomatal responses, water loss and CO_2 assimilation rates of plants in contrasting environments. *In* "Encyclopedia of Plant Physiology, Physiological Plant Ecology, Vol. 12B" (O. L. Lange, P. S. Nobel, C. B. Osmond, and H. Zeigler, eds.), pp. 180–230. Springer-Verlag, New York.

Shackel, K. A., and Hall, A. E. (1979). Reversible leaflet movements in relation to drought adaptation of cowpeas, *Vigna unguiculata* (L) Walp. *Aust. J. Plant Physiol.* **6**, 265–276.

Stahl, R. S., and McCree, K. J. (1988). Ontogenetic changes in the respiration coefficients of grain sorghum. *Crop Sci.* **28**, 111–113.

Tanner, C. B., and Sinclair, T. R. (1983). Efficient water use in crop production: Research or re-search? *In* "Limitations to Efficient Water Use in Crop Production" (H. M. Taylor, W. R. Jordan, and T. R. Sinclair, eds.), pp.1–27. ASA-CSSA-SSSA, Madison, WI.

Virgona, J. M., and Farquhar, G. D. (1996). Genotypic variation in relative growth rate and carbon isotope discrimination in sunflower is related to photosynthetic capacity. *Aust. J. Plant Physiol.* **23**, 227–236.

Virgona, J. M., Hubick, K. T., Rawson, H. M., Farquhar, G. D., and Downes, R. W. (1990). Genotypic variation in transpiration efficiency, carbon-isotope discrimination and carbon allocation during early growth in sunflower. *Aust. J. Plant Physiol.* **17,** 207–214.

White, J. W. (1993). Implications of carbon isotope discrimination studies for breeding common bean under water deficits. *In* "Stable Isotopes and Plant Carbon–Water Relations" (J. R. Ehleringer, A. E. Hall, and G. D. Farquhar, eds.), pp. 387-398. Academic Press, San Diego.

White, J. W., Castillo, J. A., and Ehleringer, J. R. (1990). Associations between productivity, root growth and carbon isotope discrimination in *Phaseolus vulgaris* under water deficit. *Aust. J. Plant Physiol.* **17,** 189–198.

Wilson, D. (1975). Variation in leaf respiration in relation to growth and photosynthesis of *Lolium. Ann. Appl. Biol.* **80,** 323–328.

Wilson, D., and Jones, J. G. (1982). Effects of selection for dark respiration rate of mature leaves on crop yields of *Lolium perenne* cv. S.23. *Ann. Bot.* **49,** 313–320.

Winzeler, M., McCullough, D. E., and Hunt, L. A. (1989). Leaf gas exchange and plant growth of winter rye, triticale and wheat under contrasting temperature extremes. *Crop Sci.* **29,** 1256–1260.

Woodruff, D. R., and Tonks, J. (1983). Relationship between time of anthesis and grain yield of wheat genotypes with differing developmental patterns. *Aust. J. Agric. Res.* **34,** 1–11.

Wright, G. C., Hubick, K. T., and Farquhar, G. D. (1988). Discrimination in carbon isotopes of leaves correlates with water-use efficiency of field-grown peanut cultivars. *Aust. J. Plant Physiol.* **15,** 815–825.

Wright, G. C., Hubick, K. T., Farquhar, G. D., and Nageswara Rao, R. C. (1993). Genetic and environmental variation in transpiration efficiency and its correlation with carbon isotope discrimination and specific leaf area in peanut. *In* "Stable Isotopes and Plant Carbon–Water Relations" (J. R. Ehleringer, A. E. Hall, and G. D. Farquhar, eds.), pp. 245–267. Academic Press, San Diego.

Wright, G. C., Nageswara Rao, R. C., and Farquhar, G. D. (1994). Water-use efficiency and carbon isotope discrimination in peanut under water deficit conditions. *Crop Sci.* **34,** 92–97.

4

Productivity in Water-Limited Environments: Dryland Agricultural Systems

C. Gimenez, F. Orgaz, and E. Fereres

I. Introduction

Dryland agricultural systems (DASs) are inherently fragile. The variability in precipitation and at low quality of many arid-zone soils leads to variable and low yields and large fluctuations in farmers' income. Farmers have always sought insurance against such variations, in particular against catastrophic years that could bring about famine and migration (Loomis, 1983).

Avoidance of crop failure has been the primary management objective over the years, a fact that explains the conservative nature of many of the practices adopted by dryland farmers. Management practices designed with the objective of avoiding crop failure tend to leave resources unused in the good years because they do not fully exploit the favorable conditions that periodically arise due to weather or soil variations. Nevertheless, risk avoidance strategies are needed to minimize threats to the sustainability of the system; thus, there has to be a balance between the measures aimed at increasing productivity and those designed for ensuring sustainability. Analyses of many traditional dryland farming systems tend to suggest that there is often room for increasing productivity, while enhancing, not threatening, sustainability.

Drought is a common phenomenon in the DASs. Crop productivity depends on the cumulative effects of water deficits on plant performance. Therefore, it is critical to understand the processes of water supply and demand in dryland crop communities and the role of adaptation in mitigating the detrimental effects of drought on crop yields.

Crop plants require a continuous supply of water to their transpiring organs to meet the evaporative demand. Such demand arises simply because internal plant surfaces are nearly saturated with water and are exposed to a drier atmospheric environment. To sustain the water flow needed to prevent tissue dehydration, plants have evolved an elaborate water gathering and transport system. Water flows from the soil into the root system and is transported via the xylem vessels to the leaves where it replaces water that evaporates through the epidermis to the atmosphere. Thus, from a purely physical viewpoint, plants are water transport systems from a source, the soil, to a sink, the atmosphere. Such systems are capable of transporting large amounts of water, equivalent, in a typical summer day, to several times their own weight. However, a small imbalance in the transport process that is hardly detectable may occur in response to alterations in water supply or demand and create a plant water deficit. Such very mild water deficits are often detrimental to yield even though large amounts of water are being transported by the plants at the same time (Hsiao *et al.*, 1976). This extreme sensitivity of plant growth processes to water deficits is responsible for the large decreases in yields observed in drought-prone regions.

Crop performance under limited water supply can be evaluated in terms of water-use efficiency [WUE; aboveground biomass (DM) per unit of evapotranspiration (ET); i.e., g m^{-2} mm^{-1}]. Because transpiration from leaves (T) and evaporation from the soil surface (E) are jointly considered in crop ET, WUE can be written as

$$\text{WUE} = (TE)/(1 + E/T). \text{ (Cooper, 1983)} \qquad (1)$$

Transpiration efficiency (TE) is used here as the ratio of biomass produced to water transpired at the whole plant level (i.e., g m^{-2} mm^{-1}). There are two possible ways of increasing biomass WUE: improving the transpiration efficiency or reducing evaporation from the soil. Maximizing WUE does not always result in higher biomass production if crop ET is negatively affected. This conservative strategy is typical of CAM species and is more successful in natural systems as a survival strategy than as a means of improving productivity in DASs.

Crop productivity must be viewed not only in terms of DM but also in relation to harvestable yield. For that purpose, a conceptual model similar to those outlined by Passioura (1977) and Fischer and Turner (1978) may be used, where water-limited productivity (P) is expressed as

$$P = T \cdot \text{TE} \cdot \text{HI}, \qquad (2)$$

where HI is the harvest index. It has been argued that if the three variables of Eq. (2) are largely independent of each other, an improvement in one of them results in an improvement of P, but this is not always the case. For example, there are situations in which an increase in T early in the season may result in a reduction of soil water availability later on, leading to a reduction in HI. In Mediterranean-type environments, conserving water for use in T late in the season also reduces TE. Matching the crop or cultivar phenology to the prevailing environment is probably the best option to minimize the detrimental effects of a limited water supply to the HI. Therefore, management strategies in the DASs should be directed at maximizing the variables of Eq. (2) while reducing their negative interactions.

Evaporation and runoff can be major losses in the water balance of the arid and semiarid zones, where it is not infrequent to find in the subsoil a depth below which the water does not penetrate. To maximize the proportion of seasonal rainfall that is used in crop T, infiltration of rainwater should be promoted and root exploration and soil water extraction by the crop should be emphasized. Water lost directly from the soil surface by evaporation reduces the potential for crop T and should be minimized. System manipulations to maximize rainfall storage in the root zone, to extract most of the water of the profile and to reduce E losses to a minimum, not only increase the productivity of DASs but also protect the soil from erosion threats.

In this chapter, we present first the physical characterization of water-limited environments and describe recent progress in weather forecasting for improving management decisions in the DASs. We then analyze the different strategies to increase productivity in dryland cropping systems. Those directed at reducing water losses and at improving T, TE, and HI are emphasized. Finally, we highlight the importance of good agronomy and the potential that simulation models have as analytical tools for predicting the behavior of agricultural systems.

II. Characterization of Water-Limited Environments

A. Drought Patterns in Dryland Cropping Systems

In dryland systems, crops depend on the amount of growing season rainfall and on the previously stored soil water in the potential root zone. Water-limited environments can be defined as those in which, in some parts of the season or in some seasons, plant water deficits arise as discussed in the Introduction and crop yields are affected by the drought.

There is great variability in climatic conditions among water-limited environments, although a common and important characteristic is the high year-to-year and seasonal variability of rainfall. Dregne (1982) (cited in Hatfield, 1990) described four patterns of rainfall distribution in dryland areas: winter, summer, continental, and multimodal. The winter rainfall pattern is characterized by long, hot, dry summers and short, mild, wet winters. Mediterranean environments are typical examples of this pattern.

In these environments, annual rainfall can vary between 275 and 900 mm (Aschmann, 1973) with more than 65% concentrated in winter. Rainfed agricultural areas of Mediterranean-like climate are located in southern Europe, north Africa, west Asia, southwestern Australia, areas along the west coast of North and South America, and in South Africa.

The summer rainfall pattern is generally the result of monsoon activity during the warm summer months with the remainder of the year being dry. The areas that represent this pattern are located south of the Sahara, between 0 and 10°N in India and between 15 and 20°S in northern Australia. Crops in the dry season are expected to rely solely on stored soil water because rainfall is very infrequent after the monsoonal rains end.

Localized, intense storms as a result of extreme surface heating character-ize the continental rainfall pattern. Although rainfall is mainly concentrated in the summer months, some precipitation occurs in every month. This pattern is found in central North America, central China, and in Argentina. Sporadic droughts may occur at any time during the growing season in these continental environments.

In the multimodal distribution pattern, the two alternating rainy and dry seasons that occur often result from moonsonal conditions. The inten-sity and duration of droughts exhibit large year-to-year variations. This fourth rainfall pattern is characteristic of equatorial areas in east Africa and India.

Regardless of the drought pattern, crop water deficits will not develop unless the soil water supply limits the crop ET rate. Thus, it is important to assess how the rainfall patterns discussed previously relate to the actual occurrence of crop water deficits. Jordan and Miller (1980) have proposed three generalized patterns of crop water deficits in dryland agriculture. The first is termed terminal drought and is characterized by adequate water supply early in the season and increasingly severe water deficits. Terminal drought is characteristic of Mediterranean environments in which the lack of rainfall and high ET of early summer combine to hasten crop water deficits near the end of the season. An initial drought period often occurs in the second crop water-deficit pattern because planting takes place before the full start of the rainy season. It is normally found in tropical areas that have a wet season that is shorter than the crop cycle. The third pattern is observed in many humid and in semiarid areas, where annual rainfall may be sufficient but intraseasonal variability is large and intermittent deficits can occur at any time during the growth period, typical of the continental rainfall pattern.

B. Precipitation Prediction

A large proportion of the year-to-year variation in crop yields of water-limited environments is due to the variation in available soil moisture, which is determined by the balance between rainfall and evapotranspirative losses. There is normally less variability in evaporative demand than in rainfall and long-term averages of ET are useful in predicting seasonal ET demand. Average rainfall is not a good indicator, however, of expected rainfall. Seasonal rainfall predictions are needed to identify when tactical changes can be made either to take advantage of predicted above-average rainfall or to reduce losses in predicted low-rainfall situations (Sonka *et al.*, 1987).

Although managers of agricultural systems have sought to predict rainfall for many years, recently have successful, semiquantitative predictions become available in some areas. Long-term daily climatic records can be statistically analyzed to predict daily rainfall. Richardson (1985) used a combination of a first-order Markov chain and a γ distribution function to simulate rainfall for crop modeling. The Markov chain described daily rainfall and the γ distribution fit the rainfall amount in a rainy day. In this approach, long series of daily weather records are needed to estimate the model parameters, which limits its applicability. However, Gen *et al.* (1986) have overcome this limitation by estimating model parameters from monthly weather records. Different studies (Richardson and Wright, 1984; Villalobos and Fereres, 1989) have compared the simulated rainfall data against observed distributions and concluded that it can be applied to a wide range of locations. Other methods, such as response farming forecasts in Kenya (McCown *et al.*, 1991), are useful in predicting seasonal rainfall. They used two predictors. One (P_I) was based on the onset of the rainy season and the other (P_{II}) depended on the cumulative amount of rainfall received early in the season. In general terms, P_{II} reduced the uncertainty in seasonal rainfall more than P_I. Loomis and Connor (1992) described how such predictions offer promise for increasing profits to maize farmers in Kenya.

Other predictors have been developed based on empirical indices derived from readily available atmospheric measurements, such as the southern oscillation index (SOI), which show considerable promise for some regions (Clewett *et al.*, 1991; Hammer and Muchow, 1991; Nicholls, 1991). The SOI is based on atmospheric pressure oscillations in the Pacific Ocean (southern oscillation) that, together with the warming of surface waters of the eastern equatorial Pacific (El Niño), are referred to jointly as ENSO (Nicholls, 1991)

Clewett *et al.* (1991) have shown that SOI has large effects on mean summer rainfall throughout Queensland (Australia). In this region, an increase in spring value of SOI from strongly negative to strongly positive is associated with a subsequent increase in mean summer rainfall of about 200 mm in the semiarid tropics and about 100 mm in the semiarid subtropics. However, Russell *et al.* (1993), studying the relationship between the

SOI and the subsequent seasonal rainfall in four regions in the subtropics of eastern Australia, suggest the need to include additional variables to the SOI in rainfall predictions. Waylen *et al.* (1994) and Pisciottano *et al.* (1994) have provided a verification of the good relationship between rainfall and SOI in Costa Rica and Uruguay, respectively.

The SOI has also been used for estimating crop yield in areas where seasonal variability in rainfall is high. Wheat yields in Australia have been correlated with values of SOI, although the best predictors appear to be trends in SOI rather than absolute values (Rimmington and Nicholls, 1993). Wheat yield was negatively correlated with the SOI of the year prior to planting and positively correlated with the SOI during the crop season. Nicholls (1986) established that Australian average yields for grain sorghum, which is a summer crop, were predictable using the SOI. However, the development of timely prediction methods for crops planted in autumn, when the southern oscillation phase might not yet be clear, is more difficult. Kuhnel (1994) indicated that the SOI alone seems to have only limited value as a predictor of total sugarcane yields over large areas of Australia. However, on a smaller scale, the SOI was a useful indicator of yields for the northern sugarcane districts.

Important progress has been made in weather forecasting and rainfall prediction over the past decade. However, there are still many agricultural areas of the world where seasonal rainfall cannot be predicted with reasonable accuracy. When farmers' income is low, they tend to use tactics directed to minimize risks, reducing the uncertainty of crop production. However, if there were reliable predictions at the sowing time of the seasonal rainfall they would use higher-risk management tactics, achieving substantial increases in farm profits.

III. Water Use and Water-Use Efficiency

A. Crop Transpiration

Early experiments and recent analyses (Tanner and Sinclair, 1983; Monteith, 1991) have emphasized the conservativeness of the transpiration–biomass relationship for a given crop–environment combination. Thus, increasing T is probably the most important means of increasing biomass production in water-limited environments. The potential volume of water for crop T is primarily determined by the amount of rainfall since last harvest. However, there are losses due to runoff, deep percolation, evaporation from the soil surface, and weed transpiration that reduce the supply for crop transpiration below the maximum potential. Although after crop emergence, such losses are directly affected by crop growth dynamics (ground cover, root growth, etc.), the water stored before planting depends on weed growth and on the soil properties, which determine the infiltration, redistribution, and evaporation processes. Such properties can be manipulated by tillage.

1. Maximizing Stored Soil Water Runoff and infiltration are closely related. When the rate of precipitation exceeds the infiltration rate, water is lost as runoff. In an unaltered, bare soil, raindrop impact causes soil aggregate dispersion and crusting, reducing infiltration. For many decades, farmers have used tillage to improve infiltration, reduce evaporation, eliminate restrictions to root penetration (clay pans, fragipans, compaction pans, etc.), and for weed control. Crop residues were buried or burned, leaving the soil completely uncovered. However, conventional tillage practices present some disadvantages in the long run. First, the positive effects of tillage in breaking the soil crust, thereby increasing infiltration, do not last because bare soil is continuously exposed to raindrop impact. Second, tractor traffic and tillage implements create a compaction layer below the tillage zone. Third, the practice of clearing the crop residues decreases the organic matter content and the structural stability of the soil. Finally, the energy use and the costs of conventional tillage are high. Such disadvantages are not nearly as critical as the high rates of erosion that are associated with conventional tillage in the semiarid zones. Soil erosion, probably the most dramatic problem in semiarid agricultural areas, can be substantially reduced with conservation tillage systems, which were developed as alternatives to conventional tillage. Conservation tillage is based on maintaining crop residues on the surface to protect the soil and includes a wide range of practices, from the use of chisel plows instead of the moldboard plow to no-till in the zero-tillage systems.

The beneficial effect of crop residues on infiltration rate is well known and has been documented in recent reviews (Fischer, 1987; Unger, 1990; Kemper, 1993). Residues protect the soil against the destructive action of raindrops and increase the organic matter in the uppermost layers, thus improving soil aggregate stability and biopore formation. In some cases, soil cracks have a positive effect on infiltration rate, for example, in vertisols at the beginning of the rainy season in Mediterranean climates (Gonzalez *et al.*, 1988). Also, the reduction in traffic tends to eliminate the plow-compacted layer. This explains the improved water balance and crop production associated with conservation tillage systems, even on bare soils as is the case of olive groves (Pastor, 1988). Crop residues have additional effects on infiltration rate by intercepting runoff, thus increasing the opportunity time for water infiltration. Crop residues delay but do not prevent evaporation. In the long run, their effects have limited impact under field conditions (Fischer, 1987; Unger *et al.*, 1991).

It should be emphasized that tillage practices are site specific. For example, in many arid zone soils, which are typically of low structural stability, some tillage is required to prevent a substantial decline in infiltration (Henderson, 1979). An important additional limitation of conservation tillage in such zones is the small amount of residues usually available. Thus, it is likely that the optimal conservation tillage practices in soils of arid areas will require some tillage passes, contrary to the single operation used in zero-tillage systems.

Finally, crop rotation can also affect the total amount of soil water available at planting. The combination of shallow and deep-rooted crops in a rotation provides a means for complete extraction of subsoil moisture and nutrients, increasing the efficiency of water use.

2. Partitioning of ET into E and T   Once the crop is established, E reduction is the most important way of increasing crop T. Direct evaporation of soil water can be described as a three-phase process (Lemon, 1956; Philip, 1957). In the first phase following wetting, E is only limited by the energy available at the soil surface, approaching potential ET for bare soil conditions (Ritchie, 1972). When the soil can no longer supply water fast enough to use all the available energy, the soil-limiting phase (phase 2) begins, and E is progressively reduced. In phase 3, the E rate is very low, and it is mainly determined by the temperature gradients driving water vapor flows (Cary, 1966). The potential E savings that may lead to increasing T and biomass production largely depend on the rainfall patterns from planting until the crop reaches full cover (LAI > 3). Infrequent rainfall and fast canopy development limit the benefits of E reductions. Nevertheless, E losses are relatively important in a variety of situations. For wheat in Australia, French and Schultz (1984) estimated that E represented 30% of ET for a wide range of conditions. When crop growth is very poor, E can reach 75% of ET, a value estimated by Cooper et al. (1987) for unfertilized barley in Syria. It has been shown that there is a partial compensating rise in T when E is reduced (Villalobos and Fereres, 1990); nevertheless, an important fraction of E can be considered a true loss.

Manipulation of the first phase of E is most important to reduce E losses under field conditions. The objective here is to achieve full ground cover quickly so that the radiation reaching the wet soil surface is reduced. Changes in planting date, starter fertilizer applications, and the use of cultivars with rapid early growth are the most relevant practices for E reduction in dryland systems.

Early plantings of spring crops in Mediterranean climates can increase crop radiation interception when the probability of rainfall is high. Gimeno *et al.* (1989) have shown that rainfed sunflower yields in southern Spain may be substantially increased by planting in winter compared to conventional spring plantings. Although different factors may have influenced the response, the reduction in E and the concomitant increase in T for the winter planting was most important in one experiment (Orgaz *et al.*, 1990). In Mediterranean climates, high rainfall probability is normally associated with low temperatures, and early plantings may not be successful because of too slow early growth and/or frost damage. Breeding genotypes for rapid early growth under low temperatures is thus an important objective for crop improvement in such climates.

Increasing planting density and adequate fertilizer applications are additional practices that enhance early crop growth and reduce E losses. In many Mediterranean soils barley and durum wheat respond markedly to N and P (Cooper *et al.*, 1987; Anderson, 1985). The increased cereal yields observed with adequate fertilization were associated with faster early growth and reduced E losses (Gregory *et al.*, 1984; Cooper *et al.*, 1987).

Early vigor is an important physiological trait that reduces E and increases T in water-limited environments. Acevedo (1991) found it to be one of the attributes that confers an advantage to barley over wheat in low rainfall areas of the eastern Mediterranean. López-Castañeda and Richards (1994a) confirmed that the early vigor of barley was responsible for the success of this crop when compared to other cereals, such as wheat, triticale, and oats, in Mediterranean-like environments. Barley grain yield was 17% higher than that of the other species and aboveground biomass was 22% higher than that of durum wheat. Because of its early vigor, barley shaded the soil surface faster, resulting in lower E and higher T (Fig. 1). An additional

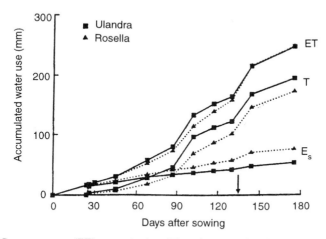

Figure 1 Crop water use (ET), transpiration (T) and evaporation from soil under the crop (Es) for Ulandra and Rosella bread wheat cultivars. Anthesis date is indicated by the arrow (reprinted from *Field Crops Res.* **39**, López-Castañeda, C., and Richards, R. A., 1994b. Variation in temperate cereals in rainfed environments. III. Water use and water-use efficiency 85–98. Copyright 1994 with kind permission of Elsevier Science-NL, Sara Burgerhartstraat 25, 1055 KV Amsterdam, The Netherlands).

advantage of its early vigor was that its faster leaf area development led to higher TE because more of its growth and transpiration occurred when vapor pressure deficit (VPD) was small (López-Castañeda and Richards, 1994b). The embryo size was identified as the most important factor accounting for the differences in early vigor among the cereal species and the width of the first seedling leaf was suggested as an indirect selection criteria in breeding programs for increasing early vigor in temperate cereals (López-Castañeda *et al.*, 1996). Within a species such as wheat, those genotypes with vigorous early growth have higher water-use efficiency, biomass production, and yield in dryland Mediterranean environments (Turner and Nicolas, 1987; Whan *et al.*, 1993).

B. Transpiration Efficiency

Most reviews dealing with the improvement of WUE attribute more importance to changes in T than in TE in the improvement of WUE, biomass production, and yield under drought (Fischer and Turner, 1978; Tanner and Sinclair, 1983; Ritchie, 1983; Turner, 1993). Little evidence of intraspecific variability in the TE of crop plant species has been found in the past (Fischer, 1981; Tanner and Sinclair, 1983). However, studies of carbon isotope discrimination (Farquhar *et al.*, 1982) have demonstrated that there are differences in TE among wheat genotypes (Farquhar and Richards, 1984), barley, peanuts, cowpea, bean, lentil, and sunflower (Condon and Hall, Chapter 3, this volume). Turner (1993) has indicated that increased TE may be inherently linked to reduced early growth. Thus, genotypes of high TE often show low potential yields and have high E/ET ratios due to slow canopy development early in the season. Such genotypes may have adaptive responses that are useful for plant survival under extreme water stress but do not show potential for high productivity under moderate water deficits. Hall *et al.* (1993) found that high TE was associated with low HI in cowpea. Breeding efforts will be needed to break such negative associations; Condon and Hall (Chapter 3, this volume) explore the genetic variation in TE and its potential for increasing crop adaptation and productivity in dryland agriculture.

In the short run, better prospects for increasing TE exist by matching the crop cycle to the season of low evaporative demand. The most important environmental factor affecting TE is the VPD (Fischer and Turner, 1978). Therefore, management practices that enhance radiation interception when VPD is low will increase TE.

Solar radiation levels also influence TE. As crop assimilation rate tends to saturate (particularly in C_3 plants) and T increases roughly linearly with increasing irradiance, TE must decline as incident radiation increases (Loomis, 1983). In agreement with this hypothesis, Kanemasu (1983) presented experimental data of the relationship between TE and irradiance

(24-hr basis) for grain sorghum. Under full radiation interception (LAI > 2.8) TE values decreased by about half as irradiance doubled. Unfortunately, the effects of VPD and of increasing irradiance were not separated in Kanemasu's (1983) study. In situations of near complete radiation interception by the crop, the VPD is the governing factor affecting TE. However, under low LAI values, irradiance must play an important role in determining TE, interacting with the fraction of total leaf area that receives radiation levels above the saturation point.

Management practices directed at increasing TE must attempt to maximize the radiation intercepted by the crop during the periods of low irradiance and VPD. Once again, early plantings and any other practice promoting early growth (planting density, fertilization, early vigor, etc.) will increase biomass production in Mediterranean-type environments not only through a reduction in *E,* as it was argued in Section III,A,2, but also through an increase in TE.

IV. Carbon Partitioning to Harvestable Yield

In many DASs, harvestable yield includes all aboveground biomass, particularly in the crop–livestock systems on drier areas. In all other cases, harvestable yield is only a plant part and the processes of assimilate partitioning and C allocation are of paramount importance for the determination of final yields.

Empirical selection for yield has increased the partitioning to harvestable organs but our understanding of the processes that determine dry matter distribution is very limited.

A. Dry Matter Distribution under Water Deficits

Water deficits affect the partitioning of dry matter to above- and belowground plant parts. The functional balance between shoot and root (Brouwer, 1983) is altered by water stress in the direction of a relative increase in carbon partitioning to the root. Another relevant partitioning process that is affected by water deficits is the distribution of dry matter between vegetative and reproductive organs. Both responses have an effect on the productivity of DASs as discussed below.

1. Leaf Area and Root Growth   Yield decreases caused by water deficits are primarily mediated by the reduction in leaf area. Numerous studies in various crops, such as maize and sorghum (Hsiao *et al.,* 1976), wheat (Robertson and Giunta, 1994), sunflower (Connor and Jones, 1985; Gimenez and Fereres, 1986), faba bean (Karamanos, 1984), cotton (McMichael *et al.,* 1973), soybean and sorghum (Constable and Hearn, 1978), and barley

(Day *et al.*, 1987), have shown how water deficits effect large reductions in leaf area that are directly linked to a decrease in intercepted radiation and in yield. Passioura (1994) has emphasized the role of leaf area in the modulation of yield under water deficits.

Leaf expansion rates (LER) are immediately reduced by water deficits (Acevedo *et al.*, 1971). In the longer term, however, it appears that leaf area expansion under water deficits is modulated via chemical signals sent by the roots (Passioura, 1994). However, the role of the shoot water status has been shown to be critical in at least one field experiment (Sadras *et al.*, 1993) in which variations in VPD affected the relation between LER and soil water status. It is probable that both hydraulic and hormonal signals interact somehow in determining the patterns of leaf expansion under water deficits (Tardieu, 1993).

Mild water deficits, while reducing leaf area, must make available additional C for other plant processes, including root growth. This is because photosynthetic processes are less sensitive to water deficits than is LER. Root extension into new soil is less affected by water deficits than LER, and that provides a feedback mechanism for the improvement of growth and water status of the shoot.

The traditional view is that large, vigorous root systems are advantageous in water-limited environments. Actually, rooting depth is directly related to water extraction (Ehlers *et al.*, 1991), whereas there are substantial differences in rooting density (Lv) among crops (Hamblin and Tennant, 1987) and cultivars (Gregory, 1994) often not related to differences in water uptake. Although differences in Lv among species are of interest for complete exploitation of available resources in a given environment, genotypic differences are mostly linked to maturity date (Blum and Arkin, 1984). For example, long-season sunflower varieties depleted the deeper layers of the profile much more than early cultivars (Fig. 2; Gimenez and Fereres, 1986). The additional water captured by the deeper root systems of late-maturing cultivars occurs near the end of the season, when evaporative demand is highest in Mediterranean-type environments. Thus, TE values are low and that may reduce the benefits of the higher *T* of late cultivars (Fig. 3; Fereres, 1987). Such cultivars are best suited to areas of deep soils that must be consistently refilled by seasonal rainfall if their use is to be sustainable.

The maintenance of green leaf area and root growth under water deficits is an important feature contributing to mitigate drought impacts on yield. Osmotic adjustment (Hsiao *et al.*, 1976) maintains root growth, slows down leaf senescence (Hsiao *et al.*, 1984), and promotes assimilate translocation to the grains (Ludlow *et al.*, 1990). Despite earlier criticism (Munns, 1988), it is now firmly established that osmotic adjustment is a valuable adaptive feature of crops and cultivars adapted to water-limited environments (Ludlow and Muchow, 1990; Tangpremsri *et al.*, 1995).

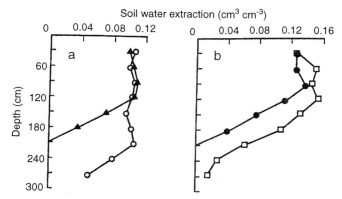

Figure 2   Soil water extraction for (a) $A_1 \times R_2$ (▲) and Contiflor (○) in 1981 and (b) $A_1 \times R_7$ (●) and Sungro (□) in 1983. Measurements are average of four replicates. $A_1 \times R_2$ and $A_1 \times R_7$ are short-season genotypes and Contiflor and Sungro are long-season genotypes (from Gimenez and Fereres, 1986).

2. Mobilization of Reserves   Mobilization of reserves to harvestable organs in response to water deficits changes the distribution of dry matter in the plant and diminishes the impact of water stress on yields. Such adaptive response has been thoroughly documented in winter cereals, in which the mobilization of preanthesis assimilates stored in leaves and stem is enhanced by water deficits.

Gallagher *et al.* (1975) estimated in barley that preanthesis assimilation

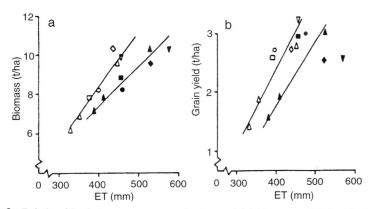

Figure 3   Relationships between evapotranspiration and (a) biomass and (b) grain yield for five early (open symbols) and five late (closed symbols) sunflower cultivars. Data obtained at Cordoba, Spain, between 1981 and 1985 (from Fereres, 1987).

contributed 2% of grain yield in a wet year but up to 74% in a very dry year. This contribution was calculated as the dry weight difference between anthesis and maturity of the vegetative aboveground organs. This methodology overestimates the mobilization of reserves because the dry weight change in the nongrain parts includes tissue losses (mainly leaves) due to wind, decay, etc. Bidinger *et al.* (1977) used labeled $^{14}CO_2$ at frequent intervals before and after anthesis to estimate that 20% of preanthesis assimilates were transferred to grain in water-stressed wheat. Austin *et al.* (1980) estimated that the mobilization of preanthesis reserves in barley was 44% of grain yield in a dry treatment and just 11% when water was not limiting. Not only does the relative proportion of preanthesis assimilation increase but also some studies (Palta *et al.*, 1994) showed that water deficits effected an absolute increase in the contribution of preanthesis reserves to wheat yield. Breeders have searched for genetic variability in this favorable character. Blum *et al.* (1983), using magnesium chlorate as a dessicant, attributed the difference among four spring wheat cultivars in sustaining kernel growth in the absence of photosynthesis to differences in stem reserves mobilization. Magnesium chlorate has been found to be too severe a dessicant to simulate drought effects by Turner and Nicolas (1987), who proposed the use of potassium iodide as a more suitable chemical. Mahalakshmi *et al.* (1994) have succesfully used this senescing agent in pearl-millet by spraying with KI at anthesis as a screening method to evaluate the genetic variability in reserve mobilization under water deficits during grain filling.

The general conclusion is that assimilate mobilization increases yield stability in drought situations by acting as a buffer against the detrimental effects of water deficits on postanthesis assimilation. Genetic variability in reserve mobilization has also been reported in soybean (Constable and Hearn, 1978) and sorghum (Ludlow *et al.*, 1990). This adaptive trait could be incorporated into new varieties to increase crop yield under water-limited environments. Attempts to use chemical desiccation to select lines under simulated postanthesis stress have been relatively successful in cereals (Nicolas and Turner, 1993).

B. Effects of Water Deficits on the Harvest Index

1. Harvest Index   The effects of water deficits on HI depend on the timing and intensity of water stress. Mild water deficits either do not alter the HI value or, exceptionally, improve it relative to situations of ample water supply. As water deficits become more severe or occur at a critical stage, HI tends to decrease below the nonstress value. Terminal drought patterns tend to reduce HI in winter cereals (Gifford, 1979; Giunta *et al.*, 1993). In a few crops, such as sorghum (Faci and Fereres, 1980) and cotton (Orgaz *et al.*, 1992), mild water stress increased HI relative to irrigated controls.

In drought situations, it is common to find a correlation between HI and yield as dry matter production is limited by the water available and the maintenance of HI becomes critical. For instance, Bolaños and Edmeades (1993) found that yield increases in maize after eight cycles of selection for drought tolerance were correlated to an increase in HI in wet and in dry environments. Improved drought tolerance was due to increased partitioning of biomass toward the developing ear during a severe drought stress that coincided with flowering rather than to a change in plant water status. Positive correlations between HI and yield have been reported for many other crops such as wheat (Loffler *et al.,* 1985; Slafer and Andrade, 1989; Siddique *et al.,* 1989), rice (Cook and Evans, 1983), sunflower (Fereres *et al.,* 1986), and chickpeas (Silim and Saxena, 1993).

2. Yield Components The harvestable part of many crops is the product of fruit number times the individual fruit weight. Fruit number and weight—the components of yield—are both affected by water deficits, particularly by its timing and its severity. If stress occurs in the early developmental stages, there could be opportunities for compensation later on, and HI and yields may not be affected. This has been shown, for example, in maize (Abrecht and Carberry, 1993), rice (Lilley and Fukai, 1994), and sorghum (Craufurd and Peacok, 1993), in which water stress in the vegetative phase did not reduce yields. On the contrary, water stress during the stages of fruit number determination or during grain filling induces the largest yield reductions. When water deficits reduce either fruit numbers or weight, there is normally some compensation between yield components. In many crops, there is more plasticity in seed number than in seed weight. For instance, Fereres *et al.* (1986) attributed most of the genotypic differences in the drought susceptibility of sunflower cultivars to adjustments in seed number. The ratio of seed number between a dryland and an irrigated treatment ranged from 0.39 to 0.56, whereas the grain weight ratio varied between only 0.73 and 0.77. Simane *et al.* (1993) studied the effect of early, mid-season, and terminal stress on durum wheat yield components. Wheat grain yield was related to the effects of early stress on kernels per spike and on duration of the grain-filling period and kernels per spike under terminal stress. When the duration of grain growth is significantly shortened by water stress, grain weight is reduced (Simane *et al.,* 1993; Savin and Nicholas, 1996).

C. Water Conservation Strategies for Optimum Allocation

We believe that one of the best strategies to obtain high yields consistently under dryland conditions is to match the patterns of crop growth and development to the patterns of water supply and use. This has largely been achieved by genetic selection in the case of winter cereals in Mediterranean-type environments (Loss and Siddique, 1994), a case that is used in the subsequent discussion. Additionally, because of the possible interactions among water supply, water use, and crop growth, crop management offers options to optimally distribute the available water among the different crop stages in order to maximize yields in water-limited environments.

In Section III, we emphasized that there are opportunities for increasing biomass production in the DASs. Such an increase may be mostly achieved through management practices aimed at increasing radiation interception by the crop when the soil surface is wet and the evaporative demand (radiation and VPD) is low. At those times, carbon may be acquired at a very low water cost, whereas the water not used in T is lost via E anyway.

The desirable management practices that can increase biomass production include early plantings, cultivars with high early vigor, adequate starter fertilizer, and high planting densities. Such practices affect favorably both T and TE, resulting in increased biomass production at anthesis and, hence, in increased yield potential. Whether such yield potential is realized will depend on the maintenance of HI at levels similar to those observed in nonlimiting water conditions. Most of the increases in grain yield during recent decades have been attributed to the increase in HI, which is now approaching its potential value in many crop species. In water-limited environments, further yield increases through genetically improved HI may be small (Siddique *et al.*, 1990); thus, crop management offers the greatest potential to achieve high HI and yields in DASs.

In Passioura's (1977) expression [Eq. (2)], HI cannot be considered independent of T and TE. On the one hand, higher biomass at anthesis can substantially contribute to grain yield by assimilate mobilization. On the other hand, such an increase in biomass could represent more water use during the vegetative phase, reducing the fraction available for grain filling and, consequently, affecting HI. In his analysis, Fischer (1979) proposed for wheat that, in water-limiting environments, there is a negative association between crop biomass at anthesis and water availability for grain filling, and an optimum biomass level at anthesis is proposed. Yield and HI reductions would occur when biomass exceeds such a level as a consequence of insufficient water available for grain filling (Fig. 4). This would be especially true when the differences in crop biomass at anthesis are associated to differences in preanthesis ET, as is the case of genotypes of different maturity date. However, when the higher crop biomass at anthesis is caused by higher T and TE without influencing preanthesis ET, the model in Fig. 2 does not apply. In fact, it is difficult to find experimental data showing reductions in HI and yield when radiation interception in the first crop stages and, hence, TE are improved. The few reports indicating such negative responses (Fischer and Kohn, 1966; Cantero-Martínez *et al.*,

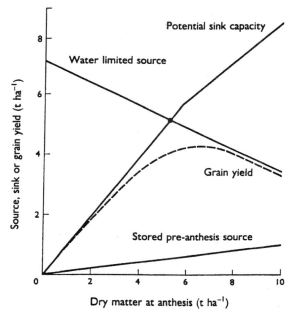

Figure 4 The relationship between growth, water use, and grain filling in wheat under postanthesis drought (from Fischer, 1979).

1995) are associated with very severe terminal stress environments and long-season genotypes. We propose that practices directed to shift a higher fraction of the available water to grain filling, at the expense of restricting early crop growth, may not be suitable for improving productivity in many water-limited environments. On the contrary, the optimum strategy should be directed at increasing biomass production through management practices that allow rapid early growth, whereas the optimum distribution of water supply between pre- and postanthesis should be determined by choosing a genotype of adequate season length. In terminal stress environments of limited water supply, early maturing cultivars should be chosen.

In the very severe stress environments, flowering cannot take place soon enough, due to either genetic or environmental factors, as is the case of frost risks in winter cereals, to conserve enough water for grain filling. Only in such situations are management practices that reduce preanthesis ET desirable. Among them, reducing planting density and low fertilization rates are the most common. When the water supply limitations are even more drastic, growing forage crops for biomass or even the use of fallow are the obvious strategies (Loomis, 1983).

V. Integration of Responses in Dryland Agricultural Systems

Crop yields in the DASs are not always constrained by water deficits. In fact, other limiting factors, such as nutrition, weed competition, etc., often reduce yields below the potential for the amount of water available. An example of the importance of other agronomic factors in limiting wheat yields in the dryland areas of south Australia was presented by French and Schultz (1984). Their analysis of a series of observations shown in Fig. 5 suggests that the primary yield constraint in many areas may be incorrect crop management rather than insufficient water supply.

Agronomy and plant breeding are responsible for the increases in crop productivity observed over the past decades. Analyses performed for wheat in several environments showed that the rate of yield increase in dryland areas has been much less than that observed in more favorable environments (Perry and D'Antuono, 1989). Evidently, the slower rate of yield

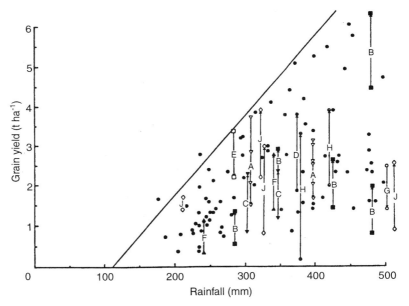

Figure 5 The relation between grain yield of wheat and April–October rainfall for experimental sites and farmers paddocks. The sloping line indicates the potential yield relation. The responses to different treatments are shown by lettered lines linking points. Yield increases were obtained by the application of nitrogen (points linked by a B line), phosporus (C line), copper (D line), control of eelworms (F line), and multifactor research (J line). Yield reductions occurred because of delayed time of sowing (A line), effects of weeds (E line), and waterlogging (G line). Variation in yields in districts are shown by the H lines (from French and Schultz, 1984).

improvement in the DASs reflects the constraint that the water limitation imposes on carbon assimilation. It also offers considerable scope for further improvement if management is optimized. A comparison in wheat trends under dryland and irrigated conditions performed by Stanhill (1986; Fig. 6) shows very similar yield improvement rates in both situations, with irrigated yields being higher than dryland yields on the average. The outstanding feature of the data in Fig. 6 is the large fluctuations in dryland yields relative to the much more stable irrigated yield trend. Improved agronomy should focus on mitigating the impact of the very bad years and on maximizing profits in the good years. If farmers could anticipate the nature of the season, their tactical decisions would reduce yield fluctuations substantially.

Among agronomic factors, nitrogen fertilization is most critical. Fast canopy development after sowing is essential to increase the T fraction in the partitioning of ET. The traditional view that nitrogen supply to dryland crops should be restricted to prevent excessive water use in the vegetative phase, so that water supply during grain filling is not exhausted, should be questioned. Three response types to N supply have been observed in

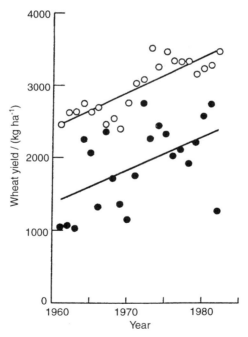

Figure 6 Trends in the level and variation of wheat yields under irrigation in Egypt (○) compared with dryland conditions in Israel (●) (from Stanhill, G., 1986. Irrigation in arid lands. *Phil. Trans. R. Soc. London A* **316**, 261–273, Fig. 2, by the Royal Society of London).

dryland crops; normally, the overall yield constraint in areas or years of low rainfall limits a positive response to N to moderate rates of fertilization, beyond which there is no response to increased N. When water supply conditions improve, crops exhibit a positive N response over a wide range of fertilization rates. Finally, only in areas (or years) of very low precipitation (less than 250 mm annual rainfall) have modest yield reductions been observed in response to increased N fertilization (i.e., Cantero-Martínez *et al.*, 1995). It seems that, in most situations, the benefits of higher *T* and higher photosynthetic rates associated with increased N supply balance or outweigh the possible losses induced by the more severe water stress in the last stages of crop development caused by increased N fertilization under dryland conditions.

Other management tactics, such as early plantings and use of cultivars with early vigor, should contribute to increasing *T* without having negative impacts on HI. There is an urgent need for breeding efforts in the development of cultivars that can achieve high initial growth rates under the low temperatures characteristic of seasons of low evaporative demand.

To optimize agronomic practices, it is essential to develop and use analytical techniques that have some predictive value. Crop simulation models are ideal tools for risk analyses (Loomis *et al.*, 1979; Boote *et al.*, 1996). Model outputs are used for generating probability distributions of yield, income, or profit as a function of the environment and management considered (Fereres *et al.*, 1993). Cumulative probability functions are developed under several scenarios to determine the distribution that is most favorable for the objective chosen (maximum average yields, highest yield in the drier years, lowest risk, etc.). Simulation models can also be a valuable tool for matching genotypes to environments (Boote *et al.*, 1996) and for evaluating management strategies to improve productivity in water-limiting conditions (Muchow and Bellamy, 1991; Villalobos *et al.*, 1994).

As an example of an application to design an optimal cropping sequence, Fereres *et al.* (1993) compared three crop sequences using wheat and sunflower in a Mediterranean climate of about 600 mm annual rainfall. Under such conditions, wheat does not use all the winter rainfall and there is substantial residual moisture in the subsoil at harvest. On the contrary, continuous sunflower cropping, a spring crop, can exhaust all available water well before harvest and even fails in the dry years. The current wheat–sunflower (W–S) rotation was compared in a simulation to continuous wheat (W), continuous sunflower (S), and a wheat–wheat–sunflower (W–W–S) rotation (Fereres *et al.*, 1993). The probability distributions for each case showed that the W–W–S sequence had higher average wheat yields than in the W–S rotation and also had higher sunflower yields in 9 of 10 years, thus it is apparently more sustainable than the W–S sequence used in the area.

Management tactics, such as planting dates, fertilization rates and timing, and cultivars of different season length, can be evaluated in long-term scenarios to select the best strategy. Another related application of models is their use for tactical variations of the selected strategies at the start of the season or within a season. This is very useful to fully exploit the resources available when the year conditions are more favorable than on the average or to avoid crop failure in the very dry years. Such applications are becoming realistic and feasible as weather predictions are improved. For instance, Hammer *et al.* (1996) evaluated the value of seasonal forecasts based on the SOI for tactical management decisions in wheat production in northeastern Australia. They showed that, by using forecasts to make decisions on N fertilizer and cultivar maturity, profits increased by 20% and the risk of a loss decreased by 35%. In this case, a crop simulation model combined with seasonal forecasts illustrates the feasibility of using such tools for improving the agricultural productivity of that area.

Both seasonal weather forecasts and simulation models need substantial improvements to increase the precision and accuracy of their predictions. Research efforts are essential for such improvements if these new tools are to play an important role in the management of DASs in the future.

VI. Conclusions

Productivity of DASs is low and all efforts to improve it have had only modest success when compared to the improvement of yield in the agricultural systems in which water is not so limiting. In the short term, we believe that strategic and tactical modifications of crop management offer scope for increasing the productivity in water-limited environments.

References

Abrecht, D. G., and Carberry, P. S. (1993). The influence of water deficits prior to tassel initiation on maize growth, development and yield. *Field Crops Res.* **31**, 55–69.

Acevedo, E. (1991). Morphophysiological traits of adaptation of cereals to Mediterranean environments. *In* "Improvement and Management of Winter Cereals under Temperature, Drought and Salinity Stresses" (E. Acevedo, E. Fereres, C. Gimenez and J. P. Srivastava, eds.), pp. 85–96. ICARDA-INIA, Madrid, Spain.

Acevedo, E., Hsiao, T. C., and Henderson, D. W. (1971). Immediate and subsequent growth responses of maize leaves to changes in water status. *Plant Physiol.* **48**, 631–636.

Anderson, W. K. (1985). Grain yield responses of barley and durum wheat to split nitrogen applications under rainfed conditions in a Mediterranean environment. *Field Crops Res.* **12**, 191–202.

Aschmann, H. (1973). Distribution and peculiarity of Mediterranean ecosystems. *In* "Mediterranean Type Ecosystem: Origin and Structure" (F. di Castri and D. Mooney, eds.). Springer-Verlag, New York.

Austin, R. B., Morgan, C. L., Ford, M. A., and Blackwell, R. D. (1980). Contributions to grain yield from pre-anthesis assimilation in tall and dwarf barley phenotypes in two contrasting seasons. *Ann. Bot.* **45,** 309–319.

Bidinger, F., Musgrave, R. B., and Fischer, R. A. (1977). Contribution of stored pre-anthesis assimilate to grain yield in wheat and barley. *Nature* **270,** 431–433.

Blum, A., and Arkin, G. F. (1984). Sorghum root growth and water use as affected by water supply and growth duration. *Field Crops Res.* **9,** 131–138.

Blum, A., Poiarkova, H., Golan, G., and Mayer, J. (1983). Chemical dessication of wheat plants as a simulator of post-anthesis stress. I. Effects on translocation and kernel growth. *Field Crops Res.* **6,** 51–58.

Bolaños, J., and Edmeades, G. O. (1993). Eight cycles of selection for drought tolerance in lowland tropical maize. I. Responses in grain yield, biomass, and radiation utilization. *Field Crops Res.* **31,** 233–252.

Boote, K. J., Jones, J. W., and Pickering, N. B. (1996). Potential uses and limitations of crop models. *Agron. J.* **88,** 704–716.

Brouwer, R. (1983). Functional equilibrium: Sense or nonsense? *Netherlands J. Agric. Sci.* **31,** 335–348.

Cantero-Martínez, C., Villar, J. M., Romagosa, I., and Fereres, E. (1995). Nitrogen fertilization of barley under semi-arid rainfed conditions. *Eur. J. Agron.* **4,** 309–316.

Cary, J. W. (1966). Soil moisture transport due to thermal gradients: Practical aspects. *Soil Sci. Soc. Am. Proc.* **30,** 428–433.

Clewett, J. F., Howden, S. M., McKeon, G. M., and Rose, C. W. (1991). Optimising farm dam irrigation in response to climatic risk. *In* "Climatic Risk in Crop Production: Models and Management for the Semiarid Tropics and Subtropics" (R. C. Muchow and J. A. Bellamy, eds.), pp. 307–328. CAB International, Wallingford, UK.

Connor, D. J., and Jones, T. R. (1985). Response of sunflower to strategies of irrigation. II. Morphological and physiological responses to water stress. *Field Crops Res.* **12,** 91–103.

Constable, G. A., and Hearn, A. B. (1978). Agronomic and physiological responses of soybean and sorghum crops to water deficits. *Aust. J. Plant Physiol.* **5,** 159–167.

Cook, M. G., and Evans, L. T. (1983). Some physiological aspects of the domestication and improvement of rice (*Oryza* spp.). *Field Crops Res.* **6,** 219–238.

Cooper, P. J., Gregory, P. J., Tully, D., and Harris, H. (1987). Improving water use efficiency of annual crops in the rainfed farming systems of west Asia and north Africa. *Exp. Agric.* **23,** 113–158.

Cooper, P. J. M. (1983). Crop management in rainfed agriculture with special reference to water use efficiency. Proc. 17th Colloquium Int. Potash Inst., Bern, Switzerland, pp. 63–79.

Craufurd, P. Q., and Peacock, J. M. (1993). Effect of heat and drought stress on sorghum (*Sorghum bicolor*). 2. Grain yield. *Exp. Agric.* **29,** 77–86.

Day, W., Lawlor, D. W., and Day, A. T. (1987). The effect of drought on barley yield and water use in two contrasting years. *Irrig. Sci.* **8,** 115–130.

Dregne, H. E. (1982). Dryland soil resources. Science and Technology Agriculture Report. Agency for International Development, Washington, DC.

Ehlers, W., Hamblin, A. P., Tennant, D., and van der Ploeg, R. R. (1991). Root system parameters determining water uptake of field crops. *Irrig. Sci.* **12,** 115–124.

Faci, J. M., and Fereres, E. (1980). Responses of grain sorghum to variable water supply under two irrigation frequencies. *Irrig. Sci.* **1,** 149–159.

Farquhar, G. D., and Richards, R. A. (1984). Isotopic composition of plant carbon correlates with water-use efficiency of wheat genotypes. *Aust. J. Plant Physiol.* **11,** 539–552.

Farquhar, G. D., O'Leary, M. H., and Berry, J. A. (1982). On the relationship between carbon isotope discrimination and intercellular carbon dioxide concentration in leaves. *Aust. J. Plant Physiol.* **9,** 121–137.

Fereres, E. (1987). Responses to water deficits in relation to breeding for drought resistance. *In* "Drought Tolerance in Winter Cereals" (P. Srivastava, E. Porceddu, E. Acevedo, and S. Varma, eds.), Proceedings of an International Workshop, ICARDA, Capri, Italy, pp. 263–273. Wiley, Chichester, UK.

Fereres, E., Gimenez, C., and Fernandez, J. M. (1986). Genetic variability in sunflower cultivars under drought. I. Yield relationships. *Aust. J. Agric. Res.* **37,** 573–582.

Fereres, E., Orgaz, F., and Villalobos, F. J. (1993). Water use efficiency in sustainable agricultural systems. *In* "International Crop Science. I" (D. R. Buxton, R. Shibles, R. A. Forsberg, B. L. Blad, K. H. Asay, G. M. Paulsen, and R. F. Wilson, eds.), pp. 83–89. CSSA, Madison, WI.

Fischer, R. A. (1979). Growth and water limitation to dryland wheat yield in Australia: A physiological framework. *J. Aust. Inst. Agric. Sci.* **45,** 83–94.

Fischer, R. A. (1981). Optimizing the use of water and nitrogen through breeding of crops. *Plant Soil* **58,** 249–278.

Fischer, R. A. (1987). Responses of soil and crop water relations to tillage. *In* "Tillage" (P. S. Cornish and J. E. Pratley, eds.), pp. 194–221. Inkata Press, Melbourne, Australia.

Fischer, R. A., and Kohn, G. D. (1966). The relationship of grain yield to vegetative growth and post-flowering leaf area in the wheat crop under conditions of limited soil moisture. *Aust. J. Agric. Res.* **17,** 281–295.

Fischer, R. A., and Turner, N. C. (1978). Plant productivity in the arid and semiarid zones. *Annu. Rev. Plant Physiol.* **29,** 277–317.

French, R. J., and Schultz, J. E. (1984). Water use efficiency of wheat in a mediterranean-type environment. II. Some limitations to efficiency. *Aust. J. Agric. Res.* **35,** 765–775.

Gallagher, J. N., Biscoe, P. V., and Scott, R. K. (1975). Barley and its environment. V. Stability of grain weight. *J. Appl. Ecol.* **12,** 319–336.

Gen, S., Penning de Vries, F. W. T., and Supit, I. (1986). A simple method for generating daily rainfall data. *Agric. For. Meteorol.* **36,** 363–376.

Gifford, R. M. (1979). Growth and yield of CO_2-enriched wheat under water-limited conditions. *Aust. J. Plant Physiol.* **6,** 367–378.

Gimenez, C., and Fereres, E. (1986). Genetic variability in sunflower cultivars under drought. II. Growth and water relations. *Aust. J. Agric. Res.* **37,** 583–597.

Gimeno, V., Fernandez, J. M., and Fereres, E. (1989). Winter plantings as a means of drought escape in sunflower. *Field Crops Res.* **22,** 307–316.

Giunta, F., Motzo, R., and Deidda, M. (1993). Effect of drought on yield and yield components of durum-wheat and triticale in a Mediterranean environment. *Field Crops Res.* **33,** 399–409.

Gonzalez, P., Fereres, E., Giraldez, J. V., Martin, I., Garcia, M., Gil, J., and Agüera, J. (1988). Non tillage dry farming in heavy clay soils under Mediterranean climate. Proc. 11th Int. Conf. of the Int. Soil Tillage Res. Organization, Edinburgh, Scotland, Vol. 2, pp. 661–666.

Gregory, P. J. (1994). Root growth and activity. *In* "Physiology and Determination of Crop Yield" (K. J. Boote, J. M. Bennett, T. R. Sinclair, and G. M. Paulsen, eds.), pp. 65–93. ASA,CSSA,SSSA, Madison, WI.

Gregory, P. J., Shepherd, K. D., and Cooper, P. J. M. (1984). Effects of fertilizer on root growth and water use of barley in Northern Syria. *J. Agric. Sci. Cambridge* **103,** 429–438.

Hall, A. E., Ismail, A. M., and Menendez, C. (1993). Implications for plant breeding of genotypic differences in water use efficiency, carbon isotope discrimination and gas exchange. *In* "Stable Isotopes and Plant Carbon–Water Relations" (J. R. Ehleringer, A. E. Hall, and G. D. Farquhar, eds.), pp. 349–370. Academic Press, San Diego.

Hamblin, A., and Tennant, D. (1987). Root length density and water uptake in cereals and grain legumes: How well are they correlated? *Aust. J. Agric. Res.* **38,** 513–527.

Hammer, G. L., and Muchow, R. C. (1991). Quantifying climatic risk to sorghum in Australia's semiarid tropics and subtropics: Model development and simulation. *In* "Climatic Risk in Crop Production: Models and Management for the Semiarids Tropics and Subtropics" (R. C. Muchow and J. A. Bellamy, eds.), pp. 205–232. CAB International, Wallingford, UK.

Hammer, G. L., Holzworth, D. P., and Stone, R. (1996). The value of skill in seasonal climate forecasting to wheat crop management in a region with high climatic variability. *Aust. J. Agric. Res.* **47**, 717–737.

Hatfield, J. L. (1990). Agroclimatology of semiarid lands. *In* "Dryland Agriculture. Strategies for Sustainability" (R. P. Singh, J. F. Parr, and B. A. Stewart, eds.), Advances in Soil Science, Vol. 13, pp. 9–26. Springer-Verlag, New York.

Henderson, D. W. (1979). Soil management in semi-arid environments. *In* "Crop Productivity in Arid and Semiarid Environments" (A. E. Hall, ed.), Ecological Studies, Vol. 35, pp. 224–237. Springer-Verlag, Berlin.

Hsiao, T. C., Fereres, E., Acevedo, E., and Henderson, D. W. (1976). Water stress and dynamics of growth and yield of crop plants. *In* "Water and Plant Life: Problems and Modern Approaches" (O. L. Lange, L. Kappen, and E. D. Schulze, eds.), Ecological Studies, Vol. 19, pp. 281–305. Springer-Verlag, Berlin.

Hsiao, T. C., O'Toole, J. C., Yambo, E. B., and Turner, N. C. (1984). Influence of osmotic adjustment on leaf rolling and tissue death in rice (*Oryza sativa* L.). *Plant Physiol.* **75**, 338–341.

Jordan, W. R., and Miller, F. R. (1980). Genetic variability in sorghum root systems: Implications for drought tolerance. *In* "Adaptations of Plants to Water and High Temperature Stress" (N. C. Turner and P. J. Kramer, eds.), pp. 383–399. Wiley, New York.

Kanemasu, E. T. (1983). Yield and water-use relationships: Some problems of relating grain yield to transpiration. *In* "Limitations to Efficient Water Use in Crop Production" (H. M. Taylor, W. R. Jordan, and T. R. Sinclair, eds.), pp. 413–417. ASA, CSSA, SSSA, Madison, WI.

Karamanos, A. J. (1984). Effects of water stress on some growth parameters and yields of field bean crops. *In* "*Vicia faba*: Agronomy, Physiology and Breeding" (P. D. Hebblethwaite, T. C. K. Dawkins, M. C. Heath, and G. Lockwood, eds.), pp. 47–59. Nijhoff/Junk, Dordrecht.

Kemper, W. D. (1993). Effects of soil properties on precipitation use efficiency. *Irrig. Sci.* **14**, 65–73.

Kuhnel, I. (1994). Relationship between the southern oscillation index and Australian sugarcane yields. *Aust. J. Agric. Res.* **45**, 1557–1568.

Lemon, E. R. (1956). The potential for decreasing soil moisture evaporation loss. *Soil Sci. Soc. Am. Proc.* **20**, 120–125.

Lilley, J. M., and Fukai, S. (1994). Effect of timing and severity of water deficits on 4 diverse rice cultivars. 3. Phenological development, crop growth and grain yield. *Field Crops Res.* **37**, 225–234.

Loffler, C. M., Rauch, T. L., and Busch, R. H. (1985). Grain and plant protein relationships in hard red spring wheat. *Crop Sci.* **25**, 521–524.

Loomis, R. S. (1983). Crop manipulations for efficient use of water: An overview. *In* "Limitations to Efficient Water Use in Crop Production" (H. M. Taylor, W. R. Jordan, and T. R. Sinclair, eds.), pp. 345–374. ASA, CSSA, SSSA, Madison, WI.

Loomis, R. S., and Connor, D. J. (1992). "Crop Ecology. Productivity and Management in Agricultural Systems," pp. 538. Cambridge Univ. Press, Cambridge, UK.

Loomis, R. S., Rabbinge, R., and Ng, E. (1979). Explanatory models in crop physiology. *Annu. Rev. Plant Physiol.* **30**, 339–367.

López-Castañeda, C., and Richards, R. A. (1994a). Variation in temperate cereals in rainfed environments. II. Phasic development and growth. *Field Crops Res.* **37**, 63–75.

López-Castañeda, C., and Richards, R. A. (1994b). Variation in temperate cereals in rainfed environments. III. Water use and water-use efficiency. *Field Crops Res.* **39**, 85–98.

López-Castañeda, C., Richards, R. A., Farquhar, G. D., and Williamson, R. E. (1996). Seed and seedling characteristics contributing to variation in early vigor among temperate cereals. *Crop Sci.* **36**, 1257–1266.

Loss, S. P., and Siddique, K. H. M. (1994). Morphological and physiological traits associated with wheat yield increases in Mediterranean environments. *Adv. Agron.* **52**, 229–276.

Ludlow, M. M., and Muchow, R. C. (1990). A critical evaluation of traits for improving crop yields in water-limited environments. *Adv. Agron.* **43,** 107–153.

Ludlow, M. M., Santamaria, J. M., and Fukai, S. (1990). Contribution of osmotic adjustment to grain yield in *Sorghum bicolor* (L.) Moench under water-limited conditions. II. Water stress after anthesis. *Aust. J. Agric. Res.* **41,** 67–78.

Mahalakshmi, V., Bidinger, F. R., Rao, K. P., and Wani, S. P. (1994). Use of the senescing agent potassium iodide to simulate water deficit during flowering and grain filling in pearl-millet. *Field Crops Res.* **36,** 103–111.

McCown, R. L., Wafula, B. M., Mohammed, L., Ryan, J. G., and Hargreaves, J. N. G. (1991). Assesing the value of a seasonal rainfall predictor to agronomic decisions: The case of response farming in Kenya. *In* "Climatic Risk in Crop Production: Models and Management for the Semiarids Tropics and Subtropics " (R. C. Muchow and J. A. Bellamy, eds.), pp. 383–409. CAB International, Wallingford, UK.

McMichael, B. L., Jordan, W. R., and Powell, R. D. (1973). Abcission process in cotton: Induction by plant water deficit. *Agron. J.* **65,** 202–204.

Monteith, J. L. (1991). Conservative behaviour in the response of crops to water and light. *In* "Theoretical Production Ecology: Reflections and Prospects" (R. Rabbinge *et al.,* eds.), pp. 3–16. PUDOC, Wageningen, The Neherlands.

Muchow, R. C., and Bellamy, J. A. (1991). "Climatic Risk in Crop Production: Models and Management for the Semiarid Tropics and Subtropics," pp. 548. CAB International, Wallingford, UK.

Munns, R. (1988). Why measure osmotic adjustment? *Aust. J. Plant Physiol.* **15,** 717–726.

Nicholls, N. (1986). Use of the Southern Oscillation to predict Australian sorghum yield. *Agric. For. Meteorol.* **38,** 9–15.

Nicholls, N. (1991). Advances in long-term weather forecasting. *In* "Climatic Risk in Crop Production: Models and Management for the Semiarid Tropics and Subtropics" (R. C. Muchow and J. A. Bellamy, eds.), pp. 427–444. CAB International, Wallingford, UK.

Nicolas, M. E., and Turner, N. C. (1993). Use of chemical desiccants and senescing agents to select wheat lines maintaining stable grain size during post-anthesis drought. *Field Crops Res.* **31,** 155–171.

Orgaz, F., Gimenez, C., and Fereres, E. (1990). Efficiency of water use in winter plantings of sunflower in a Mediterranean climate. Proc. 1st Congr. Eur. Soc. Agron., Paris, France.

Orgaz, F., Mateos, L., and Fereres, E. (1992). Season length and cultivar determine the optimum evapotranspiration deficit in cotton. *Agron. J.* **84,** 700–706.

Palta, J. A., Kobata, T., Turner, N. C., and Fillery, I. R. (1994). Remobilization of carbon and nitrogen in wheat as influenced by postanthesis water deficits. *Crop Sci.* **34,** 118–124.

Passioura, J. B. (1977). Grain yield, harvest index, and water use of wheat. *J. Aust. Inst. Agric. Sci.* **43,** 117–120.

Passioura, J. B. (1994). The yield of crops in relation to drought. *In* "Physiology and Determination of Crop Yield" (K. J. Boote, J. M. Bennett, T. R. Sinclair, and G. M. Paulsen, eds.), pp. 343–359. ASA,CSSA,SSSA, Madison, WI.

Pastor, M. (1988). Sistemas de manejo del suelo en olivar. Ph.D. dissertation, University of Cordoba, Spain, pp. 280.

Perry, M. W., and D'Antuono, M. F. (1989). Yield improvement and associated characteristics of some australian spring wheat cultivars introduced between 1860 and 1982. *Aust. J. Agric. Res.* **40,** 457–472.

Philip, J. R. (1957). Evaporation, and moisture and heat fields in the soil. *J. Meteorol.* **14,** 354–366.

Pisciottano, G., Diaz, A., Cazes, G., and Mechoso, C. R. (1994). El-Niño Southern-Oscillation impact on rainfall in Uruguay. *J. Climate* **7,** 1286–1302.

Richardson, C. W., and Wright, D. A. (1984). WGEN: A model for generating daily weather variables. ARS-8. Department of Agriculture, Agriculture Research Service, Washington, DC.

Richardson, C. W. (1985). Weather simulation for crop management models. *Trans. ASAE* **28,** 1602–1606.

Rimmington, G. M., and Nicholls, N. (1993). Forecasting wheat yields in Australia with the Southern Oscillation Index. *Aust. J. Agric. Res.* **44,** 625–632.

Ritchie, J. T. (1972). Model for predicting evaporation from a row crop with incomplete cover. *Water Resour. Res.* **8,** 1204–1213.

Ritchie, J. T. (1983). Efficient water use in crop production: Discussion on the generality of relations between biomass production and evapotranspiration. *In* "Limitations to Efficient Water Use in Crop Production" (H. M. Taylor, W. R. Jordan, and T. R. Sinclair, eds.), pp. 29–44. ASA, CSSA, SSSA, Madison, WI.

Robertson, M. J., and Giunta, F. (1994). Responses of spring wheat exposed to pre-anthesis water stress. *Aust. J. Agric. Res.* **45,** 19–35.

Russell, J. S., McLeod, I. M., Dale, M. B., and Valentine, T. R. (1993). The Southern Oscillation Index as a predictor of seasonal rainfall in the arable areas of the inland Australian subtropics. *Aust. J. Agric. Res.* **44,** 1337–1349.

Sadras, V. O., Villalobos, F. J., and Fereres, E. (1993). Leaf expansion in field-grown sunflower in response to soil and leaf water status. *Agron. J.* **85,** 564–570.

Savin, R., and Nicolas, M. E. (1996). Effects of short periods of drought and high temperature on grain growth and starch accumulation of 2 malting barley cultivars. *Aust. J. Plant Physiol.* **23,** 201–210.

Siddique, K. H. M., Kirby, E. J. M., and Perry, M. W. (1989). Ear : stem ratio in old and modern wheat varieties; Relationship with improvement in number of grains per ear and yield. *Field Crops Res.* **21,** 59–78.

Siddique, K. H. M., Tennant, D., Perry, M. W., and Belford, R. K. (1990). Water use and water use efficiency of old and modern wheat cultivars in a Mediterranean-type environment. *Aust. J. Agric. Res.* **41,** 431–447.

Silim, S. N., and Saxena, M. C. (1993). Adaptation of spring-sown chickpea to the Mediterranean basin. 2. Factors influencing yield under drought. *Field Crops Res.* **34,** 137–146.

Simane, B., Struik, P. C., Nachit, M. M., and Peacock, J. M. (1993). Ontogenic analysis of yield components and yield stability of durum-wheat in water-limited environments. *Euphytica* **71,** 211–219.

Slafer, G. A., and Andrade, F. H. (1989). Genetic improvement in bread wheat (*Triticum aestivum*) yield in Argentina. *Field Crops Res.* **21,** 289–296.

Sonka, S. T., Mjelde, J. W., Lamb, P. J., Hollinger, S. E., and Dixon, B. L. (1987). Valuing climate forecast information. *J. Climate Appl. Meteorol.* **26,** 1080–1091.

Stanhill, G. (1986). Irrigation in arid lands. *Phil. Trans. R. Soc. London A* **316,** 261–273.

Tangpremsri, T., Fukai, S., and Fischer, K. S. (1995). Growth and yield of sorghum lines extracted from a population for differences in osmotic adjustment. *Aust. J. Agric. Res.* **46,** 61–74.

Tanner, C. B., and Sinclair, T. R. (1983). Efficient water use in crop production: Research or re-search? *In* "Limitations to Efficient Water Use in Crop Production" (H. M. Taylor, W. R. Jordan, and T. R. Sinclair, eds.), pp. 1–27. ASA, CSSA, SSSA, Madison, WI.

Tardieu, F. (1993). Will progresses in understanding soil–root relations and root signalling substantially alter water flux models? *Phil. Trans. R. Soc. London* **341,** 57–66.

Turner, N. C. (1993). Water use efficiency of crop plants: Potential for improvement. *In* "International Crop Science. I" (D. R. Buxton, R. Shibles, R. A. Forsberg, B. L. Blad, K. H. Asay, G. M. Paulsen, and R. F. Wilson, eds.), pp. 75–82. CSSA, Madison, WI.

Turner, N. C., and Nicolas, M. E. (1987). Drought resistance of wheat for light-textured soils in a Mediterranean climate. *In* "Drought Tolerance in Winter Cereals" (J. P. Srivastava, E. Porceddu, E. Acevedo, and S. Varma, eds.), pp. 203–216. Wiley, Chichester, UK.

Unger, P. W. (1990). Conservation tillage systems. *In* "Dryland Agriculture. Strategies for Sustainability" (R. P. Singh, J. F. Parr, and B. A. Stewart, eds.), Advances in Soil Science, Vol. 13, pp. 27–68. Springer-Verlag, New York.

Unger, P. W., Stewart, B. A., Part, J. F., and Singh, R. P. (1991). Crop management and tillage methods for conserving soil and water in semi-arid regions. *Soil Tillage Res.* **20,** 219–240.

Villalobos, F. J., and Fereres, E. (1989). A simulation model for irrigation scheduling under variable rainfall. *Trans. ASAE* **32,** 181–188.

Villalobos, F. J., and Fereres, E. (1990). Evaporation measurements beneath corn, cotton, and sunflower canopies. *Agron. J.* **82,** 1153–1159.

Villalobos, F. J., Mateos, L., and Orgaz, F. (1994). Improvements of the efficiency of water use in sustainable agricultural systems. Proc. 3rd Cong. Eur. Soc. Agron., Abano-Padova, Italy, pp. 294–303.

Waylen, P. R., Quesada, M. E., and Caviedes, C. N. (1994). The effects of El-Niño-Southern Oscillation on precipitation in San Jose, Costa-Rica. *Int. J. Climatol.* **14,** 559–568.

Whan, B. R., Carlton, G. P., Siddique, K. H. M., Regan, K. L., Turner, N. C., and Anderson, W. K. (1993). Integration of breeding and physiology: Lessons from a water-limited environment. *In* "International Crop Science. I" (D. R. Buxton, R. Shibles, R. A. Forsberg, B. L. Blad, K. H. Asay, G. M. Paulsen, and R. F. Wilson, eds.), pp. 607–614. CSSA, Madison, WI.

5

Nitrogen as a Limiting Factor: Crop Acquisition of Ammonium and Nitrate

Arnold J. Bloom

I. Importance of Nitrogen Acquisition

Most natural and agricultural ecosystems show a dramatic increase in productivity when they receive nitrogen fertilizer (van Keulen and van Heemst, 1982) and, thus, nitrogen is considered to be a major factor that limits plant growth. Plants have evolved diverse mechanisms to capture nitrogen, including symbiotic relationships with nitrogen-fixing bacteria or mycorrhizal fungi, insect carnivory, root absorption of amino acids from organic soils, and leaf absorption of nitrogen from atmospheric deposition or foliar application. Despite an inherent interest in such mechanisms, their contributions to the nitrogen economy of crops are usually minor [nitrogen-fixing bacteria, Peoples and Craswell (1992); mycorrhizal fungi, Marschner (1995); insect carnivory, Dixon *et al.* (1980); amino acids, Chapin *et al.* (1993); foliar absorption, Nicoulaud and Bloom (1996)]. For crops, nitrogen acquisition primarily involves root absorption of ammonium (NH_4^+) or nitrate (NO_3^-) from the soil solution and the subsequent assimilation of these ions into amino acids within roots and shoots (Bloom, 1994).

This chapter describes the physiological processes through which crops absorb and assimilate NH_4^+ and NO_3^-. These processes are among the most energy intensive in life and, thus, determine not only the nitrogen budgets of plants but also their carbon balance. Crop performance—be it yield, fertilizer efficiency, or susceptibility to biological or environmental stress—generally hinges on NH_4^+ and NO_3^- acquisition. The chapter takes an "epidermis-in" approach, focusing on nitrogen acquisition from absorption at the root surface to the assimilation into organic nitrogen within tissues. The chapter by Robertson (Chapter 10, this volume) provides a more "epidermis-out" perspective in that it considers nitrogen cycling in ecosystems and soil nitrogen transformations.

The first half of this chapter treats NH_4^+ and NO_3^- separately. For each ion, I discuss (i) its importance to crops, (ii) the characteristics of its root absorption, (iii) its assimilation into amino acids, and (iv) the internal and external factors that regulate its acquisition. I emphasize root absorption at micromolar concentrations because NH_4^+ or NO_3^- levels in the rhizosphere—the environment immediately surrounding a root—usually fall within this range (Bloom, 1994). In the second half of the chapter, I contrast NH_4^+ and NO_3^- as nitrogen sources and describe the physiological and developmental consequences of using one form over the other.

For additional information on these topics, I refer the reader to the recent reviews of Hoff *et al.* (1994), Imsande and Touraine (1994), Lynch (1995), and, particularly, Oaks (1994), Glass and Siddiqi (1995), and Marschner (1995).

II. Physiology of Root Ammonium Acquisition

A. Importance of Ammonium as a Nitrogen Source

Ammonium is the preferred inorganic source of nitrogen for microorganisms and many plants (Kleiner 1981, 1985; Dortch, 1990; Huppe and Turpin, 1994). The availability of NH_4^+ from soils generally shows much less variation both seasonally and spatially than the availability of other nitrogen forms, particularly NO_3^-. As a consequence, NH_4^+ may be the major nitrogen form available at certain times and places. However, plant NH_4^+ acquisition has received relatively little attention. This lack of research activity stems from two common misconceptions: that passive diffusion of NH_3 through biological membranes is predominately responsible for NH_4^+ absorption (Henderson, 1971; Pitts, 1972) and that higher plants grow better with NO_3^- as the sole nitrogen source (Haynes, 1986; Lewis, 1986).

Root NH_4^+ absorption can proceed via NH_3 diffusion through the membranes of root cells because NH_3 is highly membrane permeable and NH_4^+ interconverts rapidly with NH_3. The following results, however, argue that root NH_4^+ influx depends more on carrier-mediated NH_4^+ transport than on NH_3 diffusion:

1. *Gradients of NH_3 between the rhizosphere and cytoplasm usually oppose passive NH_3 diffusion into root cells.* In agronomic soils, the solution near roots normally has NH_4^+ concentrations between 10 and 50 μM (about 0.03–

0.30 μg NH$_4^+$-N g^{-1} dry soil; Novoa and Loomis, 1981; Haynes, 1986) and a pH below 6 (*e.g.*, Jackson and Bloom, 1990; Lee and Ratcliffe, 1991). The pK_a of the reaction NH$_4^+$ $\leftrightarrow$ NH$_3$ + H$^+$ is 9.25. Consequently, rhizosphere NH$_3$ levels should vary between 0.006 and 0.03 μM. By contrast, the root cytoplasm maintains a pH of between 7.0 and 7.8 and contains from 3 to 438 μM NH$_4^+$ (Roberts and Pang, 1992), from 3 to 12 mM NH$_4^+$ (Lee and Ratcliffe, 1991), or more than 30 mM NH$_4^+$ (Wang *et al.*, 1993; Kronzucker *et al.*, 1995), depending on methodology; a calculation based on the pK_a indicates that cytoplasmic NH$_3$ should range from 0.07 to 700 μM. Thus, under most conditions, the NH$_3$ concentration gradient should oppose the inward diffusion of NH$_3$ across the root cell membrane (Kronzucker *et al.*, 1995).

2. *Exposure to* NH$_4^+$ *depolarizes the plasmalemma.* When a plant cell is first exposed to NH$_4^+$, the electrical potential within the cell initially shifts to a more positive value (*cf.* Section II,B). Diffusion of neutral molecules such as NH$_3$ into a cell would not alter this potential, whereas unidirectional influx of cations such as NH$_4^+$ would have the observed effect.

3. *Root* NH$_4^+$ *influx depends on energy metabolism.* About 2 ATP are expended per NH$_4^+$ absorbed (Fig. 1; Bloom *et al.*, 1992). Metabolic inhibitors, such as KCN or arsenate (Sasakawa and Yamamoto, 1978), hypoxia (Sasakawa

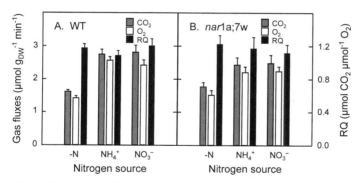

Figure 1   Root CO$_2$ evolution (hatched bars), O$_2$ consumption (open bars), and respiratory quotient (RQ; solid bars) in two genotypes of *Hordeum vulgare* L. cv. Steptoe under different nitrogen nutrition. (A) Wild-type plants and (B) a *nar*1a;*nar*7w mutant that is deficient in both NADH nitrate reductase and NAD(P)H nitrate reductase were exposed to media first devoid of nitrogen, then containing 50 μM NH$_4^+$, and last containing 50 μM NO$_3^-$. Fluxes are expressed per gram dry weight of root. Shown are the mean $\pm$ SE (n = 6–23). The wild type had lower root O$_2$ consumption when shifted from NH$_4^+$ nutrition to NO$_3^-$ nutrition (A); the *nar*1a;*nar*7w mutant, the genotype that had diminished root NO$_3^-$ assimilation, showed no change in O$_2$ consumption when shifted from NH$_4^+$ nutrition to NO$_3^-$ nutrition (B). Thus, the diminished O$_2$ consumption reflects the respiratory cost of NO$_3^-$ assimilation. (Bloom *et al.*, 1992).

and Yamamoto, 1978; Koch and Bloom, 1989), and low temperatures (Bloom *et al.*, 1993), immediately and severely diminish NH_4^+ absorption. A physical process such as NH_3 diffusion would exhibit less sensitivity to these factors.

4. *Root NH_4^+ absorption follows saturation kinetics.* Over the range of concentrations normally found in the rhizosphere, net NH_4^+ uptake approximates Michaelis–Menten kinetics (Fig. 2; Bloom, 1994). NH_4^+ influx saturates at around 200 μM NH_4^+ (Bloom, 1985; Smart and Bloom, 1988) and does not vary directly with root NH_4^+ content (Morgan and Jackson, 1988a). These observations support the contention that NH_4^+ influx is carrier mediated; NH_3 diffusion or channel-mediated NH_4^+ influx would display a more linear response to external concentration and would respond directly to root NH_4^+ content.

The second supposition—that plants prefer NO_3^-—is based on numerous solution culture experiments in which unrealistically high levels of NH_4^+ were applied or pH control was inadequate (Bloom, 1988, 1994). Nutrient solutions typically contain more than 2 mM NH_4^+, whereas most plants require less than 20 μM NH_4^+ to sustain rapid growth (Ingestad and Ågren, 1988; Smart and Bloom, 1993). High NH_4^+ levels in a nutrient solution stimulate high root absorption rates, but the low light levels found in most controlled environment chambers permit only low NH_4^+ assimilation rates: With high absorption and low assimilation, free NH_4^+ accumulates in plant tissues and becomes toxic. Ammonium toxicity probably derives from dissi-

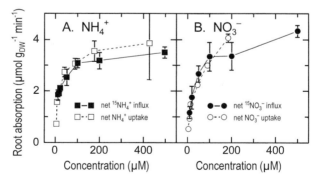

Figure 2   Kinetics of (A) NH_4^+ or (B) NO_3^- absorption for *Lycopersicon esculentum* grown under low levels of NH_4^+ and NO_3^-. Absorption was estimated either by the rate of accumulation of ^{15}N into plants exposed for 10 min to solutions containing predominantly $[^{15}N]NH_4^+$ or $[^{15}N]NO_3^-$ ($^{15}NH_4^+$ or $^{15}NO_3^-$ influx; closed symbols) or by the rate of depletion of NH_4^+ or NO_3^- from a nutrient solution after a single pass through a root cuvette (net NH_4^+ or NO_3^- uptake; open symbols). The difference between the two measures provides an estimate of efflux. (Smart and Bloom, 1988; Kosola and Bloom, 1994, 1996).

pation of the transmembrane proton gradients required both for photosynthetic and respiratory electron transport and for accumulating metabolites in the vacuole (Fig. 3). Moreover, without adequate pH control, NH_4^+-based nutrient solutions quickly become acidic as roots balance NH_4^+ absorption and assimilation through the release of H^+ into the rhizosphere (*cf.* Section II,B). Increased acidity inhibits net NH_4^+ uptake (Lycklama, 1963) so that the stunted growth often observed with NH_4^+ nutrition may result from nitrogen deficiencies at low pH, not from toxic effects (Tolley-Henry and Raper, 1986). If the experimental artifacts of NH_4^+ toxicity and media acidity are avoided, most plants grow faster when they have access to both NH_4^+ and NO_3^- (Fig. 4; Section IV,A; Cox and Reisenauer, 1973; Ganmore-Neumann and Kafkafi, 1980; Haynes, 1986; Bloom, 1988; Schortemeyer *et al.*, 1993; Smart and Bloom, 1993; Marschner, 1995).

B. Ammonium Absorption

The absorption of NH_4^+ from the root/soil interface, as outlined previously, involves a carrier-mediated membrane transport system. The characteristics of this system largely determine the capacity of a plant to obtain NH_4^+ from its environment and, thus, the fertilizer efficiency of a crop. The following sections discuss plant membrane transport systems for NH_4^+, their dependence on counterions to balance the inward movement of positive charges, and their specificity for NH_4^+.

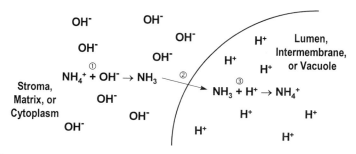

Figure 3   A schematic showing how NH_4^+ toxicity may result from the dissipation of pH gradients. Here a pH gradient across a biological membrane is depicted where the pH is higher on the left side of the membrane. The left side represents the stroma, matrix, or cytoplasm; the right side represents the lumen, intermembrane space, or vacuole; and the membrane represents the thylakoid, inner mitochondrial, or tonoplast membrane for a chloroplast, mitochondria, or root cell, respectively. (1) If NH_4^+ is present at sufficient concentrations, it will react with OH^- on the left side of the membrane to produce NH_3. (2) This NH_3 is membrane permeable and will diffuse across the membrane along its concentration gradient. (3) On the right side of the membrane, the NH_3 will react with H^+ to form NH_4^+. The net result is that both the OH^- concentration on the left side and the H^+ concentration the right side have been diminished, that is, the pH gradient has been dissipated.

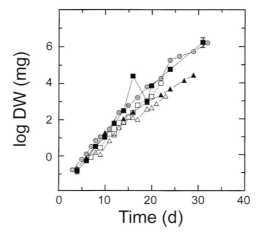

Figure 4   Growth of tomato (*Lycopersicon esculentum* Mill. cv. T5) when nitrogen was supplied as 20 μM NH$_4$Cl ($\triangle$), 50 μM KNO$_3$ ($\square$), 100 μM NH$_4$Cl ($\blacktriangle$), 200 μM KNO$_3$ ($\blacksquare$), or 20 μM NH$_4$Cl and 50 μM KNO$_3$ ($\oplus$). Log dry weight of plant (mean $\pm$ SE) is plotted against days after transplanting to the different nutrient regimes. Root temperature was 20°C. Small error bars are incorporated into the symbols. The lower concentrations of NH$_4^+$ and NO$_3^-$ were selected because they are similar to the apparent K_m's for absorption of each ion; the higher concentrations represented levels at which absorption of each ion saturates. The nutrient concentrations were maintained to within 20% of the designated concentration and pH was controlled at pH 6.0 $\pm$ 0.2. (Smart & Bloom 1993).

1. Membrane Transport   The carrier-mediated NH$_4^+$ transport system that predominates at normal rhizosphere concentrations is thought to be an uniport—a system that transports NH$_4^+$ through a membrane without requiring the movement of another ion (Ullrich, 1987; Glass, 1988). Evidence for an uniport is based on the strong depolarization of the plasmalemma (*i.e.,* a rapid increase in the electrical potential of a cell) that occurs when internodal cells of *Chara corallina* (Smith and Walker, 1978), rhizoid cells of *Riccia fluitans* (Bertl *et al.,* 1984), frond epidermal cells of *Lemna gibba* (Ullrich *et al.,* 1984), coleoptiles of *Avena sativa* (Higinbotham *et al.,* 1964), and tomato or barley roots (Ayling, 1993) are first exposed to NH$_4^+$.

2. Counterions   Although NH$_4^+$ membrane transport does not involve the immediate movement of another ion, a cell cannot sustain an inward movement of positive charges such as NH$_4^+$ without a counterion flow to restore electroneutrality. Some cations must exit the cell and/or some anions must enter the cell so that the cell membrane potential returns to a physiological favorable level. During longer exposure to NH$_4^+$, (i) cells assimilate the NH$_4^+$ they have absorbed, (ii) assimilation of NH$_4^+$ to a monocarboxylmonoamino acid generates an H$^+$ that lowers cytoplasmic pH (Allen, 1988), (iii) both the pH shift and membrane depolarization stimulate plasmalemma H^+-ATPases to pump H^+ out of the cell (Hendrich and Schroeder, 1989; Kurkdjian and Guern, 1989; Serrano, 1989), and (iv) increased acidity outside the cell stimulates anion influx (Tyerman, 1992). This H^+ efflux and anion influx balance NH_4^+ influx (Raven and Smith, 1976), and the cell recovers to its initial membrane potential. Therefore, NH_4^+-induced depolarizations are transient (Smith and Walker, 1978; Bertl *et al.*, 1984; Ullrich *et al.*, 1984; Ayling, 1993).

This scenario for root NH_4^+ absorption explains several phenomena. Many of the effects ascribed to NH_4^+ nutrition, such as diminished absorption of K^+, Cl^-, NO_3^-, and $H_2PO_4^-$ during short-term experiments, probably derive from NH_4^+-induced membrane depolarizations and, thus, are transient (Rufty *et al.*, 1982; Ullrich *et al.*, 1984; Bloom and Finazzo, 1986; Tyerman *et al.*, 1986; Vale *et al.*, 1988). An alkaline rhizosphere facilitates H^+ efflux so that the pH optimum of NH_4^+ absorption is high, above pH 8 (Lycklama, 1963; Munn and Jackson, 1978; Marcus-Wyner, 1983). The effectiveness of an anion as a counterion for NH_4^+ absorption depends not only on its concentration but also on its mobility through a membrane. Mobilities of common anions decrease in the order Cl^-, NO_3^-, PO_4^{3-}, and SO_4^{2-}; correspondingly, NH_4^+ absorption from NH_4Cl is faster than that from NH_4NO_3 or $NH_4H_2PO_4$, which, in turn, is faster than that from $(NH_4)_2SO_4$ (Lycklama, 1963).

3. Specificity   The NH_4^+ transport system that predominates under normal rhizosphere conditions is highly specific for NH_4^+. For example, $CH_3NH_3^+$ serves as a NH_4^+ transport analog in bacteria and fungi (Kleiner, 1981, 1985) and coelutes with NH_4^+ from cation-exchange columns (Kosola and Bloom, 1994); nonetheless, root affinity for NH_4^+ exceeds that for $CH_3NH_3^+$ by nearly an order of magnitude (Deane-Drummond, 1986; Kosola and Bloom, 1994). Potassium is also similar to NH_4^+—NH_4^+ competes directly with K^+ for root K^+ absorption (Rufty *et al.*, 1982; Bloom and Finazzo, 1986; Vale *et al.*, 1988; Smart and Bloom, 1991), and NH_4^+ and K^+ bind equally to soil particles (Loomis and Conner, 1992)—but K^+ has a negligible effect on root NH_4^+ absorption (Scherer *et al.*, 1984; Bloom and Finazzo, 1986; Smart and Bloom, 1988; Bloom and Sukrapanna, 1990). Crops must absorb nearly twice as much nitrogen as potassium from soils that contain a least an order of magnitude more K^+ than NH_4^+ (Epstein, 1972); clearly, a root transport system with high specificity for NH_4^+ is critical to crop performance.

C. Ammonium Assimilation

To avoid NH_4^+ toxicity, roots assimilate NH_4^+ into amino acids quickly after it is absorbed. This process is carbon limited both because the assimilation reactions are energy intensive and account for about 15% of root respiration

(Figs. 1 and 5; Bloom *et al.*, 1992) and because amino acid synthesis requires carbon skeletons, usually α-ketoglutarate generated through dark carbon fixation (Cramer *et al.*, 1993). As a consequence, NH_4^+ assimilation is central to the nitrogen and carbon budgets of roots.

Assimilation of NH_4^+ in roots occurs via a cycle involving the sequential action of glutamine synthetase (GS) and glutamate synthase (GOGAT) (Oaks and Hirel, 1985; Emes and Bowsher, 1991; Amancio and Santos, 1992). The predominant form of GS in roots is cytosolic (Suzuki *et al.*, 1981; Fentem *et al.*, 1983; Tingey *et al.*, 1987), but another form has been found in plastids (Vézina *et al.*, 1989). Probably, the cytosolic form produces glutamine for export, whereas the plastid form is the one that functions primarily for N metabolism in the root itself (Oaks, 1994). Roots may have two forms of GOGAT, one that depends on a ferredoxin-like protein as an electron donor (Suzuki *et al.*, 1982; Suzuki and Gadal, 1984) and another that can use NADH or NADPH (Suzuki and Gadal, 1984; Anderson *et al.*, 1989); both forms are found in plastids (Oaks and Hirel, 1985; Bowsher *et al.*, 1993). Altering the properties of these enzymes through genetic engineering has yielded significant changes in crop growth (Lam *et al.*, 1996), and future cultivars are likely to be modified in these reactions.

D. Regulation of Ammonium Acquisition

To meet the nitrogen demands of growth but avoid toxic NH_4^+ accumulations requires precise regulation of root NH_4^+ absorption and assimilation. Net NH_4^+ uptake, the difference between influx and efflux, changes quickly with plant nitrogen status; it decreases under high nitrogen nutrition and increases during nitrogen deprivation (Lycklama, 1963; Lee and Rudge, 1986; Morgan and Jackson, 1988a,b, 1989). Because under normal rhizosphere conditions root NH_4^+ efflux is a small fraction of influx (Fig. 2; Kosola and Bloom, 1994), net NH_4^+ uptake is probably regulated through control of influx and assimilation.

Rhizosphere NH_4^+ may serve as a positive effector and root glutamine as a negative effector of root NH_4^+ influx (Feng *et al.*, 1994). Exposure to

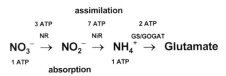

Figure 5   Energy requirements of NH_4^+ and NO_3^- absorption and assimilation. As indicated, approximately 1 ATP is required for roots to absorb either NH_4^+ or NO_3^- from the soil solution; approximately 10 ATP is required to convert NO_3^- to NH_4^+ and 2 ATP to convert NH_4^+ to glutamate. NR, nitrate reductase; NiR, nitrite reductase; and GS/GOGAT, glutamine synthetase/glutamate synthase. (Bloom *et al.*, 1992).

rhizosphere NH_4^+ increases NH_4^+ influx in nitrogen-limited plants (Goyal and Huffaker, 1986; Morgan and Jackson, 1988a), but this results more from a relief of nitrogen deficiency than from an NH_4^+-specific induction of NH_4^+ absorption (Nicoulaud and Bloom, 1994). Glutamine represses NH_4^+ influx in several microorganisms (Kleiner, 1981, 1985; Syrett and Peplinska, 1988). Data on maize indicate a positive correlation between NH_4^+ influx and the ratio of NH_4^+ to glutamine in the root (Lee *et al.*, 1992; Gojon *et al.*, 1993). Nonetheless, tissue levels of NH_4^+ and glutamine can vary independently of root NH_4^+ influx (Glass and Siddiqi, 1995).

Information on the control of root NH_4^+ assimilation is meager (Oaks and Hirel, 1985). Assimilation is an energy-intensive process (*cf.*, Section II,C) that varies with root carbohydrate levels (Reisenauer, 1978; Haynes, 1986; Bloom and Caldwell, 1988). Root carbohydrate levels are inversely correlated to the nitrogen status of a plant (Brouwer, 1967; Brouwer and DeWit, 1969; Talouizte *et al.*, 1984). Therefore, root carbohydrate levels may serve as a positive effector of NH_4^+ assimilation.

III. Physiology of Root Nitrate Acquisition

A. Importance of Nitrate as a Nitrogen Source

The availability of NO_3^- in well-aerated, temperate, agricultural soils typically exceeds that of NH_4^+ by an order of magnitude. In such soils, microbes rapidly convert organic nitrogen to NO_3^-. This NO_3^- moves relatively freely through the soil (Nye and Tinker, 1977) because NO_3^-, as an anion, does not bind to the soil cation-exchange capacity and because all NO_3^- salts are highly soluble. Plants can store relatively high concentrations of NO_3^- in their tissues without toxic effect (Lorenz, 1978; Goyal and Huffaker, 1984). For these reasons, most crops in temperate zones rely on NO_3^- to meet a major portion of their nitrogen requirements (Haynes, 1986) and to serve as a metabolically benign osmoticum (Hanson and Hitz, 1983). This does not mean that crops depend on NO_3^- as their sole nitrogen source, just that in temperate agroecosystems, NO_3^- is often the major nitrogen source.

B. Nitrate Absorption

Root absorption of NO_3^- from the soil solution depends on at least two distinct transport mechanisms: one system (mechanism I or high affinity transport system) predominates at lower concentrations (<1 mM) and appears to be carrier mediated; the other (mechanism II or low affinity transport system) predominates at higher concentrations (>1 mM) and appears to be a more general anion channel (Glass and Siddiqi, 1995). Models of nutrient depletion from the soil around roots (Nye and Tinker, 1977) and the kinetics parameters of root NO_3^- absorption (Crowley, 1975) indicate that NO_3^- concentrations near root surfaces are generally below 1 mM. Consequently, the discussion here will focus on the mechanism I system.

1. Membrane Transport   Thermodynamic conditions do not favor passive NO_3^- movement across the plasmalemma into the cytoplasm of root cells. The cytoplasm maintains a negative potential with respect to its exterior (-100 to -250 mV; Ullrich, 1991) and contains a higher NO_3^- concentration (5–30 mM; Lee and Clarkson, 1986; Zhen *et al.*, 1991; King *et al.*, 1992; Walker *et al.*, 1995) than the rhizosphere (0.1–1.0 mM; Novoa and Loomis, 1981); thus, both electrical and chemical potentials oppose NO_3^- influx. To transport NO_3^- against these potentials requires expenditure of metabolic energy (Bloom and Caldwell, 1988; Glass *et al.*, 1990; Bloom *et al.*, 1992) in a process that involves cotransport of H^+ (Ullrich, 1992; King *et al.*, 1992).

Plants commonly employ a H^+ cotransport mechanism for transport of ions or molecules against electrical and chemical potential gradients (Serrano, 1989). The plasmalemma possesses H^+-translocating ATPases that use the energy released through hydrolysis of ATP to transport H^+ out of the cytoplasm, thereby creating an electrical potential and a proton gradient across the plasmalemma. These favorable electrical and chemical potentials—the so-called protonmotive force—can drive thermodynamically unfavorable transport of ions or molecules by coupling the energetically downhill movement of H^+ to the uphill transport of the other species.

Evidence for cotransport of H^+ and NO_3^- across the plasmalemma has been based on the response of the membrane potential to varying external NO_3^- concentrations. If H^+ diffusion drives NO_3^- transport and H^+:NO_3^- stoichiometry is greater than 1, then raising external NO_3^- concentration and, presumably, increasing NO_3^- influx should produce a transient depolarization. In fronds of the aquatic higher plant *Lemna gibba,* exposure to NO_3^- caused a transient depolarization that appeared to be associated with NO_3^- influx because both increased in a similar manner with time after first exposure to NO_3^- and with increasing $[NO_3^-]_{ext}$ (Ullrich and Novacky, 1981). Intact barley roots showed a transient depolarization upon first exposure to NO_3^- that increased with $[NO_3^-]_{ext}$ (Glass *et al.*, 1992). These data support that NO_3^- influx entails the cotransport of 2 H^+.

2. Counterions   Root NO_3^- absorption is sensitive to the availability of cations (Minotti *et al.*, 1968) but not to species—with the exception of NH_4^+—of cation (Lycklama, 1963). Addition of nontoxic levels of NH_4^+ may either inhibit (Bloom and Finazzo, 1986; Glass and Siddiqi, 1995) or stimulate (Fig. 6; Bloom and Finazzo, 1986; Smart and Bloom, 1988) NO_3^- absorption. The pH optimum for root NO_3^- absorption is generally below pH 6 (Munn and Jackson, 1978; Marcus-Wyner, 1983; Vessey *et al.*, 1990), presumably because of the increased availability of H^+ for cotransport.

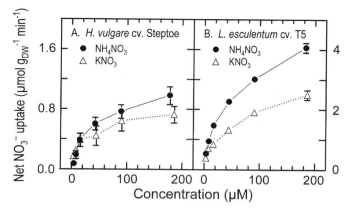

Figure 6 Concentration dependence of net NO₃ uptake when NO₃ was supplied as NH₄NO₃ (●) or as KNO₃ (△); shown are the mean ± SE with small error bars incorporated into the symbols. (A) Data for *Hordeum vulgare* L. Steptoe, a common barley cultivar in Oregon and Washington. Notice that full-scale for net NO_3^- uptake is 1.25 μmol g_{DW}^{-1} root min^{-1}. (B) Data for *Lycopersicon esculentum* Mill. T5, a fresh-market tomato cultivar grown in California. Notice that full-scale for net NO_3^- uptake is 5.0 μmol g_{DW}^{-1} root min^{-1}. (Bloom, 1994).

3. Specificity   The carrier-mediated NO_3^- transport system that operates at normal rhizosphere concentrations is highly specific for NO_3^- (Siddiqi *et al.*, 1990). Chlorate (ClO_3^-), an anion that coelutes with NO_3^- from ion-exchange columns and is readily reduced by nitrate reductase, does not compete with NO_3^- for this transport system (Siddiqi *et al.*, 1992; Kosola and Bloom, 1996). In addition, this NO_3^- system is relatively insensitive to Cl^- and SO_4^{2-}, two anions that may be present in the soil solution at high levels (Glass and Siddiqi, 1985; Deane-Drummond, 1986; Bloom and Finazzo, 1986). Such selectivity permits crops to absorb NO_3^- with minimal interference from other soil constituents.

C. Nitrate Assimilation

Assimilation of NO_3^- into NH_4^+ involves the stepwise reduction of NO_3^- to NO_2^- via nitrate reductase (NR) and NO_2^- to NH_4^+ via nitrite reductase (NiR) (Fig. 5). NR is generally believed to be cytosolic (Vaughn and Campbell, 1988; Oaks, 1994; but see Ward *et al.*, 1989; Meyerhoff *et al.*, 1994), whereas NiR is located in the stroma of chloroplasts or within plastids in roots (Oaks and Long, 1992). These reactions are energy intensive, requiring the transfer of two electrons in the form of NAD(P)H for NR and six electrons from reduced ferredoxin or a ferredoxin-like electron carrier for NiR (Fig. 5); NO_3^- assimilation consumes about 25% of the reductant generated from photosynthesis, shoot respiration, or root respiration (Figs.

1 and 7; Bloom *et al.,* 1989, 1992). Consequently, a plant may expend more than one-quarter of its energy to acquire nitrogen, a constituent that usually comprises less than 2% of total plant dry weight.

The partitioning of NO_3^- assimilation between root and shoot varies among species (Andrews, 1986), but for most species, shoot assimilation becomes increasingly important at higher light levels (Aslam and Huffaker, 1982). Under light-limited conditions, such as in a closed canopy or under elevated atmospheric CO_2 levels, NO_3^- or NO_2^- reduction is unable to compete against carbon fixation for photoreductant and must rely on reductant generated through respiration (Baysdorfer and Robinson, 1985; Bloom *et al.,* 1989). With increasing light levels, as carbon fixation becomes CO_2 limited, NO_3^- or NO_2^- reduction may exploit the surplus reductant generated from photosynthetic electron transport (Fig. 7; Robinson, 1986, 1988; Bloom *et al.,* 1989). Thus, during NO_3^- and NO_2^- photoassimilation, the high

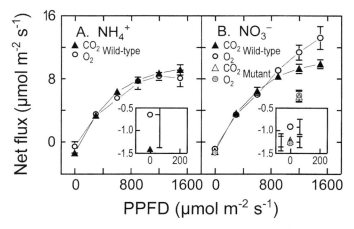

Figure 7   Influence of photosynthetic flux density (PFD) on net CO_2 influx (triangles) and net O_2 efflux (circles) for the wild-type Steptoe barley (filled symbols) and a *nar*1a;*nar*7w mutant that is deficient in both NADH and NAD(P)H NO_3^- reductases (open symbols). The external CO_2 concentration was 360 μl liter^{-1}. The inset presents the dark CO_2 and O_2 fluxes on an expanded scale. Shown are the mean $\pm$ SE ($N = 3$–8) with small error bars incorporated into the symbols. Net CO_2 influx provides an estimate of carbon fixation in the light and carbohydrate catabolism in the dark; net O_2 efflux provides an estimate of photosynthetic electron transport in the light and mitochondrial electron transport in the dark. Respiratory quotient (RQ) equals $CO_{2\ evolved}/O_{2\ consumed}$. Assimilatory quotient (AQ) equals $CO_{2\ consumed}/O_{2\ evolved}$. (A) Under NH_4^+ nutrition, RQ and AQ did not deviate significantly from unity. (B) Under NO_3^- nutrition, the wild type had a RQ and AQ that were significantly different from unity: CO_2 fluxes were identical under NH_4^+ and NO_3^-, but under NO_3^-, O_2 consumption in the dark was 25% less than CO_2 evolution and O_2 evolution at high light was 25% more than CO_2 consumption. For the mutant, RQ in the dark and AQ at 1200 μmol m^{-2} sec^{-1} equaled unity. (Bloom *et al.,* 1989).

energy demands of these reactions do not divert carbon assimilates from growth or other processes.

D. Regulation of Nitrate Acquisition

Both NO_3^- influx and NR activity increase several-fold upon first exposure to NO_3^- (Bloom and Sukrapanna, 1990; Siddiqi *et al.*, 1990, 1992; Aslam *et al.*, 1989, 1992); this induction by the substrate NO_3^- provides for a coarse control of NO_3^- absorption and assimilation. Fine control is effected through the following distinct mechanisms:

1. Root NO_3^- influx and efflux are, respectively, negatively and positively correlated with internal root NO_3^- concentration (Fig. 8; Lee and Drew, 1986; Jackson, 1978; Dean-Drummond and Glass, 1983; Teyker *et al.*, 1988). As a consequence, efflux to the rhizosphere is negligible until external concentrations reach high levels (Fig. 2), but at 1500 μM (root NO_3^- contents of about 75 mol m^{-3} H$_2$O) NO_3^- efflux may reach rates that exceed 40% of influx (Lee and Clarkson, 1986). Most of the internal NO_3^- seems to be sequestered in the vacuole (Lee and Clarkson, 1986). Therefore, only newly absorbed NO_3^- or that released from the vacuole influences NR activity (Aslam *et al.*, 1976; MacKown *et al.*, 1983).

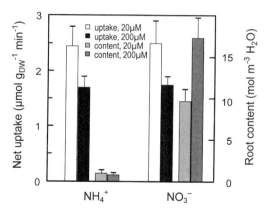

Figure 8 Net root uptake and root contents of NH_4^+ or NO_3^- for tomato (*Lycopersicon esculentum* Mill. cv. T5) grown for several weeks in medium containing 20 μM NH$_4$NO$_3$ (open bars, light hatching) or 200 μM NH$_4$NO$_3$ (closed bars, heavy hatching). Nutrient concentrations were maintained to within 20% of the designated concentration and pH was controlled at pH 6.0 ± 0.2. Steady-state root absorption of NH_4^+ or NO_3^- was monitored from a solution containing 50 μM NH$_4$NO$_3$. Root temperature was 20°C. Shown are the mean ± SE for six plants. The plants maintained relatively constant levels of NH_4^+, whereas NO_3^- contents increased significantly in the plants grown at 200 μM NH$_4$NO$_3$. These plants also absorbed NH_4^+ or NO_3^- at a slower rate when exposed to an intermediate concentration. (Smart & Bloom, 1991).

2. Glutamine and several amino acids suppress both NO_3^- absorption (Lee *et al.*, 1992) and the amount of NR protein (Shiraishi, 1992).

3. Light stimulates NO_3^- absorption and NR activity. In plants that are carbohydrate limited, NO_3^- absorption is greater during the day than at night (Clement *et al.*, 1978; Aslam *et al.*, 1979; Hansen, 1980; Pearson *et al.*, 1981; Aslam and Huffaker, 1982; Rufty *et al.*, 1984; Keltjens and Nijenstein, 1987). NR activity increases in the light both through changes in levels of the NR mRNA (Cheng *et al.*, 1992) and through posttranslational phosphorylation of existing protein (Kaiser and Spill, 1991; Huber *et al.*, 1992). A blue-light receptor and phytochrome may mediate the light responses of NO_3^- uptake in algae (Kamiya, 1989; Aparicio and Quiñones, 1991; López-Figueroa and Rüdiger, 1991; Corzo and Niell, 1992) and of leaf NR activity in higher plants (Duke and Duke, 1984; Schuster *et al.*, 1989; Appenroth *et al.*, 1992; Becker *et al.*, 1992).

All in all, internal root concentrations of NO_3^- are not as finely regulated as those of NH_4^+ (Fig. 8; Smart and Bloom, 1991), probably because NO_3^- can be stored in tissues without toxic effect (Goyal and Huffaker, 1984). Tissue NO_3^- provides plants with a storage pool to support growth when soil nitrogen availability becomes limiting (Chapin *et al.*, 1990) and with a metabolically benign osmoticant during drought (Hanson and Hitz, 1983). A common scenario is for plants to accumulate relatively high levels of NO_3^- at the beginning of a growing season and then to some degree "coast" through the rest of the season (Chapin *et al.*, 1990).

IV. Plant Growth as a Function of Ammonium or Nitrate

A. Ammonium versus Nitrate

The relative reliance of crops on NH_4^+ and NO_3^- as nitrogen sources varies not only with soil availability but also with species. The capability of a species to use one form versus the other ranges from cranberry, which cannot use NO_3^- as a major nitrogen source (Greidanus *et al.*, 1972), to radish, which cannot use NH_4^+ as the sole nitrogen source (Goyal *et al.*, 1982). Most plants—even those that are native to flooded soils where NO_3^- levels in the bulk soil are negligible (Korcak, 1989; Koch *et al.*, 1991)—perform better when they have access to both NH_4^+ and NO_3^- in the soil solution (*cf.* Section II,A). Growth on a single nitrogen form not only produces shifts in rhizosphere pH but also changes the balance between inorganic cations and anions within plant tissues (Kirkby and Mengel, 1967; Arnozis and Findenegg, 1986; MacDuff *et al.*, 1987); for example, growth on NH_4^+ as a sole nitrogen source may lead to deficiencies of cations, such as potassium or calcium, and accumulation to deleterious levels of anions such as borate or molybdate (Fig. 9; Smart and Bloom, 1993). Cation–anion imbalances may be compensated by the production of organic acids (Raven and Smith, 1976), but the long-term effects of such compensation are unclear.

Preference for one nitrogen form versus the other may change during plant development. Seedlings may be energy limited because light levels at the bottom of a canopy are usually low and the energy demands of seedling growth are great. Seedlings generally absorb more NH_4^+ than NO_3^- when both forms are available in equal amounts (A. Bloom, unpublished data) possibly because of the lower energy requirements for NH_4^+ assimilation (approximately 2 ATP for NH_4^+ vs 12 ATP for NO_3^-; cf. Sections II,C and III,C). Similar factors may come into play near the end of the life cycle when canopies are closed: NH_4^+ seems to stimulate crop reproduction (Marschner, 1995) and provide more nitrogen to seeds (Soares and Lewis, 1986) than does NO_3^-.

B. Root Growth and Soil Nitrogen Availability

Root systems must supply plants with water and nutrients under soil conditions that are extremely heterogeneous, both spatially and temporally (Jackson and Bloom, 1990). To optimize acquisition of a vital resource such as nitrogen, but minimize carbon expenditures for root structures,

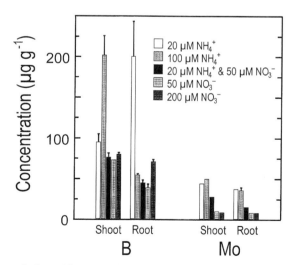

Figure 9   Accumulation of boron and molybdenum in the tissue of tomato (*Lycopersicon esculentum* Mill cv. T5) grown for 15–24 days with 20 μM NH_4^+, 100 μM NH_4^+, 20 μM NH_4^+ and 50 μM NO_3^-, 50 μM NO_3^-, or 200 μM NO_3^- as the nitrogen source. Concentration is given in μg of the element per gram dry weight root or shoot with the mean $\pm$ SE shown. (Smart and Bloom, 1993).

root production responds to local soil nitrogen levels (Hackett, 1972) in a hyperbolic fashion (Fig. 10; Bloom *et al.*, 1993). In soil zones deficient in mineral nitrogen, root growth is minimal because it is nitrogen limited. As soil nitrogen availability increases, roots tend to proliferate (Drew, 1975; Grime *et al.*, 1986; Granato and Raper, 1989; Jackson and Caldwell, 1989; Samuelson *et al.*, 1991) and form denser architecture (Fitter *et al.*, 1988; Fitter and Stickland, 1991). Where soil nitrogen exceeds an optimal level—in our study on tomatoes, 2 μg N NH_4^+ g^{-1} soil and 6 μg N NO_3^- g^{-1} soil—root growth becomes carbohydrate limited, lags behind shoot growth, and eventually ceases (Boot and Mensink, 1990; Durieux *et al.*, 1994; Marschner, 1995). With high soil nitrogen, a small proportion of the root system [3.5% in spring wheat (Robinson *et al.*, 1991) and 12% in lettuce (Burns, 1991)] can supply all the nitrogen required by a plant so that a small root system may be sufficient.

The optimization of root growth with soil nitrogen availability is consistent with the concept of a functional equilibrium between root and shoot (Brouwer, 1967; Brouwer and DeWit. 1969; Bloom *et al.*, 1985). This concept proposes that because of the distances between sources and sinks, roots

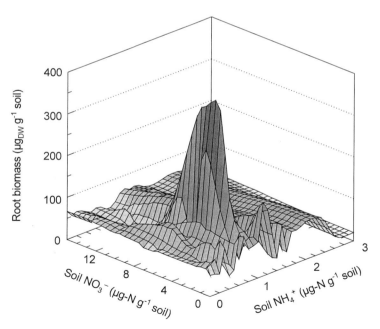

Figure 10   Root biomass (μg root dry weight g^{-1} soil) versus extractable soil NH_4^+ and NO_3^- (μg extractable N g^{-1} soil) for *Lycopersicon esculentum* L. Mill. cv T5 growing in an irrigated field that had been fallow the previous 2 years. (Bloom *et al.*, 1993).

more readily meet their requirements for nitrogen than for carbohydrates, whereas shoots more readily meet their requirements for carbohydrates than for nitrogen. Under low soil nitrogen, the root and shoot growth is nitrogen limited and carbohydrate supply in the shoot is relatively high so that carbohydrate translocation to roots is high. When a root encounters a soil patch rich in nitrogen, the nitrogen absorbed initially remains in that root and enhances growth. As the nitrogen levels in that root reach a surfeit, more nitrogen is translocated to a shoot. Vascular connections often determine that one part of the shoot receives a significant portion of its nutrients from a particular root and, in return, supplies carbohydrates first to that root (Watson and Casper, 1984; Wardlaw, 1990). This part of a shoot, now sufficient in nitrogen, becomes carbohydrate limited and translocates relatively little carbohydrate to the root; growth of this root then becomes carbohydrate limited. Such a response would explain the hyperbolic response of root production to soil nitrogen availability (Bloom *et al.*, 1993; Durieux *et al.*, 1994; Marschner, 1995).

C. Localization of Absorption and Its Immediate Effects

Root tissue diminishes in activity as it matures (Marschner, 1995). As a result, nutrient absorption per unit root surface area is greatest in the immature tissues near the apex: In particular, root NH_4^+ absorption reaches maximum rates within 1 mm of the apex (Figs. 11 and 12), whereas absorp-

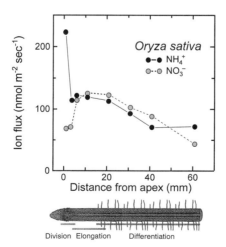

Figure 11   Absorption of NH_4^+ and NO_3^- at various distances from the apex for nodal roots in rice. The measurements localize fluxes along the root axis to within 0.8 mm of the designated distance with 95% certainty. The diagram of the root approximates the morphology at various locations. (T. Colmer and A. Bloom, unpublished data).

162   *Arnold J. Bloom*

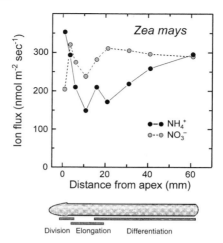

Figure 12   Absorption of NH_4^+ and NO_3^- at various distances from the apex of a maize seminal root. The measurements localize fluxes along the root axis to within 0.8 mm of the designated distance with 95% certainty. The diagram of the root approximates the morphology at various locations. (T. Colmer and A. Bloom, unpublished data).

tion of NO_3^- (Figs. 11 and 12; Henriksen *et al.*, 1992; Lazof *et al.*, 1992) and of other ions such as K^+ (Russell, 1977; Silk *et al.*, 1986) reach a maximum in zones located 3–20 mm from the apex. Presumably, this rapid nutrient absorption serves the high demands of cell division and elongation that occur in these regions. Root hairs develop and xylem matures in regions more basal (Esau, 1960). Because the root hairs present a large surface area and the mature xylem provide better vascular connections, most of NO_3^- translocated to the shoot is absorbed in these more basal root zones (Lazof *et al.*, 1992).

Rapid NH_4^+ or NO_3^- absorption in the zone of root elongation directly influences root growth and architecture. Root NH_4^+ absorption stimulates proton efflux, whereas NO_3^- absorption involves cotransport with protons (*cf.* Sections II,B and III,B; Allen, 1988); thus, these processes produce opposing pH shifts around the cell walls of elongating tissues. Protons released during NH_4^+ absorption may displace Ca^{2+} from the cell wall increasing its extensibility (Taiz and Zeiger, 1991) or stimulate wall loosening enzymes (Rayle and Cleland, 1992); in either case, root growth is stimulated in a phenomenon known as the acid growth of roots (Tanimoto *et al.*, 1989). Consequently, at physiological levels of rhizosphere nitrogen, root growth is more extensive under NH_4^+ nutrition than under NO_3^- nutrition (Table I; Cox and Reisenauer, 1973; Rufty *et al.*, 1983; Peet *et al.*, 1985; Allen and Smith, 1986; Blacquière *et al.*, 1987; MacDuff *et al.*, 1987; Anderson *et al.*, 1991; Miller and Zhen, 1991; Bloom *et al.*, 1993).

Differences in the energy requirements of NH_4^+ and NO_3^- acquisition may also influence growth under the two forms. In barley plants under NH_4^+ nutrition, 14% of carbon catabolism was coupled to NH_4^+ acquisition, whereas under NO_3^- nutrition, 5% of root carbon catabolism was coupled to NO_3^- absorption, 15% to NO_3^- assimilation, and 3% to NH_4^+ assimilation (Bloom *et al.*, 1992). Roots under NO_3^- nutrition may be more carbohydrate limited and, thus, may grow more slowly (Bloom *et al.*, 1993).

The differential response of roots to rhizosphere NH_4^+ and NO_3^- is reasonable in a ecological context. Ammonium binds strongly to cation-exchange sites on soil particles; the more extensive, finer roots that develop in the presence of NH_4^+ are well suited to the task of increasing plant access to this bound NH_4^+. Nitrate, by contrast, moves relatively freely through the soil; the thicker, perhaps less fragile, roots that develop in the presence of NO_3^- are probably adequate to access this mobile anion.

D. Research Opportunities

Despite the importance of plant nitrogen acquisition to agricultural productivity, fundamental information about plant nitrogen relations is meager. Unknown are the mechanisms through which crops acquire nitrogen from the soil or the relative dependence of crops on various forms of soil nitrogen (*i.e.*, ammonium, nitrate, or amino acids). There are only few data on how crop root systems develop in response to soil nitrogen availability or on what environmental signals roots use to be in the right place at the right time. Genetic variation in nitrogen harvest index—a measure of the ability to scavenge nitrogen from the soil and divert it into the harvested portions of the crop—remains virtually unexplored. Future efforts to manage our valuable nitrogen resources efficiently will require such information.

Table I   The Plant Parameters (Mean ± SE, *n* = 4) of Shoot and Root Biomass (mg_{DW}), Root Length (m), Root Branching (number m^{-1} root), and Root Area (cm^2) for *Lycopersicon esculentum* Mill. cv. T5 Grown in Solution Culture for 12 Days under Constant Levels of Nitrogen Nutrition and pH[a]

Treatment	Shoot biomass (mg_{DW})	Root biomass (mg_{DW})	Root length (m)	Root branching (roots m^{-1})	Root area (cm^2)
100 μM NH_4^+	34.8 ± 1.5	13.2 ± 0.6	4.27 ± 0.19	34.9 ± 1.6	17.3 ± 0.8
200 μM NO_3^-	35.4 ± 3.0	9.1 ± 0.8	3.00 ± 0.23	28.1 ± 0.8	12.9 ± 1.5

[a] Shoot and root biomass data are from Smart and Bloom (1992); the other data are from Bloom *et al.* (1992a).

References

Allen, S. (1988). Intracellular pH regulation in plants. ISI Atlas of Science. *Anim. Plant Sci.* **1**, 283–288.

Allen, S., and Smith, J. A. C. (1986). Ammonium nutrition in *Ricinus communis*: Its effect on plant growth and the chemical composition of the whole plant, xylem and ploem saps. *J. Exp. Bot.* **37**, 1599–1610.

Amancio, S., and Santos, H. (1992). Nitrate and ammonium assimilation by roots of maize (*Zea mays* l) seedlings as investigated by in vivo N-15 NMR. *J. Exp. Bot.* **43**, 633–639.

Anderson, D. S., Teyker, R. H., and Rayburn, A. L. (1991). Nitrogen form effects on early corn root morphological and anatomical development. *J. Plant Nutr.* **14**, 1255–1266.

Anderson, M. P., Vance, C. P., Heichel, G. H., and Miller, S. S. (1989). Purification and characterization of NADH-glutamate synthase form alfalfa root nodules. *Plant Physiol.* **90**, 351–358.

Andrews, M. (1986). The partioning of nitrate assimilation between roots and shoot of higher plants. *Plant Cell Environ.* **9**, 511–519.

Appenroth, K.-J., Oelmuller, R., Schuster, C., and Mohr, H. (1992). Regulation of transcript level and synthesis of nitrate reductase by phytochrome and nitrate in turions of *Spirodela polyrhiz* (L.) Schleiden. *Planta* **188**, 587–593.

Arnozis, P. A., and Findenegg, G. R. (1986). Electrical charge balance in the xylem sap of beet and sorghum plants grown with either NO_3^- or NH_4^+ nitrogen. *J. Plant Physiol.* **125**, 441–449.

Aslam, M., and Huffaker, R. C. (1982). In vivo nitrate reduction in roots and shoots of barley (*Hordeum vulgare* l.) seedlings in light and darkness. *Plant Physiol.* **70**, 1009–1013.

Aslam, M., and Huffaker, R. C. (1989). Role of nitrate and nitrite in the induction of nitrite reductase in leaves of barley seedlings. *Plant Physiol.* **91**, 1152–1156.

Aslam, M., Oaks, A., and Huffaker, R. C. (1976). Effect of light and glucose on the induction of nitrate reductase and on the distribution of nitrate in etiolated barley leaves. *Plant Physiol.* **58**, 588–591.

Aslam, M., Huffaker, R. C., Rains, D. W., and Rao, K. P. (1979). Influence of light and ambient carbon dioxide concentration on nitrate assimilation by intact barley seedlings. *Plant Physiol.* **63**, 1205–1209.

Aslam, M., Travis, R. L., and Huffaker, R. C. (1992). Comparative kinetics and reciprocal inhibition of nitrate and nitrite uptake in roots of uninduced and induced barley (*Hordeum vulgare* L.) seedlings. *Plant Physiol.* **99**, 1124–1133.

Ayling, S. M. (1993). The effect of ammonium ions on membrane potential and anion flux in roots of barley and tomato. *Plant Cell Environ.* **16**, 297–303.

Baysdorfer, C., and Robinson, M. J. (1985). Metabolic interactions between spinach leaf nitrite reductase and ferredoxin-NADP reductase. *Plant Physiol.* **77**, 318–320.

Becker, T. W., Foyer, C., and Caboche, M. (1992). Light-regulated expression of the nitrate-reductase and nitrite-reductase genes in tomato and in the phytochrome-deficient aurea mutant of tomato. *Planta* **188**, 39–47.

Bertl, A., Felle, H., and Bentrup, F. W. (1984). Amine transport in *Riccia fluitans*: Cytoplasmic and vacuolar pH recorded by a pH-sensitive microelectrode. *Plant Physiol.* **76**, 75–78.

Blacquière, T., Hofstra, R., and Stulen, I. (1987). Ammonium and nitrate nutrition in *Plantago lanceolata* and *Plantago major* L. ssp. major. I. Aspects of growth, chemical composition and root respiration. *Plant Soil* **104**, 129–141.

Bloom, A. J. (1985). Wild and cultivated barleys show similar affinities for mineral nitrogen. *Oecologia* **65**, 555–557.

Bloom, A. J. (1988). Ammonium and nitrate as nitrogen sources for plant growth. ISI Atlas of Science. *Anim. Plant Sci.* **1**, 55–59.

Bloom, A. J. (1994). Crop acquisition of ammonium and nitrate. *In* "Physiology and Determination of Crop Yield" (K. J. Boote, J. M. Bennett, T. R. Sinclair, and G. M. Paulsen, eds.), pp. 303–309. ASA, CSA, SSSA, Madison, WI.

Bloom, A. J., and Caldwell, R. M. (1988). Root excision decreases nutrient absorption and gas fluxes. *Plant Physiol.* **87,** 794–796.

Bloom, A. J., and Finazzo, J. (1986). The influence of ammonium and chloride on potassium and nitrate absorption by barley roots depends on time of exposure and cultivar. *Plant Physiol.* **81,** 67–69.

Bloom, A. J., and Sukrapanna, S. (1990). Effects of exposure to ammonium and transplant shock upon the induction of nitrate absorption. *Plant Physiol.* **94,** 85–90.

Bloom, A. J., Chapin, F. S., III, and Mooney, H. A. (1985). Resource limitation in plants—An economic analogy. *Annu. Rev. Ecol. Syst.* **16,** 363–392.

Bloom, A. J., Caldwell, R. M., Finazzo, J., Warner, R. L., and Weissbart, J. (1989). Oxygen and carbon dioxide fluxes from barley shoots depend on nitrate assimilation. *Plant Physiol.* **91,** 352–356.

Bloom, A. J., Sukrapanna, S. S., and Warner, R. L. (1992). Root respiration associated with ammonium and nitrate absorption and assimilation by barley. *Plant Physiol.* **99,** 1294–1301.

Bloom, A. J., Jackson, L. E., and Smart, D. R. (1993). Root growth as a function of ammonium and nitrate in the root zone. *Plant Cell Environ.* **16,** 199–206.

Boot, R. G. A., and Mensink, M. (1990). Size and morphology of root systems of perennial grasses from contrasting habitats as affected by nitrogen supply. *Plant Soil* **129,** 291–299.

Bowman, D. C., Paul, J. L., Davis, W. B., and Nelson, S. H. (1989). Rapid depletion of nitrgen applied to Kentucky bluegrass turf. *J. Am. Soc. Hort. Sci.* **114,** 229–233.

Bowsher, C. G., Hucklesby, D. P., and Emes, M. J. (1993). Induction of ferredoxin-NADP$^+$ oxidoreductase and ferredoxin synthesis in pea root plastids during nitrate assimilation. *Plant J.* **3,** 463–467.

Brouwer, R. (1967). Beziehungen zwischen Spross-und Wurzelwachstum. *Angew. Bot.* **41,** 244–250.

Brouwer, R., and DeWit, C. T. (1969). A simulation model of plant growth with special attention to root growth and its consequences. *In* "Root Growth" (W. J. Whittington, ed.), Proceedings of the 15th Easter School in Agricultural Science, University of Nottingham, pp. 224–244. Butterworths, London.

Brown, L. R., Chandler, W. U., Flavin, C., Jacobson, J., Pollock, C., Postel, S., Starke, L., and Wolf, E. C. (1987). "State of the World, A Worldwatch Institute Report on Progress Toward a Sustainable Society." Norton, New York.

Burns, I. G. (1991). Short- and long-term effects of a change in the spatial distribution of nitrate in the root zone on N uptake, growth and root development of young lettuce plants. *Plant Cell Environ.* **14,** 21–33.

Cameron, K. C., and Haynes, R. J. (1986). Retention and movement of nitrogen in soils. *In* "Mineral Nitrogen in the Plant–Soil System" (R. J. Haynes, ed.), pp. 166–241. Academic Press, Orlando, FL.

Chapin, F. S., Schulze, E. D., and Mooney, H. A. (1990). The ecology and economics of storage in plants. *Annu. Rev. Ecol. Syst.* **21,** 423–447.

Chapin, F. S., Moilanen, L., and Kielland, K. (1993). Preferential use of organic nitrogen for growth by a non-mycorrhizal arctic sedge. *Nature* **361,** 150–153.

Cheng, C.-L., Acedo, G. N., Cristinsin, M., and Conkling, M. A. (1992). Sucrose mimics the light induction of *Arabidopsis* nitrate reductase gene transcription. *Proc. Natl. Acad. Sci. USA* **89,** 1861–1864.

Clement, C. R., Hopper, M. J., Jones, L. H. P., and Leafe, E. L. (1978). The uptake of nitrate by *Lolium perenne* from flowing nutrient solution. II. Effect of light, defoliation, and relationship to CO_2 flux. *J. Exp. Bot.* **29,** 1173–1183.

Corzo, A., and Niell, F. X. (1992). Inorganic nitrogen metabolism in *Ulva rigida* illuminated with blue light. *Marine Biol.* **112**, 223–228.

Cox. W. J., and Reisenauer, H. M. (1973). Growth and ion uptake by wheat supplied nitrogen as nitrate or ammonium or both. *Plant Soil* **38**, 363–380.

Cramer, M. D., Lewis, O. A. M., and Lips, S. H. (1993). Inorganic carbon fixation and metabolism in maize roots as affected by nitrate and ammonium nutrition. *Physiol. Plant.* **89**, 632–639.

Crowley, P. H. (1975). Natural selection and the Michaelis constant. *J. Theor. Biol.* **50**, 461–475.

Deane-Drummond, C. E. (1986). Some regulatory aspects of ^{14}C methylamine influx into *Pisum sativum* L. cv. Feltham First seedlings. *Planta* **169**, 8–15.

Deane-Drummond, C. E., and Glass, A. D. M. (1983). Short term studies of nitrate uptake into barley plants using ion-specific electrodes and $^{36}ClO_3^-$—I. Control of net uptake by NO_3^- efflux. *Plant Physiol.* **73**, 100–104.

Delphon, P., Gojon, A., Tillard, P., and Passama, L. (1995). Diurnal regulation of NO_3^- uptake in soybean plants. 1. Changes in NO_3^- influx, efflux, and N utilization in the plant during the day night cycle. *J. Exp. Bot.* **46**, 1585–1594.

Dortch, Q. (1990). The interaction between ammonium and nitrate uptake in phytoplankton. *Marine Ecol.-Pr.* **61**, 183–201.

Drew, M. C. (1975). Comparison of the effects of a localized supply of phosphate, nitrate, ammonium and potassium on the growth of the seminal root system, and the shoot, in barley. *New Phytol.* **75**, 479–490.

Duke, S. H., and Duke, S. O. (1984). Light control of extractable nitrate reductase activity in higher plants. *Physiol. Plant.* **62**, 485–493.

Durieux, R. P., Kamprath, E. J., Jackson, W. A., and Moll, R. H. (1994). Root distribution of corn—The effect of nitrogen fertilization. *Agron. J.* **86**, 958–962.

Emes, M. J., and Bowsher, C. G. (1991). Integration and compartmentation of carbon and nitrogen metabolism in roots. *In* "Compartmentation of Plant Metabolism in Nonphotosynthetic Tissues. Soc. for Exp. Biol. Sem. Ser. 42" (Emes, M. J., ed.), pp. 147–165. Cambridge Univ. Press, Cambridge, UK.

Epstein, E. (1972). "Mineral Nutrition of Plants: Principles and Perspectives." Wiley, NY.

Esau, K. (1960). "Anatomy of Seed Plants." Wiley, New York.

Feng, J. N., Volk, R. J., and Jackson, W. A. (1994). Inward and outward transport of ammonium in roots of maize and sorghum—contrasting effects of methionine sulpoximine. *J. Exp. Bot.* **45**, 429–439.

Fentem, P. A., Lea, P. J., and Stewart, G. R. (1983). Ammonia assimilation in the roots of nitrate- and ammonia-grown *Hordeum vulgare* L. (cv Golden Promise). *Plant Physiol.* **71**, 496–501.

Fitter, A. H., and Stickland, T. R. (1991). Architectural analysis of plant root systems: 2. Influence of nutrient supply on architecture in contrasting plant species. *New Phytol.* **118**, 383–389.

Fitter, A. H., Nichols, R., and Harvey, M. L. (1988). Root system architecture in relation to life history and nutrient supply. *Funct. Ecol.* **2**, 345–351.

Ganmore-Neumann, R., and Kafkafi, U. (1980). Root temperature and percentage NO_3^-/NH_4^+ effect on tomato development. II Nutrient composition of tomato plants. *Agron. J.* **72**, 762–766.

Glass, A. D. M. (1988). Nitrogen uptake by plant roots. ISI Atlas of Science. *Anim. Plant Sci.* **1**, 151–156.

Glass, A. D. M., and Siddiqi, M. Y. (1985). Nitrate inhibition of chloride influx in barley: Implications for a proposed chloride homeostat. *J. Exp. Bot.* **36**, 556–566.

Glass, A. D. M., and Siddiqi, M. Y. (1995). Nitrogen absorption by plant roots. *In* "Nitrogen Nutrition in Higher Plants" (H. S. Srivastava and R. P. Singh, eds.), pp. 21–56. Associated, New Delhi, India.

Glass, A. D. M., Siddiqi, M. Y., Ruth, T. J., and Rufty, T. W. (1990). Studies of the uptake of nitrate in barley. 2. Energetics. *Plant Physiol.* **93,** 1585–1589.

Glass, A. D. M., Shaff, J. E., and Kochian, L. V. (1992). Studies of the uptake of nitrate in barley. IV. Electrophysiology. *Plant Physiol.* **99,** 456–463.

Goyal, S. S., and Huffaker, R. C. (1984). Nitrogen toxicity in plants. *In* "Nitrogen in Crop Production" R. D. Hauck, ed.), pp. 97–118. ASA, CSSA, SSSA, Madison, WI.

Goyal, S. S., and Huffaker, R. C. (1986). The uptake of NO_3^-, NO_2^-, and NH_4^+ by intact wheat (*Triticum aestivum*) seedlings: I. Induction and kinetics of transport systems. *Plant Physiol.* **82,** 1051–1056.

Goyal, S. S., Huffaker, R. C., and Lorenz, O. A. (1982). Inhibitory effects of ammoniacal nitrogen on growth of radish plants. II. Investigation on the possible causes of ammonium toxicity to radish plants and reversal by nitrate. *J. Am. Soc. Hort. Sci.* **107,** 130–135.

Granato, T. C., and Raper, C. D., Jr. (1989). Proliferation of maize (*Zea mays* L.) roots in response to localized supply of nitrate. *J. Exp. Bot.* **211,** 263–275.

Greidanus, T., Peterson, L. A., Schrader, L. E., and Dana, M. N. (1972). Essentiality of ammonium for cranberry nutrition. *J. Am. Soc. Hort. Sci.* **97,** 272–277.

Grime, J. P., Crick, J. C., and Rincon, J. E. (1986). The ecological significance of plasticity. *In* "Plasticity in Plants" (D. H. Jennings and A. J. Trewavas, eds.), pp. 5–29. Company of Biologists Limited, Cambridge.

Hackett, C. (1972). A method of applying nutrients locally to roots under controlled conditions, and some morphological effects of locally applied nitrate on the branching of wheat roots. *Aust. J. Biol. Sci.* **25,** 1169–1180.

Hansen, G. K. (1980). Diurnal variation of root respiration rates and nitrate uptake as influenced by nitrogen supply. *Physiol. Plant.* **48,** 421–427.

Hanson, A. D., and Hitz, W. D. (1983). Water deficits and the nitrogen economy. *In* "Limitations to Efficient Water Use in Crop Production" (H. M. Taylor, W. R. Jordan, and T. R. Sinclair, eds.), pp. 331–343. Am. Soc. Agron., Madison, WI.

Harold, F. M. (1986). "The Vital Force: A Study of Bioenergetics." Freeman, New York.

Hasenstein, K. H., and Evans, M. L. (1988). The influence of calcium and pH on growth in primary roots of *Zea mays. Physiol. Plant.* **72,** 466–470.

Haynes, R. J. (1986). Uptake and assimilation of mineral nitrogen by plants. *In* "Mineral Nitrogen in the Plant–Soil System" (R. J. Haynes, ed.), pp. 303–378. Academic Press, Orlando, FL.

Haynes, R. J., and Sherlock, R. R. (1986). Gaseous losses of nitrogen. *In* "Mineral Nitrogen in the Plant–Soil System" (R. J. Haynes, ed.), pp. 242–302. Academic Press, Orlando, FL.

Hedrich, R., and Schroeder, J. I. (1989). The physiology of ion channels and electrogenic pumps in higher plants. *Annu. Rev. Plant Physiol.* **40,** 539–569.

Henderson, P. J. F. (1971). *Annu. Rev. Microbiol.* **25,** 393–428.

Henriksen, G. H., Raman, D. R., Walker, L. P., and Spanswick, R. M. (1992). Measurement of net fluxes of ammonium and nitrate at the surface of barley roots using ion-selective microelectrodes. II. Patterns of uptake along the root axis and evaluation of the microelectrode flux estimation technique. *Plant Physiol.* **99,** 734–747.

Higinbotham, N., Etherton, B., and Foster, R. J. (1964). Effect of external K, NH4, Na, Ca, Mg, and H ions on the cell transmembrane electropotential of Avena coleoptile. *Plant Physiol.* **39,** 196–203.

Hoff, T., Truong, H. N., and Caboche, M. (1994). The use of mutants and transgenic plants to study nitrate assimilation. *Plant Cell Environ.* **17,** 489–506.

Huber, J. L., Huber, S. C., Campbell, W. H., and Redinbaugh, M. G. (1992). Reversible light dark modulation of spinach leaf nitrate reductase activity involves protein phosphorylation. *Arch. Biochem. Biophys.* **296,** 58–65.

Huppe, H. C., and Turpin, D. H. (1994). Integration of carbon and nitrogen metabolism in plant and algal cells. *Annu. Rev. Plant Physiol.* **45,** 577–607.

Imsande, J., and Touraine, B. (1994). N demand and the regulation of nitrate uptake. *Plant Physiol.* **105**, 3–7.

Ingestad, T., and Ågren, G. I. (1988). Nutrient uptake and allocation at steady-state nutrition. *Physiol. Plant.* **72**, 450–459.

Jackson, L. E., and Bloom, A. J. (1990). Root distribution in relation to soil nitrogen availability in field-grown tomatoes. *Plant Soil* **128**, 115–126.

Jackson, R. B., and Caldwell, M. M. (1989). The timing and degree of root proliferation in fertile-soil microsites for 3 cold-desert perennials. *Oecologia* **81**, 149–153.

Jackson, W. A. (1978). Nitrate acquisition and assimilation by higher plants: Processes in the root system. *In* "Nitrogen in the Environment" (D. R. Nielsen and J. G. MacDonald, eds.), Vol. 2, pp. 45–88. Academic Press, New York.

Kaiser, W. M., and Spill, D. (1991). Rapid modulation of spinach leaf nitrate reductase activity by photosynthesis. II. In vitro modulation by ATP and AMP. *Plant Physiol.* **96**, 368–375.

Kamiya, A. (1989). Effects of blue light and ammonia on nitrogen metabolism in a colorless mutant of *Chlorella*. *Plant Cell Physiol.* **30**, 513–521.

Keltjens, W. G., and Nijenstein, J. H. (1987). Diurnal variations in uptake, transport and assimilation of NO_3^- and efflux of OH^- in maize plants. *J. Plant Nutr.* **10**, 887–900.

King, B. J., Siddiqi, M. Y., and Glass, A. D. M. (1992). Studies of the uptake of nitrate in barley. V. Estimation of root cytoplasmic nitrate concentration using nitrate reductase activity—Implications for nitrate influx. *Plant Physiol.* **99**, 1582–1589.

Kirkby, E. A., and Mengel, K. (1967). Ionic balance in different tissues of the tomato plant in relation to nitrate, urea, or ammonium nutrition. *Plant Physiol.* **42**, 6–14.

Kleiner, D. (1981). The transport of NH_3 and NH_4^+ across biological membranes. *Biochim. Biophys. Acta* **639**, 41–52.

Kleiner, D. (1985). Bacterial ammonium transport. *FEMS Microbiol. Rev.* **32**, 87–100.

Knight, T. J., Durbin, R. D., and Langston-Unkefer, P. J. (1986). Effects of tabtoxinine-b-lactam on nitrogen metabolism in Avena sativa L. roots. *Plant Physiol.* **82**, 1045–1050.

Koch, G., and Bloom, A. J. (1989). Root respiration and NH_4^+ and NO_3^- absorption in relation to root hypoxia in the wild variety of tomato, Lycopersicon esculentum var. cerasiforme. *Plant Physiol.* **89**. [Abstract 717]

Koch, G. W., Bloom, A. J., and Chapin, F. S., III (1991). Ammonium and nitrate as nitrogen sources in two Eriophorum species. *Oecologia* **88**, 570–573.

Korcak, R. F. (1989). Variation in nutrient requirements of blueberries and other calcifuges. *Hort. Sci.* **24**, 573–578.

Kosola, K. R., and Bloom, A. J. (1994). Methylammonium as a transport analog for ammonium in tomato (*Lycopersicon esculentum* Mill.). *Plant Physiol.* **104**, 435–442.

Kosola, K. R., and Bloom, A. J. (1996). Chlorate as a transport analog for nitrate absorption by roots of tomato (*Lycopersicon esculentum*). *Plant Physiol.*, **110**, 1293–1299.

Kronzucker, H. J., Siddiqi, M. Y., and Glass, A. D. M. (1995). Analysis of (NH_4^+)-N^{13} efflux in spruce roots—A test case for phase identification in compartmental analysis. *Plant Physiol.* **109**, 481–490.

Kurkdjian, A., and Guern, J. (1989). Intracellular pH: Measurement and importance in cell activity. *Annu. Rev. Plant Physiol.* **40**, 271–303.

Lam, H. M., Coschigano, K. T., Oliveira, I. C., Melooliveira, R., and Coruzzi, G. M. (1996). The molecular-genetics of nitrogen assimilation into amino acids in higher plants. *Annu. Rev. Plant Physiol. Plant Mol. Biol.* **47**, 569–593.

Lazof, D. B., Rufty, T. W., Jr., and Redinbaugh, M. G. (1992). Localization of nitrate absorption and translocation within morphological regions of the corn root. *Plant Physiol.* **100**, 1251–1258.

Lee, R. B., and Clarkson, D. T. (1986). Nitrogen-13 studies on nitrate fluxes in barley roots. I. Compartmental analysis from measurements of ^{13}N efflux. *J. Exp. Bot.* **37**, 1753–1767.

Lee, R. B., and Drew, M. C. (1986). Nitrogen-13 studies of nitrate fluxes in barley roots. II. Effect of plant N-status on the kinetic parameters of nitrate influx. *J. Exp. Bot.* **37,** 1768–1779.

Lee, R. B., and Ratcliffe, R. G. (1991). Observations on the subcellular distribution of the ammonium ion in maize root tissue using *in vivo* N^{14}-nuclear-magnetic-resonance spectroscopy. *Planta* **183,** 359–367.

Lee, R. B., and Rudge, K. A. (1986). Effects of nitrogen deficiency on the absorption of nitrate and ammonium by barley plants. *Ann. Bot.* **57,** 471–486.

Lee, R. B., Purves, J. V., Ratcliffe, R. G., and Saker, L. R. (1992). Nitrogen assimilation and the control of ammonium and nitrate absorption by maize roots. *J. Exp. Bot.* **43,** 1385–1396.

Lewis, O. A. M. (1986). "Plants and Nitrogen," pp. 104. Arnold, London.

Loomis, R. S., and Conner, D. J. (1992). "Crop Ecology: Productivity and Management in Agricultural Systems." Cambridge Univ. Press, Cambridge, UK.

López-Figueroa, F., and Rüdiger, W. (1991). Stimulation of nitrate net uptake and reduction by red and blue light and reversion by far-red light in the green alga Ulva rigida. *J. Phycol.* **27,** 389–394.

Lorenz, O. A. (1978). Potential nitrate levels in edible plant parts. *In* "Nitrogen in the Environment" (D. R. Nielsen and J. G. MacDonald, eds.), Vol. 2, pp. 201–219. Academic Press, San Diego.

Lycklama, J. C. (1963). The absorption of ammonium and nitrate by perennial rye-grass. *Acta Bot. Neerl.* **12,** 361–423.

Lynch, J. (1995). Root architecture and plant productivity. *Plant Physiol.* **109,** 7–13.

MacDuff, J. H., Hopper, M. J., and Wild, A. (1987). The effect of root temperature on growth and uptake of ammonium and nitrate by Brassica napus L. in flowing solution culture. I. Growth. *J. Exp. Bot.* **38,** 42–52.

MacKown, C. T., Jackson, W. A., and Volk, R. J. (1983). Partitioning of previously-accumulated nitrate to translocation, reduction, and efflux in corn roots. *Planta* **157,** 8–14.

Marcus-Wyner, L. (1983). Influence of ambient acidity on the absorption of NO_3^- and NH_4^+ by tomato plants. *J. Plant Nutr.* **6,** 657–666.

Marschner, H. (1995). "Mineral Nutrition of Higher Plants, 2nd Ed." Academic Press, London.

McClure, P. R., Kochian, L. V., Spanswick, R. M., and Shaff, J. E. (1990). Evidence for cotransport of nitrate and protons in maize roots. I. Effects of nitrate on the membrane potential. *Plant Physiol.* **93,** 281–289.

Meyerhoff, P. A., Fox, T. C., Travis, R. L., and Huffaker, R. C. (1994). Characterization of the association of nitrate reductase with barley (*Hordeum vulgare* L.) root membranes. *Plant Physiol.* **104,** 925–936.

Miller, A. J., and Zhen, R. G. (1991). Measurement of intracellular nitrate concentrations in *Chara* using nitrate-selective microelectrodes. *Planta* **184,** 47–52.

Minotti, P. L., Williams, D. C., and Jackson, W. A. (1968). Nitrate uptake and reduction as affected by calcium and potassium. *Soil Sci. Soc. Am. Proc.* **32,** 692–698.

Morgan, M. A., and Jackson, W. A. (1988a). Inward and outward movement of ammonium in root systems: Transient responses during recovery from nitrogen deprivation in presence of ammonium. *J. Exp. Bot.* **39,** 179–191.

Morgan, M. A., and Jackson, W. A. (1988b). Suppression of ammonium uptake by nitrogen supply and its relief during nitrogen limitation. *Physiol. Plant.* **73,** 38–45.

Morgan, M. A., and Jackson, W. A. (1989). Reciprocal ammonium transport into and out of plant roots: Modifications by plant nitrogen status and elevated root ammonium concentration. *J. Exp. Bot.* **40,** 207–214.

Munn, D. A., and Jackson, W. A. (1978). Nitrate and ammonium uptake by rooted cuttings of sweet potato. *Agron. J.* **70,** 312–316.

National Research Council (1989). "Alternative Agriculture." National Academy Press, Washington, DC.

170   *Arnold J. Bloom*

Nicoulaud, B. A. L., and Bloom, A. J. (1994). Foliar applied urea as the sole nitrogen source on tomato plants to study ammonium acquisition by the roots. *Plant Physiol.* **105.** [Abstract 136]

Novoa, R., and Loomis, R. S. (1981). Nitrogen and plant production. *Plant Soil* **58,** 177–204.

Nye, P. H., and Tinker, P. B. (1977). "Solute Movement in the Soil–Root System." Univ. of California Press, Berkeley.

Nylund, J. E. (1990). Nitrogen, carbohydrate and ectomycorrhiza—The classical theories crumble. *Agric. Ecol. Environ.* **28,** 361–364.

Oaks, A. (1994). Efficiency of nitrogen utilization in C_3 and C_4 cereals. *Plant Physiol.* **106,** 1–7.

Oaks, A., and Hirel, B. (1985). Nitrogen metabolism in roots. *Annu. Rev. Plant Physiol.* **36,** 345–365.

Oaks, A., and Long, D. M. (1992). NO_3^- assimilation in root systems: With special reference to *Zea mays* (cv. W64A X W128E). *In* "Nitrogen Metabolism of Plants" (K. Mengel and D. J. Pilbeam, eds.), Proc. Phytochem. Soc. Europe, Vol. 3, pp. 91–102. Clarendon, Oxford.

Pearson, C. J., Volk, R. J., and Jackson, W. A. (1981). Daily changes in nitrate influx, efflux and metabolism in maize and pearl millet. *Planta* **152,** 319–324.

Peet, M. M., Raper, D. C., Tolley, L. C., and Robarge, W. P. (1985). Tomato responses to ammonium and nitrate nutrition under controlled root-zone pH. *J. Plant Nutr.* **8,** 787–798.

Peoples, M. B., and Craswell, E. T. (1992). Biological nitrogen fixation: Investments, expectations and actual contributions to agriculture. *Plant Soil* **141,** 13–39.

Pitts, R. F. (1972). Control of renal production of ammonia. *Kidney Int.* **1,** 297–305.

Raven, J. A., and Smith, F. A. (1976). Nitrogen assimilation and transport in vascular land plants in relation to intracellular pH regulation. *New Phytol.* **76,** 415–431.

Rayle, D. L., and Cleland, R. E. (1992). The acid growth theory of auxin-induced cell elongation is alive and well. *Plant Physiol.* **99,** 1271–1274.

Reisenauer, H. M. (1978). Absorption and utilization of ammonium nitrogen by plants. *In* "Nitrogen in the Environment" (D. R. Nielsen and J. MacDonald, eds.), Vol. 2, pp. 157–170. Academic Press, New York.

Roberts, J. K. M., and Pang, M. K. L. (1992). Estimation of ammonium ion distribution between cytoplasm and vacuole using nuclear magnetic resonance spectroscopy. *Plant Physiol.* **100,** 1571–1574.

Robinson, D. (1986). Compensatory changes in the partitioning of dry matter in relation to nitrogen uptake and optimal variations in growth. *Ann. Bot.* **58,** 841–848.

Robinson, D., Linehan, D. J., and Caul, S. (1991). What limits nitrate uptake from soil? *Plant Cell Environ.* **14,** 77–85.

Robinson, J. M. (1986). Carbon dioxide and nitrite photoassimilatory processes do not intercompete for reducing equivalents in spinach and soybean leaf chloroplasts. *Plant Physiol.* **80,** 676–684.

Robinson, J. M. (1988). Spinach leaf chloroplast CO_2 and NO_2^- photoassimilations do not compete for photogenerated reductant. Manipulation of reductant levels by quantum flux density titrations. *Plant Physiol.* **88,** 1373–1380.

Rufty, T. W., Jr., Jackson, W. A., and Raper, C. D., Jr. (1982). Inhibition of nitrate assimilation in roots in the presence of ammonium. The moderating influence of potassium. *J. Exp. Bot.* **33,** 1122–1137.

Rufty, T. W., Jr., Raper, C. D., Jr., and Jackson, W. A. (1983). Growth and nitrogen assimilation of soybeans in response to ammonium and nitrate nutrition. *Bot. Gaz.* **144,** 466–470.

Rufty, T. W., Raper, C. D., Jr., and Huber, S. C. (1984). Alterations in internal partitioning of carbon in soybean plants in response to nitrogen stress. *Can. J. Bot.* **62,** 501–508.

Russell, R. S. (1977). "Plant Root Systems. Their Function and Interaction with the Soil." McGraw-Hill, London.

Samuelson, M. E., Eliasson, L., and Larsson, C.-M. (1991). Nitrate-regulated growth and cytokinin responses in seminal roots of barley. *Plant Physiol.* **98,** 309–315.

Sasakawa, H., and Yamamoto, Y. (1978). Comparison of the uptake of nitrate and ammonium by rice seedlings. Influences of light, temperature, oxygen concentration, exogenous sucrose, and metabolic inhibitors. *Plant Physiol.* **62,** 665–669.

Scherer, H. W., MacKnown, C. T., and Leggett, J. E. (1984). Potassium–ammonium uptake interactions in tobacco seedlings. *J. Exp. Bot.* **35,** 1060–1070.

Schortemeyer, M., Feil, B., and Stamp, P. (1993). Root morphology and nitrogen uptake of maize simultaneously supplied with ammonium and nitrate in a split-root system. *Ann. Bot.* **72,** 107–115.

Schuster, C., Schmidt, S., and Mohr, H. (1989). Effect of nitrate, ammonium, light and a plastidic factor on the appearance of multiple forms of nitrate reductase in mustard (*Sinapis alba* L.) cotyledons. *Planta* **177,** 74–83.

Schwerdtfeger, W. (1976). "World Survey of Climatology. Vol. 12." Elsevier, New York.

Serrano, R. (1989). Structure and function of plasma membrane ATPase. *Annu. Rev. Plant Physiol.* **40,** 61–94.

Shiraishi, N., Sato, T., Ogura, N., and Nakagawa, H. (1992). Control by glutamine of the synthesis of nitrate reductase in cultured spinach cells. *Plant Cell Physiol.* **33,** 727–731.

Siddiqi, M. Y., Glass, A. D. M., Ruth, T. J., and Rufty, T. W. (1990). Studies of the uptake of nitrate in barley. 1. Kinetics of $^{13}NO_3^-$-influx. *Plant Physiol.* **93,** 1426–1432.

Siddiqi, M. Y., King, B. J., and Glass, A. D. M. (1992). Effects of nitrite, chlorate, and chlorite on nitrate uptake and nitrate reductase activity. *Plant Physiol.* **100,** 644–650.

Silk, W. H., Hsiao, T. C., Diedenhofen, U., and Matson, C. (1986). Spatial distribution of potassium, solutes and their deposition rates in the growth zone of the primary corn root. *Plant Physiol.* **82,** 853–858.

Smart, D. R., and Bloom, A. J. (1988). Kinetics of ammonium and nitrate uptake among wild and cultivated tomatoes. *Oecologia* **76,** 336–340.

Smart, D. R., and Bloom, A. J. (1991). Influence of root NH_4^+ and NO_3^- content on the temperature response of net NH_4^+ and NO_3^- uptake in chilling sensitive and chilling resistant *Lycopersicon* taxa. *J. Exp. Bot.* **42,** 331–338.

Smart, D. R., and Bloom, A. J. (1993). The relationship between kinetics of NH_4^+ and NO_3^- absorption and growth in the cultivated tomato (*Lycopersicon esculentum* Mill. cv. T5). *Plant Cell Environ.* **16,** 259–267.

Smith, F. A., and Walker, N. A. (1978). Entry of methylammonium and ammonium ions into *Chara* internodal cells. *J. Exp. Bot.* **29,** 107–120.

Soares, M. I. M., and Lewis, O. A. M. (1986). An investigation into nitrogen assimilation and distribution in fruiting plant of barley (*Hordeum vulgare* L. cv. Clipper) in response to nitrate, ammonium and mixed nitrate and ammonium nutrition. *New Phytol.* **104,** 385–393.

Suzuki, A., and Gadal, P. (1984). Glutamate synthase: Physiochemical and functional properties of different forms in higher plants and other organisms. *Physiol. Veg.* **22,** 471–486.

Suzuki, A., Gadal, P., and Oaks, A. (1981). Intracellular distribution of enzymes associated with nitrogen assimilation in roots. *Planta* **151,** 457–461.

Suzuki, A., Vidal, J., and Gadal, P. (1982). Glutamate synthase isoforms in rice. Immounological studies of enzymes in green leaf, etiolated leaf, and root tissues. *Plant Physiol.* **78,** 374–378.

Syrett, P. J., and Peplinska, A. M. (1988). Effects of nitrogen-deprivation, and recovery from it, on the metabolism of microalgae. *New Phytol.* **109,** 289–296.

Taiz, L., and Zeiger, E. (1991). "Plant Physiology." Benjamin/Cummings, Redwood City, CA.

Talouizte, A., Champigny, M. L., Bismuth, E., and Moyse, A. (1984). Root carbohydrate metabolism associated with nitrate assimilation in wheat previously deprived of nitrogen. *Physiol. Veg.* **22,** 19–27.

Tanimoto, E., Scott, T. K., and Masuda, Y. (1989). Inhibition of acid-enhanced elongation of *Zea mays* root segments by galactose. *Plant Physiol.* **90,** 440–444.

Tennessee Valley Authority (1991). "Commercial Fertilizers 1991." National Fertilizer and Environmental Research Center, Muscle Shoals, AL.

Teyker, R. H., Jackson, W. A., Volk, R. J., and Moll, R. H. (1988). Exogenous $^{15}NO_3^-$ influx and endogenous $^{14}NO_3^-$ efflux by two maize (*Zea mays* L.) inbreds during nitrogen deprivation. *Plant Physiol.* **86,** 778–781.

Thibaud, J. B., and Grignon, C. (1981). Mechanism of nitrate uptake in corn roots. *Plant Sci. Lett.* **22,** 279–289.

Tingey, S. V., Walker, E. L., and Coruzzi, G. M. (1987). Glutamine synthetase genes of pea encode distinct polypeptides which are differentially expressed in leaves, roots and nodules. *EMBO J.* **6,** 1–9.

Tolley-Henry, L., and Raper, C. D., Jr. (1986). Utilization of ammonium as a nitrogen source. Effects of ambient acidity on growth and nitrogen accumulation by soybean. *Plant Physiol.* **82,** 54–60.

Tyerman S. D. (1992). Anion channels in plants. *Annu. Rev. Plant Physiol. Plant Mol. Biol.* **43,** 351–373.

Tyerman, S. D., Findlay, G. P., and Paterson, G. J. (1986). Inward membrane current in Chara inflata. II. Effects of pH, Cl^- channel blockers and NH_4^+, and significance of the hyperpolarised state. *J. Membrane Biol.* **89,** 153–161.

Ullrich, W. R. (1987). Nitrate and ammonium uptake in green algae and higher plants: Mechanism and relationship with nitrate metabolism. *In* "Inorganic Nitrogen Metabolism" (W. R. Ullrich, P. J. Aparicio, P. J. Syrett, F. Castillo, eds.), pp. 32–38. Springer-Verlag, New York.

Ullrich, W. R. (1992). Transport of nitrate and ammonium through plant membranes. *In* "Nitrogen Metabolism of Plants" (K. Mengel and D. J. Pilbeam, eds.), Proc. Phytochem. Soc. Europe, Vol. 3, pp. 121–137. Clarendon, Oxford.

Ullrich, W. R., and Novacky, A. (1981). Nitrate-dependent membrane potential changes and their induction in Lemna gibba G1. *Plant Sci. Lett.* **22,** 211–217.

Ullrich, W. R., Larsson, M., Larsson, C.-M., Lesch, S., and Novacky, A. (1984). Ammonium uptake in Lemna gibba G 1, related membrane potential changes, and inhibition of anion uptake. *Physiol. Plant.* **61,** 369–376.

Vale, F. R., Volk, R. J., and Jackson, W. A. (1988). Simultaneous influx of ammonium and potassium into maize roots: Kinetics and interactions. *Planta* **173,** 424–431.

van Keulen, H., and van Heemst, H. D. J. (1982). Crop response to the supply of macronutrients. Agric. Res. Rep. No. 916. Centre for Agricultural Publishing and Documentation, Wageningen.

Vaughn, K. C., and Campbell, W. H. (1988). Immunogold localization of nitrate reductase in maize leaves. *Plant Physiol.* **88,** 1354–1357.

Vessey, J. K., Henry, L. T., Chaillou, S., and Raper, C. D. (1990). Root-zone acidity affects relative uptake of nitrate and ammonium from mixed nitrogen sources. *J. Plant Nutr.* **13,** 95–116.

Vezina, L.-P., and Langlois, J. R. (1989). Tissue and cellular distribution of glutamine synthetase in roots of pea (*Pisum sativum*) seedlings. *Plant Physiol.* **90,** 1129–1133.

Walker, D. J., Smith, S. J., and Miller, A. J. (1995). Simultaneous measurement of intracellular pH and K^+ or NO_3^- in barley root cells using triple-barreled, ion-selective microelectrodes. *Plant Physiol.* **108,** 743–751.

Wang, M. Y., Siddiqi, M. Y., Ruth, T. J., and Glass, A. D. M. (1993). Ammonium uptake by rice roots I. Fluxes and subcellular distribution of $^{13}NH_4^+$. *Plant Physiol.* **103,** 1249–1258.

Ward, M. R., Grimes, H. D., and Huffaker, R. C. (1989). Latent nitrate reductase activity is associated with the plasma membrane of corn roots. *Planta* **177,** 470–475.

Wardlaw, I. F. (1990). The control of carbon partitioning in plants. *New Phytol.* **116,** 341–381.

Watson, M. A., and Casper, B. B. (1984). Constraints on the expression of plant phenotypic plasticity. *Annu. Rev. Ecol. Syst.* **15,** 233–258.

Zhen, R. G., Koyro, H. W., Leigh, R. A., Tomos, A. D., and Miller, A. J. (1991). Compartmental nitrate concentrations in barley root cells measured with nitrate-selective microelectrodes and by single-cell sap sampling. *Planta* **185,** 356–361.

6

Trade-offs in Root Form and Function

David M. Eissenstat

I. Introduction

Agricultural research on roots has focused primarily on attributes of root systems and their mycorrhizal fungi that might be enhanced for acquisition of water and nutrients or tolerance to salinity, heavy metals, low soil water potentials, or specific root pathogens. Much less attention has been focused on the potential trade-offs between different modes of root deployment. For example, a large extensive root system of high root length density is desirable from a water or nutrient acquisition perspective but may come at a carbon cost not consistent with optimizing yield. Similarly, roots constructed to be well defended against herbivores and pathogens may be slow growing and have heavily suberized and lignified epidermal and cortical cell walls (Ryser, 1996); these traits are not consistent with roots designed to maximize water and nutrient uptake. The concepts of ecological trade-offs may aid in understanding root form and function as well as enhance opportunities for improved breeding and management of crops. In this chapter, I will review root deployment from a cost:benefit perspective, placing particular emphasis on work in citrus. I will cover how design of the optimal root system for such parameters as root:shoot ratio, root diameter, root maintenance respiration, root architecture, root hydraulic conductivity, root defense, root life span, and root symbiosis with mycorrhizal fungi must be considered in the context of relative resource availabilities. Cultivars adapted to drought, infertile soils, or defense from root-feeding organisms will likely have different root systems from those most suitable to highly intensive agriculture.

II. Citrus as a Model System

Annual crops are of great agricultural importance and their root systems, especially in cereals, have been studied extensively. Perennial dicots also have high agricultural importance but have received much less research attention. In reviewing trade-offs in roots systems, it is important to consider the unique ecological niche of annuals. For example, annual plants typically grow very fast from seed and finish their life cycle during the growing season. They are found most frequently in disturbed habitats and/or habitats with a long dry season. Perennial species, in contrast, often build root systems that can support a shoot many meters tall; survive periods of the year when minimal C is imported from the shoot; support shoot regrowth following canopy loss; tolerate wide fluctuations in soil temperature, soil water potential, oxygen availability, and herbivore/pathogen pressure; and provide the internal meristems necessary for the continuous production of new fine laterals needed to replace those that have died. Thus, although ecological trade-offs occur in annuals, many kinds of trade-offs in root form and function are most vividly expressed in perennial root systems.

Citrus is an example of a crop that exhibts many kinds of trade-offs in its root systems. Citrus is a subtropical evergreen that is believed to be an understory tree in the parts of Asia where it originated. In fruit crops like citrus, rootstocks can be specifically targeted for certain environmental conditions, such as tolerance to flooding, saline soil, or root feeding by nematodes or fungal pathogens. Trees may be grown without irrigation for sometimes greater than 30 years. Therefore, citrus growers must choose from a diverse group of rootstocks based on site conditions, scion variety, desired tree size, and other management objectives. These rootstocks have not been selected based on fruit characteristics. Indeed, many rootstocks produce an inedible fruit. Besides being adapted to a wide set of environmental conditions, an additional advantage of using rootstocks is that inferences about the role of the root system are strengthened because the roots of species of *Citrus* and related genera are compared on a common site with all the scions (shoots) genetically identical, thus providing an important internal control. Citrus rootstocks vary widely in root morphology, number and depth of structural roots, hydraulic conductivity, mycorrhizal dependency, and plant growth rates (Castle, 1987; Eissenstat, 1991, 1992; Graham and Eissenstat, 1994). Thus, they provide an excellent system to examine potential trade-offs in the structure and function of root systems.

III. Optimizing Carbon Expenditure on Roots

Physiological ecologists commonly have taken a cost: benefit or economic approach to understanding how plants cope with the availability of light, water, and nutrients (Bloom *et al.*, 1985; Chapin *et al.*, 1987). The view that plants are typically limited by several resources simultaneously is an important departure from the more traditional view of single resource limitation developed by the German chemist, Justus von Liebig, in the 19th century and that dominated the thinking of agricultural workers for a very long time (Tisdale and Nelson, 1975). Carbon has been widely used by physiological ecologists as a currency for estimating costs of allocation, tissue construction, and resource acquisition because C provides an indirect measure of energy gain and expenditure by the plant and because the rate of energy capture often limits plant growth (Chapin, 1989). Carbon limitation is most clearly demonstrated by the numerous incidences of enhanced whole plant growth when plants are exposed to elevated CO_2 (Rogers *et al.*, 1994). Other resources, however, such as N or P, can also be a useful currency in understanding costs and benefits, especially in strongly nutrient-limited environments in which C supply does not limit plant growth (Chapin, 1989; Eissenstat and Yanai, 1997).

A. The Balance between Shoot and Root Growth

An obvious example illustrating the evidence that carbon expenditure below ground can significantly affect shoot growth is the close coordination between shoot and root partitioning with nutrient supply (e.g., Ingestad and Ågren, 1991). When plants have constant relative growth rates and constant internal nutrient concentrations, the fraction of biomass allocated to the shoot is positively correlated with plant nutrient status (for N, P, and S but not K). Plants have higher relative growth rates when they are of higher nutrient status (until saturation).

Taking a simple crop modeling approach, the primary role of the shoots is to fix C for the plant, which is proportional to shoot activity (photosynthesis) and total leaf mass (or area) (e.g., Thornley and Johnson, 1990). The primary role of the roots is to acquire water and nutrients, which are a product of root activity and root mass (area). Leaf growth and photosynthesis is dependent on water and nutrients from the roots; root growth and activity is dependent on C from the shoots. This "functional equilibrium" between roots and shoots is a basic tenet of most crop growth models. For example, in sour orange seedlings grown at a range of P supply, relative growth rate was inversely related to root: shoot ratio (Eissenstat *et al.*, 1993). This principle—that when soil resources are limiting, plants must allocate a greater proportion of photosynthate to the roots, thus causing slower whole plant relative growth rates—represents a widely recognized trade-off of root production. Other aspects associated with trade-offs in root construction, such as root respiration, root morphology, and root architecture, however, are often ignored.

B. Root Construction versus Maintenance

Roots represent a considerable C cost (e.g., Caldwell, 1979). In perennial communities of natural ecosystems (e.g., grasslands, tundra, and evergreen forests), annual biomass production below ground usually exceeds that above ground (Caldwell, 1987; Fogel, 1985). In addition to C allocated for root production, C allocated for root respiration is also high (Table I). Root respiration can be partitioned into respiration used for root growth, root maintenance, and ion uptake and assimilation. Respiration used for the uptake and assimilation of nutrients is a cost associated more with whole plant function and will not be considered further. Maintenance respiration of the fine roots of a wide range of species is approximately 2 mmol C g^{-1} day^{-1} (range, 1–4 mmol C g^{-1} day^{-1}; see Amthor, 1984; Peng *et al.*, 1993). The total cost (growth respiration and root C content) to build roots, typically referred to as the construction cost, can range from 42 to 49 mmol C g^{-1} dry weight (dw) (Peng *et al.*, 1993; Amthor *et al.*, 1994). Thus, in just 20–26 days the C expended on root maintenance can

Table I Summary of Root Respiratory Costs of Volkamer Lemon Seedlings Colonized by *Glomus intraradices* (M) or Uninoculated (NM)[a]

	1P		5P	
	M	NM	M	NM
	Cost per unit root dry wt *[mmol CO_2 (g new root)$^{-1}$]*			
Construction cost	48.7	44.7	45.3	42.0
Growth respiration coefficient	11.8	10.1	10.3	8.7
	Specific respiration rates *[mmol CO_2 (g whole root system)$^{-1}$ days^{-1}]*			
Total root respiration	4.14	3.00	3.00	2.61
Ion-uptake respiration	0.83	0.72	0.40	0.48
Maintenance respiration[b]	2.46	1.77	2.05	1.62
No. of days for maintenance respiration to equal cost of root construction	19.8	25.3	22.1	25.9

[a] From Peng *et al.* (1993) and Eissenstat and Yanai (1997). Plants were grown at either low- or high-P supply (1P = 1 mM KH$_2$ PO$_4$; 5P = 5 mM P added once or twice per week, depending on plant size).

[b] Maintenance respiration represents respiration not attributable to growth or ion uptake and, thus, includes maintenance of roots and fungal biomass in the root, microbial respiration of plant exudates, and respiration of extramatrical hyphae.

exceed the C expended on root construction (Table I). Clearly, the C expended on root growth and maintenance is a considerable cost to the plant. If only a small amount of C was diverted from growth and maintenance of roots to new shoot growth then the C investment could be compounded, leading to an exponential increase in growth (under conditions in which plant growth is partially limited by C).

IV. Parallels between Roots and Leaves

When analyzing the potential trade-offs in root deployment, it is useful to draw upon the rich ecological literature on leaves. Fine roots and leaves are both generally ephemeral tissues, with a single dominant function: resource acquisition. Leaves have one primary function: acquisition of C. Roots are usually separated into fine roots and coarse roots (often based on some arbitrarily chosen root diameter, e.g., 2 mm). Coarse roots are primarily designed for extending the root system vertically and laterally, transporting water and nutrients from the fine roots to the stem, anchoring the stem in the soil, and storing carbohydrates and mineral nutrients. Although fine roots may serve minor functions in storage, anchorage, and transport, the primary function of fine roots is water and nutrient acquisition. In leaves, many traits are associated with long life span including low resource availability in the plant's preferred habitat, low maximum plant growth rates, low maximum photosynthetic rates, low dark respiration rates, low specific leaf area (area/dry wt), and high tissue density (Table II). Similar relationships may also be true in roots (Eissenstat, 1992). Long-

Table II   Characteristics Hypothesized to Be Associated with Leaf and Root Life Span[a]

Variable	Short life span	Long life span
Max. potential growth rate	High	Low
Responses to pulses or patches of nutrients	High	Low
Max. rates of photosynthesis or nutrient and water absorption	High	Low
Maintenance respiration rates	High	Low
Specific leaf area or root length	High	Low
Tissue density	Low	High
Amount of tissue construction allocated to antiherbivore defense	Low	High
Successional status	Often early	Often late
Resource availability in native habitat	High	Low

[a] From Coley *et al.* (1985), Coley (1988), Eissenstat (1992), Reich *et al.* (1992), Chapin (1993), Garnier and Laurent (1994), Grime (1994), Ryser (1996), and Eissenstat and Yanai (1996).

lived roots may tend to be coarse, invest more in immobile defenses, have low hydraulic conductivity, low specific ion absorption rates, low maintenance respiration rates, and less flexible responses to pulses in resources (Table II). Screening crop cultivars for high root growth rates or rates of ion uptake in young seedlings may represent trade-offs in root longevity and tolerance to edaphic stresses. Ecological theory suggests that resource availability (e.g., light, water, and nutrients) places a large constraint on the relative advantages of different kinds of tissue deployment. In some agricultural sectors production is moving toward lower inputs of irrigation, pesticides, and fertilizers in the hopes of higher efficiency and lower agrochemical pollution. More than 80% of the world's population is located in countries whose soils are of relatively low fertility and whose agriculturists can not afford costly agrochemical inputs (World Bank, 1993). In designing crop plants that can meet these environmental conditions, roots that are coarser, slower growing but better defended, with lower maintenance costs, may represent an important breeding objective, especially for perennial crops.

V. Specific Root Length, Root Diameter, and Tissue Density

Specific root length (SRL) is defined as the length : dry mass ratio of the fibrous root system. It has been used as a rough measure of the benefit (length) versus cost (dry wt) of the root system (e.g., Fitter, 1991). Mechanistic modeling of nutrient uptake would suggest that the most efficient deployment of root mass is to build very fine roots that maximize length and absorptive surface area per gram dry weight of root (Barber and Silberbush, 1984; Yanai *et al.,* 1995). Presumably, there are important trade-offs in building very fine roots not being considered in these nutrient uptake models (Eissenstat, 1992).

Plant species vary widely in root diameter. The diameter of the finest elements of the root system can be less than 100 μm in many graminoid species found in the Juncaceae, Cyperaceae, and Poaceae, in ericoid mycorrhizal species of the Ericaceae and Epacridaceae, and in many annual dicots such as the well-studied species, *Arabidopsis thaliana* (Harley and Smith, 1983; Fitter, 1991; Eissenstat, 1992, unpublished data). At the other extreme, many woody species found in the Magnoliaceae, Pinaceae, and Rutaceae (family of *Citrus*) and herbaceous species found in the Alliaceae produce absorptive roots that are no finer than 500–1000 μm in diameter.

It is normally assumed that SRL is proportional to the square of the root radius. This, of course, assumes the density of the tissue is largely constant. In citrus rootstock genotypes, a range of root diameters of 587–688 μm corresponded to a 44% difference in root length/root volume (Table III). Among these same genotypes, however, there was an 83% difference in SRL; the genotypes that built the finest roots also built roots with the lowest tissue density (Eissenstat, 1991). The rootstocks with the smallest diameter roots and lowest tissue density were able to proliferate roots in disturbed soil faster than the rootstock genotypes with coarse roots and higher tissue density (Fig. 1, top). Citrus rootstocks that produced greater root length in the disturbed soil also depleted water to a greater extend 8 days following a saturating rain (Fig. 1, bottom). Fitter (1994) found a similar relationship of root diameter with root proliferation when investigating root growth in nutrient patches among four herbaceous species in the Caryophyllaceae. He found a strong inverse relationship between mean root diameter and the ratio of relative extension rate (RER) in the patch to RER in the nutrient poor patch.

Ryser and Lambers (1995) identified tissue density as having a fundamental role in distinguishing traits of ecologically distinct grasses. They found that the slow-growing grass, *Brachypodium pinnatum,* had 20–106% higher root tissue density than the fast-growing grass, *Dactylis glomerata.* They suggest that production of roots at low tissue density, especially at low to moderate N supply, allows *D. glomerata* to explore much greater soil volumes than the slower-growing *B. pinnatum.* Similar relationships of high RGR with low tissue density have been found in leaves; both of these traits have been linked to shorter tissue longevity (Poorter and Remkes, 1990; Garnier and Laurent, 1994; Ryser, 1996). Thus, there is some evidence among the roots of the fine root system that dense, highly lignified and tannified

Table III   Patterns of Root Biomass Allocation among Six Citrus Rootstocks[a]

| Rootstock | Fibrous roots | | | | Pioneer roots |
	Root diameter[b] (μm)	Length/vol. (cm^{-2})	SRL or length/mass ($cm\ g^{-1}$)	Tissue density ($g\ cm^{-3}$)	(% of total root length
Sour orange	665	243	1240	0.203	5.00
Cleopatra mandarin	688	234	1280	0.184	5.59
Swingle citrumelo	673	249	1430	0.181	8.02
Rough lemon	637	293	1660	0.179	2.32
Volkamer lemon	632	287	1750	0.167	1.04
Trifoliate orange	587	338	2270	0.149	1.04
LSD (0.05)	44	40	260	0.032	3.05

[a] From Eissenstat (1991). Pioneer roots are those roots >2 mm in diameter. Only roots in the ingrowth containers were measured. Roots were collected from 0 to 14 cm depth 19 weeks after installation of ingrowth containers beneath field trees in a Valencia orange rootstock trial (n = 8 trees of each rootstock).
[b] Mean of eight trees, 20–40 0.5-mm root segments per tree.

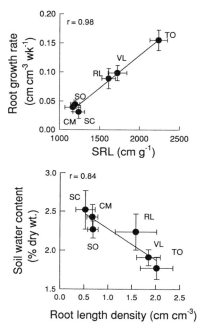

Figure 1 The relationship of SRL with root growth rate into disturbed, root-free soil under the tree of six citrus rootstocks (top). Growth rate was determined by sampling ingrowth containers at 5, 10, and 19 weeks after disturbance. Ingrowth containers were installed to a depth of 14 cm with the top flush with the soil surface. Specific root length was determined from the total root length divided by the root dry wt at each sampling. (Bottom) The relationship of root length density with gravimetric soil water content 19 weeks after disturbance. Sampling was 8 days following a saturating rain. For both curves, symbols represent rootstock means ± SE ($n = 7–8$), correlation coefficient (r) of rootstock means indicated $n = 6$. The rootstocks were sour orange (SO; *Citrus aurantium*), Swingle citrumelo (SC; *Poncirus trifoliata* × *Citrus paradisi*), Cleopatra mandarin (CM; *Citrus reshni*), rough lemon (RL; *Citrus jambhiri*), Volkamer lemon (VL; *Citrus volkameriana*), and trifoliate orange (TO; *Poncirus trifoliata*) budded to Valencia sweet orange. Trees were 13 years old (from Eissenstat, 1991).

large-diameter roots may grow slower and less readily proliferate in favorable soil patches but live longer than more fragile, small-diameter roots of higher water content.

VI. Exploiting Soil Heterogeneity

Soils are not uniform. Nutrients and soil water are usually heterogeneously distributed in space and time. A plant may optimize C expenditure on the root system by withholding carbohydrates from roots unable to acquire water and nutrients and preferentially increasing the growth and activity of roots in locations where water and nutrients are most available, thus providing greater water or nutrient gain for the C expended on the root system.

A. Preferential Carbon Allocation to Roots in Resource-Rich Patches

Proliferation of roots in localized patches rich in nutrients is a well-recognized phenomenon that occurs in most plants (e.g., Robinson, 1994). Furthermore, roots in these fertile patches often exhibit rapid shifts in their uptake kinetics, thus permitting higher specific rates of nutrient uptake at high supply rates than roots in the less fertile soil (Jackson *et al.*, 1990). The greater root growth and activity in favorable nutrient patches presumably increases the efficiency of nutrient acquisition of the root system (nutrient acquired/carbon expended). Root growth is often greatest in wetter regions of the soil (e.g., Coutts, 1982). During a growing season with progressive drying of the soil surface layers, it is quite common for roots to exhibit staggered growth with roots near the surface exhibiting the most growth early in the season, whereas those in the deeper soil have the most growth activity later in the season (e.g., Fernandez and Caldwell, 1975). Other factors, such as soil temperature and proximity to the leaves, may also contribute to these root growth patterns.

B. Reduced Carbon Expenditure to Roots in Resource-Poor Soil

Reduction of maintenance costs of roots in the infertile or dry bulk soil can occur either by decreasing maintenance respiration or by shedding roots operating at lower efficiency. There is little direct evidence to support that maintenance respiration is lower for those roots of a plant located in less fertile soil patches. Although studies have shown that *total* respiration of roots supplied with moderate N is lower than those well supplied with N (Granato and Raper, 1989), this mainly reflects respiration associated with N uptake and assimilation and not higher maintenance respiration. Indirect evidence that maintenance respiration is also affected in high-nutrient root axes, however, is indicated by greater V_{max} (maximum rate of nutrient uptake per gram root at high nutrient supply) of roots in fertile than infertile patches (Jackson *et al.*, 1990), which may cause higher rates of protein turnover and, thus, higher maintenance respiration than those roots in the less fertile bulk soil.

Reduced maintenance costs have been more clearly associated with roots in dry soil. Palta and Nobel (1989) indicate that maintenance respiration of roots of desert succulents in dry soil is only about 13% of those under wet conditions. Greatly reduced respiration of roots exposed to dry soil has also been found in other desert perennial species (e.g., Holthausen and Caldwell, 1980; Sisson, 1989) and citrus (Espeleta and Eissenstat, 1997).

Rather than reducing maintenance respiration of roots in less favorable soil, whole roots or the outer tissues (epidermis and cortex) of roots in unfavorable soil can be shed. Plant control of root longevity as a mechanism of optimizing the efficiency of the root system has not been well studied largely because of inadequate methods for tracking the fate of individual roots (Eissenstat and Yanai, 1997). Recent advances in viewing roots with clear access tubes (minirhizotrons) and extracting demographic information from videotapes with improved computer hardware and software are rapidly removing this obstacle to our understanding of root efficiency. Only a few studies have examined whether preferential shedding of roots occurs in response to nutrient heterogeneity, and these results are contradictory. Pregitzer *et al.* (1993) and Fahey and Hughes (1994) indicate that roots of northern hardwood species in low nutrient soil may die sooner than those in more fertile patches, but research with old-field herbaceous species indicated the opposite effect (Gross *et al.,* 1993).

Root shedding in dry portions of the soil profile has been more frequently reported. Root death in dry soil has been indicated in cotton (Klepper *et al.,* 1973), timothy grass (*Phleum pratense*) (Molyneux and Davies, 1983), soybean (Huck *et al.,* 1987), desert succulents (Huang and Nobel, 1992), Sitka spruce (Ferrier and Alexander, 1991), and big bluestem-dominated prairie (Hayes and Seastedt, 1987). Other species, however, such as corn (Taylor and Klepper, 1973; Stasovski and Peterson, 1991), wheat (Meyer *et al.,* 1990), orchard grass (*Dactylis glomerata*) (Molyneux and Davies, 1983), perennial ryegrass (*Lolium perenne*) (Molyneux and Davies, 1983; Jupp and Newman, 1987), onion (Stasovski and Peterson, 1993), and citrus (Kosola and Eissenstat, 1994), apparently do not readily shed roots in dry surface soil, although detailed demographic information is often lacking. Partial shedding of root tissues, however, is common. In perennial rye grass (*L. perenne*), for example, the root cortex readily dies in response to drought while the stele remains alive with a functional pericycle, thus permitting the initiation of new laterals when the soil is rewetted (Jupp and Newman, 1987). Stasovski and Peterson (1991, 1993) have examined in detail the loss of tissues of onion and corn roots exposed to drought. Maintenance costs of a root with only a functional stele are undoubtedly less than those of a root with a full cortex.

C. Citrus Root Responses in Dry Surface Soil

Carbon allocation, respiration, and death of surface roots of citrus have been studied in experiments in which deep roots have access to abundant soil moisture and in which shoots do not exhibit significant water stress. In a greenhouse study, seedlings of four citrus rootstocks that ranged widely in SRL were grown for 4 months in fine, sandy soil in a split-pot system divided into a top and bottom pot (Kosola and Eissenstat, 1994). After roots were well established in the bottom pot, water was withheld from the top pots of half of the plants. Plants were harvested every 2 weeks thereafter, 48 hr after labeling with $^{14}CO_2$. Contrary to our initial expectations, few roots died for any genotype, even after 60 days of localized soil drying. Carbon allocation to the roots, however, declined by about 80%. In a subsequent study, Espeleta and Eissenstat (1997) examined in the field the effects of dry surface soil on root mortality and root/soil respiration of Volkamer lemon seedlings and of bearing red grapefruit trees on Volkamer lemon rootstocks. Seedlings were grown adjacent to the grapefruit trees in wooden boxes buried in the soil in a similar split-pot system as described previously. Next to the seedling, a 5-mm-thick, woody lateral of the grapefruit tree was trained into a second split-pot system. Old fibrous roots were excised from the woody lateral prior to its placement into the pot system. Survivorship of new fine laterals emerging from the woody root was tracked on transparent windows on the side of the pot. Root/soil respiration from the surface of the top pot was also measured. Four weeks after withholding water, seedling root/soil respiration declined about 60% and tree root/soil respiration declined about 70%. Seedlings continued to exhibit new root growth in the dry soil after this time so that it was impossible to partition total respiration into growth, maintenance, and ion uptake. For both seedlings and adult trees, few roots died even after 6 weeks of drought. For the adults, no new root growth in the top pot was observed after about 6 weeks of withholding water. After 15 weeks of drought, respiration was 0.11 mmol C g^{-1} day^{-1} for roots of adults (Espeleta and Eissenstat, 1997) or about 4.4% of the maintenance respiration previously determined for seedlings in wet soil (Peng *et al.*, 1993).

Citrus evidently is a species that copes with dry surface soil not by shedding roots but by strongly downregulating root respiration. Because respiration rates and rates of ion uptake are downregulated in similar amounts (Espeleta and Eissenstat, 1997; Whaley, 1995), root efficiency (benefit/cost) is unchanged. Thus, simulation modeling of optimal life span predicts that citrus roots in dry surface soil should exhibit no difference in life span than those in wet soil (Eissenstat and Yanai, 1997). Moreover, it is not unreasonable that citrus exhibits less root death than many other species in response to dry surface soil because of their relatively high construction cost per unit root length compared to those of many other species.

VII. Root Architecture

Root architecture, or the spatial configuration of a root system in the soil, has several components, including root distribution with depth or distance from the stem, root topology (root branching patterns), and some aspects of root morphology such as root diameter distribution (Lynch, 1995). There is no optimal root architecture for all environmental conditions. For example, many plants have their highest concentration of roots within a few cm of the soil surface, which often corresponds to the highest concentrations of N and P (Fitter, 1994). In deserts, however, the surface of exposed soil can exceed 70°C and become very dry, leading to little root development in the top 3–5 cm of soil (Rundel and Nobel, 1991).

A. Root Distribution

Native species vary widely in the depth and lateral spread of their root system and provide insights into the trade-offs of building a deep or laterally extensive root system. In the Mediterranean-type vegetation of Western Australia, Dodd *et al.* (1984) excavated 551 root systems from 43 species of woody shrubs in sandy soil. Species that developed a root system characterized by deep taproots with few woody laterals or a dimorphic root system with deep taproots and extensive woody laterals only near the soil surface typically were plants that developed the largest shoots and were the slowest growing, longest lived, most drought tolerant and regenerated after fire mainly by resprouting. Short-lived, fast-growing species that regenerated mainly from seed generally had shallow roots and rarely invested in a distinct taproot.

Trade-offs between a root system that produces many fine roots that readily increase root length in a given soil volume and a root system that rapidly increasing the total soil volume occupied can also be examined theoretically using architectural simulation models (Berntson, 1994). Berntson parameterized the model using P as the resource acquired and root parameters based on observations of common groundsel (*Senecio vulgaris*, a fast-growing annual dicot common to disturbed soils) grown at ambient and twice-ambient CO_2 and two watering regimes. Root systems designed to occupy large soil volumes required larger average root diameters and were therefore less efficient in resource acquisition per unit soil volume than small root systems that exhibited more intensive growth around the stem. Thus, both empirical and theoretical studies indicate important trade-offs between building deep roots for drought tolerance and building a root system that maximizes root length deployment for short-term nutrient gain.

B. Root Topology

The costs and benefits of root systems of different topology have also been examined by model simulations (Fitter, 1991, 1994; Nielsen *et al.*, 1994). One extreme topology is the branching pattern of a herringbone, or one that comprises a main axis and laterals only. At the other extreme is a dichotomous branching pattern in which each lateral bifurcates. For relatively mobile resources of high diffusivity ($D > 10^{-7}$ cm^2 sec^{-1}) interroot competition within a plant is significant and a herringbone topology tends to be the most efficient in terms of resource gain per unit C expended. Herringbone topologies, however, tend to have the smallest root length per unit root volume; consequently, as a first approximation, these root systems are the most expensive in terms of C to construct. This led Fitter (1991) to predict that herringbone roots should be favored in infertile soils, whereas dichotomous branched root systems should be more prevalent in habitats where soil resources are less important.

Another area in which root architecture appears important is in the exploitation of resource-rich soil patches (Fitter, 1994). For example, roots that proliferate in patches tend to have a more dichotomous branching pattern with short, fine, laterals of high developmental order. Compared to coarse laterals in a herringbone topology, the fine roots of dichotomous branching likely are more precise in patch exploration and more readily develop a high amount of absorptive surface area for a limited C and nutrient investment. Fine, high-order laterals, however, are likely more ephemeral, less capable of nutrient and water transport, and extend less far from the parent root, leading to potentially more competition within the plant's root system than coarser laterals in a herringbone topology. Although these ideas regarding trade-offs associated with different root topologies are theoretically attractive, there is a notable lack of experimental evidence that supports that dichotomous branched roots of high-order turn over more rapidly and are less capable of transport than those with a herringbone topology.

C. Plant Breeding

Root architecture can be influenced in plant breeding programs. Much variation exists in plant root systems that has thoroughly been reviewed elsewhere (O'Toole and Bland, 1987). Here, I will provide a few examples how differences in root architecture among crop cultivars may represent a trade-off between two types of environmental stresses. Jackson (1995) compared the root architecture of cultivated (*Lactuca sativa*) and wild (*L. serriola*) lettuce. She found no difference in the root branching patterns in the two species. She did find that the cultivated species produced more laterals in the top zone along the taproot (0–5 cm) and the wild lettuce produced more laterals in the 5- to 55-cm zone. This was assumed related to the intense selection of cultivated lettuce for rapid growth under conditions of frequent water and fertilizer application at the soil surface. Similarly, Bonser *et al.* (1996) found differences in root development near the soil surface among bean (*Phaseolus vulgaris*) genotypes. Root architecture was related to the growth angle of the basal roots. Genotypes that exhibited higher relative yield (low-P yield /high-P yield) in field trials also tended to have the most horizontally growing basal roots in seedlings. For some genotypes, shallow branching was a fixed trait; for others shallow branching occurred only under P deficiency. Under drought conditions, shallow branching likely would be a serious disadvantage.

There has been extensive research aimed at selecting cultivars with deep rooting systems for drought resistance and, conversely, shallow rooted cultivars for soils of minimal water and nutrient stress (reviewed by O'Toole and Bland, 1987; Marcum *et al.*, 1995). As previously discussed, evidence based on ecological comparisons of woody species and on model simulations suggests that it may be difficult to select for cultivars that develop deep roots but still exhibit high yields under nondrought conditions. This appears to be true, for example, in turf grasses. In a comparison of cultivars of buffalo grass, "Texoka" had the highest shoot dry weight and high root density at shallow depths but was less deeply rooted than many of the other cultivars (Marcum *et al.*, 1995).

D. Absorptive versus Structural Roots

One of the potential trade-offs of investing heavily in a root system that proliferates rapidly is retaining a root system capable of rapid extensive growth. Among citrus rootstocks, those that built relatively coarse fibrous roots produced more pioneer roots (operationally defined as new roots >2 mm in diameter), a developmentally distinct group of roots characterized by a pronounced root cap and rapid, indeterminate root extension with little immediate lateral root development (Table III). These roots become the principal roots forming the structural framework of the tree and typically are not readily infected by *Phytophthora*, mycorrhizal fungi, or nematodes. Because these roots undergo secondary growth and develop large metaxylem vessels, they accommodate a root system that occupies a larger soil volume with less axial resistance (resistance to water flow in the xylem) for water transport.

The trade-offs of absorptive vs structural root development can be seen in rootstock performance. For example, trifoliate orange, the rootstock with the finest fibrous roots but fewest pioneer roots (Table III), is known for its ability to produce a small tree, especially on sandy soils (Castle, 1987). This rootstock is characterized by having shallow root systems with dense fibrous root systems but weak development of framework laterals, causing trees on this rootstock to be more susceptible to drought on sandy soils compared to other rootstocks.

VIII. Root Hydraulic Conductivity

Root hydraulic conductivity (L_P; m sec^{-1} MPa^{-1}) defines the intrinsic ability of roots to conduct water across a water potential gradient between the root surface and the xylem in the stem. Most of the resistance to water transport is normally associated with radial movement of water to the xylem of the fine roots (radial resistance; inverse of radial conductivity). Axial conductance (water movement in the xylem), however, may be an important factor affecting total water flow where xylem vessels are of small diameter such as in cereals (Greacen *et al.*, 1976), where much of the xylem is immature and xylem cross-walls persist (St. Aubin *et al.*, 1986), or where embolisms have occurred (Tyree and Cochard, 1994).

Species may have evolved high root hydraulic conductivity where rates of water extraction influence competitive success (Eissenstat and Caldwell, 1988) or where plants must quickly take advantage of small rainfall events before significant soil evaporation occurs (Rundel and Nobel, 1991). Root hydraulic conductivity can be strongly influenced by root age. For example, rain roots of *Agave deserti* and *Ferocactus acanthodes* have approximately 5-fold higher hydraulic conductivity than do 4-month-old roots and more than 15-fold higher hydraulic conductivity than do 2-year-old established roots. High hydraulic conductivity would be advantageous where an agricultural crop extracted water from the soil surface quickly following rains, thus requiring less weed management.

A. Trade-offs in Axial Conductance

High root hydraulic conductivity is not necessarily advantageous for agricultural production in dryland conditions. In areas of the world where crops must grow to maturity on a finite, limited supply of water, it may be desirable to retard water use during early growth in order to retain available water for use during reproductive growth (Taylor, 1983). For example, in Australia a breeding program in wheat was established to decrease axial conductance by selecting for smaller xylem vessel diameters in seminal roots (Richards and Passioura, 1989). Field trials established the success of the technique; narrow vessel selections yielded between 3 and 11% more than unselected controls in the driest environments. Yield differences in wetter environments were largely insignificant.

There is a positive correlation between vessel diameter and potential for cavitation of the water column in the vessel, resulting in an air embolism that greatly restricts water transport (Tyree and Ewers, 1991). Resistance of the xylem to cavitation is extremely important to drought resistance. Most research has been done with stems, but the same principles should apply to roots. For example, in a comparison of coexisting drought-deciduous and evergreen woody species in a tropical dry forest, drought-deciduous species maximized production during the wet season with higher water-transport efficiency; however, the high axial conductance led to the seasonal occurrence of embolisms (Sobrado, 1993). Evergreen species, on the other hand, had lower axial conductance but rarely developed xylem embolisms.

B. Radial Hydraulic Conductivity and SRL

In citrus seedlings, we have found that SRL is positively correlated with root hydraulic conductivity (Fig. 2, top). Note that the citrus rootstock with the highest hydraulic conductivity, *Poncirus trifoliata,* is a deciduous species, whereas true *Citrus* rootstocks are evergreen; these are patterns consistent with those observed by Sobrado (1993) in tropical trees. The positive correlation of SRL with root hydraulic conductivity in citrus rootstocks can also be seen using three parent genotypes, trifoliate orange (*P. trifoliata*), Duncan grapefruit (*Citrus paradisi*), and ridge pineapple (*Citrus sinensis*), with two of their hybrids, Swingle citrumelo (*P. trifoliata* × *C. paradisi*) and Carrizo citrange (*P. trifoliata* × *C. sinensis*) (Fig. 2, bottom). Specific root

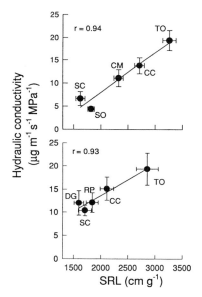

Figure 2   The relationship of SRL with root hydraulic conductivity of citrus seedlings. (Top) Original study using pressure chamber technique to pressurize entire root system and measure rates of xylem exudation from cut stem (data of Graham and Syvertsen, 1985; n = 12 seedlings per genotype ± SE; correlation coefficient (*r*) of rootstock means indicated) (SO, sour orange; SC, Swingle citrumelo; CM, Cleopatra mandarin; CC, Carrizo citrange; TO, trifoliate orange). (Bottom) Follow-up study using similar techniques that compared two citrus hybrids, SC (*Poncirus trifoliata* × *Citrus paradisi*) and CC (*P. trifoliata* × *Citrus sinensis*), with their parental genotypes, Duncan grapefruit (DG; *C. paradisi*), Ridge pineapple sweet orange (RP; *C. sinensis*), and trifoliate orange (*P. trifoliata*) (J. Barker, D. Eissenstat, and J. Syvertsen, unpublished results).

length was highest and mean root diameter and tissue density were lowest in trifoliate orange (Table III). Swingle citrumelo tended to exhibit both the SRL and hydraulic conductivity of its Duncan grapefruit parent, whereas Carrizo citrange exhibited more intermediate SRL and root hydraulic conductivity of its two parents, Ridge pineapple and trifoliate orange. Huang *et al.* (1995) examined the overall hydraulic conductivity, radial hydraulic conductivity, and axial conductance of individual primary and secondary roots of trifoliate orange, Swingle citrumelo, and sour orange. They found little difference in axial conductance but a higher radial hydraulic conductivity in trifoliate orange than in the other two rootstocks. These differences were linked to a smaller radius of the cortical tissue, lack of heavily lignified secondary walls in the hypodermis, and many more unsuberized passage cells in the secondary roots in trifoliate orange than in either Swingle citrumelo or sour orange.

IX. Root Herbivory and Root Defense

Where plants allocate C and nutrients to defensive compounds that protect roots, there should be a cost in terms of reduced maximum potential growth rates in the absence of herbivory or parasitism. Thus, plant defense theories predict a negative relationship between defense and plant growth rate. This can include phenotypic and genotypic differences in defense allocation. Phenotypic responses in relation to resource supply are illustrated in tomato (Wilkens *et al.*, 1996). At low levels of fertilization plant growth rates and production of defense compounds (rutin and chlorogenic acid) were low; at moderate rates of fertilization defense production, but not growth rate, was high; and at high rates of fertilization both defense production and growth rate were high (Wilkens *et al.*, 1996). It is also hypothesized that species that have evolved in habitats of low resource supply should allocate more to defense and have longer-lived but slower-growing tissues than those that evolved to compete in more productive habitats (Coley *et al.*, 1985). Support for this hypothesis can be found in a comparative study of 41 neotropical tree species: slow-growing shade-tolerant species had tougher, longer-lived leaves than fast-growing shade-intolerant species (Coley, 1988).

Unfortunately, there has been very little direct investigation of the trade-offs between root growth and activity and root defense. Most of the evidence is indirect. Typically, the fastest rates of mortality in minirhizotron and rhizotron studies are seen in the youngest, finest roots (Reid *et al.*, 1993; Eissenstat and Yanai, 1997). Presumably, there is a "window of vulnerability" where new roots are highly susceptible to root-feeding organisms, as found in citrus with the fungal pathogen, *Phytophthora nicotianae* (Graham,

1995). As roots age, they develop heavily suberized and lignified cell walls filled with tannins. This process can be accelerated under herbivore or pathogen pressure (Gerson, 1996). However, the accumulation of hydrophobic secondary compounds associated with aging or pathogens can also impede water and ion absorption, suggesting an important trade-off in root anatomy between absorption and defense (Rundel and Nobel, 1991; Peterson and Enstone, 1996). Other evidence of trade-offs between growth and defense comes from work on tissue density (Ryser, 1996), which will be discussed later.

Plant tolerance of root-feeding organisms may not involve production of defense compounds but rather production of "surplus" roots so that root losses to herbivores/pathogens will have only a minimal impact on the aboveground growth. For example, in lettuce, a cultivar that is resistant to corky-root (*Rhizomonas suberifaciens*) disease has thinner roots than a susceptible cultivar, presumably allowing the former to outgrow the pathogen more easily (L. E. Jackson, unpublished data). This strategy, which should commonly occur in high-resource environments such as typical agricultural conditions, also represents a trade-off because in the absence of root-feeding organisms, the excessive root growth should limit maximum potential growth rates of the plant.

Not all interactions of root herbivores or pathogens with roots should be considered trade-offs. Root morphological or architectural responses to pathogen infection may have little benefit to the plant, in terms of either defense or growth. For example, nematodes can cause gross alterations in root system morphology and architecture (Cohn *et al.*, 1996). *Pythium* spp. can restrict root branching and growth (Larkin *et al.*, 1995). In most cases, these root responses likely provide little benefit to the plant.

X. Mycorrhizae

A. Costs and Benefits of Mycorrhizae

The parallel between roots and leaves breaks down when mycorrhizal fungi are considered. Coarse roots may greatly increase their absorptive surface area by investing in the fine extramatrical hyphae of mycorrhizal fungi. Diameters of hyphal strands are typically only 2–4 μm in diameter so that a root can greatly increase its absorptive surface area by forming a symbiosis with mycorrhizal fungi. The extent that a root invests in mycorrhizal fungi can also be considered from a cost/benefit perspective.

Plants vary greatly in both the extent they are colonized by mycorrhizal fungi and their dependency on mycorrhizal fungi for P uptake and growth. Baylis (1975) first suggested that among vesicular–arbuscular mycorrhizal plants in low- to moderate-P soils, species that tend to produce large-diameter roots with few root hairs tend to be much more dependent on mycorrhizal fungi for P uptake and growth than those that produce fine roots with abundant root hairs. Subsequent studies have generally supported Baylis (Pope *et al.,* 1983; Manjunath and Habte, 1991; Jasper and Davy, 1993), although plant P demand (Hall, 1975; Graham *et al.,* 1991) and extent of root branching may also contribute to mycorrhizal dependency (Hetrick *et al.,* 1992). As one might expect, species with coarse root systems (high mycorrhizal dependency) tend to be colonized faster or to a greater extent than those of low dependency (Reinhardt and Miller, 1990; Graham and Eissenstat, 1994).

More mycorrhizal colonization is not a panacea for better plant nutrition or more favorable growth. For plants of similar P status, approximately 10–20% more current photosynthate is expended on mycorrhizal than nonmycorrhizal roots (see references in Peng *et al.,* 1993). Thus, mycorrhizal fungi may represent an appreciable carbon drain for the plant if they are not providing a nutritional benefit. Many plants are effective at limiting colonization and carbon expenditure toward the mycorrhizal fungus under conditions in which the fungus can provide little benefit. Reduced mycorrhizal colonization is well documented in environments in which P availability is high or in which C availability is low (e.g., Son and Smith, 1988). Nonetheless, the somewhat limited mycorrhizal colonization of plants in high-P soils can still consume significant quantities of carbohydrates and lead to growth depressions.

B. Citrus and Mycorrhizae

The trade-offs of mycorrhizal colonization have been critically examined in citrus. Citrus roots are quite coarse (ca. 600–700 μm). Much of their root system is devoid of root hairs. Where root hairs are present, they tend to be very short and extend less than 100 μm, which would not tend to expand appreciably the root's depletion zone (Fig. 3). Consequently, citrus plants are generally considered highly dependent on mycorrhizal fungi in low- to moderate-P soil. For example, P inflow (fmol cm^{-1} sec^{-1}) was more than three times higher in mycorrhizal than nonmycorrhizal sour orange seedlings at low-P supply (Eissenstat *et al.,* 1993). At high-P supply, however, mycorrhizal fungi may provide little or no P benefit. For example, Peng *et al.* (1993) grew mycorrhizal and nonmycorrhizal lemon seedlings at low- and high-P supply for 90 days. At the end of the experiment, mycorrhizal plants were nearly twice the dry weight of nonmycorrhizal plants at low-P supply. However, at high-P supply, mycorrhizal plants were 8% smaller than nonmycorrhizal plants. They found that the mycorrhizal fungus, *Glomus intraradices,* established numerous lipid-rich vesicles in the roots of the lemon seedlings, causing increased root fatty acid content and increased root construction cost (g glucose/g dry wt). Not surprisingly, fatty acid

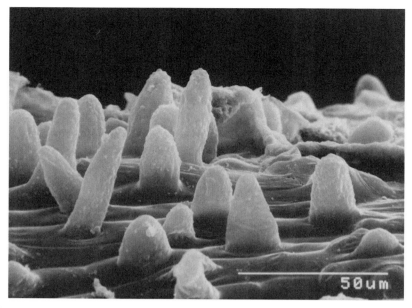

Figure 3 Scanning electron micrograph of root hairs of the citrus rootstock genotype, Cleopatra mandarin. Samples were fixed in glutaraldehyde and dehydrated in ethanol series, then critical-point dried with CO_2 and sputter coated with $80:20$ gold:palladium for 90 sec. Microscopy was performed with a Hitachi Model S530 at $200\times$ (micrograph kindly provided by Richard Crawford).

content and root construction costs were highly correlated (Fig. 4). Among high-P plants, mycorrhizal colonization increased root biomass allocation compared to nonmycorrhizal plants. At 52 days after transplanting, high-P nonmycorrhizal plants had 20% higher daily specific carbon gain (mmol g^{-1} day^{-1}) than the mycorrhizal plant because of 37% higher root–soil respiration in the mycorrhizal plants. Of the 37% higher root–soil respiration, 10% was directly attributable to building lipid-rich roots, 51% to greater mycorrhizal root biomass allocation, and the remaining 39% to maintenance of the fungal tissue in the roots and growth and maintenance of the extramatrical hyphae. Although mycorrhizal costs probably would decline as the roots aged, the results nonetheless illustrate the trade-offs of mycorrhizal colonization depending on P supply. In soils in which P availability is a problem, there is little doubt that citrus benefits from being mycorrhizal. In contrast, under intensive management conditions in which water and nutrients are supplied in abundance (as is common in Florida citrus production), mycorrhizal fungi should not be considered a symbiont but a weak parasite that may influence maximum potential growth rates and

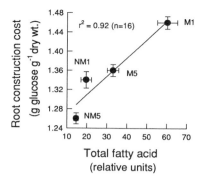

Figure 4 Relationship of construction cost of fibrous roots with their relative total fatty acid content in nonmycorrhizal (NM) and mycorrhizal (M) Volkamer lemon seedlings colonized by *Glomus intraradices* at low- (1) and high-P (5) supply (redrawn from Figure 5 in Peng, S., Eissenstat, D. M., Graham, J. H., Williams, K., and Hodge, N. C. (1993). Growth depression in mycorrhizal citrus at high-phosphorus supply. *Plant Physiol.* **101,** 1063–1071. © American Society of Plant Physiologists).

yield (Graham and Eissenstat, 1994). The costs of trying to limit mycorrhizal fungal infection, however, would likely far outweigh the potential benefits associated with increased production.

XI. Root Life Span

As in leaves, there is a wide range in median root life span, from about 14 days in apple and strawberry to more than 2 years in some pine forests (Eissenstat and Yanai, 1997). Simulation modeling can help identify factors that may influence optimal root life span. Root efficiency can be defined as the total resource benefit acquired by a root over its lifetime divided by the total carbon expended on the root (Yanai *et al.*, 1995). The carbon cost would include the initial cost of root construction and the lifetime cost of root maintenance respiration, which could also include such factors as carbon expended on mycorrhizal fungi and on root exudates. Uptake for solutes would include the uptake kinetics and water influx of the root, SRL, and root diameter, as well as soil factors affecting solute diffusion and mass flow to the root. Optimal root life span in this context would be to retain a root until its lifetime efficiency is maximized. Roots with high rates of maintenance respiration would be expected to be shed sooner than roots with low maintenance respiration. Aspects of root efficiency and root life span have been more thoroughly reviewed elsewhere (Eissenstat and Yanai, 1997).

There has been little experimental investigation of the trade-offs in root form and function that affect root life span. An exception is a study of perennial grasses, in which there is evidence of a link between root construction and root life span (Ryser, 1996). In a common garden experiment, root life span and root density were contrasted in five grass species (*Arrhenatherum elatuis, Dactylis glomerata, Holcus lanatus, Bromus erectus,* and *Festuca ovina*), which are found in a range of habitats that differed widely in nutrient availability. Tissue density of roots ranged from 0.17 to 0.28 g cm^{-3} with the species from the sites with the lowest nutrient availability having the highest tissue density. These species grew slowest and had the longest root longevity, as indicated by the percentage of dead root to total root length after approximately 1.5 years of growth. The relationships of tissue density, growth, and life span of leaves were similar to those in roots. Ryser convincingly argues that short organ life span should lead to greater nutrient losses, which in nutrient-poor habitats would be more disadvantageous than greater acquisition capacity.

One would predict that citrus roots with low SRL should live longer than those with high SRL; otherwise, it is not clear what would be the advantage of building a root that had more dry mass per unit length. However, minirhizotron investigations in a citrus rootstock trial have indicated that median life spans of trifoliate orange roots (90 days) are similar to those of Swingle citrumelo (99 days) and sour orange (90 days) (roots born between April 1992 and December 1992; $n = 565–1107$; D. Eissenstat, unpublished data), even though SRL varies by more than 80% (Table III). Root constructions costs per unit mass (g glucose g^{-1} dry wt) of these rootstocks are very similar (D. Eissenstat, unpublished data). Other costs not measured, such as root respiration and expenditure on mycorrhizal fungi, may be needed to better account for rootstock differences in life span. The factor most linked to life span in these rootstocks was resistance to *Phytophthora* infection (Kosola *et al.*, 1995), although the energetic cost of this resistance has not been quantified.

XII. Summary and Conclusions

Root construction represents a series of trade-offs. Plants that allocate more carbon on the root system and less on shoot growth normally have slower relative growth rates. Plants that build mostly fine roots may grow faster initially but do not lay down the belowground framework to support a large plant with access to deep water. Roots of high hydraulic conductivity may be more susceptible to embolism. Roots constructed for maximizing short-term benefits in water and nutrient absorption, such as small diameter, low tissue density (high water content), and high capacity to proliferate in favorable soil patches, may have a higher maintenance respiration rate and a shorter life span. These traits may be typical of roots in annual crops and fast-growing woody species of fertile soils. On the other hand, roots can be constructed to be well defended from belowground herbivores, pathogens, and soil desiccation, but at the costs of slower growth rates, higher carbon investment per unit length, lower specific absorption rates, and less ability to proliferate in favorable soil patches. Coarse fibrous roots may also invest more heavily in the mycorrhizal symbiosis, which can also represent an appreciable cost.

The concept that optimal root design represents a compromise of competing root functions needs to be more formally included in the selection and management of crops. This can be seen, for example, in the lack of drought tolerance in trifoliate orange because of its lack of deep structural roots and preponderance of fine, absorptive roots that readily proliferate and exhibit high hydraulic conductivity. Breeding for more drought tolerance in crops, however, may mean selecting for plants that invest more in deep roots or roots of smaller xylem vessel diameters but that exhibit lower yield under nondrought conditions. Root systems of crops intended for agricultural systems with low inputs of fertilizer, irrigation, and pesticides may have additional costs that would not be advantageous for crops under intensive management conditions.

Acknowledgments

Some of this work was supported by Grant NRICGP-94-37107-1024 from the United States Department of Agriculture and by Grants BSR-911824 and IBN-956050 from the National Science Foundation. Helpful comments were provided by David Bryla, Louise Jackson, Jonathan Lynch, and two anonymous reviewers.

References

Amthor, J. S. (1984). The role of maintenance respiration in plant growth. *Plant Cell Enviorn.* **7**, 561–569.

Amthor, J. S., Mitchell, R. J., Runion, G. B., Rogers, H. H., Prior, S. A., and Wood, C. W. (1994). Energy content, construction cost and phytomass accumulation of *Glycine max* (l.) Merr. and *Sorghum bicolor* (L.) Moench grown in elevated CO_2 in the field. *New Phytol.* **128**, 443–450.

Barber, S. A., and Silberbush, M. (1984). Plant root morphology and nutrient uptake. *In* "Roots, Nutrient and Water Influx, and Plant Growth" (S. A. Barber and D. R. Bouldin, eds.), pp. 65–87. American Society of Agronomy, Madison, WI.

Baylis, G. T. S. (1975). The magnolioid mycorrhiza and mycotrophy in root systems derived from it. *In* "Endomycorrhizas" (F. E. Sanders, B. Mosse, and P. B. Tinker, eds.), pp. 373–389. Academic Press, New York.

196   *David M. Eissenstat*

Berntson, G. M. (1994). Modelling root architecture: Are there tradeoffs between efficiency and potential of resource acquisition? *New Phytol.* **127,** 483–493.

Bloom, A. J., Chapin, F. S., III, and Mooney, H. A. (1985). Resource limitation in plants: An economic analogy. *Annu. Rev. Ecol. Syst.* **16,** 363–392.

Bonser, A. M., Lynch, J., and Snapp, S. (1996). Effect of phosphorus deficiency on growth angle of basal roots in *Phaseolus vulgaris. New Phytol.* **132,** 281–288.

Caldwell, M. M. (1979). Root structure: The considerable cost of belowground function. *In* "Topics in Plant Population Biology" (O. T. Solbrig, S. Jain, G. B. Johnson, and P. H. Raven, eds.), pp. 408–427. Columbia Univ. Press, New York.

Caldwell, M. M. (1987). Competition between root systems in natural communities. *In* "Root Development and Function" (P. J. Gregory, J. V. Lake, and D. A. Rose, eds.), pp. 167–185. Cambridge Univ. Press, New York.

Castle, W. S. (1987). Citrus rootstocks. *In* "Rootstocks for Fruit Crops" (R. C. Rom and R. F. Carlson, eds.), pp. 361–397. Wiley, New York.

Chapin, F. S., III (1987). Plant response to multiple environmental factors. *BioScience* **37,** 49–57.

Chapin, F. S., III (1989). The cost of tundra plant structures: Evaluation of concepts and currencies. *Am. Nat.* **133,** 1–19.

Chapin, F. S., III (1993). Evolution of suites of traits in response to environmental stress. *Am. Nat.* **142,** S78–S92.

Cohn, E., Koltai, H., Sharon, E., and Spiegel, Y. (1996). Root–nematode interactions: Recognition and pathogenicity. *In* "Plant Roots: The Hidden Half" (Y. Waisel, A. Eshel, and U. Kafkafi, eds.), 2nd ed., pp. 797–809. Dekker, New York.

Coley, P. D. (1988). Effects of plant growth rate and leaf lifetime on the amount and type of anti-herbivore defense. *Oecologia* **74,** 531–536.

Coley, P. D., Bryant, J. P., and Chapin, F. S., III (1985). Resource availability and plant antiherbivore defense. *Science* **230,** 895–899.

Coutts, M. P. (1982). Growth of Sitka spruce seedlings with roots divided between soils of unequal matric potential. *New Phytol.* **92,** 49–61.

Dodd, J., Heddle, E. M., Pate, J. S., and Dixon, K. W. (1984). Rooting patterns of sandplain plants and their functional significance. *In* "Kwongan: Plant Life of the Sandplain" (J. S. Pate and J. S. Beard, eds.), pp. 146–177. Univ. of Western Australia Press, Nedlands.

Eissenstat, D. M. (1991). On the relationship between specific root length and the rate of root proliferation: A field study using citrus rootstocks. *New Phytol.* **118,** 63–68.

Eissenstat, D. M. (1992). Costs and benefits of constructing roots of small diameter. *J. Plant Nutr.* **15,** 763–782.

Eissenstat, D. M., and Caldwell, M. M. (1988). Competitive ability is linked to rates of water extraction: A field study of two aridland tussock grasses. *Oecologia* **75,** 1–7.

Eissenstat, D. M., and Van Rees, K. C. J. (1994). The growth and function of pine roots. *Ecol. Bull.* **43,** 76–91.

Eissenstat, D. M., and Yanai, R. D. (1997). The ecology of root lifespan. *Adv. Ecol. Res.,* **27,** 1–60.

Eissenstat, D. M., Graham, J. H., Syvertsen, J. P., and Drouillard, D. L. (1993). Carbon economy of sour orange in relation to mycorrhizal colonization and phosphorus status. *Ann. Bot.* **71,** 1–10.

Espeleta, J. F., and Eissenstat, D. M. (1997). Effects of tree juvenility on citrus fine root mortality in dry surface soil. *Tree Physiol.,* in press.

Fahey, T. J., and Hughes, J. W. (1994). Fine root dynamics in a northern hardwood forest ecosystem, Hubbard Brook Experimental Forest, NH. *J. Ecol.* **82,** 533–548.

Fernandez, O. A., and Caldwell, M. M. (1975). Phenology and dynamics of root growth of three cool semi-desert shrubs under field conditions. *J. Ecol.* **63,** 703–714.

Ferrier, R. C., and Alexander, I. J. (1991). Internal redistribution in N in Sitka spruce seedlings with partly droughted root systems. *For. Sci.* **37,** 860–870.

Fitter, A. H. (1991). Characteristics and functions of root systems. *In* "Plant Roots: The Hidden Half" (Y. Waisel, A. Eshel, and U. Kafkafi, eds.), pp. 3–26. Dekker, New York.

Fitter, A. H. (1994). Architecture and biomass allocation as components of the plastic response of root systems to soil heterogeneity. *In* "Exploitation of Environmental Heterogeneity of Plants: Ecophysiological Processes Above- and Belowground" (M. M. Caldwell and R. W. Pearcy, eds.), pp. 305–323, Academic Press, New York.

Fogel, R. (1985). Roots as primary producers in below-ground ecosystems. *In* "Ecological Interactions in Soil" (A. H. Fitter, D. Atkinson, D. J. Read, and M. Usher, eds.), Spec. Publ. Ser. of the British Ecological Society No. 4, pp. 23–36. Blackwell, Oxford.

Garnier, E., and Laurent, G. (1994). Leaf anatomy, specific mass and water content in congeneric annual and perennial grass species. *New Phytol.* **128,** 725–736.

Gerson, U. (1996). Arthropod root pests. *In* "Plant Roots: The Hidden Half" (Y. Waisel, A. Eshel, and U. Kafkafi, eds.), 2nd ed., pp. 797–809. Dekker, New York.

Graham, J. H. (1995). Root regeneration and tolerance of citrus rootstocks to root rot caused by *Phytophthora nicotianae. Phytopathology* **85,** 111–117.

Graham, J. H., and Eissenstat, D. M. (1994). Host genotype and the formation and function of VA mycorrhizae. *Plant Soil* **159,** 179–185.

Graham, J. H., and Syvertsen, J. P. (1985). Host determinants of mycorrhizal dependency of citrus rootstock seedlings. *New Phytol.* **101,** 667–676.

Graham, J. H., Eissenstat, D. M., and Drouillard, D. L. (1991). On the relationship between a plant's mycorrhizal dependency and rate of vesicular–arbuscular mycorrhizal colonization. *Funct. Ecol.* **5,** 773–779.

Granato, T. C., and Raper, D., Jr. (1989). Proliferation of maize (*Zea mays* L.) roots in response to localized supply of nitrate. *J. Exp. Bot.* **40,** 263–275.

Greacen, E. L., Posana, P., and Barley, K. P. (1976). Resistance to water flow in the roots of cereals. *In* "Water and Plant Life. Problems and Modern Approaches" (O. L. Lange, L. Kappen, and E.-D. Schulze, eds.), pp. 86–100. Springer-Verlag, Berlin.

Grime, J. P. (1994). The role of plasticity in exploiting environmental heterogeneity. *In* "Exploitation of Environmental Heterogeneity of Plants: Ecophysiological Processes Above- and Belowground" (M. M. Caldwell and R. W. Pearcy, eds.), pp. 1–20. Academic Press, New York.

Gross, K. L., Peters, A., and Pregitzer, K. S. (1993). Fine root growth and demographic responses to nutrient patches in four old-field plant species. *Oecologia* **95,** 61–64.

Hall, I. R. (1975). Endomycorrhizas of *Meterosideros umbellata* and *Weinmannia racemosa. New Zealand J. Bot.* **13,** 463–472.

Harley, J. L., and Smith, S. E. (1983). "Mycorrhizal Symbiosis." Academic Press, New York.

Hayes, D. C., and Seastedt, T. R. (1987). Root dynamics of tallgrass prairie in wet and dry years. *Can. J. Bot.* **65,** 787–791.

Hetrick, B. A. D., Wilson, G. W. T., and Todd, T. C. (1992). Relationship of mycorrhizal symbiosis, rooting strategy, and phenology among tallgrass prairie forbs. *Can. J. Bot.* **70,** 1521–1528.

Holthausen, R. S., and Caldwell, M. M. (1980). Seasonal dynamics of root system respiration in *Atriplex confertifolia. Plant Soil* **55,** 307–317.

Huang, B., and Nobel, P. S. (1992). Hydraulic conductivity and anatomy for lateral roots of *Agave deserti* during root growth and drought-induced abscission. *J. Exp. Bot.* **43,** 1441–1449.

Huang, B., Eissenstat, D. M., and Achor, D. (1995). Root hydraulic conductivity in relation to its morphological and anatomical characteristics for citrus rootstocks. International Symposium: Dynamics of Physiological Processes in Roots, October 8–12, Ithaca, NY. [Abstract]

Huck, M. G., Hoogenboom, G., and Peterson, C. M. (1987). Soybean root senescence under drought stress. *In* "Minirhizotron Observation Tubes: Methods and Applications for Measuring Rhizosphere Dynamics'' (H. M. Taylor, ed.), ASA Spec. Publ. No. 50, pp. 109–121. Agronomy Society of America, Madison, WI.

Ingestad, T. and Ågren, G. I. (1991). The influence of plant nutrition on biomass allocation. *Ecol. Appl.* **1**, 168–174.

Jackson, L. E. (1995). Root architecture in cultivated and wild lettuce (*Lactuca* spp.). *Plant Cell Environ.* **18**, 885–894.

Jackson, R. B., Manwaring, J. H., and Caldwell, M. M. (1990). Rapid physiological adjustment of roots to localized soil enrichment. *Nature (London)* **344**, 58–60.

Jasper, D. A., and Davy, J. A. (1993). Root characteristics of native plant species in relation to the benefit of mycorrhizal colonization for phosphorus uptake. *Plant Soil* **155/156**, 281–284.

Jupp, A. P., and Newman, E. I. (1987). Morphological and anatomical effects of severe drought on the roots of *Lolium perenne* L. *New Phytol.* **105**, 393–402.

Klepper, B., Taylor, H. M., Huck, M. G., and Fiscus, E. L. (1973). Water relations and growth of cotton in drying soils. *Agron. J.* **54**, 307–310.

Kosola, K. R., and Eissenstat, D. M. (1994). The fate of citrus seedlings in dry soil. *J. Exp. Bot.* **45**, 1639–1645.

Kosola, K. R., Eissenstat, D. M., and Graham, J. H. (1995). Root demography of mature citrus trees: The influence of *Phytophthora nicotianae*. *Plant Soil* **171**, 283–288.

Larkin, R. P., English, J. T., and Mihail, J. D. (1995). Effects of infection by *Pithium* spp. on root system morphology of alfalfa seedlings. *Phytophathology* **85**, 430–435.

Lynch, J. (1995). Root architecture and plant productivity. *Plant Physiol.* **109**, 7–13.

Manjunath, A., and Habte, M. (1991). Root morphological characteristics of host species having distinct mycorrhizal dependency. *Can. J. Bot.* **69**, 671–676.

Marcum, K. B., Engelke, M. C., and Morton, S. J. (1995). Rooting characteristics of buffalo-grasses grown in flexible plastic tubes. *HortScience* **30**, 1390–1392.

Meyer, W. S., Tan, C. S., Barrs, H. D., and Smith, R. C. G. (1990). Root growth and water uptake by wheat during drying of undisturbed and repacked soil in drainage lysimeters. *Aust. J. Agric. Res.* **41**, 253–265.

Molyneux, D. E., and Davies, W. J. (1983). Rooting patterns and water relations of three pasture grasses growing in drying soil. *Oecologia* **58**, 220–224.

Nielsen, K. L., Lynch, J. P., Jablokow, A. G., and Curtis, P. S. (1994). Carbon cost of root systems: An architectural approach. *Plant Soil* **165**, 161–169.

O'Toole, J. C., and Bland, W. L. (1987). Genotypic variation in crop plant root systems. *Adv. Agron.* **41**, 91–145.

Palta, J. A., and Nobel, P. S. (1989). Influences of water status, temperature, and root age on daily patterns of root respiration of two cactus species. *Ann. Bot.* **63**, 651–662.

Peng, S. B., Eissenstat, D. M., Graham, J. H., Williams, K., and Hodge, N. C. (1993). Growth depression in mycorrhizal citrus at high-phosphorus supply: Analysis of carbon costs. *Plant Physiol.* **101**, 1063–1071.

Peterson, C. A., and Enstone, D. E. (1996). Functions of passage cells in the endodermis and exodermis of roots. *Physiol. Plant.* **97**, 592–598.

Poorter, H., and Remkes, C. (1990). Leaf area ratio and net assimilation rate of 24 wild species differing in relative growth rate. *Oecologia* **83**, 553–559.

Pope, P. E., Chaney, W. R., Rhodes, J. D., and Woodhead, S. H. (1983). The mycorrhizal dependency of four hardwood tree species. *Can. J. Bot.* **61**, 412–417.

Pregitzer, K. S., Hendrick, R. L., and Fogel, R. (1993). The demography of fine roots in response to patches of water and nitrogen. *New Phytol.* **125**, 575–580.

Reich, P. B., Walters, M. B., and Ellsworth, D. S. (1992). Leaf life-span in relation to leaf, plant, and stand characteristics among diverse ecosystems. *Ecol. Monogr.* **62**, 365–392.

Reid, J. B., Sorensen, I., and Petrie, R. A. (1993). Root demography in kiwifruit (*Actinidia deliciosa*). *Plant Cell Environ.* **16**, 949–957.

Reinhardt, D. R., and Miller, R. M. (1990). Size classes of root diameter and mycorrhizal fungal colonization in two temperate grassland communities. *New Phytol.* **116,** 129–136.

Richards, R. A., and Passioura, J. B (1989). A breeding program to reduce the diameter of the major xylem vessel in the seminal roots of wheat and its effect on grain yield in rain-fed environments. *Aust. J. Agric. Res.* **40,** 943–950.

Robinson, D. (1994). The responses of plants to non-uniform supplies of nutrients. *New Phytol.* **127,** 635–674.

Rogers, H. H., Runion, G. B., and Sagar, V. K. (1994). Plant responses to atmospheric CO_2 enrichment with emphasis on roots and the rhizosphere. *Environ. Pollut.* **83,** 155–189.

Rundel, P. W., and Nobel, P. S. (1991). Structure and function in desert root systems. *In* "Plant Root Growth: An Ecological Perspective" (D. Atkinson, ed.), pp. 349–378. Black-well, Oxford.

Ryser, P. (1996). The importance of tissue density for growth and life span of leaves and roots: A comparison of five ecologically contrasting grasses. *Funct. Ecol.,* **10,** 717–723.

Ryser, P., and Lambers, H. (1995). Root and leaf attributes accounting for the performance of fast- and slow-growing grasses at different nutrient supply. *Plant Soil* **170,** 251–265.

Sisson, W. B. (1989). Carbon balance of *Panicum coloratum* during drought and non-drought in the northern Chihuahuan desert. *J. Ecol.* **77,** 799–810.

Sobrado, M. A. (1993). Trade-off between water transport efficiency and leaf life-span in a tropical dry forest. *Oecologia* **96,** 19–23.

Son, C. L., and Smith, S. E. (1988). Mycorrhizal growth responses: Interactions between photon irradiance and phosphorus nutrition. *New Phytol.* **108,** 305–314.

St. Aubin, G., Canny, M. J., and McCully, M. E. (1986). Living vessel elements in the late metaxylem of sheathed maize roots. *Ann. Bot.* **58,** 577–588.

Stasovski, E., and Peterson, C. A. (1991). The effects of drought and subsequent rehydration on the structure and vitality of *Zea mays* seedling roots. *Can. J. Bot.* **69,** 1170–1178.

Stasovski, E., and Peterson, C. A. (1993). Effects of drought and subsequent rehydration on the structure, vitality and permeability of *Allium cepa* adventitious roots. *Can. J. Bot.* **71,** 700–707.

Taylor, H. M. (1983). Managing root systems for efficient water use: An overview. *In* "Limitations to Efficient Water Use in Crop Production" (H. M. Taylor, W. R. Jordan, and T. R. Sinclair, eds.), pp. 87–113. Agronomy Society of America, Madison, WI.

Taylor, H. M., and Klepper, B. (1973). Rooting density and water extraction patterns for corn (*Zea mays* L.). *Agron. J.* **65,** 965–968.

Thornley, J. H. M., and Johnson, I. R. (1990). "Plant and Crop Modelling: A Mathematical Approach to Plant and Crop Physiology." Clarendon, Oxford.

Tisdale, S. L., and Nelson, W. L. (1975). "Soil Fertility and Fertilizers." Macmillian, New York.

Tyree, M. T., and Cochard, H. (1994). Water relations of a tropical vine-like bamboo (*Rhipiocla-dum racemiflorum*): Root pressures, vulnerability to cavitation and seasonal changes in embolism. *J. Exp. Bot.* **45,** 1085–1089.

Tyree, M. T., and Ewers, F. W. (1991). Tansley Review No. 34. The hydraulic architecture of trees and other woody plants. *New Phytol.* **119,** 345–360.

Whaley, E. L. (1995). Uptake of phosphorus by citrus roots in dry surface soil. Thesis, University of Florida, Gainesville.

Wilkins, R. T., Spoerke, J. M., and Stamp, N. E. (1996). Differential responses of growth and two soluble phenolics of tomato to resource availability. *Ecology* **77,** 247–258.

World Bank (1993). "World Development Report 1993: Investing in Health." Oxford Univ. Press, Oxford.

Yanai, R. D., Fahey, T. J., and Miller, S. L. (1995). Efficiency of nutrient acquisition by fine roots and mycorrhizae. *In* "Resource Physiology of Conifers" (W. K. Smith and T. M. Hinckley, eds.), pp. 75–103. Academic Press, New York.

II

Biotic Interactions
and Processes

7

The Use of Biodiversity to Restrict Plant Diseases and Some Consequences for Farmers and Society

Maria R. Finckh and Martin S. Wolfe

I. Introduction

Until the past few hundred years, agriculture developed on the basis of biodiverse systems, with greater amounts of both interspecific and intraspecific variation than are seen today in the developed world. Some of that diversity was disadvantageous (barberry and buckthorn around wheat and oat fields) and some was advantageous, leading to the development of many useful systems of plant production that helped to constrain plant diseases (Thurston, 1992) and that are still evident in a number of indigenous systems, mostly in tropical regions. One universal advantage was almost certainly compensation: By producing a wide range of crops and animals, and by using "naturally" occurring materials, farm operations were buffered against the ravages of individual pathogens and pests.

During the past 300 years or so, correlated with the increase in the growth of the world's human population, there has been a massive increase in food production brought about by many factors including the development of plant breeding and production-orientated intensification of agriculture. This was boosted particularly in the 20th century by governmental policies aimed at increasing self-sufficiency in food production at any cost. A major development, seen all over the developed world, was specialization among few crops and the consequent massive monoculture of, for example, small and coarse grain cereals.

Recently, as experience with monoculture has developed, we now see more clearly some of the major disadvantages that follow, for example, in the way in which this encourages massive and rapid development of particular diseases, pests, and weeds. There are two general solutions to the problem. The first is to improve the monoculture technology by breeding more durably resistant crop varieties and by relying on pesticides. The second is to develop rationally the use of the biodiversity that is already available. In our view, this latter approach can help not only to deal with disease, pest, and weed problems but also provide a new framework for agricultural production that is more sustainable than the systems currently practiced in the developed world. A remarkable confirmation of the possible positive correlation between biodiversity, on the one hand, and production and sustainability, on the other hand, has been demonstrated by Tilman *et al.* (1996) for a natural ecosystem.

Genetic diversity for disease resistance has been advocated and developed as a means to control plant diseases for more than half a century by plant pathologists and breeders (e.g., Browning *et al.*, 1969; Jensen, 1952, 1988; Harlan, 1972; Stevens, 1942) and the effects of diversity on disease have been demonstrated in a large number of studies in numerous crop–pathogen systems. Our aim in this chapter is to provide an overview of these developments and to try to indicate in a more general social context some of the advantages and constraints in such an approach. We also comment on some of the difficulties involved in changing toward exploitation of biodiversity because of the resistance to change that has already become built into the large-scale specialist approach.

II. The Pathologist's View

A. The Status Quo: Monoculture

The massive increase in the use of a few crop species (crop monoculture) over the past few hundred years and reliance on an often narrow genetic base for the varieties has led to many severe and chronic disease problems, well known for cereals but also for many other crops (Adams *et al.*, 1971; Browning *et al.*, 1969; Harlan, 1972; Ullstrup, 1972).

Crop monoculture was established first in cereals and then developed widely. Within each cereal monoculture, and later in other crops, there developed further the monoculture of individual varieties and then, on a larger scale, the monoculture of individual resistance genes (we may refer, therefore, to crop monoculture, variety monoculture, and resistance monoculture; some degree of crop monoculture is now inevitable because of the high demand for wheat, rice, and maize; variety or resistance monoculture is not inevitable).

Genetic uniformity has been identified as the cause of many devastating epidemics, for example, the outbreak of late blight (caused by *Phytophthora infestans*) in 1845–1847, which destroyed the Irish potato crop and led to the starvation of 2 million people and the mass emigration of another 2 million from Ireland (Adams *et al.,* 1971; Harlan, 1972). The pathogen spread throughout Europe in the following years, destroying much of the crop, because the potato varieties at the time were uniformly susceptible to the disease because there had been no previous selection for resistance. Uniform susceptibility of coffee (*Coffea arabica*) to rust (caused by *Hemileia vastatrix*) led to massive losses in coffee production in Sri Lanka in the 1870s and to widespread epidemics in South America following the introduction of the rust into Brazil in 1970 (Schumann, 1991). In 1970, southern corn leaf blight (caused by *Cochliobolus carbonum*) destroyed much of the U.S. corn crop causing an economic loss of several billion dollars (Ullstrup, 1972).

Early breeding for resistance relied heavily on the use of simply inherited race-specific resistance (Johnson, 1961). Hopes were high that the use of such resistance would end devastating epidemics, but they were soon shattered. Often, new pathogen races selected that were virulent on previously resistant varieties soon after the varieties became popular and occupied large areas (Adams *et al.,* 1971; Harlan, 1972; Johnson, 1961; Van der Plank, 1963), resulting in what has been referred to as "boom and bust cycles" (e.g., Wolfe, 1973). For example, it took an average of 2 or 3 years for the powdery mildew pathogen (*Erysiphe graminis* f.sp. *hordei*) to overcome previously highly effective resistances in barley once they were introduced on a large scale (Brown *et al.,* 1997) (Fig. 1). Similar boom and bust cycles could be followed in the rice–rice bacterial blight (caused by *Xanthomonas oryzae* pv. *oryzae*) system (Mew *et al.,* 1992) and the wheat–stripe rust (caused by *Puccinia striiformis*) system (Danial *et al.,* 1994; Wellings and McIntosh, 1990). The epidemics that "bust" popular varieties are usually due to huge increases in population size of the relevant pathogen genotypes, providing a source population from which inoculum of virulent races migrates to new regions.

In intensive agriculture, the current method of avoiding disaster in the bust phase is the "replacement strategy": As one resistance is overcome, it is replaced by another (Duvick, 1977). This has considerable disadvantages for the farmer and for society in terms of sustainability (cost of seed, difficulty of keeping up to date, problems of marketing, loss of valuable genetic resources, and environmental and economic costs of fungicide use during the changeover period while pesticides are used). The main alternative within monocultures is to improve durability of resistance. **Durable resistance** is defined as "resistance that remains effective in a cultivar that is widely grown for a long period of time in an environment favorable to the disease" (Johnson, 1993, p. 284). Note that both the test for and exploitation of durability usually refer to large-scale monoculture. One

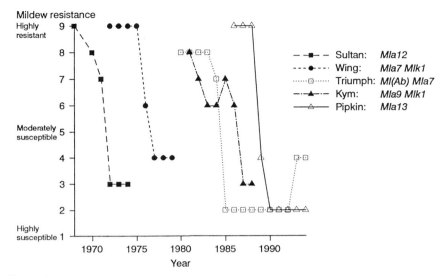

Figure 1   The "breakdown" of some barley powdery mildew resistance genes. The varieties named were the first in the United Kingdom to carry the resistance genes *Mla12*, *Mla7* + *Mlk1*, *Ml(Ab)*, *Mla9*, or *Mla13;* Triumph and Kym also had *Mla7* and *Mlk1*, respectively. The powdery mildew resistance ratings are those given in *Recommended Varieties of Cereals* by the National Institute of Agricultural Botany. 9 or 8, good resistance; 7 or 6, moderate resistance; 5 or 4, moderate susceptibility; 3, 2, or 1, high susceptibility (from Brown *et al.*, 1997).

strategy that is thought to improve durability is by the use of resistance gene combinations that reduce the probability of occurrence of a new virulent pathogen genotype (see Section III,A for discussion).

B. Host–Pathogen Coevolution

An aspect not fully considered in the "monoculture with replacement" strategy is the form of **coevolution** that occurs between host and pathogen. We define coevolution as the partial coordination of nonmixing gene pools through biotic selection caused by competition, predation, or parasitism (Ehrlich and Raven, 1964; Futuyma and Slatkin, 1983; Harper, 1982; Thompson, 1982); this depends on the availability of relevant genetic variation. As a host population develops, its changing structure results from interactions among the host components, which can strongly influence the structure and dynamics of the pathogen population. Simultaneously, the host interactions can be influenced by the development of the pathogen population (the epidemic), which may have differential effects on the host components, affecting their final reproductive output and thus fitness, where we define fitness according to Harper (1977, p. 770): "The fitness of an organism is measured by the descendents that it leaves, measured over a number of generations.''

To understand coevolutionary interactions among hosts and their pathogens, it is important to keep in mind the following: (i) Although host plants do not depend on the existence of their pathogens for survival, many pathogens, especially fungi and viruses, cannot survive and/or reproduce for long in the absence of their hosts. Thus, we may expect pathogen populations to respond more strongly to their host's genetic composition than vice versa; (ii) pathogens differ in the degree of their dependence on specific hosts, i.e., there is variation in host specificity; (iii) coevolutionary interactions depend strongly on the balance among the reproductive systems that are involved [e.g., sexual, asexual, and parasexual (i.e., recombination without meiosis)], alternate hosts, and generation time, which is generally much shorter for pathogens than for their hosts; (iv) the scale at which host and pathogen populations interact, which may have major effects on the outcome. For example, relative to the mode of dispersal of the pathogen, small separated fields may act as islands, whereas large-scale monoculture tends to join the islands (see Chapter 8, this volume) making migration for the pathogen easier, lessening population subdivision, and enhancing regional epidemics. For example, the use of different stem rust (caused by *Puccinia graminis* f.sp. *tritici*) resistance genes in wheat breeding programs in North America greatly affected the population dynamics of the rust. Overwintering inoculum from the South was selected to attack a set of resistance genes different from those used in the central and northern regions that depend on southern inoculum for the start of new epidemics each year. The southern inoculum first had to be reselected on the central and later on the northern varieties, greatly slowing down overall epidemic development (Knott, 1972)

Compared to natural systems, coevolution in agricultural systems differs principally in relation to the factors affecting "fitness" over generations of the host. In agriculture, humans decide which variety to grow on what scale and where. Although these decisions are influenced by the environmental conditions (including disease pressure), the environment can be ameliorated, for example, by the use of soil amendments and pesticides. No such interference happens in natural ecosystems and a wider range of factors may influence significantly the life history of host plants (see Section II,B,2).

1. Host–Pathogen Coevolution in Agricultural Ecosystems The direct, strong, and mutual influence of host and pathogen genetic composition has long been recognized in agroecosystems (e.g., Flor, 1953; Harlan, 1976; Johnson, 1961; Johnson *et al.*, 1967; Leonard and Czochor, 1980; McDonald *et al.*, 1989b; Stakman, 1947; Stevens, 1942; Wolfe, 1973; Wolfe and Barrett, 1977,

1980). Most of these interactions are three sided, involving host, pathogen, and humans. In some cases, in which no conscious selection is exerted on the crop by humans, the process is more truly two sided.

Flor (1953) was the first to uncover the large effects that humans can have in driving the coevolution of hosts and pathogens. He revealed the close interrelation between the resistance genes of flax (*Linum usitatissimum*) and the virulence genes of *Melampsora lini* (causing flax rust). From 20 years of race surveys, he found that the prevalence of different races of the rust was determined largely by the resistances in the common flax varieties. Based on the analysis of many crosses, Flor (1956) then hypothesized that during the coevolution of the rust and the host, complementary genic systems in host and pathogen had evolved, and that for each gene controlling resistance/susceptibility in the host there is a corresponding gene controlling avirulence/virulence in the pathogen. This was termed the gene-for-gene hypothesis. The resistance genes in the host were often found to be dominant multiple alleles involving a few loci, whereas the virulence genes in the rust were recessive and not allelic.

Johnson (1961) and Johnson *et al.* (1967) described the "man-guided evolution in plant rusts." The cereal rusts had changed their racial composition from a large number of "standard races" in earlier epidemics in central north America toward one or a group of related races that could overcome the predominantly used resistance genes. In a more detailed study covering 56 years of data, Kolmer (1991) found a close correlation between host and pathogen genetic composition in the wheat (*Triticum aestivum*)–wheat leaf rust (caused by *Puccinia recondita* f.sp. *tritici*) system in Canada. With the change to the use of markedly distinct wheat varieties in the eastern and western regions, the rust population became subdivided.

The first resistance gene used consciously in barley (*Hordeum vulgare*) to control powdery mildew (caused by *E. graminis* f.sp. *hordei*) was *Mlg*, introduced in Germany in the 1930s. Since then, successive waves of simply inherited resistances have been introduced in Europe as the pathogen overcame each in turn (Brown *et al.*, 1997; Brown, 1995; Jorgensen, 1994; Wolfe and Schwarzbach, 1978; Wolfe, 1984; Wolfe *et al.*, 1992; Wolfe and McDermott, 1994). The recent example of the *Mla13* resistance confirms the common process of an initial explosion of the newly selected virulence in one region followed by rapid and large-scale migration over a much larger area (Brown and Wolfe, 1990; Wolfe and McDermott, 1994; Wolfe *et al.*, 1992). It is important to note that, although there is a tendency for European races of the pathogen to become more complex (accumulating more of the selected virulence genes in single genotypes), the European pathogen population remains highly diverse, even to the extent of different subpopulations on winter and spring barley (Caffier *et al.*, 1996).

However, massive changes in population structure may be due to migration alone. For example, Goodwin *et al.* (1995) found that the population of *P. infestans* in New York state was replaced within a year by migrants, presumably from Mexico. As a general principle, migration (gene flow of new virulence alleles) slows the loss of diversity in highly selected pathogen populations. Other examples are the apparent homogeneity of *Stagonospora nodorum* populations worldwide, which might be due to seed-borne migration (McDonald *et al.*, 1994; Shah *et al.*, 1995), and the rapid changes in cereal rust populations in Australasia due to new migrants (Burdon *et al.*, 1982; Wellings and McIntosh, 1990).

Where host–pathogen systems are left for long periods without selective interference by humans, two-sided coevolution between host and pathogen populations can occur. Evidence for such direct coevolution in an experimental agricultural ecosystem has been found in synthetic barley populations (composite crosses; CC) that were generated by intercrossing barley varieties from around the world (see Section III,B) (Harlan and Martini, 1929; Suneson *et al.*, 1953) and that were then grown for more than 70 generations without conscious selection. From saved seed it was possible to track changes in gene frequencies over time, including resistance to *Rhynchosporium secalis* (causing scald). The interactions between host and pathogen populations were complex; the CC II and V populations, which were started at different times and from different sets of parents, did not evolve to the same level of resistance to common pathotypes of the scald pathogen, despite being grown in the same part of the field for many generations. However, it was possible to track changes in resistance gene frequencies over time that were consistent with selection and not genetic drift (Jackson *et al.*, 1978; Muona *et al.*, 1982; Saghai-Maroof *et al.*, 1983; Webster *et al.*, 1986). Levels of pathogenicity in "natural" *R. secalis* populations isolated from CC II and CC V corresponded to both the relative levels of resistance and the resistance genes in the host populations. Considerable genetic variation was maintained in both host and pathogen populations, presumably as a result of some form of frequency-dependent selection. The results were consistent with a dynamic coevolutionary process between *R. secalis* and the barley composite cross populations in which selection plays a prominent role (Allard, 1990; McDonald *et al.*, 1989a,b; McDermott *et al.*, 1989). The CC populations also maintained diversity for resistance to barley powdery mildew although this disease rarely occurs in California. Population resistance to powdery mildew increased rapidly when one of the composite crosses was grown in a mildew-prone environment in the United Kingdom (Knight *et al.*, 1988). Accumulation of resistances for barley mildew was also observed in a CC composed specifically for mildew resistance (CC XLII) (Reinhold *et al.*, 1990).

2. Host–Pathogen Coevolution in Nonagricultural Ecosystems   In contrast to agroecosystems, there is usually a high degree of interspecific and intraspecific variability in other ecosystems. Also, the biotic and abiotic environments are more diverse, the spatial distribution of plants is usually more irregular, and natural evolutionary processes, such as selection, migration, and drift, are operating (Alexander, 1988; Anikster and Wahl, 1979; Browning, 1974; Harlan, 1976; Mundt and Browning, 1985).

The differences between natural and agricultural ecosystems have consequences for the detection of coevolutionary interactions. Humans are often more sensitive than host plants to the effects of plant diseases and react by replacing whole and extensive populations of genotypes within short periods. For coevolutionary interactions to be discernible in nature, much longer timescales may be necessary. For example, over a period of 6 years, Manisterski (1987) found pronounced shifts in race composition of oat stem rust (caused by *P. graminis* f.sp. *avenae*) on wild oats (*Avena sterilis*) in Israel. However, these shifts could not be related to shifts in the resistance gene composition of the host population. The host effects might have been wiped out by the annual import of inoculum from Africa which has been documented (Dinoor, 1981).

Natural host populations are generally much smaller than agricultural populations, and this, combined with the occurrence of unfavorable environmental conditions, may reduce the size of the pathogen population to the point where random genetic drift becomes a major force. For example, Burdon and Jarosz (1992) found evidence for frequent migration of the flax rust pathogen in Australia, probably due to the frequent occurrence of bush fires that eradicate the rust but not the host population, which can survive as seed. Frequent extinction and recolonization can lead to a geographically extensive adaptation of pathogens and hosts; for the wild flax–flax rust system, the scale extends to more than several hundred kilometers (Lawrence and Burdon, 1989).

It has been claimed that disease levels in natural ecosystems are typically low (Antonovics and Levin, 1980; Borlaug, 1981; Burdon, 1978,1982; Burdon and Shattock, 1980; Dinoor and Eshed, 1984), though recent observations do not support this view (Burdon and Jarosz, 1992; Jarosz and Burdon, 1991; Kranz, 1990). One of the reasons for these conflicting viewpoints is the variability of disease pressure on the host population among seasons and locations. The fitnesses of plant and pathogen may be determined independently by factors such as competition, microclimatic effects, spatial arrangements, and additional trophic levels. Genetic interactions between pathogen and host populations may thus be less pronounced and the correlation between the genetic composition of host and pathogen population may not be detectable at the time and spatial scales that are usually possible to study. Cyclical or chaotic behavior in population dynamics may be a further factor (May, 1976).

Because of these difficulties, indirect measures are used to infer the occurrence of coevolution. For example, in environments conducive to oat crown rust (caused by *Puccinia coronata*), the genetic diversity for host resistance and for pathogen virulence was greater than in nonconducive environments (Burdon *et al.*, 1983; Oates *et al.*, 1983; Wahl, 1970). However, effects of the host on the pathogen population might be overridden by other environmental effects and additional fitness factors in the pathogen population. For example, in Israel an oat crown rust race possessing many different virulence genes allowing it to attack almost all host genotypes in the region (race 263) was still less fit than race 276, which possessed fewer virulences, because race 276 was able to reproduce and survive under a wider range of environmental conditions (e.g., Brodny *et al.*, 1988). Similar observations were made by Katsuya and Green (1967) with wheat stem rust races.

The array of possible interactive effects among hosts and pathogens in natural ecosystems is extremely complex. Coexistence of *Silene alba* and *Ustilago violaceae*, for example, may be possible only because the pioneer weed *S. alba* establishes in newly colonized areas before arrival of the pathogen, which can almost completely inhibit seed set (Alexander and Antonovics, 1988).

Although it usually causes damage, disease might also give a selective advantage to a host, for example, by making the host unattractive to herbivores or by conveying greater disease and/or drought resistance. For example, the grass *Panicum agrostides* when infected with *Balansia henningsiana,* an endophyte, produced heavier plants with more tillers than uninfected plants. Although many of these tillers were sterile, the absolute numbers of inflorescences were similar in infected and uninfected groups. Infected plants were also significantly less damaged by a leaf spot fungus (*Alternaria solani*) (Clay *et al.*, 1989). With some endophytes, therefore, the line between pathogenesis and symbiosis cannot be clearly drawn (Clay, 1988, 1990).

Pathogens can also affect host community composition (Augspurger, 1984; Zadoks, 1987), indicating the importance that higher levels of diversity, especially species mixtures, could have in disease (and insect; see Section II,C,1) management in agriculture.

In almost all studies involving nonagricultural host–pathogen systems, only the effects of pathogens on the genetics of the host but not vice versa have been considered (e.g., Alexander and Antonovics, 1988; Paul, 1990), leaving the question of the occurrence and especially of the timescale of coevolution partially unanswered. Clearly, nonagricultural ecosystems show us that, despite the evidence for host–pathogen coevolution in agricultural systems, its importance should not be overestimated. Factors such as plant competition, environmental effects, and herbivory may be as important or even more so than pathogens and act on a much shorter timescale as demonstrated, for example, by the work of Burdon and Jarosz (1992).

3. Interactions between Agricultural and Nonagricultural Ecosystems   The separation of nonagricultural and agricultural ecosystems is not clear-cut because many plants may exist in both systems and pathogens do not respect the borders. For example, many rust pathogens spend part of their life cycle on agricultural and part on wild hosts indicating the close relationship between the two systems. The eradication of barberry (*Berberis vulgare*), the alternate host for *P. graminis* f. sp. *tritici* (the wheat stem rust pathogen) from the American Midwest (fields, forests, and house gardens) in the 1920s deprived the pathogen of the possibility of sexual recombination and overwintering in the north. This had far-reaching consequences for the epidemiology, resistance breeding, and gene deployment strategies (see Section IV). Incidentally, the barberry plants had served for fencing and the introduction of barbed wire allowed their removal. Similarly, wild and cultivated *Ribes* species act as alternate hosts for the uredial and telial stages of *Cronartium ribicola* (blister rust of *Pinus cembra, P. strobus,* and *P. monticola*). The introduction of *P. strobus* (Eastern white pine) from the United States, where blister rust was absent, into Eurasia in the early 18th century allowed for the rust to jump the species barrier from *P. cembra* to *P. strobus* in Siberia, which was then reexported to the United States in 1909 on infested seedlings causing huge epidemics in white pine stands in North America (Gaeumann, 1959). It is well known that susceptible pines cannot be grown near currant bushes in gardens in rust-conducive environments (M. Finckh, personal observation; Gaeumann, 1959). A similar example occurs with the alternation of the *Gymnosporangium sabinae* (pear rust) pathogen between pear and juniper in central Europe. Weedy wild barleys have also been shown to be of importance in the evolution of the Australian population of *R. secalis* (Burdon *et al.,* 1994).

Pathogen interactions among agricultural and other environments may be particularly common where crops are grown within or in close proximity to their centers of origin and diversity and where crop relatives occur in the natural and weed flora.

C. Exploiting the Coevolution of Host and Pathogen

The understanding of the interactions between host and pathogen population genetics has led some workers to suggest alternative strategies for the use of resistance genes. The goals of such alternative strategies are as follows:

• to achieve an acceptable level of disease;

- to achieve durable disease control; and
- to control simultaneously all important diseases (and pests and possibly weeds), i.e., the development of a systems approach.

Diversity can play an important role in the control of pests and pathogens (Wolfe, 1985; Wolfe and Finckh, 1997) provided such diversity is functional. Functional diversity is diversity that limits pathogen and pest expansion and that is designed to make use of knowledge about host–pest/pathogen interactions to direct pathogen evolution (Mundt and Browning, 1985; Schmidt, 1978). The approaches to diversification can be at different taxonomic, geographic, and time levels. Such schemes span from simple cultivar mixtures to species mixtures, agroforestry, and permaculture systems.

1. Within-Field Host Diversification: Intraspecific Diversification   During the past 30–40 years, management systems for diseases and insects based on intraspecific diversification have been studied and proposed that are practicable in mechanized agricultural systems. All that is required to gain from intraspecific diversity within a field is to mix desirable but disease-susceptible crop plant genotypes with genotypes that are resistant, or at least less susceptible. In the presence of disease, there will then be a strong tendency for the overall disease level to approach that of the more resistant component rather than that of the more susceptible component (e.g., Jeger *et al.*, 1981; Wolfe *et al.*, 1981). In addition, if the components show differential susceptibility to the pathogen in question, then further advantages in disease control may be obtained; this is the area in which most of the recent research has been directed.

Intraspecific diversity in agricultural crops has been achieved through the use of near-isogenic line mixtures (NILs), multilines, or cultivar mixtures. NILs are produced by introducing different race-specific resistances into a recurrent parent (usually a commercial cultivar that has proven itself in commercial production) usually through five backcrosses that result in differentially resistant or susceptible lines of near-isogenic background (Jensen, 1988). Cultivar mixtures contain cultivars that are agronomically similar for traits such as harvest time and quality but diversified for properties such as resistance to various stresses. Multilines lie between NILs and cultivar mixtures in terms of the degree of genetic variation among the components and may be generated in different ways, e.g., by combining breeding lines from different backgrounds or by allowing for fewer backcrosses. Advantages and disadvantages of NILs, multilines, and cultivar mixtures have been summarized (Wolfe, 1985; Wolfe and Finckh, 1997).

Several mechanisms may contribute to changes in disease incidence or severity (usually a reduction) in host populations that are diverse for resistance (e.g., Barrett, 1978; Browning and Frey, 1969; Burdon and Chilvers, 1982; Mundt and Browning, 1985; Van der Plank, 1968; Wolfe, 1985; Wolfe and Finckh, 1997). The first few of these apply to all systems, whether or not there is pathogen specialization to the host in question: (i) increased distance between plants of the most susceptible component in the mixture, (ii) further restriction of pathogen spread by resistant plants that act as barriers, (iii) selection in the host population for the more competitive and/or more resistant genotypes can reduce (or sometimes increase) overall disease severity, and (iv) competitive interactions among host plants may affect plant susceptibility.

Where pathogen specialization for host genotypes does occur, then, in addition, (v) pathogens nonvirulent on a host genotype may induce resistance reactions that work against virulent races, (vi) interactions among pathogen races (e.g., competition for available host tissue) may reduce disease severity, (vii) increased distance between host plants possessing the same resistance, and (viii) barrier effects are reciprocal, i.e., plants of one host genotype will act as a barrier for the pathogen specialized to a different genotype and plants of the latter will act as a barrier for the pathogen specialized to the first genotype.

Because of the universality of these mechanisms with respect to airborne, splashborne, and some soilborne diseases, mixtures of host genotypes that vary in response to a wide range of plant diseases will tend to show an overall response to the range of diseases that is correlated with the disease levels of the components most resistant to each of the diseases. In addition, where particular components are affected by disease, there is a tendency for less affected components to compensate for them in terms of yield.

Substantial reductions of several foliar diseases have been reported in experiments with multilines and cultivar mixtures of cereals (Alexander *et al.*, 1986; Chin and Ajimilah, 1982; Bonman *et al.*, 1986; Finckh and Mundt, 1992a,b; Koizumi and Kato, 1987; McDonald *et al.*, 1989b; Wolfe, 1987; Wolfe and Barrett, 1980). Cultivar mixtures and multilines are being used on a commercial scale in the United States (wheat, oats), Denmark, Germany, Poland, Scotland, and Switzerland (barley), and Colombia (coffee) to control, respectively, wheat yellow rust, oat crown rust, barley mildew, and coffee rust (Wolfe and Finckh, 1996). When used on more than 300,000 ha in the former German Democratic Republic, powdery mildew of barley was reduced by 80% in barley cultivar mixtures within 5 years (Wolfe, 1992).

A potentially important aspect of disease restriction in diversified host populations could also be threshold density, which plays an important role in animal populations. Below a certain host density, infectious diseases do not spread: Factors such as hygiene are important here. This is analogous to plant diseases, where sanitation is especially useful when plant density is relatively low. However, if there are many host plants concentrated together, both within and among fields, the potential for inoculum production is large so that plants even in a well-managed field are likely to become diseased. If the host density is low, inoculum pressure might be so low that good management for microclimatic conditions, nutrient status, and other sanitation measures may suffice to prevent or delay epidemics.

Beyond the inherent durability of the components of the host population, durability of resistance in diversified systems depends primarily on the ability of the pathogen to evolve so-called **"super races,"** i.e., genotypes able to attack all resistance genes in the host population (Mundt, 1994), although we know of no such occurrences. This question has led to much theoretical modeling and debate together with some small-scale field experiments and data collection from areas where mixtures were used on a large scale—in former East Germany and Poland (Mundt, 1994; Wolfe and Finckh, 1997). Wolfe and Finckh concluded:

> The indications from the small-scale field experiments, models and other data, are that complex races can be selected more or less quickly leading to a reduction in effectiveness of mixtures, though this is likely to be slower than the dramatic "breakdowns" that are common in monoculture. [However,] The rates of change that do occur suggest that there should be time to allow for planned changes in the composition of host mixtures (p. 386).

Intraspecific diversification also offers possibilities for insect pest control. For example, van Emden (1966) drew attention to possibly beneficial effects of greater intravarietal diversity in the oat–frit fly (*Oscinella frit* L.) system. The flies attack the host plants only at a particular growth stage and phenological variation within an oat crop could allow for escape from attack and for subsequent compensation in growth. Varietal mixtures have been tested for the control of various insects on diverse crops; results from such experiments were generally encouraging though variable with reductions in pest attacks in some cases and increases in others (see Andow, 1991, for a comprehensive review). The mechanisms altering insect attacks in diverse systems are more complex than for the pathogen–plant interactions because, in addition to the host plant and the insect pest, predators and parasites of the insects have to be considered. As with pathogens, modes of dispersal, reproduction, and recombination are highly variable among insects. Differences in ability to disperse and to select a landing place will greatly influence the relative importance of any specific mechanism (Andow, 1991). The use of resistant hosts might lead to the concentration of insects on susceptible hosts of the same species (in adjacent fields or plants) or of alternate host species (Kennedy *et al.*, 1987). Such potential side effects have to be considered carefully and benefits and costs need to be balanced within the agroecosystem and not just within a given crop. Because of the generally less pronounced effects of varietal mixtures on insects, intrafield diversification for the management of insect pests has focused more on the use of polycultures or intercropping (e.g., Bach, 1980;

for reviews see Andow, 1991; Altieri, 1994; Altieri and Letourneau, 1982; Risch *et al.*, 1983).

Smithson and Lenne (1996) point out the large amount of research but relatively little use of intraspecific diversification in the developed world, in contrast to the little research and very extensive use in the developing world. They concluded that the overall advantages of the approach demand development of appropriate research, particularly for sustainable, subsistence agriculture.

2. Within-Field Host Diversification: Interspecific Diversification   Species mixtures have a greater potential in diversifying resistance because each component used may often be a nonhost for most if not all the pathogens that can infect the other components. This means that the level of restriction may be high for each disease and there is no response to selection for races of pathogens able to attack more than one component. However, there are some cases in which pathogens infect more than one species, particularly among soilborne pathogens. Among airborne pathogens, an example is the stem rust pathogen that attacks both barley and wheat in North America (Liu *et al.*, 1996). In this case, susceptible barley and wheat cultivars will represent a monoculture with respect to resistance to stem rust The combinations and permutations for species mixtures are almost endless and at different systems levels, including incorporation of cultivar mixtures. Some species mixtures have been widely used for many years, for example, in temperate and tropical, short-term and long-term pastures. Cereal species mixtures for feed production are currently grown on more than 1.4×10^6 ha in Poland and have been shown consistently to restrict diseases (Czembor and Gacek, 1996). Recent experiments in Poland also demonstrate possible effects on weeds in cereal and cereal–legume species mixtures (E. Gacek, personal communication; see also Bulson *et al.*, 1990). In Switzerland, the "maize-ley" system (i.e., maize planted without tillage into established leys), which is being promoted to reduce soil losses and nutrient leaching, has been shown to reduce smut (caused by *Ustilago maydis*) and attack by European stem borer (*Ostrinia nubilalis*) and aphids (Bigler *et al.*, 1995). A newly developed low-input system for growing wheat with a permanent understory of white clover (*Trifolium repens*) greatly reduces the major pest aphid species and slugs (Jones and Clements, 1993). In addition, we speculate that there should also be substantial reductions in splash-dispersed diseases, such as those caused by *Septoria* spp.

Some of the most intensive and diversified mixed-species cropping is practiced in China. Harwood and Plucknett (1981) described a multitude of different intercropping systems and remarked that diseases and insect pests were, if at all, of only minor importance. Systematic experimental research on optimizing intercropping is being carried out in China.

The gardening literature refers to numerous examples of companion planting for the limitation or prevention of a wide range of pests and pathogens; some of these appear to be generally effective and others much less so. Authors usually point out that companion planting is site specific, also, the effects may be quite specific and time dependent. Thus, although Riotte (1983) claims that marigolds (*Tagetes patula*) repell certain nematodes and thus also protect its neighbors, French (1994) warns that marigolds will push the nematodes toward their neighbors when planted together, they are beneficial only if planted before the susceptible plants.

In addition to effects on pathogens, pests, and weeds, inter- and intraspecific diversification may play an important role in buffering crops against unpredictable abiotic stresses (e.g., weather and weather-induced factors). This may be part of the reason for the often poor correlation between the disease levels in crop mixtures and the effects of those diseases on crop yields.

3. Systems Diversification: Agroforestry and Permaculture   The forms of diversification described previously refer mainly to the within-field level. The next higher level of diversification is at the whole farm or the agricultural system. There are now many examples of agroforestry systems (intercropping trees and field crops or silvopastural production systems) in the tropics (e.g., Dwivedi, 1992; Singh *et al.,* 1994), but there is also a considerable potential for integrating tree production in temperate agriculture through farm forestry (trees separated from field crops), agroforestry, or, indeed, "woody agriculture," where food, energy, and raw materials are produced by the tree crop (Rutter, 1989). One of the most holistic approaches to diversified agriculture is encompassed in the concept of permaculture (Mollison, 1990). Permaculture (from permanent agriculture) encompasses the whole living environment, which is carefully designed to produce as much as possible food, raw materials, and energy that are locally required with a minimum of external inputs; this now includes social and economic interactions as part of the concept (Whitefield, 1993). In such diversified, sustainable agricultural systems, the nature of pest and disease management may no longer be well defined because the most important strategy is avoidance of problems through system design (Bird *et al.,* 1990). Many disease control methods in traditional systems are indeed no more than strategic design (Thurston, 1992). For example, density of maize (*Zea mays*) plantings in Indonesia is adjusted by farmers to provide enough ventilation to reduce infection by downy mildew (caused by *Peronosclerospora maydis*) (Thurston, 1992). Planting depth and timing, together with site selection, are other strategic decisions often made by growers.

III. The Breeder

A. Conventional Breeding Approaches to Durable Resistance: Inherent Durability

The identification of durable resistance is a large-scale process; such resistance can only be identified *post hoc* and it cannot be predicted in a breeding program. The only possibility to increase the probability of obtaining durably resistant progeny is to use durably resistant parents. This may limit progress in other directions in the breeding program.

Flor (1956) proposed the use of resistance gene combinations or pyramids in hosts to decrease the probability of a new pathogen race occurring. The probability hypothesis relies on a lack or very low rate of recombination in the pathogen population and pyramids have therefore only a limited scope of application. Also, the greater durability that has been observed with some resistance gene combinations may be a function of the particular genes used and not of reduced probabilities (Mundt, 1990). The most sophisticated approach to the breeding of resistance gene pyramids has been taken by rice (*Oryza sativa*) pathologists and breeders who are applying detailed knowledge of the blast pathogens' (*Pyricularia grisea*) population structure and the fact that certain resistance genes appear to provide resistance to whole pathogen lineages (Zeigler *et al.*, 1994, 1995). This approach relies on a lack of migration and sexual and other forms of recombination among fungal lineages. Recent findings suggest, however, that parasexual recombination may be common in this pathogen (Zeigler *et al.*, 1997). Also, in some of the centers of diversity of rice blast, the pathogen population is in gametic equilibrium (i.e., all tested genetic markers are randomly associated, indicating that random recombination is taking place) with no evidence of lineage subdivision.

Monogenic resistances have been durable (e.g., Van Ginkel and Rajaram, 1993) and race nonspecific in some cases (Parlevliet, 1993) but are considered more likely to be overcome than oligo- or polygenically inherited resistances. Breeding for partial resistance of barley to a number of diseases has also been successful (Parlevliet, 1989). However, race-non-specific and/or partial resistance is not always available and/or is difficult to identify and incorporate into a crop (Johnson, 1984; Leonard and Mundt, 1984; Parlevliet, 1989; Van der Plank, 1968). Another difficulty is that resistances reported earlier to be race nonspecific have later been shown to be race specific (see Parlevliet, 1989, for examples).

A selection scheme for partial resistance in oats to crown rust has been in place since 1953 in Minnesota, where the rust's sexual host (*Rhamnus cathartica*) is grown in close proximity to the test plots of differentials, cultivars, and breeding lines to facilitate recombination in the pathogen

(Moore, 1966). Some durably resistant oat lines have resulted from this approach despite difficulties in selection due to induced resistance and other confounding factors (K. Leonard, personal communication).

Some progress has been made in the analysis of traits contributing to durable resistance. For example, recombinant inbred lines of the rice cultivar Moroberekan were analyzed with RFLPs to identify markers linked to resistance. A large number of quantitative trait loci (QTLs) were identified in addition to a few major resistance genes. The identified markers allow the pyramiding of QTLs together with major genes into cultivars through marker-assisted selection (Wang *et al.*, 1994).

B. Improving Intraspecific Diversity

There is a considerable literature on the analysis of genotypes for their mixing ability (e.g., Allard, 1960; Baker and Briggs, 1984; Federer *et al.*, 1982; Gacek *et al.*, 1996; Knott and Mundt, 1990). One important result is that high-yielding cultivars bred for production in pure stands are often effective in mixtures despite earlier contrary predictions (e.g., Hamblin and Donald, 1974). However, mixing ability analyses require experimentation with pure stands and mixtures and conclusions about the mixing ability of a set of cultivars cannot be extended to cultivars that have not been tested. Instead of expensive experimentation, breeding of mixtures for high performance, for example, via bulk selection, in composite crosses or by top crosses (see below) could reduce costs and increase genetic diversity without sacrificing other advantages of modern high-yielding varieties and at the same time improve ecological combining ability (see below).

For a composite cross, diverse varieties or lines are intercrossed in all combinations. The F_1 plants are then grown as a bulked hybrid population for subsequent generations without conscious selection. This has been termed an "evolutionary plant breeding method" (Harlan and Martini, 1929; Jensen, 1988; Suneson, 1956). Allard and Adams (1969) found that genotypes from such a composite cross of barley that had not been exposed to conscious selection by humans over 18 generations showed significantly higher positive yield synergisms when grown in mixtures than did a mixture of cultivars that had not coevolved. A parallel analysis of the effect of neighboring genotypes on each other led the authors to the conclusion that, in the composite cross, there had been selection both for good competitors, i.e., a genotype yields more when bordered by another genotype than by itself, and for good neighbors, i.e., a genotype induces a mean increase in yield when bordering another genotype. The "ecological combining ability" to which Allard and Adams refer is the combination of good competitive ability and neighborhood effects. In addition to greater ease of selecting for ecological combining ability, composite crosses can play an important role in genetic resource conservation (Allard, 1990).

Kilen and Keeling (1990) tested whether labor-saving bulk selection in the field could be used to improve host population resistance. They exposed a bulked soybean (*Glycine max*) population produced through a single cross with known initial frequencies of the genes *Rps1-b* and *Rps3*, conferring resistance to races 1 and 3 of Phytophthora rot (caused by *Phytophthora megasperma*), respectively, for 5 years to race 1 of the pathogen in the field. They found that the *Rps1-b* gene increased significantly in frequency, whereas the *Rps3* gene did not.

In pearl millet (*Pennisetum glaucum*) Mahalakshmi *et al.* (1992) used groups of selected landraces as pollinators for high-yielding lines (i.e., top crossing). From the progeny, they were able to greatly increase yield responsiveness to improved environmental conditions, compared with the landraces, while at the same time maintaining the high yield stability of the original landraces. The authors found that not only did the top cross hybrids always outyield their pollinators but also yield increases were greater for the low-yielding landraces. They concluded that "if this is generally the case, breeding efforts could thus be concentrated on adaptive traits such as disease and insect resistance, and their combining ability and fertility restoration in the breeding of topcross pollinators rather than on yield potential" (p. 932).

Maintenance of locally adapted genetic diversity might also be facilitated by promoting only partial replacement of farmer's seeds mixtures with improved varieties. For example, Trutmann and Pyndji (1994), in Zaire, replaced various proportions of local bean varieties with high-yielding lines resistant to angular leaf spot (caused by *Colletotrichum lindemuthianum*). At replacement rates of 25–50%, disease on the local varieties was greatly reduced and yields of both the local and the high-yielding mixture components were increased between 16 and 24%. Thus, although the susceptible varieties benefited greatly from the epidemiological effects, the resistant lines also benefited from the plant–plant interactions. Because seed is selected with great care by farmers to maintain variability of traits in subsistence agricultural systems (Sperling *et al.*, 1993), there is reason to expect that amendment of seed sources with new high-yielding resistant material and then leaving the mixtures to farmers' selection should protect them from undue loss of genetic diversity.

In summary, intraspecific diversity could be improved both by directed selection for mixture performance and by using novel approaches to the generation of diversity.

C. Interspecific Mixing and Selection

Farmers grow species mixtures in many parts of the world even though, with the exception of pastures, almost no breeding effort has been invested in such systems. For example, in Poland, despite earlier official discouragement, the area grown with cereal species mixtures increased steadily over the past 30 years. Farmers tended to maintain older varieties for this purpose that were known to perform well in mixtures (Czembor and Gacek, 1996). Formal efforts by public breeders to improve species mixtures have now started (Daellenbach *et al.*, 1996; E. Gacek, personal communication).

Selection for interspecific mixing ability is a complex issue and there is no general concept on how this should be done. Wright (1983, 1985) developed statistical methods for the selection of components for interspecific or intergenotypic mixtures and discussed long-term effects of different selection methods. The greatest problem with any mixture trial is the exponential increase of treatments required when many possible combinations are to be tested.

There has been little research on the genetic and/or physiological factors necessary for a good mixing partner. Common sense suggests that genotypes that complement each other in root and shoot architecture may compete least for below- and aboveground resources. However, competitive abilities of plants are always measured relative to others and there is still no measure for the absolute competitive ability of individual plants, thus making predictions exceedingly difficult. It would be of great interest to determine if physiological measures, such as the water and nutrient extraction potential of plants, could be used to predict their competitive ability when grown together.

There is only scant literature on breeding for multiple cropping and mixtures (see reviews by Francis, 1990; Jensen, 1988; Smith and Francis, 1986). The most important resource for breeders in breeding for mixtures may be farmers who have long-standing experience in growing mixtures. This was demonstrated by Sperling *et al.* (1993), who found that the success of heterogeneous bean varieties in Rwanda was greater when selected by local farmers than when selected by breeders. It is the interaction of farmers and breeders that produces the best results.

D. Breeder Protection

Varietal and species mixtures may consist of seed produced by different breeders leading to possible difficulties in the distribution of royalties, particularly if one or two generations of seed increase are necessary before mixtures are sold. Inevitably, the component frequencies in such seed mixtures will differ from the original frequencies, which poses legal difficulties because accurate labeling is not possible. Breeders to whom we have spoken feel that sharing of royalties proportionally to the initial mixture frequencies should be acceptable; there should also be the advantage of durable seed sales. To help avoid the problem of limiting mixture diversity to the products of a single company, it would be preferable to separate the breeding and marketing of seeds.

Registration of diversified varieties currently presents legal difficulties both in Europe and in the United States because of the Union for the Protection of New Varieties (UPOV) guidelines in Europe and the Plant Varietal Protection Act (PVPA) in the United States, which require that a variety must be uniform genetically and in appearance and be readily distinguishable from other varieties in order to be accepted for registration. This ensures that newly registered varieties are not old varieties renamed and provides protection for a genuinely novel product. However, varieties registered after the PVPA came into effect often differed only in agronomically minor and unimportant traits that could be used for differentiation and description (Kloppenburg, 1988).

Moreover, there are serious biological constraints associated with the requirement for homogeneity. Innumerable varieties and landraces have disappeared from the market because they did not fulfill the laws' requirements. For example, it has been estimated that approximately 75% of all vegetable varieties in Europe have disappeared within 10 years since the inception of UPOV in 1961 (Mooney, 1979). Indeed, great concern has been voiced worldwide about the precipitous genetic erosion in agriculture due to legislative measures, general breeding methods, and genetic engineering technology (e.g., Fowler and Mooney, 1990; Kloppenburg, 1988).

The first legislative attempt to allow for the legal maintenance and sale of genetically diverse and nonhomogeneous cereal landraces and so-called "population varieties" was initiated in Switzerland through the "Ordinance on Production and Circulation of Cereal Seed" (Verordnung über die Produktion und das Inverkehrbringen von Getreidesaatgut) from December 23, 1994, by the Department of National Economy. Varieties are here defined as: *"a totality of plants within the lowest level of a botanical taxon independent of the criteria that need to be fulfilled for varietal protection"* (translated by the authors). Under that definition, the criteria of distinguishability, homogeneity, and constancy that are necessary for breeders' rights are uncoupled from the definition of a variety (Articles 4 and 5). It is then allowed explicitly to sell and circulate nonhomogeneous materials in order to maintain the biodiversity of a species. Specific rules as to the labeling and amount of trade allowed apply and only Swiss landraces can be grown in Switzerland (Article 22). The ordinance also provides for the registration of breeding lines for the sole purpose of being used in line mixtures for disease and other stress control. Such breeding lines are subject to the same protection rules as varieties. Although the ordinance points in the right direction, there is still no provision for the release of diversified varieties, such as composite crosses or top crosses, because they do not comply with UPOV rules.

It is understandable that breeders, especially in the private sector, must be rewarded for their work and one way is through royalties. However, in the current situation, the PVPA and the UPOV convention prevent the inclusion of genetic diversity for disease, pest, and other abiotic stress resistances into population varieties. There is a need to find new solutions that will allow breeders to be compensated for their efforts without constraining the potential for improving yield stability and the durability of disease and pest resistance in practical agriculture through intracrop diversity.

IV. The Farmer

Farmers who want to be successful need to be good agriculturists, economists, and ecologists. They are confronted constantly with economical and ethical/philosophical problems requiring decisions that often are, or appear to be, mutually exclusive in their consequences.

Ecological and economical concerns can lead to the same conclusion, however. For example, during the 1980s, the East German government imposed a change in spring barley management from production in large fields of single varieties to cultivar mixtures to preserve the varieties' resistance and to save on fungicide use that had become necessary to prevent major losses from powdery mildew. Fungicide needs for mildew control decreased by as much as 80% without loss of yield and with consequent large savings over a 6-year period (Wolfe, 1992). Polish farmers have been growing cereal species mixtures on an increasing scale over the past 30 years because the mixtures greatly reduced risk (Czembor and Gacek, 1996). It is also to minimize risk that subsistence farmers in many parts of the world rely on some form of mixed cropping and often they deliberately select mixtures (Bonman *et al.,* 1986; Smithson and Lenne, 1996; Sperling *et al.,* 1993).

If it is economically attractive for farmers to produce within a diversified system, they will choose that option (M. Wolfe, unpublished data). Farmers, however, are dependent on land distribution patterns, consumer preferences, the economic system, plant variety legislation, political decisions, and society at large because it influences agriculture and food policies.

V. The Agronomist/Adviser

Agricultural advisers need to consider diversification strategies for disease control not only at the farm level but also at the interfarm level. This includes the development of strategies for the constant rotation of mixture composition over time to delay pathogen adaptation (gene deployment in time), diversification on regional or continental scales (regional gene deployment), and the integration of different available management methods. Often a conflict of interest among farming methods has to be dealt with. For example, no or reduced tillage, which is an integral part of integrated farming practices (IFP), may reduce erosion but at the same time increase certain disease problems. However, many of these problems can be taken care of by appropriate crop rotations and green manure treatments that are also part of IFP (e.g., Davis *et al.,* 1996; Teich, 1994). For example, mulches and minimum tillage are known to increase earthworm populations and these have been shown to positively affect Rhizoctonia bare patch (caused by *Rhizoctonia solani*) and take-all (caused by *Gaeumannomyces graminis*) of wheat, two diseases for which no genetic resistance is available (Stephens *et al.,* 1994 a,b,c). In cereal production in Germany, Odoerfer *et al.* (1994) showed that the use of pesticides alone was not sufficient to achieve maximum yield. This was only possible when rotations were used, indicating that factors in addition to disease play a crucial role in determining yield.

Deployment in time of resistance to the green leaf hopper (*Nephotettix virescens*), the vector of Tungro disease of rice, in combination with synchronous planting has been shown to contribute to the control of the disease in the province of South Sulawesi in Indonesia. Rice varieties with different green leafhopper resistance genes are rotated between seasons to prevent the buildup of hopper populations highly virulent on any one variety (Manwan and Sama, 1985; Sama *et al.,* 1991).

There is a possibility for "recycling" defeated resistance genes after the previously selected virulent pathogen or insect strains have decreased in frequency (Wolfe, 1992). However, this will work only if, in the absence of selective host resistance, there is selection against the corresponding virulence, leading to a drastic decrease in its frequency. This approach has been successful for the control of Hessian fly (*Mayeticola destructor*) on wheat (Gallun, 1977; Hare, 1994).

For regional gene deployment, host genotypes diverse for their resistances are planted in different fields or geographic areas. This is most useful against pathogens that are dispersed from areas where they overwinter or oversummer to areas where they usually cannot persist and the epidemic is dependent on the introduction of outside inoculum, e.g., by wind (Browning *et al.,* 1969; Van der Plank, 1968). If the resistances in the host populations differ among the regions, then a pathogen race on its way along its usual seasonal pathway will be effectively stopped when encountering only resistant hosts in an area. However, if inoculum virulent on the hosts grown in that region can persist in that region, i.e., the pathogen is residual, then the strategy will fail (Van der Plank, 1963).

Regional gene deployment strategies have been proposed for potato late blight (Van der Plank, 1963), for breaking the "Puccinia path" of oat crown rust (Browning *et al.*, 1969) and wheat stem rust (Knott, 1972; Van der Plank, 1968) in central North America, and to control barley powdery mildew and wheat stripe rust in the United Kingdom and Europe in general (Priestley and Bayles, 1982; Wolfe and Barrett, 1977, Wolfe, 1992). None of these have been or are likely to be directly successful because other priorities of farmers and breeders are deemed more important.

VI. The Consumer

A. Product Quality

Concerns over product quality are said to originate from the consumer but they are often led by the industry. For example, a longer shelf-life is arguably of greater benefit to the supermarket than to the individual consumer. It has also been questioned if consumers really are so concerned about every last blemish on fruit and vegetables or if they are conditioned to be this way (Clunies-Ross and Hildyard, 1992).

It appears that consumers, and farmers, are often led to believe that field mixtures would reduce the product quality but without real evidence. Indeed, maltsters usually mix different varieties together for malting to achieve a desired malting quality and recent data have shown that the malting quality of a batch of barley is dependent much more on the conditions under which it was grown than on the individual variety. Furthermore, variety mixtures of the same quality class can provide a more stable quality across environments than the single component varieties (Baumer and Wybranietz, 1995).

The case is similar for wheat processing, in which millers mix together different batches of grain that were all tested separately, according to the quality requirements of processors. This could also be achieved with batches of cultivar mixtures. Although, in general, it is difficult to market mixtures in Europe, no such difficulty has been found in the Pacific Northwest of the United States (C. Mundt, personal communication). Indeed, a local cooperative of IFP farmers in southern Germany who are producing on the basis of cultivar mixtures are sometimes unable to provide sufficient grain for the demands of millers and bakers (El Titi, personal communication).

Coffee is often regarded as being a particularly sensitive crop from the point of view of quality. However, beans of uniform high quality are produced in Colombia from mixtures that have been selected to be variable in terms of rust resistance but uniform for quality characteristics (Browning, 1997; Moreno-Ruiz and Castillo-Zapata, 1990); such mixtures are now produced on some 350,000 ha.

Mixtures of cereals and legumes, for example, wheat and beans, can be used directly for animal feed; planting the mixture requires much less energy than mixing the components after harvest. Any necessary adjustment to the composition can be easily checked and made.

B. Technical Aspects

A common argument against crop diversification is that such practices are inappropriate for mechanised cultivation. However, there are many simple technological changes that could make mixed cropping more attractive. For example, it would not be difficult to divide the hopper of a cereal seed drill into sections to be filled with different varieties, thus avoiding the need for premixing seed. For harvesting, modern technology for mechanical or electronic separation of seeds can be easily exploited, particularly where there are large differences among the component seeds, for example, between cereals and legumes. Electronic sorting could also be applied to the separation of different fruit varieties. In vegetable production, mixed cropping might be generally more feasible because the manual labor input is very high to start with.

Besides possible positive effects on disease, strip cropping is also used as an effective tool in crop management. Vegetables, field crops, and fruit trees are planted on high beds to enhance drainage during the wet season in China, Taiwan, Thailand, Indonesia, and the Philippines (Sorjan farming system, Hoque, 1984). In some of these Southeast Asian countries drainage ditches are used to grow additional crops of rice, water hyacinth, and edible snails (Luo and Lin, 1991), a system that has been adapted for experimental research at the Asian Vegetable Research and Development Center in Taiwan (C. Thönnissen, personal communication). Strips of cultivars rather than intimate mixtures may help reduce unwanted intergenotypic competitive effects while still providing adequate disease control (e.g., Brophy and Mundt, 1991; Huang *et al.*, 1994).

Technical constraints may often be more a problem of perception than of reality. For example, production of seed mixtures does not appear to pose an insurmountable problem to seed suppliers in the Pacific Northwest of the United States. There, variety mixtures are in great demand by wheat growers who now grow more than 100,000 ha of mixtures (C. Mundt, personal communication).

In vegetable production, which is much more labor-intensive, diversified production (e.g., strip planting) is commonly practiced. Reasons for this include the need to continuously supply the markets with fresh produce and to reduce risks from crop losses. Interestingly, in the past few years, the extension of diversity among salad leaf crops, encouraged by the consumer, has enabled the grower to spread his/her risk among a wider range of species, thus reducing dependence on lettuce (*Lactuca sativa*) and the associated problem of lettuce downy mildew (*Bremia lactucae*), which is expensive and difficult to deal with because of the genetic variation in the pathogen allowing it to overcome resistant varieties and fungicides in intensive production.

VII. Politicians and Economists

Short-term market considerations and positioning are often the most important criteria for political and economical actions rather than the long-term sustainability of systems. In addition, the global economic framework, such as the General Agreement on Trade and Tariffs and public debts, often places limits on locally and especially on environmentally useful actions. Government policies have been and must continue to be concerned with food production to meet the demands of the burgeoning world population. However, we now recognize that policies concerned solely with production are not sustainable, leading, in particular, to loss of soil by mismanagement, which is compounded by other policies leading to extension of urbanization. A large contributory aspect is dependence on monoculture (particularly at the level of variety or resistance) with consequent loss of biodiversity, which requires massive support from external inputs, particularly pesticides. It is often claimed that even higher inputs will be required to produce the food necessary to meet future demands. Clearly, some inputs will always be needed for food production in comparison to a natural ecosystem. The question is, what is agronomically, ecologically, ethically, and economically viable and acceptable? For example, there is growing evidence that the increase in production is often much smaller than expected, based on inputs such as fertilizers (Brown and Kane, 1994; Cassman *et al.*, 1995; Diwan and Kallianpur, 1985).

In our view, changes in policy toward improved methods of production, of the kind already instituted in countries such as Switzerland and Austria, can help to change the balance, rewarding strategies that lead to increased exploitation of biodiversity while effectively punishing those strategies that lead to nonsustainable exploitation of soil and monoculture. With sufficient backing from appropriate research and development (which also requires considerable changes in policy), such policies will, in the long term, through their impact on greater sustainability of production be less expensive for governments by providing the same or greater production at less cost. In this sense, crop diversification can deliver many benefits in terms of soil maintenance and enhancement; disease, pest, and weed control; and more efficient employment of labor through, for example, reduced seasonality.

We already discussed in section III,D some of the implications of the current system of breeder protection for agricultural production based on diversification. The juxtaposition of breeders' rights and interests versus the current and future needs of society at large is a serious dilemma. Plant breeders are rendering a service to society and deserve to be recognized and compensated for this service. However, the compensation must not compromise the genetic diversity and thus the future ability of our crops to defend themselves from biotic and abiotic stresses (see also Clunies-Ross, 1995). It may therefore be necessary to separate plant breeding from other agribusiness undertakings such as seed and agrochemical production.

VIII. Society at Large and Future Perspectives

Whereas agricultural production has risen steadily and at times explosively between 1950 and 1990, we cannot expect it to continue to rise at the same rate. This is because agricultural land is being lost at an ever-increasing rate due to environmental degradation, industrialization, and settlement in a world with a yet unchecked population growth (Brown and Kane, 1994). Stable agricultural systems together with a massive increase in tree planting, which reduce or eliminate erosion, salination, and other factors leading to the loss of productive land, need to be developed in order to ensure the necessary food security for human society to be stable (Brown and Kane, 1994; Mollison, 1990). An important contribution already comes from exploitation of biodiversity in cropping systems in the developing world. We believe, with Smithson and Lenne (1996), that this should be developed much further in the developed world.

To place this in a broader view, we know that ecological interactions are highly dynamic and subject to change, but that changes occur at an evolutionary timescale, which is usually far longer than the relevant human timescale of planning and operation. Normally, therefore, we have to take ecological interactions as given parameters that we can only try to learn about and to make use of to improve agricultural systems. We cannot change the basic interactions in our timescale.

Human society is also based on interactions, but changes in our perceptions of how systems need to change can be much faster than adaptation in ecosystems. Thus, although we can learn about ecology and make use of what is there, we can also learn about human society so as to be able to change and modify important factors in agriculture. In this sense, we have tried to point out that we should not allow our attitudes and laws to constrain us from rational and novel exploitation of biodiversity in our agricultural ecosystems.

Acknowledgment

We thank J. A. Browning for providing helpful comments on the manuscript.

References

Adams, M. W., Ellingboe, A. H., and Rossman, E. C. (1971). Biological uniformity and disease epidemics. *BioScience* **21**, 1067–1070.
Alexander, H. M. (1988). Spatial heterogeneity and disease in natural populations. *In* "Spatial Components of Epidemics" (M. J. Jeger, ed.), pp. 144–164. Prentice-Hall, Englewood Cliffs, NJ.
Alexander, H. M., and Antonovics, J. (1988). Disease spread and population dynamics of anther-smut infection of *Silene alba* caused by the fungus *Ustilago violacea*. *J. Ecol.* **76**, 91–104.
Alexander, H. M., Roelfs, A. P., and Cobbs, G. (1986). Effects of disease and plant competition in monocultures and mixtures of two wheat cultivars. *Plant Pathol.* **35**, 457–465.
Allard, R. W. (1960). Relationship between genetic diversity and consistency of performance in different environments. *Crop Sci.* **1**, 127–133.
Allard, R. W. (1990). The genetics of host–pathogen coevolution: Implications for genetic resource conservation. *Heredity* **81**, 1–6.
Allard, R. W., and Adams, J. (1969). Population studies in predominantly self-pollinating species. XIII. Intergenotypic competition and population structure in barley and wheat. *Am. Nat.* **103**, 620–645.
Altieri, M. A. (1994). "Biodiversity and Pest Management in Agroecosystems." Haworth, New York.
Altieri, M. A., and Letourneau, D. K. (1982). Vegetation management and biological control in agroecosystems. *Crop Prot.* **1**, 405–430.
Andow, D. A. (1991). Vegetational diversity and arthropod population responses. *Annu. Rev. Entomol.* **36**, 561–586.
Anikster, Y., and Wahl, I. (1979). Coevolution of the rust fungi on Gramineae and Liliaceae and their hosts. *Annu. Rev. Phytopathol.* **17**, 367–403.
Antonovics, J., and Levin, D. A. (1980). The ecological and genetic consequences of density-dependent regulation in plants. *Annu. Rev. Ecol. Syst.* **11**, 411–452.
Augspurger, C. K. (1984). Seedling survival of tropical tree species: Interactions of dispersal distance, light-gaps, and pathogens. *Ecology* **65**, 1705–1712.
Bach, C. E. (1980). Effects of plant density and diversity on a specialist herbivore, the striped cucumber beetle, *Acalymma vittata* (Fab.). *Ecology* **61**, 1515–1530.
Baker, R. J., and Briggs, K. G. (1984). Comparison of grain yield of uniblends and biblends of 10 spring barley cultivars. *Crop Sci.* **24**, 85–87.
Barrett, J. A. (1978). A model of epidemic development in variety mixtures. *In* "Plant Disease Epidemiology" (P. R. Scott and A. Bainbridge, eds.), pp.129–137. Blackwell, Oxford.
Baumer, M., and Wybranietz, J. (1995). Einfluss von Sortenmischungen auf die Malzqualitaet der Sommergerste. Presentation at the Workshop on Seed Production and Varietal Matters by the German Breeder Union in Hannover, Germany, 15–17 March 1995.
Bigler, F., Waldburger, M., and Frei, G. (1995). Vier Maisanbauverfahren 1990 bis 1993. Krankheiten und Schaedlinge. *Agrarforschung* **2**, 380–382.
Bird, G. W., Edens, T., Drummond, F., and Groden, E. (1990). Design of pest management systems for sustainable agriculture. *In* "Sustainable Agriculture in Temperate Zones" (C. A. Francis, C. B. Flora, and L. D. King, eds.), pp. 55–110. Wiley, New York.

Bonman, J. M., Estrada, B. A., and Denton, R. I. (1986). Blast management with upland rice cultivar mixtures. *In* "Proceedings of the Symposium on Progress in Upland Rice Research," pp. 375–382. International Rice Research Institute, Los Banos, Laguna, Philippines.

Borlaug, N. E. (1981). Increasing and stabilizing food production. *In* "Plant Breeding II" (K. J. Frey, ed.), pp. 467–492. Iowa State Univ. Press, Ames.

Brodny, U., Wahl, I., and Rotem, J. (1988). Factors conditioning dominance of race 276 of *Puccinia coronata avenae* on *Avena sterilis* populations in Israel. *Phytopathology* **78**, 135–139.

Brophy, L. S., and Mundt, C. C. (1991). Influence of plant spatial patterns on disease dynamics, plant competition and grain yield in genetically diverse wheat populations. *Agric. Ecosyst. Environ.* **35**, 1–12.

Brown, J. K. M. (1995). Pathogen's responses to the management of disease resistance genes. *Adv. Plant Pathol.* **11**, 75–102.

Brown, J. K. M., and Wolfe, M. S. (1990). Structure and evolution of a population of *Erysiphe graminis* f. sp. *hordei*. *Plant Pathol.* **39**, 376–390.

Brown, J. K. M., Foster, E. M., and O'Hara, R. B. (1997). Adaptation of pathogen populations to durable and non-durable resistance: The example of cereal powdery mildew. *In* "The Gene-for-Gene Relationship in Plant Parasite Interactions" (I. R. Crute, E. Holub, and J. J. Burdon, eds.), in press. CAB International, Wallingford, UK.

Brown, L. R., and Kane, H. (1994). "Full House. Reassessing the Earth's Population Carrying Capacity." Norton, New York.

Browning, J. A. (1974). Relevance of knowing about natural ecosystems to development of pest management programs in agroecosystems. *Proc. Am. Phytopathol. Soc.* **1**, 191–199.

Browning, J. A. (1997). A unifying theory of the genetic protection of crop plant populations from diseases. *In* "Disease Resistance from Crop Progenitors and Other Wild Relatives" (I. Wahl, G. Fischbeck, and J. A. Browning, eds.). Springer-Verlag, Berlin/New York.

Browning, J. A., and Frey, K. J. (1969). Multiline cultivars as a means of disease control. *Annu. Rev. Phytopathol.* **7**, 355–382.

Browning, J. A., Simons, M. D., Frey, K. J., and Murphy, H. C. (1969). Regional deployment for conservation of oat crown rust resistance genes. *Spec. Rep. Iowa Agric. Home Econ. Exp. Stn.* **64**, 49–56.

Bulson, H. A. J., Snaydon, R. W., and Stopes, C. E. (1990). Intercropping autumn-sown field beans and wheat: Effects on weeds under organic farming conditions. *In* "Crop Protection in Organic and Low Input Agriculture" (R. J. Unwin, ed.), pp. 55–62. The British Crop Protection Council, Farnham, Surrey, UK.

Burdon, J. J. (1978). Mechanisms of disease control in heterogeneous populations—An ecologists view. *In* "Plant Disease Epidemiology" (P. R. Scott and A. Bainbridge, eds.), pp.193–200. Blackwell, Oxford.

Burdon, J. J. (1982). The effect of fungal pathogens on plant communities. *In* "The Plant Community as a Working Mechanism" (E. I. Newman, ed.), pp. 99–112. Blackwell, Oxford.

Burdon, J. J., and Chilvers, G. A. (1982). Host density as a factor in disease ecology. *Annu. Rev. Phytopathol.* **20**, 143–166.

Burdon, J. J., and Jarosz, A. M. (1992). Temporal variation in the racial structure of flax rust (*Melampsora lini*) populations growing on natural stands of wild flax (*Linum marginale*): Local versus metapopulation dynamics. *Plant Pathol.* **41**, 165–179.

Burdon, J. J., and Shattock, R. C. (1980). Disease in plant communities. *Appl. Biol.* **5**, 145–219.

Burdon, J. J., Marshall, D. R., Luig, N. H., and Gow, D. J. S. (1982). Isozyme studies on the origin and evolution of *Puccinia graminis* f. sp. *tritici* in Australia. *Aust. J. Biol. Sci.* **35**, 231–238.

Burdon, J. J., Oates, J. D., and Marshall, D. R. (1983). Interactions between *Avena* and *Puccinia* species. I. The wild hosts: *Avena barbata* Pott ex Link, *A. fatua* L. *A. ludoviciana* Durieu. *J. Appl. Ecol.* **20**, 571–584.

Burdon, J. J., Abbott, D. C., Brown, A. H. D., and Brown, J. S.(1994). Genetic structure of the scald pathogen (*Rhynchosporium secalis*) in South East Australia: Implications for control strategies. *Aust. J Agr. Res.* **45**, 1445–1454.

Caffier, V., Hoffstadt, T., Leconte, M., and de Vallavieille-Pope, C. (1996). Seasonal changes in French populations of barley powdery mildew. *Plant Pathol.* **45**, 454–468.

Cassman, K. G., De Datta, S. K., Olk, D. C., Alcantara, J., Samson, M., Descalsota, J., and Dizon, M. (1995). Yield decline and the nitrogen economy of long-term experiments on continuous, irrigated rice systems in the tropics. *In* "Soil Management. Experimental Basis for Sustainability and Environmental Quality" (R. Lal and B. A. Stewart, eds.), pp. 181–222. Lewis, Boca Raton, FL.

Chin, K. M., and Ajimilah, N. H. (1982). Rice variety mixtures. *In* "Proceedings of the Padi Workshop, 5–6 January 1982, ED–PB," pp. 203–216. Bumbong Lima, Province Wellesly, Malaysia.

Clay, K. (1988). Fungal endophytes of grasses: A defensive mutalism between plants and fungi. *Ecology* **69**, 10–16.

Clay, K. (1990). The impact of parasitic and mutualistic fungi on competitive interactions among plants. *In* "Perspectives on Plant Competition" (J. B. Grace and D. Tilman, eds.), pp. 391–413. Academic Press, San Diego.

Clay, K., Cheplick, P., and Marks, S. (1989). Impact of the fungus *Balansia henningsiana* on *Panicum agrostoides:* Frequency of infection, plant growth and reproduction, and resistance to pests. *Oecologia* **80**, 374–380.

Clunies-Ross, T. (1995). Mangolds, manure and mixtures. The importance of crop diversity on British farms. *Ecologist* **25**, 181–187.

Clunies-Ross, T., and Hildyard, N. (1992). "The Politics of Industrial Agriculture." Earthscan, London.

Czembor, H. J., and Gacek, E. S. (1996). The use of cultivar and species mixtures to control diseases and for yield improvement in cereals in Poland. *In* "Proceedings of the 3rd Workshop on Integrated Control of Cereal Mildews across Europe, Nov. 5–10 1994, Kappel a. Albis, Switzerland" (E. Limpert, M. R. Finckh, and M. S. Wolfe, eds.), pp. 177–184. Office for Official Publications of the EC, Brussels, Belgium.

Daellenbach, G. C., Finckh, M. R., Gacek, E. S., and Wolfe, M. S. (1996). Competitive interactions in mixtures of barley, oat and wheat in the presence and absence of powdery mildew in field and greenhouse experiments. *In* "Proceedings of the Third Workshop on Integrated Control of Cereal Mildews Across Europe. 5–9 Nov. 1994, Kappel a. Albis, Switzerland" (E. Limpert, M. R. Finckh, and M. S. Wolfe, eds.), pp. 197–202. Office for Official Publications of the EC, Brussels, Belgium.

Danial, D. L., Stubbs, R. W., and Parlevliet, J. E. (1994). Evolution of virulence patterns in yellow rust races and its implications for breeding for resistance in wheat in Kenya. *Euphytica* **80**, 165–170.

Davis, J. R., Huisman, O. C., Westermann, D. T., Hafez, S. L., Everson, D. O., Sorensen, L. H., and Schneider, A. T. (1996). Effects of green manures on Verticillium wilt of potato. *Phytopathology* **86**, 444–453.

Dinoor, A. (1981). Epidemics caused by fungal pathogens in wild and crop plants. *In* "Pests, Pathogens, and Vegetation" (J. M. Thresh, ed.), pp. 143–158. Pitman, Boston.

Dinoor, A., and Eshed, N. (1984). The role and importance of pathogens in natural plant communities. *Annu. Rev. Phytopathol.* **22**, 443–466.

Diwan, R., and Kallianpur, R. (1985). Biological technology and land productivity: Fertilizers and food production in India. *World Dev.* **13**, 627–638.

Duvick, D. N. (1977). United States crops. *Ann. N.Y. Acad. Sci.* **287**, 86–96.

Dwivedi, A. P. (1992). "Agroforestry Principles and Practices." IBH, New Delhi.

Ehrlich, P. R., and Raven, P. H. (1964). Butterflies and plants: A study in coevolution. *Evolution* **18**, 586–608.

Federer, W. T., Conningale, J. C., Rutger, J. N., and Wijesinha, A. (1982). Statistical analyses of yields from uniblends and biblends of eight dry bean cultivars. *Crop Sci.* **22**, 111–115.

Finckh, M. R., and Mundt, C. C. (1992a). Stripe rust, yield, and plant competition in wheat cultivar mixtures. *Phytopathology* **82**, 905–913.

Finckh, M. R., and Mundt, C. C. (1992b). Plant competition and disease in genetically diverse wheat populations. *Oecologia* **91**, 82–92.

Flor, H. H. (1953). Epidemiology of flax rust in the North Central States. *Phytopathology* **43**, 624–628.

Flor, H. H. (1956). The complementary genic systems in flax and flax rust. *Adv. Genet.* **8**, 29–54.

Fowler, C., and Mooney, P. R. (1990). "Shattering: Food Politics, and the Loss of Genetic Diversity." Univ. of Arizona Press, Tucson.

Francis, C. A. (1990). Breeding hybrids and varieties for sustainable systems. *In* "Sustainable Agriculture in Temperate Zones" (C. A. Francis, C. B. Flora, and L. D. King, eds.), pp. 24–54. Wiley, New York.

French, J. (1994). "The Organic Garden Problem Solver." Angus & Robertson, Sydney.

Futuyma, D. J., and Slatkin, M. (1983). Introduction. *In* "Coevolution" (D. J. Futuyma and M. Slatkin, eds.). Sinauer, Sunderland, MA.

Gacek, E. S., Czembor, H. J., and Nadziak, J. (1996). Disease restriction, grain yield and its stability in winter barley cultivar mixtures. *In* "Proceedings of the Third Workshop on Integrated Control of Cereal Mildews across Europe. 5–9 Nov. 1994 Kappel a. Albis, Switzerland" (E. Limpert, M. R. Finckh, and M. S. Wolfe, eds.), pp. 185–190. Office for Official Publications of the EC, Brussels, Belgium.

Gaeumann, E. (1959). "Rostpilze Mitteleuropas." Büchler, Bern.

Gallun, R. L. (1977). Genetic basis of Hessian fly epidemics. *Ann. N. Y. Acad. Sci.* **287**, 223–229.

Goodwin, S. B., Sujkowski, L. S., Dyer, A. T., Fry, B. A., and Fry, W. E. (1995). Direct detection of gene flow and probable sexual reproduction of *Phytophthora infestans* in northern North America. *Phytopathology* **85**, 473–479.

Hamblin, J., and Donald, C. M. (1974). The relationship between plant form, competitive ability and grain yield in a barley cross. *Euphytica* **23**, 535–542.

Hare, J. D. (1994). Status and prospects for an integrated approach to the control of rice planthoppers. *In* "Planthoppers: Their Ecology and Management" (R. F. Denno and T. J. Perfect, eds.), pp. 614–632. Chapman & Hall, New York.

Harlan, H. V., and Martini, M. L. (1929). A composite hybrid mixture. *J. Am. Soc. Agron.* **21**, 487–490.

Harlan, J. R. (1972). Genetics of disaster. *J. Environ. Qual.* **1**, 212–215.

Harlan, J. R. (1976). Disease as a factor in plant evolution. *Annu. Rev. Phytopathol.* **14**, 31–51.

Harper, J. L. (1977). "Population Biology of Plants." Academic Press, New York.

Harper, J. L. (1982). After description. *In* "The Plant Community as a Working Mechanism" (E. I. P. Newman, ed.), pp. 11–25. Blackwell, Oxford.

Harwood, R. R., and Plucknett, D. L. (1981). Vegetable cropping systems. *In* "Vegetable Farming Systems in China" (D. Plucknett, ed.), pp. 45–118. Westview, Boulder, CO.

Hoque, M. Z. (1984). On farm research and management. *In* "Cropping Systems in Asia." International Rice Research Institute, P.O. Box 933, 1099 Manila, Philippines.

Huang, R., Kranz, J., and Welz, H. G. (1994). Selection of pathotypes of *Erysiphe graminis* f. sp. *hordei* in pure and mixed stands of spring barley. *Plant Pathol.* **43**, 458–470.

Jackson, L. F., Kahler, A. L., and Webster, R. K. (1978). Conservation of scald resistance in barley composite cross populations. *Phytopathology* **68**, 645–650.

Jarosz, A. M., and Burdon, J. J. (1991). Host–pathogen interactions in natural population of *Linum marginale* and *Melampsora lini:* II. Local and regional variation in patterns of resistance and racial structure. *Evolution* **45**, 1618–1627.

Jeger, M. J., Jones, D. G., and Griffiths, E. (1981). Disease progress of non-specialized fungal pathogens in intraspecific mixed stands of cereal cultivars. II. Field experiment. *Ann. Appl. Biol.* **98**, 199–210.

Jensen, N. F. (1952). Intra-varietal diversification in oat breeding. *Agron. J.* **44,** 30–34.

Jensen, N. F. (1988). "Plant Breeding Methodology." Wiley-Interscience, New York.

Johnson, R. (1984). A critical analysis of durable resistance. *Annu. Rev. Phytopathol.* **22,** 309–330.

Johnson, R. (1993). Durability of disease resistance in crops: Some closing remarks about the topic and the symposium. *In* "Durability of Disease Resistance" (T. Jacobs and J. E. Parlevliet, eds.), pp. 283–300. Kluwer, Dordrecht.

Johnson, T. (1961). Man-guided evolution in plant rusts. *Science* **133,** 357–362.

Johnson, T., Green, G. J., and Samborski, D. J. (1967). The world situation of the cereal rusts. *Annu. Rev. Phytopathol.* **5,** 183–200.

Jones, L., and Clements, R. O. (1993). Development of a low-input system for growing wheat (*Triticum vulgare*) in a permanent understorey of white clover (*Trifolium repens*). *Ann. Appl. Biol.* **123,** 109–119.

Jorgensen, J. H. (1994). Genetics of powdery mildew resistance in barley. *Crit. Rev. Plant Sci.* **13,** 97–119.

Katsuya, K., and Green, G. T. (1967). Reproductive potentials of races 15B and 56 of wheat stem rust. *Can. J Bot.* **45,** 1077–1091.

Kennedy, G. G., Gould, F., dePonti, O. M. B., and Stinner, R. E. (1987). Ecological, agricultural, genetic, and commercial considerations in the deployment of insect-resistant germplasm. *Environ. Entomol.* **16,** 327–338.

Kilen, T. C., and Keeling, B. L. (1990). Gene frequency changes in soybean bulk populations exposed to Phytophthora rot. *Crop Sci.* **30,** 575–578.

Kiley-Worthington, M. (1993). "Ecological Food First Farming." Souvenir Press, UK.

Kloppenburg, J. R. (1988). "First the Seed. The Political Economy of Plant Biotechnology." Cambridge Univ. Press, Cambridge, UK.

Knight, S. C., Barrett, J. A., Spencer, J. L., Hayler, J. B. R., and Ibrahim, K. M. (1988). Coevolution of composite cross XLII of barley and the powdery mildew fungus *Erysiphe graminis*. "Abstracts of Papers, Fifth International Congress of Plant Pathology, Kyoto, Japan, 20–27 August 1988, p. 273.

Knott, D. R. (1972). Using race-specific resistance to manage the evolution of plant pathogens. *J. Environ. Qual.* **1,** 227–231.

Knott, E. A., and Mundt, C. C. (1990). Mixing ability analysis of wheat cultivar mixtures under diseased and non-diseased conditions. *Theor. Appl. Genet.* **80,** 313–320.

Koizumi, S., and Kato, H. (1987). Effect of mixed plantings of susceptible and resistant rice cultivars on leaf blast development. *Ann. Phytopathol. Soc. Jpn.* **53,** 28–38.

Kolmer, J. A. (1991). Phenotypic diversity in two populations of *Puccinia recondita* f.sp. *tritici* in Canada during 1931–1987. *Phytopathology* **81,** 311–315.

Kranz, J. (1990). Fungal disease in multispecies plant communities. Tansley review No. 28. *New Phytol.* **116,** 383–405.

Lawrence, G. J., and Burdon, J. J. (1989). Flax rust from *Linum marginale*. Variation in a natural host–pathogen interaction. *Can. J. Bot.* **67,** 3192–3198.

Leonard, K. J., and Czochor, R. J. (1980). Theory of genetic interactions among populations of plants and their pathogens. *Annu. Rev. Phytopathol.* **18,** 337–358.

Leonard, K. J., and Mundt, C. C. (1984). Methods for estimating epidemiological effects of quantitative resistance to plant diseases. *Theor. Appl. Genet.* **67,** 219–230.

Liu, J. Q., Harder, D. E., and Kolmer, J. A. (1996). Competitive ability of races of *Puccinia graminis* f.sp. *tritici* on three barley cultivars and a susceptible wheat cultivar. *Phytopathology* **76,** 627–632.

Luo, S. M., and Lin, R. J. (1991). High bed–low ditch system in the Pearl river delta, south China. *Agric. Ecosyst. Environ.* **36,** 101–109.

Mahalakshmi, V., Bidinger, F. R., Rao, K., and Raju, D. S. (1992). Performance and stability of pearl millet topcross hybrids and their variety pollinators. *Crop Sci.* **32,** 928–932.

Manisterski, J. (1987). Parasitic specialisation of *Puccinia graminis* f. sp. *avenae* in Israel during 1971–1977. *Plant Dis.* **71**, 842–844.

Manwan, I., and Sama, S. (1985). Use of varietal rotation in the management of tungro disease in Indonesia. *Indonesia Agric. Res. Dev. J.* **7**, 43–48.

May, R. M. (1976). Simple mathematical models with very complicated dynamics. *Nature* **261**, 459–467.

McDermott, J. M., McDonald, B. A., Allard, R. W., and Webster, R. K. (1989). Genetic variability for pathogenicity, isozyme, ribosomal DNA and colony color variants in populations of *Rhynchosporium secalis*. *Genetics* **122**, 561–565.

McDonald, B. A., McDermott, J. M., Allard, R. W., and Webster, R. K. (1989a). Coevolution of host and pathogen populations in the *Hordeum vulgare–Rhynchosporium secalis* pathosystem. *Proc. Nat. Acad. Sci. USA* **86**, 3924–3927.

McDonald, B. A., McDermott, J. M., Goodwin, S. B., and Allard, R. W. (1989b). The population biology of host–parasite interactions. *Annu. Rev. Phytopathol.* **27**, 77–94.

McDonald, B. A., Miles, J., Nelson, L. R., and Pettyway, R. E. (1994). Genetic variability in nuclear DNA of field populations of *Stagonospora nodorum*. *Phytopathology* **84**, 250–255.

Mew, T. W., Vera Cruz, C. M., and Medalla, E. S. (1992). Changes in race frequencies of *Xanthomonas oryzae* pv. *oryzae* in response to the planting of rice cultivars. *Plant Dis.* **76**, 1029–1032.

Mollison, B. (1990). ''Permaculture. A Practical Guide for a Sustainable Future.'' Island Press, Washington, DC.

Mooney, P. R. (1979). ''Seeds of the Earth. A Public or Private Resource. Published by Inter Pares (Ottawa) for the Canadian Council for Intern. Co-operation and the Intern. Coalition for Development Action (London).'' Mutual Press, Ottawa, Canada.

Moore, M. B. (1966). Buckthorns and oat breeding for resistance to crown rust. *Phytopathology* **56**, 891. [Abstract]

Moreno-Ruiz, G., and Castillo-Zapata, J. (1990). The variety Colombia: A variety of coffee with resistance to rust (*Hemileia vastatrix* Berk. & Br.). *Cenicafe Chinchiná–Caldas–Colombia Tech. Bull.* **9**, 1–27.

Mundt, C. C. (1990). Probability of mutation to multiple virulence and durability of resistance gene pyramids. *Phytopathology* **80**, 221–223.

Mundt, C. C. (1994). Techniques for managing pathogen co-evolution with host plants to prolong resistance. *In* ''Rice Pest Science and Management'' (P. S. Teng, K. L. Heong, and K. Moody, eds.), pp. 193–205. International Rice Research Institute, P.O. Box 933, 1099 Manila, Philippines.

Mundt, C. C., and Browning, J. A. (1985). Genetic diversity and cereal rust management. *In* ''The Cereal Rusts Vol. II'' (A. P. Roelfs and W. R. Bushnell, eds.), pp. 527–559. Academic Press, Orlando, FL.

Muona, O., Allard, R. W., and Webster, R. K. (1982). Evolution of resistance to *Rhynchosporium secalis* (Oud.) Davis in Barley Composite Cross II. *Theor. Appl. Genet.* **61**, 209–214.

Oates, J. D., Burdon, J. J., and Brouwer, J. D. (1983). Interactions between *Avena* and *Puccinia* species. II. The pathogens, *Puccinia coronata* CDA and *Puccinia graminis* Pers. f. sp. *avenae* Eriks. & Henn. *J. Appl. Ecol.* **20**, 585–596.

Odoerfer, A., Obst, A., and Pommer, G. (1994). The effects of different leaf crops in a long lasting monoculture with winter wheat. 2. Disease development and effects of phytosanitary measures. *Agribiol. Res.* **47**, 56–66.

Parlevliet, J. E. (1989). Identification and evaluation of quantitative resistance. *In* ''Plant Disease Epidemiology. Volume 2: Genetics, Resistance, and Management'' (K. J. Leonard and W. E. Fry, eds.), pp. 215–245. McGraw-Hill, New York.

Parlevliet, J. E. (1993). What is durable resistance, a general outline. *In* ''Durability of Disease Resistance'' (T. Jacobs and J. E. Parlevliet, eds.), pp. 23–40. Kluwer, Dordrecht.

Paul, N. D. (1990). Modification of the effects of plant pathogens by other components of natural ecosystems. *In* "Pests, Pathogens, and Plant Communities" (J. J. Burdon and S. R. Leather, eds.). Blackwell, Oxford.

Priestley, R. H., and Bayles, R. A. (1982). Evidence that varietal diversification can reduce the spread of cereal diseases. *J. Natl. Inst. Agric. Bot.* **16**, 31–38.

Reinhold, M., Bjarko, M. E., Sands, D. C., and Bockelman, H. E. (1990). Changes in resistance to powdery mildew in a barley composite cross. *Can. J. Bot.* **68**, 916–919.

Riotte, L. (1983). "Roses Love Garlic. Secrets of Companion Planting with Flowers." Storey Communications, Pownal, VT.

Risch, S. J., Andow, D., and Altieri, M. A. (1983). Agroecosystem diversity and pest control: Data, tentative conclusions and new research directions. *Environ. Entomol.* **12**, 625–629.

Rutter, P. A. (1989). Reducing earth's "greenhouse" CO_2 through shifting staples production to woody plants. *In* "Proceedings of the Second North American Conference on Preparing for Climate Change, Dec. 6–8, 1988, Washington DC." The Climate Institute, Washington.

Saghai-Maroof, M. A., Webster, R. K., and Allard, R. W. (1983). Evolution of resistance to scald, powdery mildew, and net blotch in barley composite cross II populations. *Theor. Appl. Genet.* **66**, 279–283.

Sama, S., Hasanuddin, A., Manwan, I., Cabunagan, R., and Hibino, H. (1991). Integrated management of rice tungro disease in South Sulawesi, Indonesia. *Crop Prot.* **10**, 34–40.

Schmidt, R. A. (1978). Diseases in forest ecosystems: The importance of functional diversity. *In* "Plant Disease: An Advanced Treatise, Vol 2" (J. G. Horsfall and E. B. Cowling, eds.), pp. 287–315. Academic Press, New York.

Schumann, G. L. (1991). "Plant Diseases: Their Biology and Social Impact." APS Press, St. Paul, MN.

Shah, D., Bergstrom, G. C., and Ueng, P. P. (1995). Initiation of septoria blotch epidemics in winter wheat by seedborne *Stagonospora nodorum*. *Phytopathology* **85**, 452–457.

Singh, P., Pathak, P. S., and Roy, M. M. (1994). "Agroforestry Systems for Sustainable Land Use." IBH, New Delhi.

Smith, M. E., and Francis, C. A.(1986). Breeding for multiple cropping systems. *In* "Multiple Cropping Systems" (C. A. Francis, ed.), pp. 219–249. Macmillan, New York.

Smithson, J. B., and Lenne, J. M. (1996). Varietal mixtures: A viable strategy for sustainable productivity in subsistence agriculture. *Ann. Appl. Biol.* **128**, 127–158.

Sperling, L., Loevinsohn, M. E., and Ntabomvura, B. (1993). Rethinking the farmer's role in plant breeding: Local bean experts and on-station selection in Rwanda. *Exp. Agric.* **29**, 509–519.

Stakman, E. C. (1947). Plant diseases are shifting enemies. *Am. Scientist* **35**, 321–350.

Stephens, P. M., Davoren, C. W., Doube, B. M., and Ryder, M. H. (1994a). Ability of the lumbricid earthworms *Aporrectodea rosea* and *Aporrectodea trapezoides* to reduce the severity of take-all under greenhouse and field conditions. *Soil Biol. Biochem.* **26**, 1291–1297.

Stephens, P. M., Davoren, C. W., Ryder, M. H., and Doube, B. M. (1994b). Influence of the earthworms *Aporrectodea rosea* and *Aporrectodea trapezoides* on *Rhizoctonia solani* disease of wheat seedlings and the interaction with a surface mulch of cereal-pea straw. *Soil Biol. Biochem.* **26**, 1285–1287.

Stephens, P. M., Davoren, C. W., Ryder, M. H., Doube, B. M., and Correll, R. L. (1994c). Field evidence for reduced severity of Rhizoctonia bare- patch disease of wheat, due to the presence of the earthworms *Aporrectodea rosea* and *Aporrectodea trapezoides*. *Soil Biol. Biochem.* **26**, 1495–1500.

Stevens, N. E. (1942). How plant breeding programs complicate plant disease problems. *Science* **95**, 313–316.

Suneson, C. A. (1956). An evolutionary plant breeding method. *Agron. J.* **48**, 188–191.

Suneson, C. A., Stevens, H. (1953). Studies with bulked hybrid populations of barley. *Tech. Bull. U.S. Dept. Agric.* **1067**, 1–14.

Teich, A. H. (1994). Disease control in wheat using ecological principles. *Genet. Pol.* **35B**, 127–135.

Thompson, J. N. (1982). Evolutionary ecology, interaction, and coevolution. *In* "Interaction and Coevolution" (J. N. Thompson, ed.). Wiley, New York.

Thurston, H. D. (1992). "Sustainable Practices for Plant Disease Management in Traditional Farming Systems." Westview, Boulder, CO.

Tilman, D., Wedin, D., and Knos, J. (1996). Productivity and sustainability influenced by biodiversity in grassland ecosystems. *Nature* **379**, 718–720.

Trutmann, P., and Pyndji, M. M. (1994). Partial replacement of local common bean mixtures by high yielding angular leaf spot resistant varieties to conserve local genetic diversity while increasing yield. *Ann. Appl. Biol.* **125**, 45–52.

Ullstrup, A. J. (1972). The impacts of the Southern com leaf blight epidemics of 1970–1971. *Annu. Rev. Phytopathol.* **10**, 37–50.

Van der Plank, J. E. (1963). "Plant Diseases: Epidemics and Control." Academic Press, New York.

Van der Plank, J. E. (1968). "Disease Resistance in Plants." Academic Press, New York.

van Emden, H. F. (1966). Plant insect relationships and pest control. *World Rev. Pest Control* **5**, 115–123.

Van Ginkel, M., and Rajaram, S. (1993). Breeding for durable resistance to diseases in wheat: An international perspective. *In* "Durability of Disease Resistance" (T. Jacobs and J. E. Parlevliet, eds.), pp. 259–272. Kluwer, Dordrecht.

Wahl, I. (1970). Prevalence and geographic distribution of resistance to crown rust in *Avena sterilis*. *Phytopathology* **60**, 746–749.

Wang, G. L., Mackill, D. J., Bonman, J. M., McCouch, S. R., Champoux, M. C., and Nelson, R. J. (1994). RFLP mapping of genes conferring complete and partial resistance to blast in a durably resistant rice cultivar. *Genetics* **136**, 1421–1434.

Webster, R. K., Saghai-Maroof, M. A., and Allard, R. W. (1986). Evolutionary response of Barley Composite Cross II to *Rhynchosporium secalis* analyzed by pathogenic complexity and by gene-by-race relationships. *Phytopathology* **76**, 661–668.

Wellings, C. R., and McIntosh, R. A. (1990). *Puccinia striiformis* f. sp. *tritici* in Australasia: Pathogenic changes during the first 10 years. *Plant Pathol.* **39**, 316–325.

Whitefield, P. (1993). "Permaculture in a Nutshell." Permanent, Clanfield Hampshire, UK.

Wolfe, M. S. (1973). Changes and diversity in populations of fungal pathogens. *Ann. Appl. Biol.* **75**, 132–136.

Wolfe, M. S. (1984). Trying to understand and control powdery mildew. *Plant Pathol.* **33**, 451–466.

Wolfe, M. S. (1985). The current status and prospects of multiline cultivars and variety mixtures for disease resistance. *Annu. Rev. Phytopathol.* **23**, 251–273.

Wolfe, M. S. (1987). The use of variety mixtures to control diseases and stabilize yield. *In* "Breeding Strategies for Resistance to the Rusts of Wheat," pp.91–99. CIMMYT, Mexico D.F.

Wolfe, M. S. (1992). Barley diseases: Maintaining the value of our varieties. *In* "Barley Genetics VI" (L. Munk, ed.), pp. 1055–1067. Munksgaard, Copenhagen.

Wolfe, M. S., and Barrett, J. A. (1977). Population genetics of powdery mildew epidemics. *Ann. N.Y. Acad. Sci.* **287**, 151–163.

Wolfe, M. S., and Barrett, J. A. (1980). Can we lead the pathogen astray? *Plant Dis.* **64**, 148–155.

Wolfe, M. S., and Finckh, M. R. (1997). Diversity of host resistance within the crop: Effects on host, pathogen and disease. *In* "Plant Resistance to Fungal Diseases" (H. Hartleb, R. Heitefuss, and H. H. Hoppe, eds.), pp. 378–400. Fischer-Verlag, Stuttgart.

Wolfe, M. S., and McDermott, J. M. (1994). Population genetics of plant pathogen interactions: The example of the *Erysiphe graminis–Hordeum vulgare* pathosystem. *Annu. Rev. Phytopathol.* **32**, 89–113.

Wolfe, M. S., and Schwarzbach, E. (1978). Patterns of race changes in powdery mildews. *Annu. Rev. Phytopathol.* **16,** 159–180.

Wolfe, M. S., Barrett, J. A., and Jenkins, J. E. E. (1981). The use of mixtures for disease control. *In* "Stategies for the Control of Cereal Diseases" (J. F. Jenkyn and R. T. Plumb, eds.), pp.73–80. Blackwell, Oxford.

Wolfe, M. S., Braendle, U. E., Koller, B., Limpert, E., McDermott, J. M., Mueller, K., and Schaffner, D. (1992). Barley mildew in Europe: Population biology and host resistance. *Euphytica* **63,** 125–139.

Wright, A. J. (1983). The expected efficiencies of some methods of selection of components for inter-genotypic mixtures. *Theor. Appl. Genet.* **67,** 45–52.

Wright, A. J. (1985). Selection for improved yield in inter-specific mixtures or intercrops. *Theor. Appl. Genet.* **69,** 399–407.

Zadoks, J. C. (1987). The function of plant pathogenic fungi in natural communities. *In* "Disturbance in Grasslands" (J. Van Andel, ed.), pp.201–207. Junk, Dordrecht.

Zeigler, R. S., Tohme, J., Nelson, R. J., Levy, M., and Correa-Victoria, F. J. (1994). Lineage exclusion: A proposal for linking blast population analysis to resistance breeding. *In* "Rice Blast Disease" (R. S. Zeigler, S. A. Leong, and P. S. Teng, eds.), pp. 267–292. CAB International International Rice Research Institute, Wallingford, UK P.O. Box 933, 1099 Manila, Philippines.

Zeigler, R. S., Cuoc, L. X., Scott, R. P., Bernardo, M. A., Chen, D. H., Valent, B., and Nelson, R. J. (1995). The relationship between lineage and virulence in *Pyricularia grisea* in the Philippines. *Phytopathology* **85,** 443–451.

Zeigler, R. S., Scott, R. P., Leung, H., Bordeos, A. A., Kumar, J., and Nelson, R. J. (1997). Evidence of parasexual exchange of DNA in the rice blast fungus challenges its exclusive clonality. *Phytopathology* **87,** 284–294.

8

Plant–Arthropod Interactions in Agroecosystems

D. K. Letourneau

I. Introduction: Crops as Resources

As growers plan the temporal and spatial pattern of crops in their fields and make decisions about management practices to optimize the productivity of these crops, they are inadvertently designing the local resource base for the arthropod community, including herbivores and their natural enemies. This chapter focuses on ecological principles that can be applied within agroecosystem management schemes to change the quality of the crop as a resource. The aim is to reduce the colonization, growth, survival, and reproduction of phytophagous arthropods as consumers. Many of these ecological concepts of plant–arthropod interactions and consumer dynamics are derived from research over the past three decades. The 1970s produced a rich body of theory on predator–prey, parasitoid–host, and herbivore–plant interactions. The 1980s brought challenges of theory based on two trophic levels and the emergence of research on multitrophic level systems, involving interactions among plants, their herbivores, and the natural enemies of the herbivores. Critical tests of these hypotheses in the 1980s and 1990s and of island biogeography and trophic cascades theory from the 1960s have clarified the extent to which general patterns apply to specific conditions and species assemblages. This ecological knowledge of plant–arthropod interactions offers explanations for why agroecosystems are relatively vulnerable to pest irruption and can be applied in agriculture to create prophylactic conditions against pest outbreaks. The reward for

Ecology in Agriculture 239

Copyright © 1997 by Academic Press.
All rights of reproduction in any form reserved.

effective preventative measures is an increase in marketable yields without the monetary cost, the health risks (e.g., to farmworkers and consumers), and the environmental disruption (e.g., groundwater contamination and nontarget species kills) resulting from reactive applications of insecticides and miticides. Pimentel (1986, 1993) estimates that the return per dollar invested in ecologically based cultural controls of insect pests ranges from $30 to $300 compared to an estimated $4 per $1 invested in pesticides.

The extension of ecological principles to pest regulations in agroecosystems, however, is hampered in two fundamental ways. First, the nature of a "principle" is generality, but the staggering amount of diversity in habits and life histories among phytophagous arthropods defies any strict adherence to generalities. Perhaps it is time to envision, given technological advances in data management and expert systems, the collection and organization of detailed results from ecological studies on plant–arthropod interactions as a basis for making specific decisions in pest management. However, the second caveat involves the degree to which results from ecological studies in natural systems can be transferred to understanding plant–arthropod interactions in agroecosystems. Comparative studies of herbivore abundance and diversity between natural and managed habitats are rare (Barbosa, 1993). Thus, assumptions about parallels in natural and managed systems are, for the most part, untested. Therefore, I have included counterexamples and critical analyses of the evidence for applicability and transferability among systems, and I recommend the use of ecological principles to guide, not to prescribe. I have organized the chapter to emphasize plant–arthropod interactions from the plant perspective and have included explicit statements of ecological concepts, patterns, or hypotheses that have predictive value for crop plant defense. The main concepts for achieving pest management are biodiversity, integrated systems of ecological control, and heterogeneity.

II. Ecological Consequences of Economically Driven Decisions: The Grower's Perspective

Growers must understand, prioritize, and juggle a complex array of information "bits" to strategically aim at maximum profits by balancing incremental costs of inputs with expected gains in marketable yields. Crop production decisions are made as growers consider a range of factors as disparate as seed prices, mechanical equipment options, marketing trends, and soil pH levels. Meanwhile, these same factors determine the attributes of the resource base for insects and mites (e.g., Zadoks, 1993), which account for approximately 13% of the species reducing yields in agricultural crops worldwide and 13% of total crop losses (Pimentel, 1993). As the grower considers seed prices, he/she is contemplating the selection of a particular crop and crop cultivar, which in turn determines the food resource being presented to the local arthropod community. Decisions about mechanized production tools may be made in response to labor availability and other factors not directly connected to pest management, but may indirectly affect crop vulnerability to pest outbreaks in various ways, such as their effects on soil compaction, tillage schemes, crop rotation, or even dust levels on crop foliage. Marketing trends may determine harvest dates (and thus planting dates) and the phenology of the crop at harvest (e.g., harvests for fully matured corn ears or "baby" corn ears or vine-ripened versus green tomato fruits), which determines the length of time resources are available for pest populations; and soil pH levels will guide the grower in his/her decision about soil amendments, which can alter crop quality as resources for growth and survival of insects and mites.

If effective chemical insecticides and miticides were as inexpensive and available as they once were, growers could continue to pay little attention to the connections between management decisions and pest irruption. However, the past decade or so has been a time of economic stress for tens of thousands of U.S. farmers who, at a time of highest output per unit of input, have still suffered severe financial problems (Duffy, 1991). Recent surveys have shown that the majority of farmers are concerned, in the current atmosphere of incentives for reduced input costs, about reliance on chemical inputs (Duffy, 1991). One way to prevent unnecessary use of agrichemicals is to add pest outbreak prevention to the considerations used in the grower's strategic approach to crop production.

III. Exploitation Strategies: The Herbivore's Perspective

Phytophagous arthropods can be found within mite families (Order Acarina) and within nine orders of insects (Coleoptera, Collembola, Diptera, Hymenoptera, Thysanoptera, Phasmida, Lepidoptera, Orthoptera, and Hemiptera) and make up more than 90% of the species in the latter four orders (Strong *et al.,* 1984). These herbivores comprise four cross-cutting categories of host plant use: (i) those that derive all of their resources from a single crop, completing their entire life history in the crop field (such as two-spotted mites, corn rootworms, and cabbage aphids); (ii) those that require diverse resources not offered by a single crop for their survival and reproduction; and, as subsets of the first two categories, (iii) arthropods that exhibit a wide host range such as the corn earworm and two-spotted mite, which can forage on many crops including corn, tomato, soybean, and cotton; and (iv) those species that are restricted to particular crops and related genera or families, such as cabbage aphids or cucumber beetles.

Feeding biologies, dispersal ranges, host plant location mechanisms, and generation times vary considerably within these diverse groups of arthropod herbivores. Phytophagous mites and insects feed on foliage, stems, roots, fruit, and seeds in numerous ways, including chewing, rasping, sucking out the contents of plant cells, tapping into the vascular system, mining leaves or stems, and producing plant galls. Some arthropods are relatively sessile. For example, European corn borers overwinter in corn stalks and codling moths stay under the bark of apple trees, building resident populations in fields or orchards. On the other hand, many arthropods are extremely vagile, dispersing regularly over long distances and across wide areas by exploiting the ephemeral but very structured nature of air movement (Wellington, 1983). Robust herbivores and their minute parasitoids can exhibit coupled displacement in long-distance migration, as shown by the Australian plague locust, *Chortoicetes terminifera* Walker, and its egg parasitoid, *Scelio fulgidus* Crawford, which disperse independently on wind currents to the same location (Farrow, 1981). Whether an herbivore moves long or short distances to encounter its host plant, different mechanisms are used for host habitat finding, host plant location, host assessment, and host plant use. For many insects, visual cues and chemical cues are both important in this complex process of colonizing host plant patches. Once arthropods develop resident populations in crops, they can exploit the resource within a single generation per year (univoltine) or they may complete 2–12 or more generations (multivoltine) during a growing season. The diversity in feeding, movement, host finding, and life histories is underscored by the fact that ecological patterns of crop exploitation are dynamic. Each herbivore species has an evolutionary history that interacts with ecological events to produce adaptations to changes in their host plant resources.

Herbivores that exploit cultivated varieties often encounter resources that differ in fundamental ways from the plants' wild relatives (see Chapter 2, this volume). Attributes that have arisen out of the selection process for high productivity and palatability of specific plant parts tend to increase the crop's suitability as a host for phytophagous arthropods. First, compared to their progenitors, crop plants can contain lower levels and simplified suites of antiherbivore defenses (Kennedy and Barbour, 1992), can possess a more uniform genetic composition, and may experience lower levels of plant stress. Second, the presentation of these plants to herbivores differs from the conditions found in most of the communities of their wild relatives in its tendency for uniformity in species composition, age distribution of the population, and structural pattern. The application of management technologies to crop fields constitutes the third level of intervention that changes the quality of crops as resources to herbivores. For example, agrichemicals effect changes in the host plant and its microenvironment. Such emergent qualities of crop production enterprises directly or indirectly impact feeding, reproduction, migration, survival, and the behaviors associated with these arthropod life processes. Many of these aspects of herbivore attributes and crop habitats will be expanded in the following sections to better understand plant–arthropod interactions and the opportunities and constraints to pest management in agricultural systems.

IV. Ecological and Evolutionary Antiherbivore Strategies: The Plant's Perspective

Although outbreaks of phytophagous arthropods and widespread plant injury do occur in pristine, native habitats (e.g., Wolda, 1978), plants in their native habitats have a relatively low probability of being devastated by herbivores (Mattson and Addy, 1975). The mechanisms by which plants gain protection from herbivore attack are both passive and active. I use the term "passive" to describe ways that the number of potential herbivore colonists are limited through the escape of plants in time and space. This term also serves to describe resistance to herbivores that is gained through association with other plants. Crops growing in vegetational assemblages can incur reduced herbivore attack when herbivores are drawn to an associated, preferred host or when associated plants mask the presence of the host plant to colonizing herbivores. Another form of passive defense occurs through tolerance mechanisms, which do not serve to reduce herbivory but allow the plant to sustain herbivory with relatively little impact on plant fitness (or yield). In contrast, "active" mechanisms of plant defense include plant resistance traits that directly reduce herbivore performance as it uses the plant as a resource. These include antibiotic characteristics that inhibit feeding, growth, survival, or reproduction of herbivores and can be present continuously (constitutive resistance) or can occur facultatively (induced resistance) in the plant.

A key characteristic of natural plant populations is their genetic and phenotypic heterogeneity. Whether plants derive protection from herbivores through passive or active mechanisms or a combination of both, individual plants tend to occur in natural habitats as a mosaic of resistance factors due to genetic variability (Whitham *et al.*, 1984) and induced responses. Such heterogeneity inherently reduces the probability of counter-resistance in rapidly evolving phytophagous arthropods and increases the durability of plant defenses over time.

A. Escape in Time

Most herbaceous, early successional plants are relatively ephemeral and therefore escape discovery by herbivores and/or the buildup of damaging herbivore populations that would occur if plants were more predictable or persistent.

For plants in general, the expectancy of being found by herbivores depends largely on patch size and temporal (or durational) stability. In natural plant communities, patch size and temporal stability for plants tend to increase with successional stage (Southwood, 1987). Feeny (1976) and Rhoades and Cates (1976) proposed that early successional plants, as relatively less apparent plants, may escape herbivore damage in space and time. However, many plants from which annual crops have been derived are early successional species, having traits characteristic to those of colonists (e.g., cole crops, carrots, and lettuce). Thus, when annual plants are selected for agricultural production, set in large monocultures, and maintained for one or more growing seasons, their "apparency" is artificially increased. Although experimental tests of the role of apparency *per se* in annual crop vulnerability are not available, a quick comparison of the number of major pests on 19 annuals and 15 perennials by Hill (1987) suggests a twofold difference, on average ($\bar{X} = 7.9 \pm 4.4$ SD on annuals vs $\bar{X} = 3.3 \pm 2.7$ SD on perennials; Student's $t = 2.25$, $P = 0.0314$), between the variety of severe pest pressures on crops derived from ephemeral stock and on perennial crops.

Price (1984a,b) developed a conceptual framework of resource availability categories that includes continuous, steadily renewed, and pulsing resources. Although the aim of his study was to express different probabilities of interspecific competition among insect herbivores, it serves as a useful framework for understanding the dynamics of arthropod–crop synchrony (Fig. 1). For many crops in most geographical locations, growers have some flexibility in (i) crop selection from one growing season to the next, (ii) planting dates, and (iii) timing of harvest, all of which can be used to the detriment of arthropod population growth on crops as resources.

1. Crop–Pest Phenologies   The number of herbivores colonizing the crop at any given time depends in part on the absolute abundance of herbivores in the habitat and their vagility. A knowledge of pest life cycles can be used in some cropping plans to shift the synchrony of the crop, or vulnerable stages of the crop, away from peak dispersing periods for the arthropods. If a key pest is univoltine, one tactic would be to sow the crop after monitored adult flight patterns are on the wane (Fig. 1, late pulsed resource). Hessian fly damage to wheat is reduced substantially when planting is timed to follow the decline in fly populations (Buntin *et al.*, 1990). Similarly, for lowest preharvest infestation of dent corn by Angoumois grain moths (Lepidoptera: Gelechiidae), late planting (June) is recommended (Weston *et al.*, 1993).

In contrast, multivoltine pest species tend to increase in abundance with each generation over the growing season. It may be possible to time the

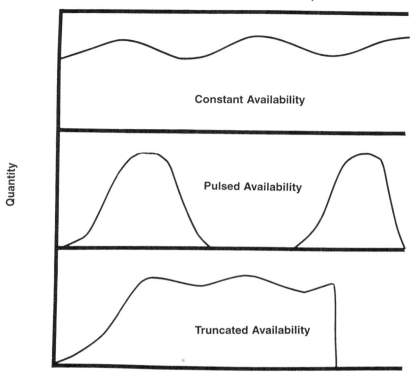

Figure 1   Hypothetical options for managing resource availability (quantity of suitable crop material) for phytophagous arthropods during the growing season, contrasting long season cropping or constantly renewed resources with either early or late season (pulsed) crops or with early crop harvest (truncated).

initiation of annual crops to coincide with low pest numbers, early in the season (Fig. 1, early pulsed resource). Mack and Backman (1990) showed that early planting of peanut crops reduced the incidence of lesser cornstalk borer. Likewise, early sowing of sugar beet allowed the plants to reach a more resistant growth stage before virus-vectoring aphids colonized the crop (Heathcote, 1970).

In a recent comparative study of commercial tomato production on organic and conventional farms in the Sacramento Valley in California, the timing of crop initiation was the single most explanatory factor for crop damage by foliage-feeding insects (Drinkwater *et al.*, 1995; Letourneau *et al.,* 1996). Reduction of photosynthetic tissue by thrips (primarily western flower thrips, *Frankliniella occidentalis*) was greatest in early sown tomato and damage by strip-feeding insects, primarily lepidopterans, was heaviest in late-planted fields, despite the wide range of management practices used on the 18 organic and conventional farms (Fig. 2).

Different optimal planting dates were also found for different pests in dryland cotton, with early (April) plantings suffering disproportionately high levels of thrips and boll weevil damage and later plantings having high populations of cotton leafhoppers and cotton aphids (Slosser, 1993). When pests respond differentially to planting date, the grower must devise a strategy for achieving host plant escape from herbivores based on the severity of different pests, but may also take into account the likelihood of early colonizers acting as alternate hosts for natural enemies of the main pests (Perring *et al.,* 1988).

2. Crop Rotations   Outbreak levels of crop-damaging arthropods that remain in or near the crop field from one growing season to the next can be avoided by disrupting population growth with sequential planting of host and nonhost plants. For example, the northern and western corn rootworms (*Diabrotica* spp., Coleoptera: Chrysomelidae), which have one generation per year in the U.S. corn belt, have become serious pests in nonrotated corn. Populations are not supported on broadleafed weeds and crops or even on grassy weeds or sorghum (Gray and Luckmann, 1994); and corn–soybean rotations have been a very effective control of these

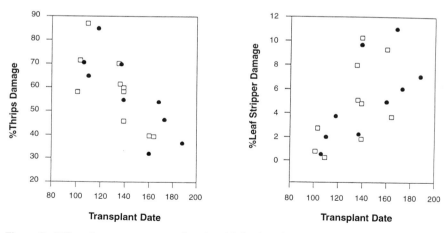

Figure 2   Effect of tomato crop transplant date (Julian) on the percentage of foliage damaged by thrips and leaf strippers (mostly caterpillars), with early tomatoes (April) accruing more thrips damage and less leaf stripper damage by the pink fruit stage (when 10% of the fruit is ripening). □, Conventional farm; ●, organic farm.

rootworms for more than six decades. Although rotation has been effective for soil-dwelling corn rootworms and can be a pivotal control tactic for soilborne pathogens and nematodes, most arthropod pests that can disperse long distances may be suppressed only if crop rotations are implemented on a regional basis (Finch, 1989). However, strong and widespread selection pressures, even with cultural practices such as rotation, can drive undesirable adaptations in pest populations. Levine *et al.* (1992) have found evidence that rotation has selected for longer dormancy periods in corn rootworm. Whereas most of the rootworms starve upon emergence in a field lacking a host plant, a portion of the population remains in diapause through the soybean crop and emerges when corn is sown again.

3. Resource Truncation   Plants in natural habitats must complete their development and produce seed to maintain fitness. However, yield for many crops is not dependent on seed maturation and is not subject to the same constraints that influence the progenitors of crops in nature. Standard practice in irrigated cotton, for example, includes the application of chemical defoliants, dessicants, or plant growth regulators before the end of the growing season to cause the crop to drop the leaves and squares (buds) that act as food and shelter resources for pink bollworms, bollworms, and the boll weevil (Adkisson, 1962; Jones *et al.,* 1990). In a similar way, early termination of the cotton crop (before November frost) in nonirrigated cotton may provide an ecologically based alternative to multiple fall pesticide applications currently mandated by the Texas Boll Weevil Eradication Program. Cotton squares and small green bolls remaining on the crop after mid-September will not mature before the final harvest, and boll weevils using these resources after mid-September have a lower incidence of mortality through the winter diapause (Montandon *et al.,* 1994). Therefore, early harvested cotton under these conditions has comparable yields while boll weevil controls and chemical input costs (capital and environmental) are reduced.

Pest phenologies and life histories determine appropriate intervention times for disrupting the resource base for pests. An example of a much shorter cycle of pulsed resources (Fig. 1) is shown by the lucerne leafroller *Merophyas divulsana* in New South Wales. The generation time of this multivoltine herbivore approximates that of the normal harvesting cycle of the crop, so early harvesting of the crop is a key control tactic in the management of this pest (Bishop and McKenzie, 1986). Also in Australia, sugarcane presents an example of resource truncation over a period of several years. Populations of the sugarcane soldier fly, *Inopus rubriceps,* increase locally from year to year on a continuous supply of stubble left in the field to produce new rattoon shoots after harvest (Samson, 1993). Periodic plowing is necessary to disrupt the resource before crop failure occurs from resident fly populations.

B. Escape in Space

There exists a species–area relationship such that the equilibrium number of species supported by an island increases with the size of the island relative to its distance from the mainland.

The insular nature of cultivated areas has motivated several analogies regarding crops as islands available for colonization by arthropods (Price and Waldbauer, 1994; Strong, 1979; Simberloff, 1985). MacArthur and Wilson's (1967) dynamic equilibrium theory of island biogeography allows the prediction of colonization rates and mortality/emigration rates on a comparative basis, with respect to crop area and distance from the sources of colonizers. Although this body of theory organizes thought on the colonization process in crop fields, some impediments to its application include (i) the notion that temporal dynamics of crop fields, as frequently disturbed communities, may cause responses different from those predicted by equilibrium models; (ii) the fact that biogeographic theory does not distinguish among types of colonizers, which is critical when considering natural biological control and arrival rates of pests versus their predators (Price, 1976; Stenseth, 1981); and (iii) the current status, in which the few empirical data available on diversity, size, and distance relationships provide an insufficient basis for agricultural planning recommendations (Simberloff, 1985). Nevertheless, general predictions from theory may be extremely important in anticipating and managing plant–arthropod interactions on the landscape level in agroecosystems.

1. Isolation from Colonizer Pools   On a global scale, crops have experienced a far more dramatic level of release from key herbivores than have indigenous plants growing in their natural habitats. Release from herbivore pressure in native habitats would be a relatively rare, temporary, and localized circumstance due to stochastic events. In contrast, there tends to be a gradual development of herbivore communities on crops introduced to new continents. Strong (1974) and Strong *et al.* (1977) traced the development of pest assemblages through time for cacao and sugarcane, respectively. In both crops, the initial numbers of herbivore species on these crops after introduction was much lower than the number of species that had exploited these crops previously in their place of origin. On a regional level, the placement of crops distant from other crop fields or closely related species can also result in reduced colonization by herbivore species, but much of the evidence for fields that are free from certain pests is anecdotal. Understandably, many ecologists and entomologists study herbivores where and when they are abundant, so data are relatively lacking for locations in which herbivore populations are absent or at extremely low levels (e.g., Nothenagle and Schultz, 1987; Price *et al.*, 1984).

However, relatively rapid colonization rates of pests from nearby sources (e.g., adjacent fields) has been documented frequently. Examples include the regional movement of exploding populations of whiteflies from cotton to melon and from melon to lettuce as these crops are harvested sequentially (Blua *et al.,* 1994); the movement of chinchbugs, grasshoppers, and armyworms from small grains to nearby corn fields (Price and Waldbauer, 1994); and rapid population growth of cabbage seedpod weevils in rape within close proximity of alternate hosts and overwintering sites (Boyd and Lentz, 1994). In an explicit test of distant versus near "crop islands," Kruess and Tscharntke (1994) showed that most herbivore species colonized small red clover patches but that the abundance of these herbivores declined with greater distances from a source of colonizers. Furthermore, the number of species of parasitoids colonizing the red clover patches was significantly lower, through increased extinction rates and decreased colonization rates, as distance from the source increased.

Economies of scale that dominate the agricultural sector in developed countries allow farmers to reduce unit production costs by increasing farm size and becoming more specialized. The resultant changes in crop island size over time as part of a matrix of wildlands, fallow fields, land planted in other crops, and urban areas will have widespread effects on pest–crop dynamics as crop fields expand, as particular crops intensify as profitable enterprises, and as wildlands become more fragmented. Rabb (1978) addressed appropriate research scales almost two decades ago when he criticized the propensity of single-commodity, closed system approaches to pest management in research and decision making as deficient for problems that demand attention to "large unit ecosystem heterogeneity." Ryszkowski *et al.* (1993) showed that landscape heterogeneity was a strong determinant of both insect biomass and species diversity. Research at the landscape level on animal movement and metapopulation dynamics using GIS will allow rigorous analyses of the effects of host plant (habitat) isolation on herbivore colonization (e.g., Kalkhoven, 1993). As theory develops in the field of conservation biology on habitat linkages for conserving endangered species and key factors in extinction, applications will accrue in the conservation of natural enemies and the suppression of pests, respectively.

2. Size-Discovery Patterns   Species richness is positively correlated to size on oceanic islands. Similarly, in mainland communities, large, isolated patches of a single plant species may support more species than do small, isolated areas (Strong, 1979). Larger host islands can collect more individuals by random probability of encounter (Connor and McCoy, 1979). Furthermore, patch detection by dispersing specialist herbivores increases, in general, with size and homogeneity. Ecological tenets on island biogeography, once again, suggest that extensive monocultural production units can favor pest outbreaks.

Extinction rates depend on resource availability in the system. Because crop plants are supplied to the system, or "reset" at certain intervals (Levins and Wilson, 1980), the resource base is predictable, at least for herbivores colonizing sufficiently early in the crop cycle. Because of resource availability for herbivores and delayed resource (prey/hosts) availability for natural enemies, extinction rates will be greater for parasitoids and predators discovering crop islands early in the season. In Kruess and Tscharntke's (1994) comparison of 16 red clover meadow fragments ranging in size from 300 to 800,000 m², the species richness of herbivores and of parasitoids and the proportion of parasitized seed weevils (*Apion ochropus*) were all significantly and positively correlated with "island" size. Differential resource availability for colonizing species on different trophic levels is not an explicit part of MacArthur and Wilson's model (1967), but disparities become important in pest management. Also, the equilibrium theory of biogeography does not allow for comparisons of single, large crop fields versus a network of several small fields of the same total area, but the contrasting designs are likely to differ in terms of suitability for both herbivores and biological control (Price, 1976; Altieri, 1994).

C. Reduce Palatability or Suitability

The array of environmental factors that affect the resistance of the host to herbivore attack, in concert with genetic components of intraspecific variation in plant morphology, allelochemistry, and nutrition, produce heterogeneity that may be central to the regulation of herbivore populations.

If the "world is green," with plants making up the most obvious and readily available resource in terrestrial communities, why don't herbivores reduce it to stubble? The answer, in part, is that plants are less suitable as food than they appear to be (Strong *et al.*, 1984). In natural systems, physical and chemical defenses as well as nutrient deficiencies present major hurdles for phytophagous arthropods in colonizing and using plants as resources (Feeny, 1990). Our current body of knowledge represents almost three decades in ecological research and provides the fundamental tools for selecting and maintaining relatively resistant crops.

Like their wild progenitors, crops possessing physical or biochemical properties that reduce their attractiveness or adequacy as a food resource are relatively resistant to exploitation by herbivores. Resistance can be achieved through inherent plant traits and/or modified by environmental conditions.

Ecological theories of plant defense suggest that resources are allocated optimally among growth, reproduction, and protection. In evolutionary time, plants have evolved physical and chemical defensive mechanisms against herbivores, pathogens, and other plants, depending on the strength of these organisms as selective factors. If energy and materials are used exclusively for defense, then the cost of these resources is perceived as the concomitant reduction in resources available for other plant functions related to fitness. If defensive structures or compounds are beneficial in other ways, such as trichomes as antidessication features, the costs of these defenses are mitigated by gains derived from multiple functions (Zangerl and Bazzaz, 1992). Theoretically, costs and benefits of antiherbivore defense are not uniform among plant parts but vary as a function of the relative fitness value of the tissue and the relative probability that the tissue will be attacked by herbivores. Therefore, genetic defense systems in wild progenitors may be out of phase for crops whose agricultural value (yields) differs fundamentally from their fitness value (e.g., leaves or fruit versus seeds).

Even when plant resistance traits are used effectively as tools for the regulation of crop pests, the combination of rapid generation time, genetic variation, and natural selection has led to the development of resistance to plant defenses among populations and species of phytophagous arthropods (Via, 1990). Because resistant crop varieties are designed to increase mortality or decrease fecundity of herbivores, they eventually select for better adapted pests that are more difficult to control. For example, the notorious greenbugs (*Schizaphus graminum* aphids) have been controlled since the 1940s using various lines of resistant wheat, but decade after decade new counterresistances have developed in aphid populations, making it necessary to establish alternative varieties with novel resistance traits (Wilhoit, 1992). Such counterresistance can be managed, to some extent, using knowledge of population dynamics and natural selection. An increasing amount of research effort on resistance management tactics has occurred during the past decade and prompted, in 1988, the creation of a new categorical heading: "Insecticide Resistance and Resistance Management" for indexing articles in *The Journal of Economic Entomology*. Recommendations for maintaining susceptible individuals in the population are a direct result of population ecology.

1. Physical Attributes   Crop varieties with morphological barriers to arthropod movement, feeding, and oviposition may incur lower levels of damage from phytophagous arthropods and produce increased yields, depending on the energetic cost of the defensive structures and resource allocations in the crop. Although morphological barriers themselves, such as spines, toughness of tissues, pubescence, and waxes (Kogan, 1994), are often derived from chemical bases, such as the overproduction of silica or lignin or the cross-linkage of lipids, the actual effect on herbivores is only through physical deterrence rather than through the olfaction or gustatory processes that characterize chemical defenses. To my knowledge there is no general theory contrasting the occurrence of physical versus chemical defenses among plant taxa. Neither are there predictive guidelines regarding target herbivore groups that respond to physical versus chemical plant defenses. Both chemical and physical resistance traits are variable in occurrence and widespread among plant taxa. Although physical and chemical resistance traits are usually treated separately, they in fact act as a package in many cases. For example, the trichomes of wild tomato (*Lycopersicum hirsutum*) produce 2-tridecanone and zingiberene, which add toxicity to the basic physical deterrence of plant hairs for several arthropod pest species (Hare, 1990). Both types of defense are linked to other plant functions as part of the plant's structural attributes (e.g., stalk strength, water retention, or metabolic functions). Physical defenses may be less subject to environmental variability than are chemical constituents and are generally detectable only in close proximity when the arthropod is already in contact with the plant. Because they generally require different mechanisms of counterresistance, a combination of physical and chemical defenses can increase the durability of resistance in plants.

a. Toughness   Sclerophyllous leaves and bracts are tough; they resist fracturing and bending. A relatively thick cuticle and a high cell wall : cytoplasm ratio (and thus a high crude fiber : crude protein ratio) reduces their suitability for food (Specht and Rundel, 1990). They are relatively expensive to the plant but effectively deter herbivores (Turner, 1994). Although most crop plants have malacophylls (soft textured leaves), varieties with relatively tougher tissues can confer resistance against some phytophagous arthropods (Norris and Kogan, 1980). The role of leaf toughness as a significant impediment to herbivore feeding has been shown, for example, for the mustard beetle (*Phaedon cochleariae*) on cole crops (Tanton, 1962), the cereal leaf beetle (*Oulema malanopius*) on grains (Shade and Wilson, 1967), and for a wide variety of lepidopteran herbivores on maize (Bergvinson *et al.,* 1994a).

b. Trichomes   Leaf surface texture of crop plants can be an extremely important determinant of its suitability for an herbivore. Plant hairs, whether multicellular or unicellular, glandular or nonglandular, straight, spiral, hooked, simple, or stellate, occur on plants in almost every family (Johnson, 1975; Levin, 1973). Leaf trichomes vary geographically within species in natural habitats, and the patterns of variation involving glandularity, which has little effect on biophysical properties of the leaf surface, suggest that herbivory may be an important selective factor for this variation among natural populations. Hooked trichomes on wild *Passiflora* plants successfully impale butterfly larvae as they forage (Gilbert, 1971) in much the same way that adult leafminers are trapped on resistant bean varieties (Quiring *et al.,* 1992). Plants with dense trichomes can also deter insects by preventing the proboscis from entering the tissue, by inhibiting attachment to the plant surface (Lee *et al.,* 1986), or by disrupting oviposition (Dimock and Tingey, 1988). Movement can be impeded, especially with sticky exudates from glandular hairs (Gibson, 1976), and some trichomes are toxic (Casagrande, 1982) or produce pheromone mimics that signal insects to drop from the leaf surface (Gibson and Pickett, 1983).

Trichomes are not restricted to plant foliage. Their presence on fruits can also reduce damage, as has been shown for peaches versus nectarines infested with thrips and mites (Cravedi and Molinari, 1984), and may explain differences in insect and mite infestation of kiwi fruit varieties (Tomkins and Steven, 1993).

Recent studies indicate that trichome density can be increased faculta-tively in response to herbivore damage on leaves (Baur *et al.,* 1991), climatic conditions (Gianfagna *et al.,* 1992), and possibly by stem-borer damage to the meristem (Letourneau and Barbosa, 1997). I am not aware of any studies on induced trichome development in response to herbivore damage in crop plants; however, if trichome production depends on sufficient nutrient uptake (Hagen and Anderson, 1967) and compromises yields in some crops (see Agren and Schemske, 1993), facultative production may be an efficient resistance mechanism.

As in virtually all plant defenses, the vast array of herbivores that come in contact with the plant include species that tolerate or prefer the trait that confers resistance generally (e.g., Webster *et al.,* 1994). For example, *Heliothis zea* (corn earworm) oviposits preferentially on pubescent soybean varieties (Panda, 1979). Trichomes can also protect herbivores from their parasitoids or predators (Table I). However, in contrast to the examples of trichomes acting as barriers to natural enemy searching efficiency, Shah (1981) showed that rapid turning rates in predatory lady beetle on hirsute brassicaceous crops resulted in increased encounters with their aphid prey.

c. Other Physical Barriers   Vital functions in ectopic herbivores (e.g., feed-ing and reproduction) require attachment to the plant surface. In this regard, waxy blooms on foliage can make a plant an unsuitable host for herbivores that lack special adaptations for stability (Bodnaryk, 1992). For internal feeders, entry into the preferred tissues for feeding is critical. Structural barriers such as silicon-reinforced vascular tissue can prevent cane pith feeding by early instar lepidopterans; extremely tight husks can reduce the incidence of corn earworm damage on some corn varieties (Kogan, 1994). These types of physical barriers to herbivores may interact favorably with biological control. The herbivore's time of exposure to natu-ral enemies, for example, can be extended by delaying herbivore entry to refugia. Structural barriers are also likely to be differentially effective for arthropods with different morphological traits (e.g., a caterpillar versus a wasp).

2. Chemical Attributes   Two parallel lines of investigation of chemical de-fenses in plants have advanced independently: (i) Ecologists and evolution-ary biologists have developed concepts of plant ecophysiology, plant–

Table 1 Examples of Mechanisms by which Physical or Chemical Plant Defenses Can Inhibit Biological Control

Enemy	Herbivore	Plant	Mechanism	Reference
Anthocorids	*Brevicoryne brassicae*	Brussels sprouts (*Brassica oleracea gemmifera* (Brassicaceae)	Deterred by waxy leaves	Way and Murdie (1965)
Encarsia formosa Gahan (Hymenoptera: Aphelinidae)	*Trialurodes vaporariorum* (Westwood) (Homoptera: Aleurodidae)	Hirsute cucumber varieties	Trichome mat hinders movement and searching efficiency	Hulspas-Jordaan and van Lenteren (1978)
Trichogramma pretiosum and *Chrysopa rufilabris* Burm. (Neuroptera: Chrysomelidae)	*Heliothis zea*	Hirsute cotton varieties	Increased trichome density reduced ability for small predators to locate and destroy host/prey	Treacy *et al.* (1985)
Predaceous mites	Herbivorous mites	*Ficus carica* L. (Moraceae)	Glandular hairs correlated with increased or decreased rate of prey consumption depending on predator species	Rasmy (1977), Rasmy and Elbanhawy (1974)
Coccinellids	Aphids	Hirsute potato varieties	Reduction in adult searching time; larval mobility impaired; but these effects were attenuated in the field	Obrycki and Tauber (1984)

Predator	Prey	Crop	Effect	Reference
Predaceous mites	Tetranychid mites	Apple and peach	Possible role of pubescence in determining predator–prey contact rates on different plants	Putman and Herne (1966)
Hyposoter exiguae (Viereck; Ichneumonidae)	*Heliothis zea*	Tomato	Increased tomatine in host plant reduced adult parasitoid longevity, size, and survival	Campbell and Duffey (1979)
Geocoris punctipes	*Anticarsia gemmatalis*	Soybeans	Predators attacking prey on resistant varieties had increased mortality rates	Rogers and Sullivan (1986)
Archytas marmoratus	*Heliothis zea*	Cornsilks	Lower percentage of emergent adults	Mannion *et al.* (1994)
Copidosoma truncatellum	*Pseudoplusia includens*	Soybean (*Glycine max* L.)	Increased parasitoid development time	Orr and Boethel (1986)
Pediobius foveolatus	*Epilachna varivestis*	Soybean	Male-biased sex ratio	Dover *et al.* (1987)
Campoletis sonorensis	*Spodoptera frugiperda*	Maize (*Zea mays* L.)	Increased parasitoid development time	Wiseman and Isenhour (1990), Pair *et al.* (1985)
Geocoris punctipes	*Heliothis zea*	Hypothetically, transgenic crops	β-Exotoxin	Herbert and Harper (1986)
Chrysoperla carnea; *Trichogramma* sp.; *Hippodaemia convergens*	—	Tobacco	Trichomes and sticky exudates caused reduced searching speeds	Belcher and Thurston (1982), Elsey (1974), Rabb and Bradley (1968)

animal interactions, and theories of coevolution among taxonomic groups, and (ii) crop scientists have established cultivars resistant to various arthropod pests and pathogens (Kogan, 1994). However, recent linkages among these complementary fields have produced insights into mechanisms of resistance, plant phylogenetic constraints on resistance traits, behavioral and physiological mechanisms of herbivore counterresistance, the role of heterogeneity in the durability of resistance, and environmental influences on plant chemical defenses.

Soil management practices in agroecosystems can have cascading effects on plant chemical composition and herbivore damage. Resource allocation due to different soil amendments accounts for many aspects of primary and secondary metabolites in plants and can influence oviposition, growth rates, survival, and reproduction in the arthropods that use them as hosts. Agricultural studies that integrate soil management or fertilizer regimes with host plant resistance expression provide practical extensions of theoretical work in ecology concerning, for example, carbon:nitrogen balance, consumer dynamics, and trophic cascades.

a. Secondary Metabolites   Although many resistant varieties of crop plants have been developed without knowledge of the mechanisms underlying their effects on pests, plant secondary compounds are implicated in many cases. Chemical defenses are ubiquitous in plants and result from the interplay between phylogenetic constraints, past selection on ancestral types, the severity of stress caused by herbivores, pathogens and other plants, the energetic cost of defense, past and current nutrient availability to plants, and the effectiveness (value) of the defense.

Compared to phytochemicals, which can act as attractants or repellents for host finding, oviposition, and feeding, visual and mechanical stimuli are often secondary (Hanson, 1983; Visser, 1986). Once feeding ensues, active resistance can occur through mechanisms in at least four categories: feeding deterrence, toxicity, digestibility reduction, and phytohormonal influences. 2,4-Dihydroxy-7-methoxy-1,4-benzoaxazin-3-1 (DIMBOA) acts as a feeding deterrent to the European corn borer (Lepidoptera: Pyralidae) and serves as a genetic constitutive defense of corn plants against this key pest (Barry *et al.,* 1994). Resistant varieties can have an order of magnitude higher concentrations than susceptible varieties, and standard visual feeding assays are used by breeders to detect reduced feeding of first-instar larvae on whorl leaves. Alkaloids such as nicotine, glucosinolates such as sinigrin, and cyanogenic glycosides, found in tobacco, cabbage, and cassava, respectively, are acutely toxic to most herbivores (Rosenthal and Janzen, 1979). Digestibility reducing agents, in the form of condensed tannins, are found in many cultivars of cotton and are extremely potent in reducing the larval growth of *H. zea* and *H. virescens* on this crop. Phytoecdysones, which can disrupt moulting processes in arthropods, are mainly restricted to ferns and gymnosperms. Of the four categories of chemical defense compounds, toxins and feeding deterrents are most common in plants from which crops have been derived.

Unfortunately, no matter how effectively the antifeedant and toxic compounds function to reduce damage from most herbivores, particular arthropod species, in many cases, are adapted to tolerate the compounds and even to prefer feeding on plants that contain them. For example, one of the most bitter compounds known, cucurbitacin, acts both as a feeding deterrent and as a toxin for most insects and mites. However, cucurbitacin induces compulsive feeding behavior in adult cucumber beetles (*Diabrotica* spp.). Deheer and Tallamy (1991) showed that the larvae also feed more extensively on roots of the bitter varieties or when roots from nonbitter varieties are coated with cucurbitacin. Due to the presence of aromatic compounds, including terpenoids and phenylpropanoids (Berenbaum, 1990), which are active against most generalist feeders and have insecticidal properties (Lichtenstein and Casida, 1963), most species in the Apiaceae (e.g., carrot, poison hemlock, and wild parsnip) support a relatively depauperate herbivore fauna. Specialist herbivores on these plants are restricted to this family and a few closely related plant families. However, crops selected from wild progenitors in this family can have substantially lower levels of these compounds than their ancestors, allowing generalist herbivores to inflict relatively more damage in cultivation (Kennedy and Barbour, 1992). Because some of the most toxic compounds (furanocoumarins) are photoactivated and contained in vessels adjoining the major leaf veins, generalist herbivores can reduce exposure by feeding on the shaded portion of celery or skeletonizing the leaves, thus acquiring still lower concentrations of these toxic compounds (Zangerl, 1990). Thus, costs and benefits of chemically based resistance in crop plants depend on the balance between potential economic losses caused by herbivores that are controlled versus herbivores that are attracted and any physiological (yield) costs to the production or storage of defensive compounds (Kogan, 1986).

Environmental factors, many of which are controlled in agricultural fields, sometimes modify the production of secondary metabolites in plants in complex ways. For example, nitrogen fertilizers can increase the net production of alkaloids, glucosinolates, and cyanogenic glycosides. However, additional potassium can decrease alkaloid production, low phosphorus encourages alkaloid production, and N only increases glucosinolates if sufficient sulfur is present in the soil (Waterman and Mole, 1989). Glucosinolates, which are a characteristic constituent of brassicaceous plants (Cruciferae), are toxic to herbivores not specialized to tolerate these secondary metabolites (Mattson and Haack, 1987). Although optimal defense theories suggest that concentration of these nitrogen-based defensive compounds may be enhanced when plants are grown in N-rich soils (Finch, 1988), they may be "diluted" by extranormal plant productivity in fertilized soil (Wolfson, 1980). Water stress and salinity also have effects on herbivores that are difficult to predict. These conditions tend to increase alkaloids in plants but decrease other compounds, such as terpenes.

In addition to soil characteristics, light levels and light quality influence C : N ratios and secondary metabolites. For example, glycoalkaloid production in potato leaves was increased with N fertilizer when the weather was cool and cloudy, but under warm and sunny conditions the crop has lower levels in fertilized than in unfertilized field plots (Nowacki *et al.*, 1976). In another study, DIMBOA levels were higher in conditions of low UV light, but plants were more susceptible to corn borer larvae despite this increase because other phenolic compounds that control cell wall strength were inhibited (Bergvinson *et al.*, 1994a,b). Thus, light levels may explain an increased susceptibility to folivores of greenhouse-grown corn plants or plants grown in overcast weather, despite optimal soil and moisture conditions.

Crop plants often have a diverse set of chemical constituents that affect herbivores, and the flux of these chemicals is dynamic. Therefore, the results of studies showing that alkaloids are decreased with N fertilizer under warm, sunny conditions may not mean that the plants are more susceptible to herbivores under these conditions. Rhoades (1983) speculated that these plants are resource rich and may be producing an alternative, more costly defense such as proteinase inhibitors. The interaction with UV light, chemical defense (DIMBOA), and structural defense (involving phenolics) also suggests that trade-offs occur in defense mechanisms in response to extrinsic factors.

Dynamic fluctuations can also occur temporally and in response to herbivore attack. Induced resistance refers to increased protective responses by plants after the plant is attacked or wounded. The reaction occurs in a relatively short time, and the elevated levels of secondary chemicals can be local or systemic (Ryan, 1983). In terms of optimality theory, these facultative defenses may incur less overall cost, especially if the probability of herbivore attack is low or extremely variable. The benefit of inducible defenses is that they do not incur direct costs unless the plant is actually attacked by an herbivore; the cost is that some degree of damage must be tolerated until the defensive compounds are produced or translocated (Zangerl and Bazzaz, 1992). Karban and Myers (1989) document numerous examples of induced host response to stress or injury, including the concentration of proteinase inhibitors in solanaceous plants and of latex in damaged cucurbits. Although the physiological mechanisms of induced resistance in grapevines are not yet known, the concept of induced resistance has been used in the management of Pacific spider mites in vineyards

(Hougen-Eitzman and Karban, 1995). Inoculation of economically unimportant, herbivorous Willamette mites during shoot expansion causes a systemic resistance to subsequent infestations by Pacific spider mite pests. The amount of inducible resistance varies among cultivars. In cotton, levels of induced resistance tended to correlate directly with levels of constitutive resistance (Brody and Karban, 1992). Resistance to mites can be induced by nonspecific injury in cotton, and cross-resistance was demonstrated by McIntyre *et al.* (1981). However, some induced resistance mechanisms can be quite specific, as shown by attempts to induce resistance to herbivores with inoculation of pathogens (e.g., Ajlan and Potter, 1991, 1992).

Resistance traits (especially constitutive traits) that cause extremely virulent and effective resistance against a broad spectrum of potential pests may be desirable and highly marketable. Using molecular techniques, such attributes are now more possible to achieve. However, they are also the least durable option for pest management. Variably susceptible target species with many generations per year (and rapid turnover of symbionts in many arthropod herbivores) can produce counterresistant biotypes within a few years (Gallun and Khush, 1980; Dowd, 1991). Just as it is critical to manage resistance to pesticides for long-term use, we must also use our ecological and evolutionary knowledge base to manage single-factored genetic selection forces of chemical plant defenses in crops. Although current techniques in genetic modification of crop plants to incorporate resistance traits hold potential for the production of durable, multifaceted plant resistance (Gould, 1988), current research and, certainly, the first marketable products of this technology are based on strong, vertical resistance (Meeusen and Warren, 1989).

Currently in the stage of seed production for commercial release in the next year, various crops, including corn, cotton, tobacco, and potato, will have resistant cultivars that contain the bacterial gene responsible for producing the toxic glycoprotein of *Bacillus thuringiensis,* a pathogen of arthropods. This δ endotoxin is extremely effective in reducing growth and survivorship in lepidopterans (e.g., Jenkins *et al.,* 1993), and virulent strains of *B. thuringiensis* have been sought for gene extraction in this process. Benedict *et al.* (1993) showed <2% survival of larval *H. virescens* and *H. zea* on all six transgenic lines of cotton. The high level of virulence, the production of the δ endotoxin by all plant tissues all the time, and the widespread deployment of this uniform resistance factor will maximize the probability of counterresistance in target pests. Although marketed as a more durable resistance, single genes incorporated into these cultivars to produce additional types of endotoxins effective against dipterans and coleopterans are functionally just as subject to resistance by each target group.

General rules from research on pesticide resistance, cross-resistance, and population genetics are (i) to maximize compatibility with other pest management tactics, such as biological control, and to minimize reliance on a single resistance factor; (ii) to reduce virulence as much as possible to minimize selective pressure for counterresistance or to maximize toxicity to make expression of counterresistance alleles more recessive; (iii) to avoid widespread use over long periods of time or to provide refuges for susceptible subpopulations of arthropod herbivores; and (iv) to encourage heterogeneity in the host plant population such that selective pressures (resistance mechanisms) are not uniform or predictable [e.g., variety mixtures (Wilhoit, 1991)]. Finally, the notion of gene transfer across related plant species is an emergent risk, not considered in resistance management, by pesticides that cannot reproduce themselves (Klinger and Ellstrand, 1994; Abbott, 1994).

b. Nutritional Suitability Primary plant metabolites and their products are converted to arthropod biomass and used for growth and reproduction. Although arthropods can usually survive on the nutrients provided by their host plants, relative growth rates and fecundity levels can be influenced significantly by the absence, paucity, or imbalance of essential nutrients such as vitamins, amino acids, sugars, or mineral nutrients (Kalode and Pant, 1967; Maxwell, 1972). Many nutritional factors can influence the level of damage that the crop ultimately incurs [including water content and different sources or levels of calcium, magnesium, phosphorus, potassium, or sulfur (e.g., Kindler and Staples, 1970; Culliney and Pimentel, 1986; Shah *et al.*, 1986; Manuwoto and Scriber, 1985)]; however, total N has been implicated as a critical, limiting nutrient for both plants and their consumers (Mattson, 1980; Scriber, 1984a; Slansky and Rodriguez, 1987). Because herbivores contain between 7 and 14% N by weight and plants range from 0.03 to 7% dry weight, Mattson (1980) argues that foliage N level can be a major regulator of herbivory rates.

In a review of 100 years of research, Scriber (1984a) counted a minimum of 135 studies that showed increased damage and/or enhanced growth of leaf-chewing insects or mites versus fewer than 50 studies in which herbivore damage was reduced on plants with higher N. In the aggregate, these studies support the nitrogen limitation hypothesis and suggest a corollary nitrogen-damage hypothesis with practical implications for fertilizer use in agriculture: High N inputs can lead to greater rates of herbivore damage to crops.

It is not clear, however, if the nitrogen-damage hypothesis or the results of Scriber's (1984a) review can be extrapolated as a warning about fertilizer inputs and pest exacerbation in agroecosystems (Scriber, 1984b). To assess the applicability of the nitrogen-damage hypothesis to agroecosystems, I critically reviewed the literature used to generate the hypothesis. Using 70 original papers, either cited in major reviews (Scriber, 1984a; Slansky and

Rodriguez, 1987; Leathe and Ratcliffe, 1974; Lipke and Fraenkel, 1956; Singh, 1970; Jones, 1976) or conducted in the decade since Scriber's (1984a) review, I investigated the potential role of experimental design in generating the general assumption that increased plant N increases herbivore populations or damage.

Based on 100 comparisons of insects and mites (some of the 70 papers included more than one species) on plants treated experimentally with high and low N fertilizer levels, two-thirds of the insects and mites studied showed an increase in growth, survival, reproductive rate (or reproductive potential as indicated by responses such as increased pupal weight), population densities, or plant damage levels in response to increased N fertilizer. The remaining third of arthropods studied showed either a decrease in performance or damage with fertilizer N ($n = 11$) or no significant change ($n = 22$). Of the 100 arthropod species studied, the authors reported actual measurements of plant tissue N concentration for 35; the other two-thirds reported fertilizer rates but not relative plant effects. Of chewing (i.e., leaf-stripping) insects, mites, mining and boring insects, and sucking insects included in these experiments, only chewing insects did not exhibit a strong positive response to N in approximately 75% of the studies.

However, experimental design played heavily in this skewed response to plant N, and probably did so in Scriber's (1984a) review in which approximately 75% of more than 180 studies of insects and mites showed a positive response to increased plant N. In fact, the experiments that generated these data do not seem to be immediately useful in predicting the response of phytophagous arthropods to N fertilizer practices in functioning agroecosystems. First, more than 50% of the studies were conducted with potted plants (Table II) versus less than 10% conducted in large-scale crop fields, which provide a more realistic set of conditions of both N uptake by the plant and herbivore response.

Second, the studies conducted in large crop fields do not clearly support the nitrogen-damage hypothesis. Although the sample size is very small, the majority of comparisons showed no significant increase in arthropod performance or damage with increased N (Table II). Even in field plot experiments the results were less than 60:40 in support of the N-damage hypothesis. Only in greenhouse studies was there a distinct basis of support for the nitrogen-damage hypothesis.

Third, actual damage to plants was measured in only one-fifth of the studies. Population levels (which include different age classes) may be the next most important indicator of damage, but the studies measuring these parameters were not nearly as likely to support the nitrogen-damage hypothesis as were those measuring some indicator of growth, survival, or reproductive rate of individual insects (Table II).

Table II Assessment of 100 Experimental Studies of the Nitrogen-Damage Hypothesis (Taken from the Primary Literature between 1940 and 1995) with Respect to Scale of Experiment, Parameters Measured, and Feeding Guild of Arthropod Herbivores

		% of studies with	
	% of 100 studies	Increased damage or performance	Decrease or no change in damage or performance
Experimental scale			
Large crop fields	7	29	71
Small field plots	36	58	42
Potted plants	57	77	23
Parameter measured			
Damage	21	52	48
Population	39	59	41
Reproduction	32	78	22
Growth	4	100	0
Survival	4	100	0
Feeding guild			
Chewers	48	58	42
Mites (cell content feeders)	13	77	23
Miners or borers	11	73	27
Suckers	28	75	25

Experimental scale has been an important factor in nitrogen studies of *Pieris rapae* oviposition preference (Letourneau and Fox, 1989). The authors found orders of magnitude higher egg placement on high N plants compared to low N plants on potted collards exposed to cabbage butterflies in the field, but this difference was not apparent in plants grown in small-scale field plots with a similar range of fertilizer levels. Particular arthropod species, even congenerics, can react to N fertilization rate differently. Damage by the beanfly *Ophiomyia spencerella* in Malawi was significantly greater in fertilized bean than in unfertilized bean plots; however, *O. phaseoli* showed no significant effect of fertilizer treatment on density or damage (Letourneau, 1994). Finally, Letourneau *et al.*'s (1996) study of herbivore damage on tomato on 20 commercial farms in California showed a wide range of tissue N concentrations in tomato foliage among farms but no differences in damage.

The explanation for a muted effect of N on arthropods in large-scale field studies compared to potted plants may include (i) actual fertilizer rates and/or plant physiological processes affecting the rate of N uptake, (ii) soil heterogeneity effects at the level of individual or small patches of plants, (iii) microclimate effects on herbivore behavior and physiological response, and (iv) interactions with the biotic community including natural enemies that are affected by changes in plant quality and concomitant changes in the herbivores. As examples of the latter, the fecundity of the herbivorous mite *Panonychus ulmi* on apples treated with three levels of nitrogen fertilizer (Huffaker *et al.*, 1970) increased up to fourfold in density with N level when *Amblyseius potentillae* predators were not present; however, in the presense of predaceous mites, phytophagous mite levels were similar among N treatments (but see Walde, 1995). In contrast, fertilized cotton plots (NPK) exhibited higher levels of *H. zea* (Boddie) than did check plots despite significantly higher population densities of *Hippodamia convergens* G-M, *Coleomegilla maculata langi* Timberlake, and *Orius insiduosus* (Say) in fertilized cotton (Adkisson, 1958). Chiang (1970) demonstrated that fertilized midwestern corn fields had significantly fewer (half) corn rootworms than did unfertilized control plots. Although ground beetles and spiders were not affected, the populations of mites, both predaceous and herbivorous, were three times higher in fertilized plots. Through three seasons of field and lab experiments, Chiang concluded that predation by mites accounted for a 20% control of corn rootworm under unfertilized field conditions and a 63% level of control when manure was applied.

D. Tritrophic Interactions with Resistance Mechanisms

Plant–herbivore interactions cannot be understood fully without incorporating direct and cascading effects of the third trophic level.

Although in evolutionary history, host range and feeding preferences (relative acceptance) of phytophagous arthropods have been determined by plant characteristics, by various aspects of the habitat in which they search, and by the herbivore's behavioral and physiological plasticity (Fox and Morrow, 1981; Scriber, 1983), special attention has been paid only recently to selection pressures for host plants imposed by natural enemies (Lawton and McNeill, 1979; Price *et al.*, 1980; Bernays and Graham, 1988). Likewise, ecologists have only recently concentrated research efforts on the consequences of plant physical defenses and phytochemistry for predators and parasitoids of herbivores (Boethel and Eikenbarry, 1986).

Early practioners of integrated pest management assumed that host plant resistance and biological control were compatible, and even synergistic. Theoretical support for this assumption was derived from deterministic population models predicting more effective regulation of pest populations by their parasitoids if the host's rate of increase (r_m) was suppressed (Hare, 1992). The rate of increase of herbivores may indeed decrease in response to plant resistance in most cases. However, the acceptibility and suitability of herbivores feeding on relatively resistant versus susceptible hosts can differ in qualities distinct from their rate of increase and in ways that directly affect their enemies (Campbell and Duffey, 1979) (Table I). For example,

Manduca sexta (Johannson) (Lepidoptera: Sphingidae) survives equally well on tobacco cultivars with high and low nicotine content, but the gregarious parasitoid *Cotesia congregata* (Say) (Hymenoptera: Braconidae) exhibits lower survival and, for females, slower development time on the high nicotine varieties (Thorpe and Barbosa, 1986).

E. Reduce Resource Concentration

Herbivores, especially those with restricted host ranges, are more likely to find and remain on hosts in pure stands of their host plant than in mixed assemblages of hosts and nonhosts.

To mitigate vulnerability to herbivores caused by growing evolutionarily adapted, nonapparent plants in "apparent" conditions, it is often possible to mask some of the attributes that aid in their location by arthropod herbivores. For specialized feeders using chemical cues to colonize host plants, a reduction in characteristic volatiles emanating from the host plant may reduce host plant apparency. Trap crops associated with the target crop may also function to concentrate herbivores and reduce economic damage. In addition, mixtures of host and nonhost plants can result in lower herbivore colonization and reduced tenure time or increased emigration from the target crop (Risch, 1981).

1. Chemical Attributes   Specific plant secondary compounds are known to attract some herbivores and to repel others. To increase the palatability of crop plants for human consumption, some of these compounds have been either reduced to lower levels or eliminated through selective breeding. For some crop–pest combinations, plant apparency is reduced, and plants are subjected to lower levels of damage. For example, cucurbitacins, which afford a bitter taste to cucumbers, are chemical cues for specialized herbivores, acting as behavioral arrestants and feeding stimulants (Howe *et al.,* 1976). Nonbitter varieties of cucumber may be less attractive to cucurbit-feeding arthropods and are certainly damaged less by these specialists (DaCosta and Jones, 1971a) than are cultivars or wild relatives with the full complement of cucurbitacins. Similarly, isothiocyanates in *Brassica* crops and their wild relatives provide critical host location and acceptance cues for specialist herbivores. Altieri and Gliessman (1983) showed how wild mustard could be used as a crop associate to draw key pests, as specialist herbivores, away from the less chemically apparent broccoli crop.

The distinction between specialist and generalist phytophages, however, becomes an important consideration when secondary chemistry of the crop plants is used as a way to reduce herbivore damage. Crops selected to have reduced concentrations of attractive compounds for monophagous herbivores can yield higher populations of ubiquitous generalist feeders that can become major pests. DaCosta and Jones (1971b) found that reduced attack by cucumber beetles on nonbitter cucurbit varieties was accompanied by serious damage by a generalist mite that normally avoids cucurbitacins as defensive compounds in cucurbitaceous crops.

2. Vegetational Diversity   Root (1973) proposed the resource concentration hypothesis, which predicts that many herbivores, especially those with a narrow host range, are more likely to find, survive, and reproduce on hosts that occur in pure or nearly pure stands. Subsequent reviews of documented experimental tests (Risch *et al.,* 1983; Andow, 1986, 1991) have shown that in the majority of cases, a reduced concentration of host plant resources results in lower herbivore abundance. Tonhasca and Byrne (1994), using a meta-analysis approach on data from 21 studies, found that crop diversity caused a moderate reduction in herbivorous insect populations compared to monocultures (but see Tonhasca, 1994). My own research on corn–bean–squash intercropping in Mexico and California supports these conclusions (Table III), with pure stands of squash supporting significantly greater levels of herbivores, on average, than did squash in the crop assemblage (Letourneau and Altieri, 1983; Letourneau, 1986, 1990a). Andow (1986) found that specialist herbivores were more likely to be disrupted in plant mixtures than were generalist feeders. This result is consistent with the resource concentration hypothesis because, for generalist herbivores, plant mixtures may constitute a varied resource, but if all plants serve as hosts the mixture is functionally no less concentrated than a pure stand of one host. Herbivore abundance on squash (Table III) was almost always greater in squash monoculture than on squash in polyculture, even for some generalists and mixed taxa. The notable exception to the rule was the squash specialist, *Anasa tristis* (squash bug), which showed a propensity to deposit its egg masses on squash in polyculture perhaps because of the shaded habitat or differences in plant quality (Table III).

Mechanisms of herbivore/pest reduction within host–nonhost mixtures are not well understood and probably vary among arthropod–plant combinations. Cromartie (1975) suggested that crucifer-feeding insects were disoriented by way of visual or gustatory stimuli in host–nonhost mixtures compared to pure stands. Tahvanainen and Root (1972) showed that the presence of associated nonhost tomato or ragweed (*Ambrosia artemisiifolia*) foliage or even the odors of these plants resulted in lower feeding damage on collard leaves by flea beetles. However, increased emigration rates of flea beetles rather than colonization rates accounted for lower numbers in intercropped broccoli (Garcia and Altieri, 1992).

Although it is possible to generalize about effects of resource concentration and pest damage, the details of these interactions vary for different crops, crop mixtures, and key pests. For example, Weiss *et al.* (1994) found an increase in flea beetle densities with increased oilseed rape (*Brassica*

Table III   Mean Densities of Specialist (*) Herbivores,
Generalist Herbivores, and Groups Including Generalists and
Specialists on Squash in Monoculture and Polyculture in Mexico
and California Experimental Plots

Taxon	Mean seasonal density ($\pm$SE) per 21 squash leaves	
	Monoculture	Polyculture
Tabasco		
Coleoptera		
Acalymma spp.*	18.1 $\pm$ 0.4[a]	4.6 $\pm$ 0.1[b]
Diabrotica spp.*	6.1 $\pm$ 0.4[a]	0.7 $\pm$ 0.1[b]
Lepidoptera		
Diaphania hyalinata*		
Eggs	10.9 $\pm$ 1.9[a]	2.5 $\pm$ 0.9[b]
Larvae	24.5 $\pm$ 1.7[a]	15.0 $\pm$ 3.1[b]
Diaphania nitidalis*		
Eggs	19.1 $\pm$ 0.3[a]	2.4 $\pm$ 0.9[b]
Homoptera		
Aphidae	30.1 $\pm$ 2.5[a]	2.9 $\pm$ 0.3[b]
Cicadellidae	7.2 $\pm$ 0.4[a]	3.1 $\pm$ 0.2[b]
Hemiptera		
Anasa tristis*	9.9 $\pm$ 0.7[a]	30.4 $\pm$ 1.3[b]
California		
Thysanoptera		
Frankliniella occidentalis	726.3 $\pm$ 34[a]	308.2 $\pm$ 18[b]
Homoptera		
Empoasca spp.	74.3 $\pm$ 12[a]	93.8 $\pm$ 16[a]

Note. Means with different letters in the same row are significantly different
(ANOVA, $P < 0.05$).

napus) concentration across mixtures of field pea and rape. However, flea beetle loads per plant did not differ significantly from pure stands to stands that were highly "diluted" with field pea. It may be that oilseed rape cues are not disrupted by the presence of field pea, as they may be with plants that have stronger volatile components. Also, they were attempting to achieve control during the first 2 weeks of the crop, when pea plants were small and perhaps less effective as disruptants (Weiss *et al.*, 1994). Finally, under field conditions, the flea beetles may have produced an aggregation phero-mone capable of overriding any effects due to intercropping (Peng and Weiss, 1992).

F. Recruit Natural Enemies

Mixed vegetation supports a greater abundance and diversity of predators and parasitoids than simple stands of a single plant because of the provision of varied resources such as attractants, refugia, alternate hosts, and food.

Plants interact with the third trophic level in various ways individually and in association such that phytophagous arthropods using those plants as resources are subject to heightened levels of attack (Table IV). Pemberton and Turner (1989) found that tiny pockets on plant leaves can act as domatia for predaceous mites. Bentley (1977) has established that ants, as "pugnaceous bodyguards," are maintained as resident protectors when energy-rich nectar is secreted from plant glands. Atsatt and O'Dowd (1976) described "insectary plants" as those providing food or shelter for predators or parasitoids that then cause mortality to herbivores on associated plant species. The number of ecological studies incorporating three trophic levels has increased substantially in the past two decades compared to early research that focused on species pairs: plant–herbivore, predator–prey, or parasitoid–host interactions (Letourneau, 1988). Many data (e.g., Clark and Dallwitz, 1975; Gilbert, 1978; McClure, 1980; Kareiva, 1982; Murdoch *et al.*, 1995) and much theory (e.g., Murdoch and Oaten, 1975; May and Anderson, 1978; Clark and Holling, 1979) on consumer dynamics in two level systems demonstrate the regulation of populations at the lower trophic level (plant, prey, or host) by natural enemies. Thus, theoretical considerations of multitrophic level interactions will provide important applications in agroecosystems as the field develops.

1. Increased Natural Enemy Activity via Plant Rewards   Factors such as the physical properties of plant surfaces, structural attributes of plants, chemical production by plants, and microclimatic conditions become vital aspects of "habitat suitability" for the natural enemies of herbivores and determine the efficiency of a particular parasitoid or predator as a biological control agent. Energetically expensive products that attract natural enemies afford plants a selective advantage if the cost of production is less than the cost of tissue loss and injury due to herbivores.

Plant species in at least 95 families (more than 1000 species) produce energy-rich substances that attract entomophagous insects and increase their residence time and foraging efficiency with respect to plant defense (Whitman, 1994). A classic example of plant rewards that produce a net gain in plant fitness via biotic plant protection is the bull's thorn *Acacia* tree, which provides lipid-rich food bodies, extrafloral nectar, and houses colonies of stinging ants (Janzen, 1966). I know of no cultivated varieties that provide as elaborate an array of incentives for natural enemies (Way and Khoo, 1992), but direct interactions do occur. For example, cotton, fava bean, and vetch varieties that retain extrafloral nectaries may incur lower levels of damage because foraging ants use these herbivores as food for the colony. Olson and Nechols (1995) discovered that hair-like trichomes on squash leaves produce glucose and galactose, and peg-like trichomes secrete both glucose and protein; both exudates are imbibed by *Gryon pennsylvanicum,* an egg parasitoid of the squash bug.

Table IV Examples of Mechanisms by which Plants Attract Natural Enemies That Attack Phytophagous Herbivores

Enemy	Herbivore	Plant	Mechanism	Reference
Chrysoperla carnea	Aphids	Cotton	Detection of caryophyllene from foliage; detection of indole acetaldehyde from honeydew and phenology-specific plant volatile by gravid females	Flint *et al.* (1979), van Emden and Hagen (1976), Hagen (1986, and references therein)
Diaeretiella rapae	Cabbage aphid	Crucifers	Attraction to allyl isothiocyanate; increased parasitism	Read *et al.* (1970)
Campoletis sonorensis (Cam.) (Hymenoptera: Ichneumonidae)	*Heliothis virescens*	Cotton, sorghum	Plant volatiles attract parasitoid to host habitat	Elzen *et al.* (1983)
Collops vittatus Say (Coleoptera: Melyridae)	Aphids	Cotton	Male response to caryophyllene in foliage	Flint *et al.* (1981)
Microplitis demolitor (Hymenoptera: Braconidae)	*Pseydoplusia includens*	Soybean	Seven- and eight-carbon hydrocarbons, aldenydes, and ketones attract parasitoid to plant	Ramachandran and Norris (1991)
Amblyseius limonicus Garman and McGregor, *A. californicus* (McGregor), *A. anonymus* Chant and Baker, *Cydrodromella pilosa* (Chang) (Acari: Phytoseiidae)	*Monochychellus tanajoa* (Bondar) (Acari: Tetranychidae)	Cassava	Attraction to mite-infested leaves	Janssen *et al.* (1990)
Phytoseiulus persimilis	*Tetranychus urticae*	Lima bean (*Phaseolus lunatus*)	Damaged host plant produces volatiles attractive to predators	Dicke *et al.* (1990)
Cotesia marginiventris	*Spodoptera exigua*	Corn seedlings	Volatiles from damaged leaves attract female wasps	Turlings *et al.* (1991)

Comparative studies demonstrate that other factors can also cause ento-mophagous insects to be more abundant in the presence of particular plants, even in the absence of host or prey. They may be attracted/arrested by chemicals released by the herbivore's host plant (Table IV). Some ento-mophagous insects are more abundant in the presence of particular plants, even in the absence of their hosts or prey, or they are attracted to or arrested by chemicals released by the herbivore's host plant or other associ-ated plants (Monteith, 1960; Shahjahan, 1974; Nettles, 1979; Martin *et al.*, 1990). Parasitism of a pest can be higher on some crops than on others (Read *et al.*, 1970; Martin *et al.*, 1976; Nordlund *et al.*, 1988; Johnson and Hara, 1987; Gerard, 1989; Lewis and Gross, 1989) and different parasitoids can be associated with particular crops or crop habitats even if their hosts are widespread (Lewis and Gross, 1989; Felland, 1990; van den Berg *et al.*, 1990).

2. Increased Natural Enemy Activity via Association with Other Plants Root's (1973) enemies hypothesis predicts that complex vegetation will provide more resources for natural enemies (alternate hosts, refugia, nectar, and pollen) than will monocultures, and, as a result, herbivore irruption will be rapidly checked by a high diversity and abundance of enemies. Andow's (1986, 1991) reviews of the literature on pest population densities in mixed cropping versus monoculture showed that 56% of the herbivores had lower population densities, 16% had higher population densities, and 28% had similar or variable densities in polyculture compared to monoculture.

Studies in temperate and tropical habitats on corn–bean–squash or corn–bean mixtures versus monocultures of these crops illustrate some additional mechanisms by which natural enemy populations are enhanced in vegetation mixtures. First, patterns of overall host–parasitoid or prey–predator densities differ in monoculture versus polyculture in ways that can enhance biological control. For example, egg parasitism of *D. hyalinata* (Lepidoptera: Pyralidae), a common herbivore on squash, was uncommon at low egg densities in squash monocultures but showed a density-dependent increase (Letourneau, 1987). In polycultures, where presumably the densities of other hosts on corn and bean contributed to the overall egg density, low *D. hyalinata* egg densities exhibited four times greater parasitism rates. Similarly, Coll and Bottrell (1995) found that mortality rates of Mexican bean beetle larvae (*Epilachna varivestis* Mulsant) were density independent in bean monocultures compared to being inversely density dependent in corn–bean dicultures. The shift in functional re-sponse caused enhanced biological regulation of the herbivore in mixed vegetation.

Second, structural complexity, as a component of vegetational diversity, is associated with increased densities of natural enemies in mixed crop systems. In Mexico, parasitoids were more abundant in corn–bean–squash tricultures than in squash monocultures, similar to their densities in corn monocultures, suggesting that the structural characteristics of corn as an intercrop may have been responsible for increased numbers (Letourneau, 1987). In California, the components of triculture were experimentally separated to test for the effects of structural complexity on natural enemies (Letourneau, 1990a). Early season colonization rates of the anthocorid *Orius tristicolor* (White) to squash monocultures made structurally complex with artificial plants were greater than colonization rates to squash in simple monoculture but were comparable to the high rates of colonization to corn–bean–squash polyculture. In the same polyculture system, *Erigone* spiders (Aranae: Micryphantidae) accumulated in higher densities on squash in polyculture probably because ballooning individuals were intercepted by tall corn plants (Letourneau, 1990b). Russell (1989) provides other examples of studies testing the enemies hypothesis, and I have listed some mechanisms in Table V.

In contrast to the majority of studies, Nafus and Schreiner (1986) found lower parasitism rates in corn intercropped with squash; the addition of squash decreased the abundance of the coccinellid *Coleomegilla maculata* (De Geer) on squash because of a nonuniform distribution of prey. Andow and Risch (1985) and Williams *et al.* (1995) found no effect of strip intercropping on green cloverworm parasitism rates. Rämert (1996) documented lower damage of carrots by *Psila rosae* (carrot fly) when lucerne was added as an intercrop. However, carabid and staphylinid predators were not always more abundant in the intercrop than in carrot monocultures and probably did not contribute to the lower incidence of carrot flies in the intercrop.

Sheehan (1986) offers a possible explanation for variable responses of natural enemies. He extends the resource concentration concept to predict that specialist enemies will respond to mixed vegetation differently, and probably less favorably, than will generalist predators and parasitoids because of the importance of alternate prey for generalists. The designation of host/prey specialization categories, however, tends to rely on only one aspect of the resource spectrum of parasitoids and predators. A more comprehensive set of predictions would include a range of species (or even individual) characteristics, such as relative vagility, resource needs, and habitat location cues as determinants of parasitoid and predator response to vegetational diversity.

Noncrop plants can be employed, as are intercrops, to foster natural enemy populations and community development (Norris, 1986; Barney *et al.*, 1984). Rapidly colonizing, fast-growing plants offer resources for natural enemies such as alternate prey/hosts, pollen, or nectar, as well as microhabitats that are not available in weed-free monocultures (van Emden 1965b).

Crop fields with a dense weed cover and high plant diversity usually have more predaceous arthropods than do weed-free fields (Pimentel, 1961; Dempster, 1969; Flaherty, 1969; Pollard, 1971; Root, 1973; Smith, 1976; Speight and Lawton, 1976). For example, carabid ground beetles (Dempster, 1969; Speight and Lawton, 1976; Thiele, 1977), syrphid flies (Pollard, 1971; Smith, 1976), and coccinellid lady beetles (Bombosch, 1966; Perrin, 1975) are abundant in weed-diversified systems. As a result, outbreaks of certain types of crop pests are more likely to occur in weed-free fields than in weed-diversified crop systems (Dempster, 1969; Flaherty, 1969; Root, 1973; Altieri *et al.,* 1977).

Flaherty (1969) showed enhancement in the control of herbivorous mites on grapevines with Johnson grass ground cover. The grass acted as a source of predaceous mites in temporal synchrony with growing herbivorous mite populations in vineyards. Other examples of cropping systems in which the presence of specific weeds has enhanced the biological control of particular pests are listed in Table V. However, weeds can have detrimental effects on arthropod pest damage to crops and do not always promote biological control (Altieri and Letourneau, 1982; Powell *et al.,* 1986).

V. Crop Protection: An Ecological Perspective

Principles of ecology and evolution can contribute vital knowledge for designing and integrating methods to enhance, rather than mitigate, the ecological factors that cause herbivores "to be caught between the devil" (natural enemies) "and the deep blue sea" (host plant defense/nutrient deficiency) (Lawton and McNeill, 1979). Unfortunately, because agricultural land use and management practices are dictated mainly by short-term economic forces, pest control plans are often reactive rather than preventative.

Because farming systems in a region are managed over a range of energy inputs, levels of crop diversity, and successional stages, variations in insect dynamics are likely to occur that are difficult to predict. However, based on current ecological and agronomic theory, low pest potentials may be expected in agroecosystems that exhibit the following characteristics as a result of specific crop and vegetation management regimes:

1. small, scattered fields creating a structural mosaic of adjoining crops and uncultivated land that potentially provide shelter and alternate food for natural enemies (van Emden, 1965b; Altieri and Letourneau, 1982). Although pests may also proliferate in these environments, depending on plant species composition (Altieri and Letourneau, 1984; Collins and Johnson, 1985; Levine, 1985; Slosser *et al.,* 1985; Lasack and Pedigo, 1986),

Table V Examples of Mixed Crop or Crop–Noncrop Associations That Enhance the Biological Control of Key Pests on Different Target Crops

Crop	Associated plants	Pest(s) regulated	Factor(s) involved	Reference
Apple	Natural weed complex	Tent caterpillar (*Malacosoma americanum*) and codling moth [*Laspeyresia pomonella* (L.) (Lepidoptera: Olethreutidae)]	Increased activity and abundance of parasitic wasps	Leius (1967)
Beans	Natural weed complex	*Epilachna varivestis* L.	Increased predation on eggs	Andow (1990)
Beans	Corn	Leafhoppers (*Empoasca kraemeri*), leaf beetle (*Diabrotica balteata*), and fall armyworm (*Spodoptera frugiperda*)	Increase in beneficial insects and interference with colonization	Altieri *et al.* (1978)
Brussels sprouts	Natural weed complex	Imported cabbage butterfly [*Pieris rapae* (L.) (Lepidoptera; Pieridae)] and aphids [*Brevicoryne brassicae* (L.) (Homoptera: Aphididae)]	Alteration of crop background and increase of predator colonization	Smith (1976), Dempster (1969)
Brussels sprouts	*Brassica kaber* (DC.) (Brassicaceae)	*B. brassicae*	Attraction of syphids (*Allograpta obliqua*) (Say) (Diptera: Syrphidae) to *B. kaber* flowers	Altieri (1984)
Cabbage	Mixed clover, *Trifolium repens* L. and *T. praetense* L.	*Erioischia brassicae*, cabbage aphids, and imported cabbage butterfly (*Pieris rapae*)	Interference with colonization and increase in ground beetles	Dempster and Coaker (1974)
Collards	*Amaranthus retroflexus* L. (Amaranthaceae), *Chenopodium album*, *Xanthium strumarium* L. (Asteraceae)	Green peach aphid (*Myzus persicae*)	Increased abundance of predators (*Chrysoperla carnea*, Coccinellidae, and Syrphidae)	Horn (1981)

Cotton	Alfalfa	Plant bugs (*Lygus hesperus* and *L. elisus*)	Prevention of emigration and synchrony in the relationship between pests and natural enemies	van den Bosch and Stern (1969), Godfrey and Leigh (1994)
Cotton	Cowpea	Boll weevil (*Anthomomus grandis*)	Population increase of parasitic wasps (*Eurytoma* sp.)	DeLoach (1970)
Cotton	Sorghum or maize	Corn earworm (*Heliothis zea*)	Increased abundance of predators	Altieri *et al.* (1981)
Soybean	*Cassia obtusifolia* L. (Fabaceae)	*Nezara viridula* (L.) (Heteroptera: Pentatomidae), *Anticarsia gemmatalis* (Lepidoptera: Noctuidae)	Increased abundance of predators	
Sugar beet	*Matricaria chamomilla* L., *Lamium purpureum* L., and others	*Aphis fabae* L., *Atomaria linearis* Stephen, *Diplopoda* spp.	Increased abundance of parasites (Hymenoptera, carabids, and spiders)	Bosch (1987)
Sweet potatoes	Corn	Leaf beetles (*Diabrotica* spp.) and leafhoppers (*Agallia lingula*)	Increase in parasitic wasps	Risch (1979)
Winter wheat	Grassland	Cereal aphids	Colonization source for *Aphidius rhopalosiphi* DeStefani Perez females	Vorley and Wratten (1987)
Squash	Maize and bean	*Diaphania hyalinata* (L.) (Lepidoptera: Pyralidae)	Habitat more suitable for parasitoids/alternate hosts for *Trichogramma* spp.	Letourneau (1987)
Cacao	Coconut	*Helopeltis theobromae* Miller	Provision of nest sites for ants [*Oecophylla smaragdina* (F.) and *Dolichoderus thoracicus* (Smith) (Hymenoptera: Formicidae)]	Way and Khoo (1991)

the presence of these pests and/or alternate hosts may be critical for maintaining natural enemies in the area;

2. discontinuity of monoculture in time through rotations, use of short maturing varieties, use of crop-free or preferred host-free periods (Stern, 1981; Lashomb and Ng, 1984; Desander and Alderweireldt, 1990);

3. resistant crop varieties designed or managed to maximize compatibility with other control tactics and to minimize the potential for counterresistance in arthropod herbivores (Gould, 1988; Kogan, 1994; Alstad and Andow, 1995);

4. soil management practices that optimize and complement other control methods and avoid conditions that exacerbate pest buildup for those species that are directly affected [e.g., fertilized bean and stemborers in Malaŵi (Letourneau 1994)];

5. high crop diversity through mixtures in time and space (Cromartie, 1981; Altieri and Letourneau, 1982; Risch *et al.*, 1983; but see Andow and Risch, 1985; Nafus and Schreiner, 1986);

6. farms with a dominant perennial crop component. Orchards are considered to be more stable, as permanent ecosystems, than are annual crop systems. Because orchards suffer less disturbance and are characterized by greater structural diversity, possibilities for the establishment of biological control agents are generally higher, especially if floral undergrowth diversity is encouraged (Huffaker and Messenger, 1976; Altieri and Schmidt, 1984);

7. high crop densities and/or the presence of tolerable levels of weed background (Shahjahan and Streams, 1973; Altieri *et al.*, 1977; Sprenkel *et al.*, 1979; Mayse, 1983; Andow, 1983a; Buschman *et al.*, 1984; Ali and Reagan, 1985; but see Perfecto *et al.*, 1986); and

8. high genetic diversity resulting from the use of variety mixtures or several lines of the same crop (Perrin, 1977; Whitham, 1983; Gould, 1986; Altieri and Schmidt, 1987).

All of the previous generalizations can serve in the planning of pest outbreak prevention strategies in agroecosystems; however, they must take into account local variations in climate, geography, crops, local vegetation, inputs, pest complexes, etc. that might increase or decrease the potential for pest development under some vegetation management conditions.

Figure 3 Synoptic guide for management options with respect to ecological characteristics of arthropod pests, with plant resistance through antibiosis targeting generalist rather than specialist herbivores, associational resistance through intercropping being most appropriate for specialist feeders, nutrient deficiencies in crops generalizing across feeding groups but likely to be more effective for reducing r-selected species than K-selected pests, and possible options for extremely K-selected as truncated or pulsed cropping and for extremely r-selected as unapparent and isolated from colonizer sources.

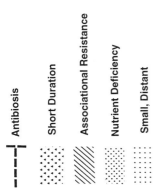

Antibiosis

Short Duration

Associational Resistance

Nutrient Deficiency

Small, Distant

Pests

Generalists ------------> Specialists

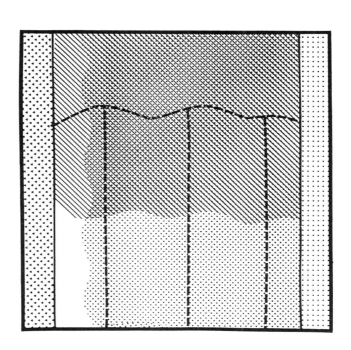

Pests

r-selected ------------> K-selected

Figure 3 represents a partial synopsis of the material in this chapter on using ecological ideas to guide the management of arthropod pests. Although there are numerous assumptions embedded in this synoptic approach, it provides a starting point for relating ecologically based pest management tactics to different types of arthropod pests. It is logical that pest-resistant cultivars (that rely on antibiosis or nonpreference) are most effective against herbivores with relatively broad host ranges (see Sections IV,D,1 and IV,D,2) and that associational resistance is a complementary tactic that is most effective for relatively specialist feeders. The importance of nitrogen deficiencies or nutrient imbalance is likely to be more pronounced for fast-growing r-selected species than for extremely K-selected pests but may affect both specialists and generalists. Obviously, reducing the duration of resource availability would be a tactic for disrupting extremely K-selected species but may not be possible for controlling rapidly colonizing/reproducing r-selected pests (but see Conway, 1976). Perhaps the most difficult pests to control ecologically are extreme r-selected species. I suggest that small, isolated crop patches may experience sufficiently reduced colonization rates and avoid pest buildup as has been seen anecdoctally when fields are extremely far from colonizer source pools.

For most pests, the ecologically based pest control practices described are not mutually exclusive and should be integrated. Tactics that reduce the suitability of a crop as a resource through the manipulation of host quality (e.g., pest resistance due to antibiosis) or vegetation management (e.g., intercropping or trap cropping) may work synergistically with the spatial and temporal crop manipulations suggested for extreme r- or K-selected pests. Similarly, crop timing or isolation techniques can enhance the effectiveness of reduced host quality or vegetation management. Finally, each of these tactics may have concomitant effects on the complex of natural enemies that attack target pests, providing an additional series of considerations relevant to the use of the synoptic guide (Fig. 3) for the prevention of pest outbreaks.

Acknowledgments

This work was supported in part by a University of California, Santa Cruz (UCSC), Division of Social Sciences Research grant and a grant from the Center for Agroecology and Sustainable Food Systems, UCSC. Helpful discussions with M. A. Altieri, P. Barbosa, J. Dlott, and L. Fox improved the quality of the manuscript, and M. Hemler managed to negotiate various sets of handwritten revisions sent from remote field stations and still meet the deadlines.

References

Abbott, R. J. (1994). Ecological risks of transgenic crops. *Trends Ecol. Evol.* **9,** 280–282.

Adkisson, P. (1962). Timing of defoliants and dessicants to reduce populations of the pink bollworm in diapause. *J. Econ. Entomol.* **55**, 949–951.

Adkisson, P. L. (1958). The influence of fertilizer applications on populations of *Heliothis zea* (Boddie), and certain insect predators. *J. Econ. Entomol.* **51**, 757–759.

Agren, J., and Schemske, D. W. (1993). The cost of defense against herbivores: An experimental study of trichome production in *Brassica rapa. Am. Nat.* **141**, 338–350.

Ajlan, A. M., and Potter, D. A. (1991). Does immunization of cucumber against anthracnose by *Colletotrichum lagenarium* affect host suitability for arthropods? *Entomol. Exp. Appl.* **58**, 83–91.

Ajlan, A. M., and Potter, D. A. (1992). Lack of effect of tobacco mosaic virus-induced systemic acquired resistance on arthropod herbivores in tobacco. *Phytopathology* **82**, 647–651.

Ali, A. D., and Reagan, T. E. (1985). Vegetation manipulation impact on predator and prey populations in Louisiana sugarcane ecosystems. *J. Econ. Entomol.* **78**, 1409–1414.

Alstad, D. N., and Andow, D. A. (1995). Managing the evolution of insect resistance to transgenic plants. *Science* **268**, 1894–1897.

Altieri, M. A. (1984). Patterns of insect diversity in monocultures and polycultures of Brussels sprouts. *Protection Ecol.* **6**, 227–232.

Altieri, M. A. (1994). "Biodiversity and Pest Management in Agroecosystems," pp. 185. Food Products Press, New York.

Altieri, M. A., and Gliessman, S. R. (1983). Effects of plant diversity on the density and herbivory of the flea beetle, *Phyllotreta cruciferae* Goeze, in California collard cropping systems. *Crop Protection* **2**(4), 497–450.

Altieri, M. A., and Letourneau, D. K. (1982). Vegetation management and biological control in agroecosystems. *Crop Protection* **1**(4), 405–430.

Altieri, M. A., and Letourneau, D. K. (1984). Vegetation diversity and insect pest outbreaks. *CRC Crit. Rev. Plant Sci.* **2**, 131–169.

Altieri, M. A., and Schmidt, L. L. (1984). Abundance patterns and foraging activity of ant communities in abandoned, organic and commercial apple orchards in Northern California. *Agric. Ecosyst. Environ.* **11**, 341–352.

Altieri, M. A., and Schmidt, L. L. (1986). Population trends, distribution patterns, and feeding preferences of flea beetles (*Phyllotreta cruciferae* Goeze) in collard-wild mustard mixtures. *Crop Protection* **5**(3), 170–175.

Altieri, M. A., and Schmidt, L. L. (1987). Mixing cultivars of broccoli reduces populations of the cabbage aphid *Brevicoryne brassicae* (L.). *California Agric.* **41**(11/12), 47–26.

Altieri, M. A., Schoonhoven, A., and Doll, J. D. (1977). The ecological role of weeds in insect pest management systems: A review illustrated with bean (*Phaseolus vulgaris* L.) cropping systems. *PANS* **23**, 185–206.

Altieri, M. A., Francis, C. A., Schoonhoven, A., and Doll, J. (1978). Insect prevalence in bean (*Phaseolus vulgaris*) and maize (*Zea mays*) polycultural systems. *Field Crops Res.* **1**, 33–49.

Altieri, M. A., Lewis, W. J., Nordlund, D. A., Gueldner, R. C., and Todd, J. W. (1981). Chemical interactions between plants and *Trichogramma* sp. wasps in Georgia soybean fields. *Protection Ecol.* **3**, 259–263.

Andow, D. A. (1983a). Plant diversity and insect populations: Interactions among beans, weeds, and insects. Ph.D. dissertation, Cornell University, Ithaca, NY.

Andow, D. A. (1983b). The extent of monoculture and its effects on insect pest populations with particular reference to wheat and cotton. *Agric. Ecosyst. Environ.* **9**, 25–35.

Andow, D. A. (1986). Plant diversification and insect population control in agroecosystems. *In* "Some Aspects of Integrated Pest Management" (D. Pimentel, ed.). Department of Entomology, Cornell University, Ithaca, NY.

Andow, D. A. (1990). Population dynamics of an insect herbivore in simple and diverse habitats. *Ecology* **72**(3), 1006–1017.

Andow, D. A. (1991). Vegetational diversity and arthropod population response. *Annu. Rev. Entomol.* **36**, 561–586.

Andow, D. W., and Risch, S. J. (1985). Predation in diversified agroecosystems: Relations between a coccinellid predator *Coleomegilla maculata* and its food. *J. Appl. Ecol.* **22,** 357–372.

Atsatt, P. R., and O'Dowd, D. J. (1976). Plant defense guilds. *Science* **193,** 24–29.

Barbosa, P. (1993). Lepidopteran foraging on plants in agroecosystems: Constraints and consequences. *In* "Caterpillars: Ecological and Evolutionary Constraints on Foraging" (N. E. Stamp and T. M. Case, eds.), pp. 526–566. Chapman & Hall, New York.

Barney, R. J., Lamp, W. O., Armbrust, E. J., and Kapusta, G. (1984). Insect predator community and its response to weed management in spring-planted alfalfa. *Protection Ecol.* **6,** 23–33.

Barry, D., Alfara, D., and Darrah, L. L. (1994). Relation of European corn borer (Lepidoptera: Pyralidae) leaf-feeding resistance and DIMBOA content in maize. *Environ. Entomol.* **23**(1), 177–182.

Baur, R., Binder, S., and Benz, G. (1991). Nonglandular leaf trichomes as short-term inducible defense of the grey alder, *Alnus incana* (L.), against the chysomelid beetle, *Agelastica alni* L. *Oecologia* **87,** 219–226.

Belcher, D. W., and Thurston, R. (1982). Inhibition of movement of larvae of the convergent lady beetle by leaf trichomes of tobacco. *Environ. Entomol.* **11,** 91–94.

Benedict, J. H., Sachs, E. S., Altman, D. W., Ring, D. R., Stone, T. B., and Sims, S. R. (1993). Impact of δ-endotoxin-producing transgenic cotton on insect–plant interactions with *Heliothis virescens* and *Helicoverpa zea* (Lepidoptera: Noctuidae). *Environ. Entomol.* **22,** 1–9.

Bentley, B. C. (1977). Extrafloral nectaries and protection by pugnacious bodyguards. *Annu. Rev. Ecol. Syst.* **8,** 407–427.

Berenbaum, M. R. (1990). Evolution of specialization in insect–umbellifer associations. *Annu. Rev. Entomol.* **35,** 319–343.

Bergvinson, D. J., Arnason, J. T., Hamilton, R. I., Mihm, J. A., and Jewell, D. C. (1994a). Determining leaf toughness and its role in maize resistance to the European corn borer (Lepidoptera: Pyralidae). *J. Econ. Entomol.* **87**(6), 1743–1748.

Bergvinson, D. J., Arnason, J. T., Hamilton, R. I., Tachibana, S., and Towers, G. H. N. (1994b). Putative role of photodimerized phenolic acids in maize resistance to *Ostrini nubilalis* (Lepidoptera: Pyralidae). *Environ. Entomol.* **23**(6), 1516–1523.

Bernays, E., and Graham, M. (1988). On the evolution of host specificity in phytophagous arthropods. *Ecology* **69**(4), 886–892.

Bishop, A. L., and McKenzie, H. J. (1986). Development of *Merophyas divulsana* (Walker) (Lepidoptera: Tortricidae) in relation to lucerne harvesting as a control strategy. *J. Aust. Entomol. Soc.* **25,** 229–233.

Blua, M. J., Perring, T. M., Nuessly, G. S., Duffus, J. E., and Toscano, N. C. (1994). Seasonal cropping pattern effects on abundance of *Bemisia tabaci* (Homoptera: Aleyrodidae) and incidence of lettuce infectious yellows virus. *Environ. Entomol.* **23**(6), 1422–1427.

Bodnaryk, R. P. (1992). Leaf epicuticular wax, an antixenotic factor in Brassicaceae that affects the rate and pattern of feeding by flea beetles *Phyllotreta cruciferae* (Goeze). *Can. J. Plant Sci.* **72,** 1295–1303.

Boethel, D. J., and Eikenbarry, R. D. (1986). "Interactions of Plant Resistance and Parasitoids and Predators of Insects," pp. 224. Ellis Horwood, New York.

Bombosch, S. (1966). Occurrence of enemies on different weeds with aphids. *In* "Ecology of Aphidophagus Insects" (I. Hodek, ed.), pp. 177–179. Academic Printing House, Prague.

Bosch, J. (1987). The influence of some dominant weeds on beneficial and pest arthropods in a sugarbeet field. *Zeit. Pflanzenkrankheit. Pflanzenschutz* **94,** 398–408.

Boyd, M. L., and Lentz, G. L. (1994). Seasonal incidence of the cabbage seedpod weevil (Coleoptera: Curculionidae) on rapeseed in West Tennessee. *Environ. Entomol.* **23**(4), 900–905.

Brody, A. K., and Karban, R. (1992). Lack of tradeoff between constitutive and induced defenses among varieties of cotton. *Oikos* **65,** 301–306.

Buntin, G. D., Bruckner, P. L., and Johnson, J. W. (1990). Management of the Hessian fly (Diptera: Cedidomyiidae) in Georgia by delayed planting of winter wheat. *J. Econ. Entomol.* **83**, 1025–1033.

Buschman, L. L., Pitre, H. N., and Hodges, H. F. (1984). Soybean cultural practices: Effects on populations of geocorids, nabids, and other soybean arthropods. *Environ. Entomol.* **13**, 305–317.

Campbell, B. C., and Duffey, S. S. (1979). Tomatine and parasitic wasps: Potential incompatibility of plant-antibiosis with biological control. *Science* **205**, 700–702.

Casagrande, R. A. (1982). Colorado potato beetle resistance in wild potato, *Solanum berthaultii*. *J. Econ. Entomol.* **75**, 368–372.

Chiang, H. C. (1970). Effects of manure applications and mite predation on corn rootworm populations in Minnesota. *J. Econ. Entomol.* **63**, 934–936.

Clark, L. R., and Dallwitz, M. J. (1975). The life system of *Cardiaspina albitextura* (Psyllidae), 1950–74. *Aust. J. Zool.* **23**, 523–561.

Clark, W. C., and Holling, C. S. (1979). Process models, equilibrium structures, and population dynamics: On the formulation and testing of realistic theory in ecology. *Fortschr. Zool.* **25**, 29–52.

Coll, M., and Bottrell, D. G. (1995). Predator–prey association in mono- and dicultures: Effects of maize and bean vegetation. *Agric. Ecosyst. Environ.* **54**, 115–125.

Collins, F. L., and Johnson, S. J. (1985). Reproductive response of caged adult velvetbean caterpillar and soybean looper to the presence of weeds. *Agric. Ecosyst. Environ.* **14**, 139–149.

Connor, E. F., and McCoy, E. D. (1979). The statistics and biology of the species-area relationship. *Am. Nat.* **113**, 791–833.

Conway, G. R. (1976). Man versus pests. *In* "Theoretical Ecology: Principles and Applications" (R. M. May ed.), pp. 257–281. Blackwell, Oxford, UK.

Cravedi, P., and Molinari, F. (1984). Thysanoptera injurious to nectarines. *Informatore-Fitopatalogico* **34**, 12–26.

Cromartie, J. J. (1975). The effect of stand size and vegetational background on the colonization of cruciferous plants by herbivorous insects. *J. Appl. Ecol.* **12**, 517–533.

Culliney, T. W., and Pimentel, D. (1986). Ecological effects of organic agricultural practices on insect populations. *Agric. Ecosyst. Environ.* **15**, 253–266.

DaCosta, C. P., and Jones, C. M. (1971a). Cucumber beetle resistance and mite susceptibility controlled by the bitter gene in *Cucumis sativus* L. *Science* **172**, 1145–1146.

DaCosta, C. P., and Jones, C. M. (1971b). Resistance in cucumber, *Cucumis sativus* L. to three species of cucumber beetles. *Hort. Sci.* **6**, 340–342.

Deheer, C. J., and Tallamy, D. W. (1991). Affinity of spotted cucumber beetle (Coleoptera: Chrysomelidae) larvae to cucurbitacins. *Environ. Entomol.* **20**(4), 1173–1175.

DeLoach, C. J. (1970). The effect of habitat diversity on predation. *Proc. Tall Timbers Conf. Ecol. Anim. Control Habitat Management* **2**, 223–241.

Dempster, J. P. (1969). Some effects of weed control on the numbers of the small cabbage white (*Pieris rapae* L.) on Brussels sprouts. *J. Appl. Ecol.* **6**, 339–345.

Dempster, J. P., and Coaker, T. H. (1974). Diversification of crop ecosystems as a means of controlling pests. *In* "Biology in Pest and Disease Control" (D. P. Jones and M. E. Solomon, eds.), pp. 106–114. Wiley, New York.

Desander, K., and Alderweireldt, M. (1990). The carabid fauna of maize fields under different rotation regimes. *Mededelingen-van-de-Faculteit-Landbouwwetenschappen Rijksuniversiteit-Gent.* **55**(2b), 493–500.

Dicke, M., Sabelis, M. W., Takabayashi, J., Bruin, J., and Posthumus, M. A. (1990). Plant structures of manipulating predator–prey interactions through allelochemicals: Prospects for application in pest control. *J. Chem. Ecol.* **16**, 3091–3118.

Dimock, M. B., and Tingey, W. M. (1988). Host acceptance behavior of the Colorado potato beetle and glandular trichomes. *Physiol. Entomol.* **13**, 399–406.

Dover, B. A., Davies, D. H., Strand, M. R., Gray, R. S., Keeley, L. L., and Vinson, S. B. (1987). Ecdysteroid-titre reduction and developmental arrest of last-instar *Heliothis virescens* larvae by calyx fluid from the parasitoid *Campoletis sonorensis. J. Insect Physiol.* **33**(5), 333–338.

Dowd, P. F. (1991). Symbiont-mediated detoxification in insect herbivores. *In* "Microbial Mediation of Plant–Herbivore Interactions" (P. Barbosa, V. A. Krischik, and C. G. Jones, eds.), pp. 411–440. Wiley, New York.

Drinkwater, L. E., Workneh, F., Letourneau, D. K., van Bruggen, A. H. C., and Shennan. C. (1995). Fundamental differences between conventional and organic tomato agroecosystems in California. *Ecol. Appl.* **5**, 1098–1112.

Duffy, M. (1991). Economic considerations of sustainable agriculture for midwestern farmers. *In* "Sustainable Agriculture Research and Education in the Field: A Proceedings, pp. 92–106. National Academy of Sciences, Washington, DC.

Elsey, K. D. (1974). Influence of plant host on searching speed of two predators. *Entomophaga* **19**, 3–6.

Elzen, G. E., Williams, H. J., and Vinson, S. B. (1983). Response by the parasitoid *Campoletis sonorensis* (Hymenoptera: Ichneumonidae) to chemicals (synomones) in plants: Implications for host habitat location. *Environ. Entomol.* **12**, 1873–1877.

Farrow, R. A. (1981). Aerial dispersal of *Scelio fulgidus* (Hym: Scelionidae), parasite of eggs of locusts and grasshoppers (Orth: Acrididae). *Entomophaga* **26**, 349–355.

Feeny, P. P. (1976). Plant apparency and chemical defense. *Rec. Adv. Phytochem.* **10**, 1–40.

Feeny, P. P. (1990). Theories of plant chemical defense: A brief historical survey. *Symp. Biol. Hung.* **39**, 163–175.

Felland, C. M. (1990). Habitat-specific parasitism of the stalk borer (Lepidoptera: Noctuidae) in northern Ohio. *Environ. Entomol.* **19**, 162–166.

Finch, S. (1988). Entomology of crucifers and agriculture—Diversification of the agroecosystem in relation to pest damage in cruciferous crops. *In* "The Entomology of Indigenous and Naturalized Systems in Agriculture" (M. K. Harris and C. E. Rogers, eds.), pp. 39–71, Westview Press, Boulder, CO.

Finch, S. (1989). Ecological considerations in the management of *Delia* pest species in vegetable crops. *Annu. Rev. Entomol.* **34**, 117–137.

Flaherty, D. (1969). Ecosystem trophic complexity and Willamette mite *Eotetranychus willametei* (Acarina: Tetranychidae) densities. *Ecology* **50**, 911–916.

Flint, H. M., Salter, S. S., and Walters, S. (1979). Caryophyllene: An attractant for the green lacewing. *Environ. Entomol.* **8**, 1123–1125.

Flint, H. M., Merkle, J. R., and Sledge, M. (1981). Attraction of male *Collops vittatus* in the field by caryophylline alcohol. *Environ. Entomol.* **10**, 301–304.

Fox, L. R., and Morrow, P. A. (1981). Specialization: Species property or local phenomenon? *Science* **211**, 887–893.

Gallum, R. L., and Khush, G. S. (1980). Genetic factors affecting expression and stability of resistance. *In* "Breeding Plants Resistant to Insects" (F. G. Maxwell and P. R. Jennings, eds.), pp. 63–85. Wiley, New York.

Garcia, M. A., and Altieri, M. A. (1992). Explaining differences in flea beetle *Phyllotreta cruciferae* Goeze densities in simple and mixed broccoli cropping systems as a function of individual behavior. *Entomol. Exp. Appl.* **62**, 201–209.

Gerard, P. J. (1989). Influence of egg depth in host plants on parasitism of *Scolypopa australis* (Homoptera: Ricaniidae) by *Centrodora scolypopa* (Hymenoptera: Aphelinidae). *New Zealand Entomol.* **12**, 30–34.

Gianfagna, T. J., Carter, C. D., and Sacalis, J. N. (1992). Temperature and photoperiod influence trichome density and sesquiterpene content of *Lycopersicon* f. *hirsutum. Plant Physiol.* **100**, 1403–1405.

Gibson, R. W. (1976). Glaundular hairs of *Solanum polyadenium* lessen damage by the Colorado potato beetle. *Ann. Appl. Biol.* **82**, 147–159.

Gibson, R. W., and Pickett, J. A. (1983). Wild potato repels aphids by release of aphid alarm pheromone. *Nature* **302**, 608–609.

Gilbert, L. E. (1971). Butterfly–plant coevolution: Has *Passiflora adenopoda* won the selectional race with heliconiine butterflies? *Science* **172**, 585–586.

Gilbert, L. E. (1978). Development of theory in the analysis of insect plant interactions. *In* "Analysis of Ecological Systems" (D. J. Horn, R. D. Mitchell, and G. R. Stairs, eds.), pp. 117–154, Ohio State Univ. Press, Columbus.

Godfrey, L. D., and Leigh, T. F. (1994). Alfalfa harvest strategy effect on Lygus bug (Hemiptera: Miridae) and insect predator population density: Implications for use as trap crop in cotton. *Environ. Entomol.* **23**, 1106–1118.

Gould, F. (1986). Simulation models for predicting durability of insect-resistant germplasm: Hessian fly (Diptera: Cecidomyiidae) in resistant winter wheat. *Environ. Entomol.* **15**(1), 11–23.

Gould, F. (1988). Evolutional biology and genetically engineered crops. *BioScience* **38**, 26–33.

Gray, M. E., and Luckmann, W. H. (1994). Integrating the cropping system for corn insect pest management. *In* "Introduction to Pest Management" (R. L. Metcalf, ed.), 3rd Ed., pp. 507–541. Wiley, New York.

Hagen, A. F., and Anderson, F. N. (1967). Nutrient imbalance and leaf pubescence in corn as factors influencing leaf injury by the adult western corn rootworm. *J. Econ. Entomol.* **60**, 1071–1073.

Hagen, K. S. (1986). Ecosystem analysis: Plant cultivars (HPR), entomophagous species and food supplements. *In* "Interactions of Plant Resistance and Parasitoids and Predators of Insects" (D. J. Boethel and R. D. Eikenbary, eds.), pp. 151–197. Ellis Horwood, Chichester, UK.

Hanson, F. E. (1983). The behavioral and neurophysiological basis of food plant selection by lepidopterous larvae. *In* "Herbivorous Insects" (S. Ahman, ed.), pp. 3–23. Academic Press, New York.

Hare, J. D. (1990). Ecology and management of the Colorado potato beetle. *Annu. Rev. Entomol.* **35**, 81–100.

Hare, J. D. (1992). Effects of plant variation on herbivore–natural enemy interactions. *In* "Plant Resistance to Herbivores and Pathogens" (R. S. Fitz and E. L. Simms, eds.), pp. 278–298. Univ. of Chicago Press, Chicago.

Heathcote, G. O. (1970). Effect of plant spacing and time of sowing of sugar beet on aphid infestations and spread of virus yellows. *Plant. Pathol.* **19**, 32–39.

Herbert, D. A., and Harper, J. D. (1986). Bioassays of a beta-exotoxin of *Bacillus thuringiensis* against *Geocoris punctipes* (Hemiptera: Lygaeidae). *J. Econ. Entomol.* **79**, 592–595.

Hill, D. S. (1987). "Agricultural Insect Pests of Temperate Regions and Their Control." Cambridge Univ. Press, Cambridge, UK.

Horn, D. J. (1981). Effects of weedy backgrounds on colonization of collards by green peach aphid, *Myzus persicae,* and its major predators. *Environ. Entomol.* **10**, 285–289.

Hougen-Eitzman, D., and Karban, R. (1995). Mechanisms of interspecific competition that result in successful control of Pacific mites following inoculations of Wilamette mites on grapevines. *Oecologia* **103**, 157–161.

Howe, W. L., Sanburn, J. R., and Rhodes, A. M. (1976). Western corn rootworm adult and spotted cucumber beetle association with *Cucurbita* and cucurbitacins. *Environ. Entomol.* **5**, 1043–1048.

Huffaker, C. B., and Messenger, P. S. (1976). "Theory and Practice of Biological Control," pp. 788. Academic Press, New York.

Huffaker, C. B., Van der Vrie, M., and McMurtry, J. A. (1970). Ecology of tetranychid mites and their natural enemies: A review. II. Tetranychid populations and their possible control by predators: An evaluation. *Hilgardia* **40**, 391–458.

Hulspas-Jordaan, P. M., and van Lenteren, J. C. (1978). The relationship between host–plant leaf structure and parasitization efficiency of the parasitic wasp, *Encarsia formosa* Gahan (Hymenoptera: Aphelinidae). *Med. Fac. Landbouww. Rijksuniv. Gent.* **43**, 431–440.

Janssen, A., Hofker, C. D., Braun, A. R., Mesa, N., Sabelis, M. W., and Bellotti, A. C. (1990). Preselecting predatory mites for biological control: The use of an olfactometer. *Bull. Entomol. Res.* **80**, 177–181.

Janzen, D. H. (1966). Coevolution of mutualism between ants and acacias in Central America. *Evolution* **20**, 249–275.

Jenkins, J. N., Parrott, W. L., McCarthy, J. C., and Callahan, F. E. (1993). Growth and survival of *Heliothis virenscens* (Lepidoptera: Noctuidae) on transgenic cotton containing a truncated form of the delta endotoxin gene from *Bacillus thuringiensis. J. Econ. Entomol.* **86**(1), 181–185.

Johnson, H. B. (1975). Plant pubescence: An ecological perspective. *Bot. Rev.* **41**, 233–258.

Johnson, M. W., and Hara, A. H. (1987). Influence of host crop of parasitoids (Hymenoptera) of *Liriomyza* spp. (Diptera: Agromyzidae). *Environ. Entomol.* **16**, 339–344.

Jones, D., and Granett, J. (1982). Feeding site preferences of seven lepidopteran pests of celery. *J. Econ. Entomol.* **75**, 449–453.

Jones, F. G. W. (1976). Pests, resistance and fertilizers. *In* "Fertilizer Use and Plant Health. Proceedings of the 12th International Potash Institute Der Bund A. G." (V. Taysi, ed.), pp. 233–258. Bern, Switzerland.

Jones, R. G., Bauer, P. J., Roof, M. E., and Langston, M. A. (1990). Effect of reduced rates of ethephon on late-season insect oviposition and feeding sites in cotton. *J. Entomol. Sci.* **25**, 246–252.

Kalkhoven, J. T. R. (1993). Survival of populations and scale of the fragmented agricultural landscape. *In* "Landscape Ecology and Agroecosystems" (R. G. H. Bunce, L. Ryszkowski, and M. G. Proletti, eds.), pp. 83–90. Lewis, Boca Raton, FL.

Kalode, M. B., and Pant, N. C. (1967). Studies on the amino acids, nitrogen, sugar and moisture content of maize and sorghum varieties and their relation to *Chilo zonellus* (Swin.) resistance. *Indian J. Entomol.* **29**(2), 139–144.

Karban, R., and Carey, J. R. (1984). Induced resistance of cotton seedlings to mites. *Science* **225**, 53–54.

Karban, R., and Myers, J. H. (1989). Induced plant responses to herbivory. *Annu. Rev. Ecol. Syst.* **20**, 331–348.

Kareiva, P. M. (1982). Experimental and mathematical analyses of herbivore movement: Quantifying the influence of plant spacing and quality on foraging discrimination. *Ecol. Monogr.* **52**(3), 261–282.

Kennedy, G. G., and Barbour, J. D. (1992). Resistance variation in natural and managed systems. *In* "Plant Resistance to Herbivores and Pathogens" (R. S. Fitz and E. L. Simms, eds.), pp. 13–41. Univ. of Chicago Press, Chicago.

Kindler, S. D., and Staples, R. (1970). Nutrients and the reaction of two alfalfa clones to the spotted alfalfa aphid. *J. Econ. Entomol.* **63**(3), 939–940.

Klinger, T., and Ellstrand, N. C. (1994). Engineered genes in wild populations: Fitness of weed–crop hybrids of *Raphanus sativus. Ecol. Appl.* **4**, 117–120.

Kogan, M. (1986). Plant defense strategies in host–plant resistance. *In* "Ecological Theory and IPM Practice" (M. Kogan, ed.), pp. 83–134. Wiley, New York.

Kogan, M. (1994). Plant resistance in pest management. *In* "Introduction to Insect Pest Management" (R. L. Metcalf and W. H. Luckmann, eds.), 3rd Ed, pp. 73–128. Wiley, New York.

Kruess, A., and Tscharntke, T. (1994). Habitat fragmentation, species loss, and biological control. *Science* **264**, 1581–1584.

Lasack, P. M., and Pedigo, L. P. (1986). Movement of stalk borer larvae (Lepidoptera: Noctuidae) from noncrop areas into corn. *J. Econ. Entomol.* **79**, 1697–1702.

Lashomb, J. H., and Ng, Y.-S. (1984). Colonization by Colorado potato beetles, *Leptinotarsa decemlineata* (Say) (Coleoptera: Chrysomelidae), in rotated and nonrotated potato fields. *Environ. Entomol.* **13,** 1352–1356.

Lawton, J. H., and McNeill, S. (1979). Between the devil and the deep blue sea: On the problem of being a herbivore. *In* "Population Dynamics" (R. M. Anderson, B. D. Turner, and L. R. Taylor, eds.), pp. 223–244. Blackwell, Oxford, UK.

Leathe, K. T., and Ratcliffe, R. H. (1974). The effect of fertilization on disease and insect resistance. *In* "Forage Fertilization" (D. Mays, ed.), pp. 481–503. Agronomy Society of America, Madison, WI.

Lee, Y. I., Kogan, M., and Larsen, J. R., Jr. (1986). Attachment of the potato leafhopper to soybean plant surfaces as affected by morphology of the pretarsus. *Entomol. Exp. Appl.* **42,** 101–107.

Leius, K. (1967). Influence of wild flowers on parasitism of tent caterpillar and codling moth. *Can. Entomol.* **99,** 444–446.

Letourneau, D. K. (1986). Associational resistance in squash monocultures and polycultures in tropical Mexico. *Environ. Entomol.* **15**(2), 285–292.

Letourneau, D. K. (1987). The enemies hypothesis: Tritrophic interactions and vegetational diversity in tropical agroecosystems. *Ecology* **68,** 1616–1622.

Letourneau, D. K. (1988). Conceptual framework of three-trophic-level interactions. *In* "Novel Aspects of Insect–Plant Interactions" (P. Barbosa and D. K. Letourneau, eds.), pp. 1–9. Wiley, New York.

Letourneau, D. K. (1990a). Mechanisms of predator accumulation in a mixed crop system. *Ecol. Entomol.* **15,** 63–69.

Letourneau, D. K. (1990b). Code of ant–plant mutualism broken by parasite. *Science* **248,** 215–217.

Letourneau, D. K. (1994). Bean fly, management practices, and biological control in Malaŵian subsistence agriculture. *Agric. Ecosyst. Environ.* **50,** 103–111.

Letourneau, D. K., and Altieri, M. A. (1983). Abundance patterns of a predator *Orius tristicolor* (Hemiptera: Anthocoridae), and its prey, *Frankliniella occidentalis* (Thysanoptera: Thripidae): Habitat attraction in polycultures versus monocultures. *Environ. Entomol.* **122,** 1464–1469.

Letourneau, D. K., and Barbosa, P. (1997). Ants, stem borers and pubescence in *Endospermum* in Papua New Guinea. Submitted for publication.

Letourneau, D. K., and Fox, L. R. (1989). Effects of experimental design and nitrogen on cabbage butterfly oviposition. *Oecologia* **80,** 211–214.

Letourneau, D. K., Drinkwater, L. E., and Shennan, C. (1996). Soil management effects on crop quality and insect damage in commercial organic and conventional tomato fields. *Agric. Ecosyst. Environ.* **57,** 179–187.

Levin, D. A. (1973). The role of trichomes in plant defense. *Quant. Rev. Biol.* **48,** 3–15.

Levine, E. (1985). Oviposition by the stalk borer, *Papaipema nebris* (Lepidoptera: Noctuidae), on weeds, plant debris, and cover crops in cage tests. *J. Econ. Entomol.* **78,** 65–68.

Levine, E., Olumi-Sadeghi, H., and Fisher, J. R. (1992). Discovery of multi-year diapause in Illinois and South Dakota northern corn rootworm (Coleoptera: Chrysomelidae) eggs and incidence of the prolonged diapause trait in Illinois. *J. Econ. Entomol.* **85,** 262–267.

Levins, R., and Wilson, M. (1980). Ecological theory and pest management. *Annu. Rev. Entomol.* **25,** 7–29.

Lewis, W. J., and Gross, H. R. (1989). Comparative studies on field performance on *Heliothis* larval parasitoids *Microplitis croceipes* and *Cardiochiles nigriceps* at varying densities and under selected host plant conditions. *Florida Entomol.* **72**(1), 6–14.

Lichtenstein, E. P., and Casida, J. E. (1963). Myristicin, an insecticide and synergist occurring naturally in the edible parts of parsnip. *J. Agric. Food Chem.* **11,** 410–415.

Lipke, H., and Fraenkel, G. (1956). Insect nutrition. *Annu. Rev. Entomol.* **1,** 17–44.

MacArthur, R. H., and Wilson, E. O. (1967). "The Theory of Island Biogeography." Princeton Univ. Press, Princeton, NJ.

Mack, T. P., and Backman, C. B. (1990). Effects of two planting dates and three tillage systems on the abundance of lesser cornstalk borer (Lepidoptera: Pyralidae), other selected insects and yield in peanut fields. *J. Econ. Entomol.* **83,** 1034–1041.

Mannion, C. M., Carpenter, J. E., Wiseman, B. R., and Gross, H. R. (1994). Host corn earworm (Lepidoptera: Noctuidae) reared on meridic diet containing silks from a resistant corn genotype on *Archytas marmoratus* (Diptera: Tachinidae) and *Ichneumon promissorius* (Hymenoptera: Ichneumonidae). *Environ. Entomol.* **23,** 837–845.

Manuwoto, S., and Scriber, J. M. (1985). Differential effects of nitrogen fertilization of three corn genotypes on biomass and nitrogen utilization by the southern armyworm, *Spodoptera eridania. Agric. Ecosyst. Environ.* **14**(1/2), 25–40.

Martin, P. B., Lingren, P. D., Greene, G. L., and Ridgway, R. L. (1976). Parasitization of two species of Plusiinae and *Heliothis* spp. after release of *Trichogramma pretiosum* in seven crops. *Environ. Entomol.* **5,** 991–995.

Martin, W. R., Jr., Nordlund, D. A., and Nettles, W. C., Jr. (1990). Response of parasitoid *Eucelatoria bryani* to selected plant material in an olfactometer. *J. Chem. Ecol.* **16**(2), 499–508.

Mattson, W. J. (1980). Herbivory in relation to plant nitrogen content. *Annu. Rev. Ecol. Syst.* **11,** 119–161.

Mattson, W. J., Jr., and Addy, N. D. (1975). Phytophagous insects as regulators of forest primary production. *Science* **190,** 515–522.

Mattson, W. J., and Haack, R. A. (1987). The role of drought stress in provoking outbreaks of phytophagous insects. *In* "Insect Outbreaks" (P. Barbosa and J. C. Schultz, eds.), pp. 365–407. Academic Press, New York.

Maxwell, F. G. (1972). Host plant resistance to insects—nutritional and pest management relationships. *In* "Insect and Mite Nutrition" (J. G. Rodriquez, ed.), pp. 599–609. North-Holland, Amsterdam.

May, R. M., and Anderson, R. M. (1978). Regulation and stability of host–parasite interactions. II. Destabilizing processes. *J. Anim. Ecol.* **47,** 267–279.

Mayse, M. A. (1983). Cultural control in crop fields: Habitat management techniques. *Environ. Management* **7,** 15.

McClure, M. S. (1980). Foliar nitrogen: A basis for host suitability for elongate hemlock seeds, *Fiorinia externa* (Homoptera: Diaspididae). *Ecology* **61,** 72–79.

McIntyre, J. L., Dodds, J. A., and Hare, J. D. (1981). Effects of localized infection of *Nictinia tabacum* by tobacco moasic virus on systemic resistance against diverse pathogens and an insect. *Phytopathology* **71,** 297–301.

Meeusen, R. L., and Warren, G. (1989). Insect control with genetically engineered crops. *Annu. Rev. Entomol.* **34,** 373–381.

Montandon, R., Slasser, J. E., and Clark, L. E. (1994). Late-season termination effects on cotton fruiting, yield, and boll weevil (Coleoptera: Curculionidae) damage in Texas dryland cotton. *J. Econ. Entomol.* **87,** 1647–1652.

Monteith, L. G. (1960). Influence of plants other than the food plants of their host on host-finding by tachinid parasites. *Can. Entomol.* **92,** 641–652.

Murdoch, W. W., and Oaten, A. (1975). Predation and population stability. *Adv. Ecol. Res.* **9,** 2–131.

Murdoch, W. W., Luck, R. F., Swarbrick, S. L., Walde, S., Yu, D. S., and Reeve, J. D. (1995). Regulation of an insect population under biological control. *Ecology* **76,** 206–217.

Nafus, D., and Schreiner, I. (1986). Intercropping maize and sweet potatoes. Effects on parasitization of *Ostrinia funnacalis* eggs by *Trichogramma chilonis. Agric. Ecosyst. Environ.* **15,** 189–200.

Nettles, W. C. (1979). *Eucelatoria* sp. females: Factors influencing responses to cotton and okra plants. *Environ. Entomol.* **8**, 619–623.

Nordlund, D. A., Lewis, W. J., and Altieri, M. A. (1988). Influences of plant produced allelochemicals on the host and prey selection behavior of entomophagous insects. *In* "Novel Aspects of Insect–Plant Interactions" (P. Barbosa and D. K. Letourneau, eds.). Wiley, New York.

Norris, D. M., and Kogan, M. (1980). Biochemical and morphological bases of resistance. *In* "Breeding Plants Resistant to Insects" (F. G. Maxwell and P. R. Jennings, eds.), pp. 23–51. Wiley, New York.

Norris, R. F. (1986). Weeds and integrated pest management systems. *Hort. Sci.* **21**, 402–410.

Nothenagle, P. J., and Schultz, J. C. (1987). What is a forest pest? *In* "Insect Outbreaks" (P. Barbosa and J. C. Schultz, eds.), pp. 59–80. Academic Press, San Diego.

Nowacki, E., Jurzysta, M., Gorski, P., Nowacka, D., and Waller, G. R. (1976). Effect of nitrogen nutrition on alkaloid metabolism in plants. *Biochem. Physiol. Pfanz.* **169**, 231–240.

Obrycki, J. J., and Tauber, M. J. (1984). Natural enemy activity on glandular pubescent potato plants in the greenhouse: An unreliable predictor of effects in the field. *Environ. Entomol.* **13**, 679–683.

Olson, D. L., and Nechols, J. R. (1995). Effects of squash leaf trichome exudates and honey on adult feeding, survival, and fecundity of the squash bug (Heteroptera: Coreidae) egg parasitoid *Gryon pennsylvanicum* (Hymenoptera: Scelionidae). *Environ. Entomol.* **24**(2), 454–458.

Orr, D. B., and Boethel, D. J. (1986). Influence of plant antibiosis through four trophic levels. *Oecologia* **70**, 242–249.

Pair, S. D., Wiseman, B. R., and Sparks, A. N. (1985). Influence of four corn cultivars on fall armyworm (Lepidoptera: Noctuidae) establishment and parasitization. *Fall Armyworm Symp.* **69** (3), 566–570.

Panda, M. (1979). "Principles of Host Plant Resistance to Insect Pests," pp. 386. Allenheld/Universe, New York.

Pemberton, R. W., and Turner, C. E. (1989). Occurrence of predatory and fungivorous mites in leaf domatia. *Am. J. Bot.* **76**, 105–112.

Peng, C., and Weiss, M. J. (1992). Evidence of an aggregation pheromone in the flea beetle, *Phyllotreta cruciferae* (Goeze) (Coleoptera: Chrysomelidae). *J. Chem. Ecol.* **18**, 875–884.

Perfecto, I., Horwith, B., Vandermeer, J., Schultz, B., McGuinness, H., and Dos Santos, A. (1986). Effects of plant diversity and density on the emigration rate of two ground beetles, *Harpalus pennsylvanicus* and *Evarthrus sodalis* (Coleoptera: Carabidae), in a system of tomatoes and beans. *Environ. Entomol.* **15**, 1028–1031.

Perrin, R. M. (1975). The role of the perennial stinging nettle *Urtica dioica,* as a reservoir of beneficial natural enemies. *Ann. Appl. Biol.* **81**, 289–297.

Perrin, R. M. (1977). Pest management in multiple cropping systems. *Agroecosystems* **3**, 93–118.

Perring, T. M., Fanar, C. A., and Toscano, N. C. (1988). Relationships among tomato planting date, potato aphids (Homoptera: Aphidae), and natural enemies. *J. Econ. Entomol.* **81**, 1107–1112.

Pimentel, D. (1961). The influence of plant spatial patterns on insect populations. *Ann. Entomol. Soc. Am.* **54**, 61–69.

Pimentel, D. (1986). Agroecology and economics. *In* "Ecological Theory and Integrated Pest Management Practice" (M. Kogan, ed.), pp. 299–319. Wiley, New York.

Pimentel, D. (1993). Cultural controls for insect pest management. *In* "Pest Control and Sustainable Agriculture" (S. A. Corey, D. J. Dall, and W. M. Milne, eds.), pp. 35–43. CSIRO, Australia.

Pollard, D. G. (1971). Hedges VI: Habitat diversity and crop pests—A study of *Brevicoryne brassicae* and its syrphid predators. *J. Appl. Ecol.* **8**, 751–780.

Powell, W., Dean, G. J., and Wilding, N. (1986). The influence of weeds on aphid-specific natural enemies in winter wheat. *Crop Protection* **4**, 182–189.

Price, P. W. (1976). Colonization of crops by arthropods: Non-equilibrium communities in soybean fields. *Environ. Entomol.* **5**, 605–611.

Price, P. W. (1984a). "Insect Ecology," 2nd Ed., pp. 607. Wiley, New York.

Price, P. W. (1984b). Alternative paradigms in community ecology. *In* "A New Ecology: Novel Approaches to Interactive Systems" (P. W. Price, C. N. Slobodchikoff, and W. S. Gaud, eds.), pp. 353–383. Wiley, New York.

Price, P. W., and Waldbauer, G. P. (1994). Ecological aspects of pest management. *In* "Introduction to Pest Management" (R. L. Metcalf and W. H. Luckmann, eds.), 3rd Ed., pp. 35–65. Wiley, New York.

Price, P. W., Bouton, C. E., Gross, P., McPheron, B. A., Thompson, J. N., and Weis, A. E. (1980). Interactions among three trophic levels: Influence of plants on interactions between insect herbivores and natural enemies. *Annu. Rev. Ecol. Syst.* **11**, 41–65.

Price, P. W., Slobodchikoff, C. N., and Gaud, W. S. (eds.) (1984). "A New Ecology: Novel Approaches to Interactive Systems." Wiley, New York.

Putman, W. L., and Herne, D. C. (1966). The role of predators and other biotic factors in regulating the population density of phytophagous mites in Ontario peach orchards. *Can. Entomol.* **98**, 808–820.

Quiring, D. T., Timmins, P. R., and Park, S. J. (1992). Effect of variations in hooked trichome densities of *Phaseolus vulgaris* on longevity of *Liriomyza trifolii* (Diptera: Agromyzidae) adults. *Environ. Entomol.* **21**, 1357–1361.

Rabb, R. L. (1978). A sharp focus on insect populations and pest management from a wide-area view. *Bull. Entomol. Soc. Am.* **24**, 55–61.

Rabb, R. L., and Bradley, J. R. (1968). The influence of host plants on parasitism of eggs on the tobacco hornworm. *J. Econ. Entomol.* **61**, 1249–1250.

Ramachandran, R., and Norris, D. M. (1991). Volatiles mediating plant–herbivore–natural enemy interactions: Electroantennogram responses of soybean looper, *Pseudoplusia includens,* and a parasitoid, *Microplitis demolitor,* to green leaf volatiles. *J. Chem. Ecol.* **17**, 1665–1690.

Rämert, B. (1996). The influence of intercropping and mulches on the occurrence of polyphagous predators in carrot fields in relation to carrot fly (*Psila rosae* (F.)) (Dipt., Psilidae) damage. *J. Appl. Entomol.* **120**, 39–46.

Rasmy, A. H. (1977). Predatory efficiency and biology of the predatory mite *Amblyseius gossipi* (Acarina: Phytoseiidae) as affected by physical surfaces of the host plant. *Entomophaga* **22**, 421–423.

Rasmy, A. H., and Elbanhawy, E. M. (1974). Behaviour and bionomics of the predatory mite *Phytoseius plumifer* (Acarina: Phytoseiidae) as affected by physical surfaces of the host plant. *Entomophaga* **19**, 255–257.

Read, D. P., Feeny, P. P., and Root, R. B. (1970). Habitat selection by the aphid parasite *Diaeretiella rapae* (Hymenoptera: Braconidae) and hyperparasite *Charips brassicae* (Hymenoptera: Cynipidae). *Can. Entomol.* **102**, 1567–1578.

Rhoades, D. F. (1983). Herbivore population dynamics and plant chemistry. *In* "Variable Plants and Herbivores in Natural and Managed Systems" (R. F. Denno and M. S. McClure, eds.), pp. 155–220. Academic Press, New York.

Rhoades, D. F., and Cates, R. G. (1976). Towards a general theory of plant antiherbivore chemistry. *Rec. Adv. Phytochem.* **10**, 168–213.

Risch, S. J. (1979). A comparison, by sweep sampling, of the insect fauna from corn and sweet potato monocultures and dicultures in Costa Rica. *Oecologia* **42**, 195–211.

Risch, S. J. (1981). Insect herbivore abundance in tropical monocultures and polycultures: An experimental test of two hypotheses. *Ecology* **62**, 1325–1340.

Risch, S. J., Andow, D., and Altieri, M. A. (1983). Agroecosystem diversity and pest control: Data, tentative conclusions, and new directions. *Environ. Entomol.* **12**(3), 625–629.

Rogers, D. J., and Sullivan, M. J. (1986). Nymphal performance of *Geocoris punctipes* (Hemiptera: Lygaeidae) on pest-resistant soybeans. *Environ. Entomol.* **15**, 1032–1036.

Root, R. B. (1973). Organization of a plant–arthropod association in simple and diverse habitats: The fauna of collards (*Brassica oleraceae*). *Ecol. Monogr.* **43**, 95–124.

Rosenthal, G. A., and Janzen, D. H. (1979). "Herbivores: Their Interactions with Secondary Plant Metabolites," pp. 718. Academic Press, New York.

Russell, E. P. (1989). Enemies hypothesis: A review of the effect of vegetational diversity on predatory insects and parasitoids. *Environ. Entomol.* **18**, 590–599.

Ryan, C. A. (1983). Insect-induced chemical signals regulating natural plant protection responses. *In* "Variable Plants and Herbivores in Natural and Managed Systems" (R. F. Denno and M. S. McClure, eds.), pp. 43–60. Academic Press, New York.

Ryszkowski, L., Karg, J., Margaret, G., Paoletti, M. G., and Zlotin, R. (1993). Above-ground insect biomass in agricultural landscapes in Europe. *In* "Landscape Ecology and Agroecosystems" (R. G. H. Bunce, L. Ryszkowski, and M. G. Paoletti, eds.), pp. 71–82. Lewis, London.

Samson, P. R. (1993). Evaluation of a procedure for screening sugarcane clones for tolerance to sugarcane soldier fly. *In* "Pest Control and Sustainability Agriculture" (S. A. Corey, D. J. Dall, and W. M. Milne, eds.), pp. 330–332. CSIRO, Australia.

Scriber, J. M. (1983). The evolution of feeding specialization, physiological efficiency, and host races. *In* "Variable Plants and Herbivores in Natural and Managed Systems" (R. F. Denno and M. S. McClure, eds.), pp. 373–412. Academic Press, New York.

Scriber, J. M. (1984a). Nitrogen nutrition of plants and insect invasion. *In* "Nitrogen in Crop Production" (R. D. Hauck, ed.), pp. 441–460. Am. Soc. Agron., Madison, WI.

Scriber, J. M. (1984b). Plant–herbivore relationships: Host plant acceptability. *In* "The Chemical Ecology of Insects" (W. Bell and R. Carde, eds.), pp. 159–202. Chapman & Hall, London.

Shade, R. E., and Wilson, M. C. (1967). Leaf-vein spacing as a factor affecting larval feeding behavior of the cereal leaf beetle, *Oulema melanopus* (Coleoptera: Chrysomelidae). *Annu. Rev. Entomol. Soc. Am.* **60**, 493–496.

Shah, D. M., Horsch, R. B., Klee, H. J., Kishore, G. M., Winter, J. A., Tumer, N. E., Hironaka, C. M., Sanders, P. R., and Gasser, C. S. (1986). Engineering herbicide tolerance in transgenic plants. *Science* **233**, 478–482.

Shah, M. A. (1981). The influence of plant surfaces on the searching behavior of Coccinellid larvae. *Entomol. Exp. Appl.* **31**, 377–380.

Shahjahan, M. (1974). *Erigeron* flowers as food and attractive odor source for *Peristenus pseudopallipes*, a braconid parasitoid of the tarnished plant bug. *Environ. Entomol.* **3**, 69–72.

Shahjahan, M., and Streams, A. S. (1973). Plant effects on host-finding by *Leiophron pseudopallipes* (Hymenopterra: Braconidae), a parasitoid of the tarnished plant bug. *Environ. Entomol.* **2**, 921–925.

Sheehan, W. (1986). Response by specialist and generalist natural enemies to agroecosystem diversification: A selective review. *Environ. Entomol.* **15**, 456–461.

Simberloff, D. (1985). Island biogeographic theory and integrated pest management. *In* "Ecological Theory and Integrated Pest Management Practice" (M. Kogan, ed.), pp. 19–35. Wiley, New York.

Singh, P. (1970). Host–plant nutrition and composition: Effects on agricultural pests. Inform. Bull. No. 6, Canada Department of Agriculture.

Slansky, F., Jr., and Rodriguez, J. G. (eds.) (1987). "Nutritional Ecology of Insects, Mites, Spiders, and Related Invertebrates," pp. 1016. Wiley, New York.

Slosser, J. E. (1993). Influence of planting date and insecticide treatment on insect pest abundance and damage in dryland cotton. *J. Econ. Entomol.* **86**, 1213–1222.

Slosser, J. E., Jacoby, P. W., and Price, J. R. (1985). Management of sand shinnery oak for control of the boll weevil (Coleoptera: Cucurlionidae) in the Texas rolling plains. *J. Econ. Entomol.* **78**, 383–389.

Smith, J. G. (1976). Influence of crop background on natural enemies of aphids on Brussels sprouts. *Ann. Appl. Biol.* **83**, 15–29.

Southwood, T. R. E. (1987). Plant variety and its interaction with herbivorous insects. *In* "Insects–Plants" (V. Labeyrie, G. Fabres, and D. Lachaise, eds.), pp. 61–69. Junk, Dordrecht.

Specht, R. L., and Rundel, P. W. (1990). Sclerophyll and foliage nutrient status of mediterranean-climate plant communities in Southern Australia. *Aust. J. Bot.* **38**, 459–474.

Speight, H. R., and Lawton, J. H. (1976). The influence of weed cover on the mortality imposed on artificial prey by predatory ground beetles in cereal fields. *Oecologia* **23**, 211–233.

Sprenkel, R. K., Brooks, W. M., van Duyn, J. W., and Deitz, L. L. (1979). The effects of three cultural variables on the incidence of *Nomuroea rileyi*, phytophagous Lepidoptera, and their predators on soybeans. *Environ. Entomol.* **8**, 337–339.

Stenseth, N. C. (1981). How to control pest species: Application of models from the theory of island biogeography in formulating pest control strategies. *J. Appl. Ecol.* **18**, 773–794.

Stern, V. M. (1981). Environmental control of insects using trap crops, sanitation, prevention and harvesting. *In* "CRC Handbook of Pest Management in Agriculture" (D. Pimentel, ed.), Vol. 1, pp. 199–207. CRC Press, Boca Raton, FL.

Strong, D. R. (1974). Rapid asymptotic species accumulation in phytophagous insect communities: The pests of cacao. *Science* **185**, 1064–1066.

Strong, D. R. (1979). Biogeographical dynamics of insect–host plant communities. *Annu. Rev. Entomol.* **24**, 89–119.

Strong, D. R., McCoy, E. D., and Rey, J. R. (1977). Time and the number of herbivore species: The pests of sugar cane. *Ecology* **58**, 167–175.

Strong, D. R., Lawton, J. H., and Southwood, T. R. E. (1984). "Insects and Plants: Community Patterns and Mechanisms," pp. 313. Blackwell, Oxford, UK.

Tahvanainen, J. O., and Root, R. B. (1972). The influence of vegetational diversity on the population ecology of a specialized herbivore, *Phyllotreta cruciferae* (Coleoptera: Chrysomelidae). *Oecologia* **10**, 321–346.

Tanton, M. T. (1962). The effect of leaf "toughness" on the feeding of larvae of the mustard beetle *Phaedon cochleariae* Fab. *Entomol. Exp. Appl.* **5**, 74–78.

Thiele, H. U. (1977). Quantitative investigations on the distribution of carabids. *In* "Carabid Beetles in Their Environments" (H. U. Thiele, ed.), pp. 13–48. Springer-Verlag, Berlin.

Thorpe, K. W., and Barbosa, P. (1986). Effects of consumption of high and low nicotine tobacco by *Manduca sexta* on the survival of the gregarious endoparasitoid *Cotesia congregata*. *J. Chem. Ecol.* **12**, 1329–1337.

Tomkins, A. R., and Steven, D. (1993). Some observations on the relative susceptibility of *Actinidia* species and cultivars to damage or infestation by insects and mites. *In* "Pest Control and Sustainability Agriculture" (S. A. Corey, D. J. Dall, and W. M. Milne, eds.), pp. 320–322. CSIRO, Australia.

Tonhasca, A., Jr. (1994). Response of soybean herbivores to two agronomic practices increasing agroecosystem diversity. *Agric. Ecosyst. Environ.* **48**, 57–65.

Tonhasca, A., Jr., and Byrne, D. N. (1994). The effects of crop diversification on herbivorous insects: A meta-analysis approach. *Ecol. Entomol.* **19**, 239–244.

Treacy, M. F., Zummo, G. R., and Benedict, J. H. (1985). Interactions of host–plant resistance in cotton with predators and parasites. *Agric. Ecosyst. Environ.* **13**, 151–157.

Turlings, T. C. J., Tumlinson, J. H., Eller, F. J., and Lewis, W. J. (1991). Larval-damaged plants: Source of volatile synomones that guide the parasitoid *Cotesia marginiventris* to the microhabitat of its hosts. *Entomol. Exp. Appl.* **58**, 75–82.

Turner, I. M. (1994). Sclerophylly: Primarily protective? *Funct. Ecol.* **8**, 669–675.

van den Berg, H., Nyambo, B. T., and Waage, J. K. (1990). Parasitism of *Helicoverpa armigera* (Lepidoptera: Noctuidae) in Tanzania: Analysis of parasitoid–crop association. *Environ. Entomol.* **19**(4), 1141–1145.

van den Bosch, R., and Stern, V. M. (1969). The effect of harvesting practices on insect populations in alfalfa. *Proc. Tall Timbers Conf. Ecol. Anim. Control Hab. Management* **1**, 47–54.

Vandermeer, J., and Andow, D. A. (1986). Prophylactic and responsive components of an integrated pest management program. *J. Econ. Entomol.* **79**, 299–302.

van Emden, H. F. (1965a). The effect of uncultivated land on the distribution of cabbage aphid (*Brevicoryne brassicae*) on an adjacent crop. *J. Appl. Ecol.* **2**, 171–196.

van Emden, H. F. (1965b). The role of uncultivated land in the biology of crop pests and beneficial insects. *Sci. Hort.* **17**, 121–136.

van Emden, H. F., and Hagen, K. S. (1976). Olfactory reactions of the green lacewing, *Chrysopa carnea*, to tryptophan and certain breakdown products. *Environ. Entomol.* **5**, 469–473.

Via, S. (1990). Ecological genetics and host adaptation in herbivorous insects: The experimental study of evolution in natural and agricultural systems. *Annu. Rev. Entomol.* **35**, 421–446.

Visser, J. H. (1986). Host odor perception in phytophagous insects. *Annu. Rev. Entomol.* **31**, 121–144.

Vogt, E. A., and Nechols, J. R. (1993). Responses of the squash bug (Hemiptera: Coreidae) and its egg parasitoid, *Gryon pennsylvanicum* (Hymenoptera: Scelionidae) to three *Cucurbita* cultivars. *Environ. Entomol.* **22**(1), 238–245.

Vorley, V. T., and Wratten, S. D. (1987). Migration of parasitoids (Hymenoptera: Braconidae) of cereal aphids (Hemiptera: Aphididae) between grassland, early-sown cereals and late-sown cereals in southern England. *Bull. Entomol. Res.* **77**, 555–568.

Walde, S. J. (1995). How quality of host plant affects predator–prey interaction in biological control. *Ecology* **76**(4), 1206–1219.

Walker, P. W., and Allsopp, P. G. (1993). Factors influencing populations of *Eumargarodes laingi* and *Promargarodes* spp. (Hemiptera: Margarodidae) in Australian sugarcane. *Environ. Entomol.* **22**(2), 362–367.

Waterman, P. G., and Mole, S. (1989). Extrinsic factors influencing production of secondary metabolites in plants. *In* "Insect–Plant Interactions" (E. A. Bernays, ed.), Vol. I, pp. 107–134. CRC Press, Boca Raton, FL.

Way, M. J., and Khoo, K. C. (1991). Colony dispersion and nesting habits of the ants, *Dolichoderus thoracicus* and *Decophylla smaragdina* (Hymenoptera: Formicidae), in relation to their success as biological control agents on cocoa. *Bull. Entomol. Res.* **81**, 341–350.

Way, M. J., and Khoo, K. C. (1992). Role of ants in pest management. *Annu. Rev. Entomol.* **37**, 479–503.

Way, M. J., and Murdie, G. (1965). An example of varietal variations in resistance of Brussels sprouts. *Ann. Appl. Biol.* **56**, 326–328.

Webster, J. A., Inayatullah, C., Hamissou, M., and Mirkes, K. A. (1994). Leaf pubescence effects in wheat on yellow sugarcane aphids and greenbugs (Homoptera: Aphididae). *J. Econ. Entomol.* **87**(1), 231–240.

Weiss, M. J., Schatz, G. G., Gardner, J. C., and Nead, B. A. (1994). Flea beetle (Coleoptera: Chrysomelidae) populations and crop yield in field pea and oilseed rape intercrops. *J. Environ. Entomol.* **23**(3), 654–658.

Wellington, W. S. (1983). Biometeorology of dispersal. *Bull. Entomol. Soc. Am.* **29**, 24–29.

Weston, P. A., Barney, R. J., and Sedlacek, J. D. (1993). Planting date influences preharvest infestation of dent corn by Angoumois grain moths (Lepidoptera: Gelechiidae). *J. Econ. Entomol.* **86**, 174–180.

Whitham, T. G. (1983). Host manipulation by parasites: Within-plant variation as a defense against rapidly evolving pests. *In* "Variable Plants and Herbivores in Natural and Managed Systems" (R. F. Denno and M. S. McClure, eds.), pp. 15–42, Academic Press, New York.

Whitham, T. G., Williams, A. G., and Robinson, A. M. (1984). The variation principle: Individual plants as temporal and spatial mosaics of resistance to rapidly evolving pests. *In* "A New Ecology: Novel Approaches to Interactive Systems" (P. W. Price, C. N. Slobodchikoff, and W. S. Gaud, eds.), pp. 15–52. Wiley, New York.

Whitham, T. G., Morrow, P. A., and Potts, B. M. (1994). Plant hybrid zones as centers of biodiversity—The herbivore community of two endemic Tasmanian eucalypts. *Oecologia* **97**, 481–490.

Whitman, D. (1994). Plant bodyguards: Mutualistic interactions between plants and the third trophic level. *In* "Functional Dynamics of Phytophagous Insects" (T. N. Ananthakrishnan, ed.), pp. 207–248. Science Publishers, Lebanon, New Hampshire.

Wilhoit, L. R. (1991). Modelling the population dynamics of different aphid genotypes in plant variety mixtures. *Ecol. Modelling* **55**, 257–283.

Wilhoit, L. R. (1992). Evolution of herbivore virulence to plant resistance: Influence of variety mixtures. *In* "Plant Resistance to Herbivores and Pathogens" (R. S. Fitz and E. L. Simms, eds.), pp. 91–119. Univ. of Chicago Press, Chicago.

Williams, C. E., Pavuk, D. M., Taylor, D. H., and Martin, T. H. (1995). Parasitism and disease incidence in the green cloverworm (Lepidoptera: Noctuidae) in strip-intercropped soybean agroecosystems. *Environ. Entomol.* **24**(2), 253–260.

Wiseman, B. R., and Isenhour, D. J. (1990). Effects of resistant maize silks on corn earworm (Lepidoptera: Noctuidae) biology—A laboratory study. *J. Econ. Entomol.* **83**(2), 614–617.

Wolda, H. (1978). Fluctuations in abundance of tropical insects. *Am. Nat.* **112**, 1017–1025.

Wolfson, J. L. (1980). Oviposition response of *Pieris rapae* to environmentally induced variation in *Brassica nigra*. *Entomol. Exp. Appl.* **27**, 223–232.

Zadoks, J. C. (1993). Crop protection: Why and how. *In* "Crop Protection and Sustainable Agriculture" (D. J. Chadwick and J. Marsh, eds.), pp. 48–60. Wiley, Chichester, UK.

Zangerl. A. R. (1990). Furanocoumarin induction in wild parsnip—Evidence for an induced defense against herbivores. *Ecology* **71**, 1926–1932.

Zangerl, A. R., and Bazzaz, F. A. (1992). Theory and pattern in plant defense allocation. *In* "Plant Resistance to Herbivores and Pathogens" (R. S. Frtiz and E. L. Simms, eds.), pp. 363–391. Univ. of Chicago Press, Chicago.

9

Many Little Hammers: Ecological Management of Crop–Weed Interactions

Matt Liebman and Eric R. Gallandt

I. Introduction

Despite the use of a considerable amount of technology and many hours of human labor, weeds cause substantial reductions in yields of crops grown in both industrialized and developing countries. In the United States, more than $6 billion is spent annually on herbicides, tillage, and cultivation to control weeds in cropland (Chandler, 1991), yet annual crop losses due to weed infestation currently exceed $4 billion (Bridges and Anderson, 1992). In many developing countries, hand labor for weed control may consume up to half of the total labor demand for crop production (Akobundu, 1991). Worldwide, Akobundu (1987) estimated that weeds reduce crop yields by 5% in the most developed countries, 10% in the less developed countries, and 25% in the least developed countries. Because weeds pose a recurrent and nearly ubiquitous threat to crop productivity and farm profitability, and because weeds respond dynamically to a wide range of farming practices, weed management has a key role in the design and function of agroecosystems.

Over the past four decades, weed management in the industrialized countries has been dominated by a focus on herbicide technology (Zimdahl, 1991; Wyse, 1992; Abernathy and Bridges, 1994). In the United States, yearly application of herbicides to agricultural land exceeds 200 million kg of active ingredients (Gianessi and Puffer, 1990) and accounts for more than 60% of the total mass of all agricultural pesticides applied (Aspelin,

1994). In many developing countries, amplification of crop production has been viewed as dependent on increased use of agrichemicals (Conway and Barbier, 1990), and government subsidy programs have often promoted the use of herbicides and other pesticides (Repetto, 1985).

Increased reliance on herbicides has generally been accompanied by marked improvements in crop productivity and farm labor efficiency. In recent years, however, several factors have led to a reappraisal of heavy emphasis on chemical weed control technology and to a growing interest in alternative management strategies that are less reliant on herbicides and more reliant on manipulations of ecological phenomena, such as competition, allelopathy, herbivory, disease, and responses to soil disturbance (Liebman and Dyck, 1993a; Wyse, 1994). We call these alternative approaches to suppressing weed survival, growth, and reproduction "ecological weed management."

In North America and northern Europe, a major factor propelling interest in ecological weed management has been recognition that herbicides, applied in the course of normal farming practices, are responsible for widespread contamination of ground and surface supplies of drinking water (Hallberg, 1989; Leistra and Boesten, 1989; NRC, 1989, pp. 101–109; Goolsby *et al.*, 1991, 1993; Thurman *et al.*, 1991; USEPA, 1992; Nelson and Jones, 1994). Public concern over this contamination has inspired strict groundwater regulations in Germany (Gassman, 1993), whereas other industrialized nations, e.g., Sweden (Weinberg, 1990; Bellinder *et al.*, 1994), the United Kingdom (Lawson, 1994), and Canada (Hamill *et al.*, 1994), have responded with policies to reduce or minimize the quantities of pesticides applied in crop production. In the United States, the Clinton adminstration announced its support for coordinated efforts to reduce pesticide use in food production in a 1993 joint statement from the Department of Agriculture, Environmental Protection Agency, and Food and Drug Adminstration (Gutfeld, 1993).

A second factor propelling interest in ecological weed management is the more than 1 million incidents of unintentional pesticide poisoning that occur throughout the world each year (WHO, 1990). Pesticide poisoning is most frequent in developing countries (WHO, 1990), although it is not uncommon in industrialized countries such as the United States (e.g., Stone *et al.*, 1988). Public health data from Costa Rica (Hilje *et al.*, 1992, p. 79; Dinham, 1993, p. 105) suggest that herbicides may contribute to a significant portion of pesticide poisonings in developing countries, where safe use is difficult because of unavailable or prohibitively expensive protective equipment, inadequate or insufficiently enforced safety standards, poor labeling, illiteracy, and insufficient knowledge of pesticide hazards by handlers and applicators (Pimentel *et al.*, 1992).

A third factor promoting interest in ecologically based weed management strategies is the likelihood that farmers will face increasingly intractable weed problems with fewer chemical control options. Herbicide resistance has been noted in more than 100 weed species (Warwick, 1991). This resistance has been measured to evolve under field conditions in 4 or 5 years (Holt, 1992) and is now appearing in new species at a rate equal to that observed for insecticide resistance in arthropod pests (Holt and LeBaron, 1990). Both older (e.g., triazine) and newer, more environmentally friendly (e.g., sulfonylurea) herbicide chemistries have been met with the evolution of resistance, and cross-resistances within single weed species to different classes of herbicides have been detected (Holt, 1992; Gill, 1995). In addition to resistance within weed populations, heavy reliance on herbicides has, in some instances, shifted the composition of weed communities toward species that are more difficult to control (e.g., nutsedges, *Cyperus* spp.; Keeley, 1987). Growing numbers of government regulations and restrictions and increasing costs of research, development, and registration are expected to reduce the availability of older herbicides and the rate at which new herbicides are introduced (Holt and LeBaron, 1990).

A final factor promoting greater attention to ecological weed management is wider recognition of farming systems that operate profitably and productively with little or no use of herbicides and other agrichemicals. Profitable reduced input and organic systems seem to be possible for most commodities; examples include production of row crops and small grains in South Dakota (Smolik *et al.*, 1993), mixed grain and livestock in the midwestern United States (Lockeretz *et al.*, 1981), mixed vegetable, fruit, and livestock in New Zealand (Reganold *et al.*, 1993), and rice in Japan (Andow and Hidaka, 1989). Although a paucity of data makes it difficult to assess the range of conditions under which reduced input and organic farming systems will be successful, it is clear that such systems offer important opportunities to study the effects of nonchemical weed management practices on the survival, reproductive success, and dispersal (in both time and space) of different weed species. The information generated by such studies is, in our view, extremely valuable for the development of broadly adaptable weed management strategies that promote adequate yields and profits while protecting human health and environmental quality.

II. Information and Approaches

Ecological weed management involves the use of diverse types of information and a variety of control tactics to develop strategies for subjecting weeds to multiple, temporally variable stresses. The goals of imposing these stresses are reductions in (i) the density of weed propagules and seedlings,

(ii) the rate of weed seedling emergence relative to crop emergence, (iii) the rate of weed dispersal between and within fields, (iv) the proportion of available resources consumed by weeds, and (v) the proportion of weed communities composed of particularly noxious genotypes and species (Aldrich, 1984, pp. 399–435). To achieve these goals there is particular value in information regarding spatial and temporal patterns of weed abundance, reductions in crop yield due to weeds, weed life histories, weed niche characteristics, mechanisms of resource competition, and predicted outcomes of simulation models.

Information about spatial and temporal patterns of weed abundance is critical for determining which species are present in different portions of a management area (field, farm, or region) and whether these species are increasing or decreasing over time in response to management. Such information is generated by continual monitoring and constitutes the foundation of any integrated pest management system (Bottrell, 1979). Information generated by monitoring forms the basis for evaluating the efficacy of previous management strategies and identifying areas within fields and across landscapes where control measures need to be intensified (Mortensen *et al.*, 1993; Johnson *et al.*, 1995).

Weed-related yield loss information describes the extent to which crop yield is reduced by weed populations and communities, with emphasis on their density, mass, or leaf area; species composition; time of emergence; and period of association with the crop (Zimdahl, 1980, pp. 29–93; Kropff and Lotz, 1993a). Although the mechanistic basis of crop–weed interactions is not addressed in yield loss–weed infestation relationships, additional information concerning environmental factors (e.g., soil moisture conditions) may be included to suggest why the outcome of crop–weed interactions may vary from year to year and place to place. Predictions of yield loss from early season measurements of weed seed and plant density might be used to indicate when and where control measures are justified by agronomic and economic criteria (Streibig *et al.*, 1989; Forcella *et al.*, 1993).

Life history information results from intensive study of weed life cycles, including stages of germination or resumption of growth, seedling establishment or shoot emergence, growth and resource use, sexual or vegetative reproduction, dispersal, and quiescence (Mortimer, 1983). Life history studies are valuable for identifying stages that are potentially the most susceptible to control measures (Maxwell *et al.*, 1993). They are also important for projection of increases or declines of weed populations over time in response to different management practices, such as tillage, cultivation, grazing, and crop rotation (Firbank *et al.*, 1985; Maxwell *et al.*, 1988; Jordan, 1993a; Jordan *et al.*, 1995).

Niche information is developed by quantifying weed and crop germination and growth responses to variations in biological, physical, and chemical factors (e.g., Kennedy *et al.*, 1991; Ogg *et al.*, 1994). Such measurements help define the range of ecological conditions to which different weed and crop species (and genotypes) are best and least adapted. Of particular interest are possible differences between crops and weeds in germination and establishment and tolerance or susceptibility to manipulable environmental and management factors, such as mechanical disturbance, crop residue additions, herbivorous insects, and plant pathogens.

Mechanistic information concerning the ecophysiological basis of competition is generated through measurements of resource capture, resource conversion, growth, and allocation processes in crop and weed species grown in single-species stands and in mixture (Berkowitz, 1988; Kropff and Lotz, 1993b). Through intensive study of the mechanisms of crop–weed resource competition, management strategies may be developed to minimize stress on crops and maximize stress on weeds at particularly vulnerable stages of development.

A sixth type of information is generated by the integration of yield loss, life history, niche, and ecophysiological data into models with which to predict the outcome of crop–weed interactions when factors such as emergence time, plant height, and leaf area expansion rate are varied (Kropff *et al.*, 1993b). Such models may also aid efforts to breed crops with improved ability to tolerate or suppress weeds.

Guided by these diverse types of information, weed management strategies can be developed that employ nonselective and selective tactics. Nonselective tactics reduce weed numbers, growth, and reproduction through general herbivory, competition from rapidly growing cover crops, mowing, tillage, or the use of a broad spectrum herbicide such as glyphosate [*N*-(phosphonomethyl) glycine]. Because these manipulations can damage or kill a growing crop, they are carried out either before the crop is established or after it has been harvested to affect the weed pressure facing a subsequent crop. In contrast, selective techniques exploit differential responses between crop and weed species to control tactics such that the crop is favored and weeds are placed at a disadvantage. Examples of commonly used selective techniques include preemergence and postemergence herbicides and cultivation. Much less time and many fewer resources have been focused on the discovery and development of selective weed-suppression techniques that are based on ecological differences between crop and weed species, such as the use of weed pathogens and herbivores as biocontrol agents and rotation sequences that generate weed-suppressive crop residues. Nonetheless, available literature indicates that ecologically selective techniques can play key roles in suppressing weeds and promoting crop growth.

The full repertoire of possible tactics useful in ecologically based weed management strategies includes (i) different types and times of soil disturbance, through tillage and cultivation; (ii) diversification of crop vegetation, through cover cropping, rotation sequences, intercropping, and green manuring; (iii) herbivory by livestock and weed-feeding insects; (iv) diseases caused by applied or indigenous weed pathogens; (v) improvements in crop competitive ability through the use of weed-tolerant and weed-suppressive varieties, and through manipulations of crop density and spatial arrangement; and (vi) manipulations of resource conditions, through irrigation and fertility management (Liebman and Janke, 1990; Regnier and Janke, 1990; Cardina, 1995). Herbicides are not excluded from the toolkit but are viewed as options rather than absolute requirements for crop production. They are used only when and where the concerted application of other control tactics fails to reduce and maintain weeds at acceptable levels, and they are used in a manner that poses minimal risks to humans, other nontarget organisms, and the environment.

III. Basic Principles Guiding Weed Management Strategies

A. Competition in Agroecosystems

Resource removal that affects another individual defines competition. The competitive interactions of greatest concern to farmers occur when resources necessary for crop growth are removed by weeds and yield is reduced (Kropff, 1993b). Allelopathy, another important plant–plant interaction in agricultural systems, describes the addition of phytotoxicants into the environment and the subsequent reduction in growth of susceptible plants. Although mechanistically distinct, competition and allelopathy are not easily distinguished in the field. Interference is used to describe the net effect of competitive (removal) and allelopathic (addition) interactions (Gliessman, 1986).

Generally speaking, crops are planted at a constant density chosen to maximize economic yield per unit area (Radosevich and Roush, 1990). Although agricultural ecosystems are highly productive, intraspecific competition commonly limits the yield of individual crop plants, demonstrating that, even in a weed-free environment, supplies of certain resources are at suboptimal levels. Add to this environment weeds, and the total demand for resources, and therefore the intensity of resource competition, is increased. Interspecific competition is thus an intense and important interaction in agroecosystems.

Weed management has as its goals (i) exclusion of weeds from crop environments, (ii) killing of weeds before they interfere with crop growth and reproduction, and (iii) suppression of any surviving weeds. Interspecific resource competition and allelopathy are pivotal processes in obtaining the last of these three goals.

B. Effects of Weed Density, Duration, and Distribution

1. Yield Loss Relationships Results of experiments with a variety of weed–crop combinations, including wild oat (*Avena fatua* L.) with barley (*Hordeum vulgare* L.) and wheat (*Triticum aestivum* L.) (Cousens *et al.*, 1987), nightshade (*Solanum* spp.) with tomato (*Lycopersicon esculentum* Miller) (Weaver *et al.*, 1987), barnyard grass [*Echinochloa crus-galli* (L.) Beauv.] with sugar beet (*Beta vulgaris* L.) (Norris, 1992), common ragweed (*Ambrosia artemisiifolia* L.) with bean (*Phaseolus vulgaris* L.) (Chikoye *et al.*, 1995), redroot pigweed (*Amaranthus retroflexus* L.) with corn (*Zea mays* L.) (Knezevic *et al.*, 1994), and velvetleaf (*Abutilon theophrasti* Medikus.) with corn (Lindquist *et al.*, 1996), indicate that the relationship between crop yield loss and weed density often conforms to a rectangular hyperbola (Aldrich, 1987). Typical yield loss relationships are shown in Fig. 1 (Blackshaw, 1993a), which illustrates the effects of different densities (and age classes) of downy brome

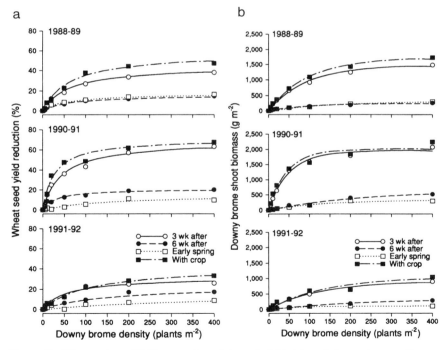

Figure 1 Winter wheat (*Triticum aestivum* L.) seed yield (a) and downy brome (*Bromus tectorum* L.) shoot biomass (b) as functions of downy brome density and times of emergence relative to wheat. Lines are fitted nonlinear regression curves for each emergence date (reproduced with permission from Blackshaw, 1993a).

(*Bromus tectorum* L.) on winter wheat. Drawn from such data sets is the first principle guiding weed management: *Reduce weed density.*

It can be seen in Fig. 1 that for a given age class of weeds, yield reduction due to weed infestation varied considerably among years: Weed competition was most damaging to wheat seed yield in 1990–1991, least damaging in 1991–1992, and intermediate in its effects in 1988–1989. Relationships between crop yield and weed density often vary considerably among environments and years, coincident with variation in temperature conditions, moisture availability, fertility source, and other factors (Aldrich, 1987; Mortensen and Coble, 1989; Bauer *et al.*, 1991; Liebman and Ohno, 1997; Lindquist *et al.*, 1996). For the experiments shown in Fig. 1, variation among years in the yield loss–weed density relationship was attributed to differences in the timing and quantity of rainfall (Blackshaw, 1993a). Thus, although general patterns of crop yield loss to varying densities and age classes of weeds appears consistent, specific outcomes are governed by a host of factors including weather, whose precise effects may be extremely difficult to predict without much larger data resources.

2. Critical Periods for Weed Control   The experimental results illustrated in Fig. 1 indicate that earlier emerging weeds produced more biomass than later emerging weeds, and crop seed yield was inversely proportional to weed density. In general, weed populations that emerge earlier in the growing season are more damaging to crop yield than are populations that emerge later (Cousens *et al.*, 1987; Wyse, 1992). Thus, a second basic principle guiding weed management is: *Delay weed emergence relative to crop emergence.*

Crop species can differ greatly in their sensitivity to different durations of weed competition, as demonstrated in experiments in which purple nutsedge (*Cyperus rotundus* L.) was allowed to grow in several vegetable crops after weed-free conditions were maintained for varying numbers of weeks (William and Warren, 1975). Maintenance of nutsedge-free conditions for 3 weeks following planting was required to prevent yield reduction of transplanted cabbage (*Brassica oleracea* L.), whereas garlic (*Allium sativum* L.) yield was reduced unless nutsedge was excluded for more than 18 weeks. Other crops tested in this experiment gave intermediate results.

To better identify the period in a crop's life cycle when it must be kept weed free to prevent unacceptable yield loss, experiments can be conducted using two types of treatments: (i) Those in which the crop is first grown without weeds for specified periods of time and then weeds are allowed to grow for the remainder of the season (as in the purple nutsedge–vegetable experiments discussed above); and (ii) those in which the crop is first grown with weeds for specified periods of time and then weeds are excluded for the rest of the season (Oliver, 1988; Zimdahl, 1988). Data generated from such experiments indicate that, in some cases, a crop may be unaffected by association with weeds during the early and late portions of the growing season if weeds are excluded during the middle period. When this pattern occurs, a clearly defined "critical weed-free period" is said to exist. For example, in a study conducted by Woolley *et al.* (1993) involving bean infested with weeds for different durations, crop yield loss did not exceed 3% if weeds were excluded during the interval of 15–60 days after the crop was planted.

In other cases, it is not possible to identify a critical weed-free period that occurs between periods of acceptable weed infestation. The important management issue then becomes how long weed exclusion efforts must be maintained before they can be relaxed. Van Acker *et al.* (1993a) noted, for example, that if weeds were prevented from growing in soybean from the time the crop was planted until it had four leaves (about 30 days after crop emergence), crop yield loss was <2.5% and weed biomass was reduced >97%. Similarly, no yield loss occurred in corn if redroot pigweed was prevented from growing until the crop reached the seven-leaf stage (Knezevic *et al.*, 1994).

As might be expected, the critical weed-free period for a given crop may vary considerably among sites and years, as was noted by Hall *et al.* (1992), who studied the effects of different durations of weed competition on corn grain yield. Factors contributing to variability in critical periods for weed control may include climatic and edaphic conditions affecting weed and crop emergence and growth rates, weed species composition, and weed density.

3. Growth Characteristics as Predictors of Weed Competition Variations between sites and years in weed and crop emergence times, growth rates, and densities would seem to preclude any unifying analysis of the effects of weed populations on crop yield or the ability to forecast the outcome of competitive interactions early enough in the season that cultivation or postemergence herbicides might still be used. A proposed solution to this problem is to use early season measurements of weed and crop leaf area to calculate a relative leaf area ratio with which to predict competitive outcomes. Theoretically, such an index could integrate variations in weed densities, emergence times, and growth rates into a quantifiable, consistent predictor of crop yield loss (Kropff and Spitters, 1991; Kropff *et al.*, 1992). Kropff and Lotz (1992) used data from five field experiments with different weed densities and times of emergence to demonstrate that leaf area measurements of sugar beet and lambsquarters (*Chenopodium album* L.) made 30 days after crop emergence could better predict crop yield loss than could weed density. Relative leaf area was also more successful than weed density in predicting soybean yield loss due to johnsongrass [*Sorghum hala-*

pense (L.) Pers.] (Vitta *et al.*, 1993), corn yield loss due to redroot pigweed (Knezevic *et al.*, 1995), and bean yield loss due to common ragweed (Chikoye and Swanton, 1995). Kropff and Lotz (1993a) provide further examples of the relative leaf area approach, including possible modifications to allow for time-dependent differences in growth characteristics such as leaf expansion rates.

The work of Lotz *et al.* (1994) is of particular interest from the standpoint of practical application of the relative leaf area approach in scouting farm fields. Because a large amount of labor is required to measure leaf area directly, the investigators tested the possibility of estimating leaf area using more rapid, nondestructive measurements of canopy cover and radiation reflectance. The study involved seven field experiments with spring wheat, sugar beet, and several weed species. Cover measurements made using a grid-quadrat frame were well correlated with direct measurements of leaf area until 3 or 4 weeks after crop emergence; reflectance measurements worked well in spring wheat but not in sugar beet. Based on results of this study, it would seem worthwhile to direct more research attention toward the development of rapid field techniques for estimating leaf area.

Other aspects of plant size that can be assessed easily early in the growing season may also prove useful for predicting the outcome of crop–weed interactions. For example, Bussler *et al.* (1995) grew corn with velvetleaf and common cocklebur (*Xanthium strumarium* L.) and found that 30–58% of the variation in corn seed production could be predicted from estimates of plant volume derived from nondestructive measurements of crop and weed densities, heights, and canopy widths made 23–28 days after planting.

Information regarding weed density and duration of competition might be used to establish thresholds and scouting methodologies but does not provide insight as to the mechanisms driving the crop–weed interactions. Mechanistic information requires a greater degree of reductionist investigation (see Section III,C,4).

4. Spatial Patterns   A certain amount of variation in yield loss–weed density relationships may derive from variation in weed spatial distributions. As the result of innate and human-related dispersal patterns, weeds often occur in clumps within fields in a manner best described by the negative binomial distribution (Cousens and Mortimer, 1995, pp. 55–85, 217–242). It is thus important to consider that the local neighborhood of most crop plants within those fields is filled with many fewer weeds than the overall field average would suggest (Cardina *et al.*, 1995; Johnson *et al.*, 1995). For a given average density over a broad area, clumped weeds are expected to be less damaging to crop yield than are randomly or evenly distributed weeds (Auld and Tisdell, 1988; Brain and Cousens, 1990).

Clumped distributions of weeds within fields have two implications for management. First, if appropriate tools were available, weed control measures might be applied only in weedy areas within fields; weed-free areas could be ignored. For example, Johnson *et al.* (1995) used multiple small quadrat samples (0.76 m^2) and geostatistical techniques to describe the spatial characteristics of weed populations in 12 commercial corn and soybean fields in Nebraska. In the interrow areas, where herbicides were not applied, 30% of the sample area was free of broadleaf weeds and 70% was free of grass weeds. Johnson *et al.* (1995) concluded that if herbicide application were linked to real-time weed-sensing technologies (Woebbecke *et al.*, 1995; Mortensen *et al.*, 1997), considerable reductions in herbicide use might be achieved. On a slightly larger scale (e.g., sections of fields), knowledge of weed distributions might aid the deployment of other site-specific weed management tactics, such as the choice of particular crop sequences and tillage practices or the use of higher seeding rates (Mortensen *et al.*, 1997).

The second, and corollary, implication of clumped weed distributions is that managing weed dispersal within fields can be an important strategy in a weed management system. Simulation models suggest that (i) if weed seed dispersal from local patches can be restricted during activities such as tillage and harvest, then the rate of weed population increase may be reduced (Ballare *et al.*, 1987); and (ii) management of weed dispersal may have greater influence on crop yield than the relative competitive ability of the weed with respect to the crop (Maxwell and Ghersa, 1992). Thus, we suggest a third principle for weed management: *Limit the dispersal of weed propagules.*

C. Mechanisms of Resource Competition

1. A Conceptual Model Resource capture, resource conversion, and biomass allocation are key processes affecting plant growth and reproduction (Mooney and Chiariello, 1984; Berkowitz, 1988). A conceptual model that incorporates these processes and provides a means of organizing information concerning resource competition between crops and weeds is shown in Fig. 2. As shown in the model, resources (e.g., light, water, and nutrients) may be either generally accessible to both crops and weeds, in which case competition for limited quantities may occur, or selectively accessible to either crop or weed species, as in mixtures of legumes, which use atmospheric N, and nonlegumes, which lack access to this N source. Because plant biomass accumulation and reproductive effort are strongly linked to resource consumption, we suggest a fourth principle for weed management: *Maximize the proportion of available resources consumed by crops and minimize the proportion consumed by weeds.*

For both generally accessible and selectively accessible resources, the capture of one resource (e.g., nitrogen) may affect capture of other resources (e.g., light) through feedback loops involving growth and biomass

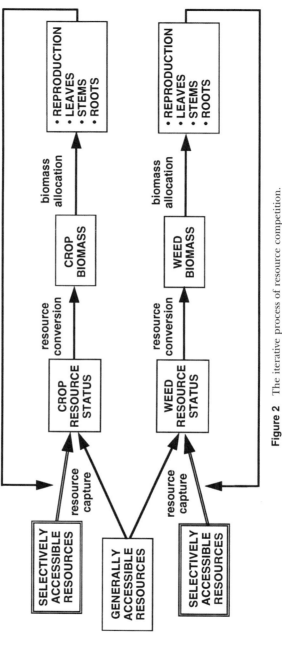

Figure 2 The iterative process of resource competition.

allocation. Biomass, representing seed or vegetative structures present at the start of crop–weed interactions, may be viewed as the starting point of the model, although its importance in determining competitive outcomes has been questioned (Wilson, 1988a).

The model shown in Fig. 2 includes state and rate variables that can be quantified (Kropff and Spitters, 1992; Kropff, 1993a). State variables, boxed in Fig. 2, might include concentration of inorganic N in the soil, plant N content, total biomass, and biomass of individual plant organs; examples of rate variables include canopy light interception per unit time, photosynthetic carbon (C) gain per unit of N absorbed, photosynthetic C gain per unit of water transpired, and biomass partitioning coefficients. Abiotic and biotic environmental factors, such as temperature, precipitation, incident photosynthetically active radiation, and pathogens, can affect resource availability and physiological processes of resource capture and conversion, but for simplicity are not included here. Kropff (1993a) describes a mechanistic, ecophysiological model of crop–weed competition that includes environmental factors at the crop–weed system's boundaries as driving variables.

Direct measurements of resource availability, capture, conversion, and allocation can be made using instruments to quantify gas exchange, radiation interception throughout the canopy, soil and plant water potentials, nutrient concentrations, and other ecophysiological parameters (e.g., Pearcy *et al.*, 1991). Growth analysis techniques may also prove useful in examining rates of biomass accumulation and canopy production and efficiencies of biomass production calculated on the basis of leaf area, nutrient content, and other state variables (e.g., Evans, 1972; Hunt, 1982; Van Acker *et al.*, 1993b). In both field and glasshouse experiments, techniques to partition shoots and roots can be used to determine whether the locus of competitive interactions is above or below ground (Snaydon, 1979; Wilson, 1988b; Exley and Snaydon, 1992; Bozsa and Oliver, 1993).

One of the advantages of using a resource-based model to organize investigations of crop–weed interactions is that it encourages acquisition of information with which to differentiate crop *tolerance* of weeds ("competitive response") from crop *suppression* of weeds ("competitive effect") (Goldberg and Landa, 1991). Tolerance involves the ability to maintain growth through (i) high resource conversion efficiency at depleted resource levels or (ii) access to a selectively available pool of resources. In contrast, suppression involves the ability to preempt the use of generally accessible growth resources from competitors (Goldberg, 1990). Strategies to minimize crop yield loss to weed infestation might emphasize improved tolerance of weeds, whereas strategies to minimize weed growth and reproduction would prioritize improved suppression of weeds (Jordan, 1993b). Because weed-tolerant crop varieties may not be those with the strongest ability to suppress weed growth, and because a variety's ability to suppress weeds may carry a yield penalty (Jordan, 1993b), a resource-based model of crop–weed competition may facilitate resolution of potentially contradictory crop breeding objectives.

2. Resource-Related Effects of Weeds on Crops   Resource-related mechanisms of weed competition against crops help to identify those resources that might be better manipulated to a crop's advantage.

Experiments conducted by Cruz *et al.* (1983) and Graham *et al.* (1988) illustrate the effects of varying weed density on crop resource use. In a study conducted under dryland conditions with rice (*Oryza sativa* L.) and a mixture of weed species, Cruz *et al.* (1983) observed that leaf water potential of the crop decreased (i.e., moisture stress increased) as weed density increased. Graham *et al.* (1988) reported that leaf area, interception of photosynthetically active radiation, net photosynthesis, total biomass, and seed production of sorghum (*Sorghum vulgare* L.) were reduced as density of pigweed (*Amaranthus* spp.) increased.

A study conducted by Young *et al.* (1983) with quackgrass [*Elytrigia repens* (L.) Nevski] and soybean illustrates how resource availability may be manipulated to help identify factors for which competition may be occurring. The investigators observed that addition of irrigation water could offset the weed's detrimental effect on crop leaf water potential but could not eliminate all of the weed's negative effects on yield. Macronutrient content of soybean leaves was not affected by the presence of quackgrass; competition for light by the weed against the crop was inferred from measurements of plant height.

Weed infestation can result in reductions in crop macronutrient uptake and yield, as shown in experiments with corn (Vengris *et al.*, 1955) and wheat (Soni and Ambasht, 1977). A study conducted by Volz (1977) demonstrates the value of monitoring nutrient availability as well as crop and weed nutrient uptake. Volz (1977) showed that corn N concentration and uptake were depressed by the presence of yellow nutsedge (*Cyperus esculentus* L.), but that nutsedge N uptake failed to account for the depression. Measurements of soil inorganic N concentration and denitrifying bacteria population densities led the investigator to suggest that nutsedge reduced N availability to corn by increasing losses through denitrification. Results of this study emphasize that interactions between crops and weeds are not necessarily limited to resource competition.

3. Resource-Related Effects of Crops on Weeds   By understanding how to create resource deficits in a weed's immediate environment, and how a weed responds to such deficits, crop and soil management strategies may be devised to reduce weed growth and reproduction. Research efforts following this approach have generally focused on weed–crop canopy relations and light deficits, but management of soil resources may offer additional options.

Shetty *et al.* (1982) used bamboo frames to manipulate light conditions in a field experiment and reported that shading reduced height, leaf area, dry matter production, and tuber production of purple nutsedge. Height, leaf area, and seed production of several other annual weeds were also reduced by shading. The investigators concluded that by manipulating crop canopies to create desired levels of shading, substantial weed suppression could be achieved.

Resource deficits may create desirable shifts in allocation patterns of weed biomass. In two field experiments in which screen cloth was used to create shaded conditions, Williams (1970) observed that shading of quackgrass resulted in an 11–16% reduction in shoot dry weight but a 47–51% decrease in rhizome weight. Williams (1970) suggested that selecting crops that shade the weed "would most help in its control."

Lotz *et al.* (1991) demonstrated that shading reduces shoot biomass and tuber production in yellow nutsedge, and that crops differ in their ability to shade the weed. Compared to a crop-free control, tuber number per nutsedge plant was reduced 25% by winter rye (*Secale cereale* L.), 69% by winter barley, 96% by corn, and 100% by hemp (*Cannabis sativa* L.). Reductions in nutsedge tuber production were reflected in decreased competition from nutsedge (i.e., higher yields) in a subsequent corn crop. Lotz *et al.* (1991) concluded that use of crops that produce dense canopy cover during a long period of the growing season may result in a considerable decrease in nutsedge population density on infested farmland.

Liebman and Robichaux (1990) examined the effects of N and light deficits on the photosynthetic performance, growth, and seed production of white mustard (*Brassica hirta* Moench) in field experiments in which the weed was grown alone, in mixture with barley sole crops, or in mixture with barley–field pea (*Pisum sativum* L.) intercrops. Barley had no effect on the amount of photosynthetically active radiation (PAR) reaching the weed, but it did reduce the weed's leaf N concentration, photosynthetic surface area, net photosynthesis, and biomass production. In contrast, barley–pea intercrops reduced both weed leaf N concentrations and the amount of PAR incident on weed leaves. Reductions in N status and PAR interception were accompanied by reductions in photosynthetic surface area, net photosynthesis, biomass, and seed production. Compared to use of a short statured pea variety with small leaves, use of a taller pea variety with larger leaves increased shading of the weed and further decreased its growth and seed production. Thus, cultivar choice may have important effects on a crop's ability to suppress growth of associated weed species through shading.

In a field competition experiment in which both the crop and the weed were tree species [Douglas fir (*Pseudotsuga menziesii* Franco.) and red alder (*Alnus rubra* Bong.), respectively], Shainsky and Radosevich (1992) demonstrated dynamic interactions between plant population density, competition for water, and competition for light. Red alder grew taller than Douglas fir, consequently reducing the fir's access to PAR and thereby its growth rate. However, when the investigators increased the density of Douglas fir, leaf water potential and leaf area of red alder decreased. The reduction in alder leaf area resulted in less shading of Douglas fir and a concomitant increase in Douglas fir's leaf area and growth rate. The combination of population and ecophysiological approaches presented in this study is exceptional and would be extremely useful in studies of annual crop and weed species.

4. Ecophysiological Simulation Models of Resource Competition   With an adequate amount of ecophysiological information, conceptual models of crop–weed competition for resources, such as that shown in Fig. 2, can be developed into quantitative simulation models. Ecophysiological simulation models of crop–weed competition typically start with characterizations of each species grown in pure stand, and then predict growth and yield in mixture based on height and canopy characteristics affecting interspecific shading and light-driven transpiration rates. Similarly, root length or mass may be used to predict the outcome of competition for nutrients.

In the INTERCOM model described by Kropff and Spitters (1992), model parameters include functions describing vertical leaf area profile; light distribution throughout the canopy; carbon dioxide assimilation and light response curves of individual leaves at different heights in the canopy; maintenance respiration rates; phenological patterns of growth and development driven by accumulated heat units; dry matter partitioning patterns; leaf area development and senescence patterns; plant height; and transpiration rates. INTERCOM and similar models have been used to study competition for light between winter wheat and wild oat (Weaver *et al.*, 1994); competition for light and nitrogen between rice and barnyard grass (Graf and Hill, 1992); and competition for light and water between soybean and common cocklebur, rice and barnyard grass, corn and barnyard grass, sugar beet and lambsquarters, and tomato and eastern black nightshade (*Solanum ptycanthum* Dun.) (Wilkerson *et al.*, 1990; Wiles and Wilkerson, 1991; Kropff *et al.*, 1993c).

Concordance between observed and predicted responses in these studies has generally been very good. Thus, it can be hoped that ecophysiological simulation models will be useful for (i) developing hypotheses concerning mechanisms of resource competition, (ii) examining variability in the outcome of crop–weed interactions between years and environments, and

(iii) evaluating potential impacts of alternative weed management practices at a cost considerably lower than that incurred by multisite, multiyear field experiments. For example, simulation models of rice–barnyard grass competition conducted by Kropff *et al.* (1993a) indicated that using a rice cultivar with 5% increases in leaf area expansion rate and height parameters could reduce crop yield loss to weed competition by 6%. Because considerable genetic variation exists in rice for these traits, "there may be potential for breeding for higher competitive ability." As this example demonstrates, simulation can be useful for suggesting useful directions and hypotheses—it must, however, be substantiated by research in the field.

IV. Managing Crop–Weed Interactions

A. Strategies to Reduce Weed Propagule and Seedling Densities

As indicated in Fig. 1, factors that reduce weed population density and that delay weed emergence relative to crop emergence can be expected to improve crop performance and increase the competitive dominance of the crop over associated weeds. A variety of existing practices can be used to achieve these objectives and numerous ecological relationships exist with potential for exploitation as components of weed management strategies.

1. Crop Rotation and Use of Cover Crops   Crop rotation is one of the most powerful cultural management techniques available to farmers for reducing weed seed and seedling densities (Liebman and Dyck, 1993b; Liebman and Ohno, 1997). When rotation sequences include crops that differ in planting and maturation dates, competitive and allelopathic characteristics, and associated management practices (e.g., tillage, cultivation, mowing, and grazing), weeds can be confronted with an unstable and frequently inhospitable environment that prevents their proliferation. Figure 3 illustrates marked reductions in weed seed and seedling densities when corn was grown in rotation with winter wheat compared with corn grown as a continuous monoculture (Covarelli and Tei, 1988). Rotation of corn with soybean was also found to reduce weed seed and seedling densities compared with continuous corn (Forcella and Lindstrom, 1988). Although crop rotation is a well-known and long-used method of maintaining or increasing crop yields, more research is needed to understand factors affecting weed demography in different rotation systems and to identify rotation strategies that are especially effective for weed management.

Cover crops are often planted to control erosion and, when leguminous species are used, to improve soil productivity through N fixation. Cover crops may also be used to suppress weeds. For example, lower weed densities were observed when various vegetable crops were planted without tillage

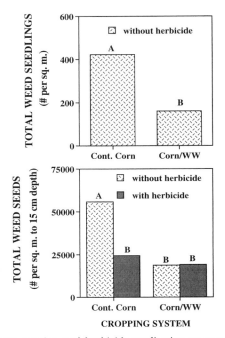

Figure 3   Effect of crop rotation and herbicide application on weed plant (upper panel) and seed densities (lower panel). Continuous corn was compared to a wheat/corn rotation, each with and without herbicides used for weed control. The predominant weed species were *Amaranthus retroflexus* L., *Chenopodium album* L., and *Echinochloa crus galli* L. Different letters above bars indicate significant differences ($P < 0.05$) in mean values (adapted with permission from Covarelli and Tei, 1988).

into the residues of gramineous cover crops compared to when the vegetables were planted into a tilled seedbed or an untilled bed without cover crop residues (Putnam and DeFrank, 1983; Putnam *et al.*, 1983). The investigators attributed much of this effect to allelopathic compounds released from the cover crops that selectively suppressed weed emergence.

Other data indicate that physical characteristics of cover crop residues may also affect weed emergence. Teasdale and Mohler (1993) reported that residues of rye and hairy vetch (*Vicia villosa* Roth.) cover crops on the soil surface reduced light transmittance and soil temperature amplitude sufficiently to reduce weed emergence. The investigators noted, however, that maintenance of soil moisture beneath cover crop residues could enhance weed emergence. Results of greenhouse and field studies indicated that hairy vetch residue used as a surface mulch reduced emergence of some weed species but not others (Teasdale *et al.*, 1991; Teasdale, 1993).

Cover crops that are incorporated into the soil may also affect weed emergence. Boydston and Hang (1995) reported that weed density was reduced 73–85% and crop yield was increased 10–18% when potato was grown after rapeseed (*Brassica napus* L.) compared with after fallow. Glucosinolate compounds released or derived from rapeseed residues were suggested as possible agents of weed suppression. In a field experiment examining weed-related effects of a crimson clover (*Trifolium incarnatum* L.) cover crop and synthetic N fertilizer, Dyck and Liebman (1994) found that soil-incorporated clover residue strongly suppressed emergence of common lambsquarters but only slightly reduced emergence of sweet corn (Table I). Nitrogen fertilizer was found to stimulate emergence of the weed but to reduce corn emergence when clover residue was present. The investigators concluded that use of crimson clover as a N source provided weed control benefits both as a direct suppressant of weed emergence and as a substitute for N fertilizer.

2. Tillage and Cultivation   Tillage, for seedbed preparation, and cultivation, for weed control, can affect established plants and reproductive structures. Tillage practices can range from nearly complete soil inversion, with moldboard plowing, to minimal soil disruption, with the use of zero-tillage (direct drilling) techniques. In addition to determining characteristics of soil disturbance and residue incorporation, tillage practices can have important effects on weed density and species composition (Buhler, 1995; Froud-Williams, 1988). Factors determining tillage effects on weeds include (i) depth of seed burial, (ii) seed survival at different soil depths, (iii) seed dormancy responses to burial, (iv) seedling ability to emerge from different burial depths, and (v) the quantity of new seeds added to the soil seedbank (Mohler, 1993). Because weed species may differ in these factors, responses to tillage are often species specific. For example, in a study comparing moldboard plow, chisel plow, and no-tillage systems for soybean production, Buhler and Oplinger (1990) observed that lambsquarters densities were not greatly influenced by tillage systems, whereas redroot pigweed densities were generally highest in the chisel plow system. Giant foxtail (*Setaria faberi* Herrm.), an annual grass species, was most abundant in the no-tillage system and least abundant in the moldboard plow system. In contrast, velvetleaf, an annual broadleaf species, was most abundant in the moldboard plow system and least abundant in the no-tillage system. Similarly in corn production systems, Buhler (1992) observed that density responses to tillage differed among weed species. Derksen *et al.* (1993) have suggested that changes in weed communities are influenced more by location and year than by tillage systems, but it appears that tillage is potentially useful for reducing weed density if choice among tillage practices is based on knowledge of the full spectrum of ecological sensitivities of different weed species.

Because seeds of many weed species require exposure to light to germi-

Table I Effect of Crimson Clover (*Trifolium incarnatum* L.) Residue and Nitrogen (N) Fertilizer on Emergence of Common Lambsquarters (*Chenopodium album* L.) and Corn (*Zea mays* L.)[a]

Fertilizer treatment (kg N ha^{-1})	Total emergence (plants m^{-2})				Time to 50% emergence (days after planting)			
	Lambsquarters		Corn		Lambsquarters		Corn	
	Without residue	With residue	Without residue	With residue	Without residue	With residue	Without residue	With residue
0	215	150	7.8	7.6	11.3	16.5	7.1	7.4
60	357	287	7.8	6.2	14.4	18.1	7.0	8.0
120	382	260	7.9	7.1	14.0	15.7	6.8	8.6
180	360	257	7.7	6.8	14.9	17.7	6.8	7.3
ANOVA								
Residue	$P < 0.001$		$P < 0.001$		$P < 0.01$		$P < 0.01$	
N fertilizer[b]	$P < 0.001$		$P < 0.001$		NS		NS	
Residue *N fertilizer	NS		$P < 0.05$		NS		NS	

[a] Adapted from Dyck and Liebman (1994), by permission of Kluwer Academic Publishers.
[b] Contrast of 0 kg N ha^{-1} vs (60 + 120 + 180 kg N ha^{-1})/3. NS, not significant.

nate, attention has been directed recently toward the possibility of preparing crop seedbeds at night or during the day using tillage equipment covered with light-excluding hoods. In field trials Ascard (1994) observed that, compared to daylight tillage, both of the aforementioned dark-tillage techniques reduced weed density.

In some farming areas, germination of common weed species may occur in predictable flushes that are driven by accumulated heat units and rainfall (e.g., Harvey and Forcella, 1993). Because weed seedlings are extremely vulnerable to tillage operations, Forcella *et al.* (1993) conducted experiments in the north central United States to determine whether synchronicity in weed emergence could be exploited for management purposes. Delaying final tillage operations—the so-called stale seedbed strategy—allowed the investigators to kill a very high percentage of weed seedlings before planting corn and soybean and, consequently, to reduce weed competition against the crops.

Following planting, cultivation can be used before and after crop emergence to reduce weed densities between and within crop rows (Terpstra and Kouwenhoven, 1981; Buhler *et al.,* 1992; Rasmussen, 1992; Rydberg, 1993; Mulder and Doll, 1994; Rasmussen and Svenningsen, 1995; Vangessel *et al.,* 1995). Increased interest in alternatives to herbicides has resulted in a number of new cultivation implements, some of which are capable of working under high residue conditions and therefore are compatible with soil conservation objectives (Eadie *et al.,* 1992). Interest has also increased in machinery for flame weeding, which kills weed seedlings through cell rupture rather than incineration (Daar, 1987). Flame weeding can be used to destroy weeds emerging before crop emergence (Ascard, 1995a,b); postemergence flaming, long practiced in cotton (*Gossypium hirsutum* L.), is also possible in certain other crops such as onion (*Allium cepa* L.) and corn (Daar, 1987).

3. Seed Predators and Pathogens Seed predation by insects, rodents, and birds can greatly reduce weed seed and seedling density, particularly in reduced tillage systems, which characteristically have greater amounts of residue cover (House and Brust, 1989). For example, Brust and House (1988) reported that, over a 5-week period, indigenous weed seed predators consumed 68% of seeds placed as baits in a zero-tillage soybean production system containing grain straw on the soil surface, but consumed only 27% of the seeds placed as bait in a clean-tilled soybean system. Weed seed predators may be indigenous (Lund and Turpin, 1977; Brust and House, 1988) or they may be cultured and released (Kremer and Spencer, 1989a,b). Weed seed predators exhibit species-specific preferences and thus may affect the relative abundance of different weed species (Brust and House, 1988; House and Brust, 1989). Pathogens, acting with (Kremer and Spencer,

1989a,b) or without (Kennedy *et al.*, 1991; Kremer, 1993) insects attacking seeds, can also be important for reducing weed seed and seedling densities. Despite data indicating weed seed predators and pathogens can be manipulated and augmented to desirable ends, optimum management strategies exploiting these interactions have not yet been developed.

4. Soil Solarization   In areas with extended periods of hot weather, use of plastic tarps to heat moist soil prior to planting (soil solarization) has proven useful as a means of killing weed seeds and reducing weed seedling densities in subsequent crops (Egley, 1983; Horowitz *et al.*, 1983; Standifer *et al.*, 1984; Elmore *et al.*, 1993). Kumar *et al.* (1993) provided data indicating the positive effects of soil solarization on both weed suppression and crop yields, although the effect on crop yield might be attributed to reductions in pathogen populations as well as reduction in weed pressure (Stapleton and DeVay, 1986).

B. Strategies to Limit Weed Dispersal

The exclusion of weed seeds and other propagules from habitats and fields where they have previously been absent has long been recognized as an important means of reducing weed problems. Sanitation methods for limiting weed dispersal into new areas include using only crop seed free of weed contamination; composting of manures to kill weed seeds; cleaning of tillage and harvest equipment between operations in different fields; and mowing, cultivating, or spraying to kill weeds in areas bordering crop fields (Walker, 1995).

More problematic is the prevention or reduction of weed dispersal on a more local scale, i.e., between patches within fields. Perimeter portions of a cropped field may contain higher densities of weeds than more central portions (Wilson and Aebischer, 1995). Thus, site-specific control methods, such as higher seeding rates, different tillage practices, and different rates or types of herbicides, might be applied to field margins to control weed propagule production and subsequent movement into the rest of the field (see Section III,B,4). Where high densities of weeds occur in scattered "hot spots", the elimination of the propagules through hand roguing and spot spraying can be important in preventing further colonization. Combine harvesters clearly have the ability to spread weed seeds throughout a field (Cousens and Mortimer, 1995; pp.79–84); their ability to act as weed "seed predators", however, has received very little attention. Maxwell and Ghersa (1992) and Cousens and Mortimer (1995, p. 290) have emphasized the need for more integrative work by weed managers and agricultural engineers to develop better machinery for harvesting and removing weed seeds.

C. Strategies to Manipulate Resource Availability and Resource Capture

Manipulating access to water, nutrients, and light can provide opportunities to manage the outcome of weed–crop interactions to a crop's advantage. Access to resources can be manipulated in spatial, temporal, and physiological dimensions.

1. Spatial Aspects   Manipulating access to water can have poweful effects on crop and weed performance in arid areas. Figure 4 illustrates how selective placement of irrigation water improved weed management in tomato production systems studied in California (Grattan *et al.*, 1988). Three irrigation systems were compared: sprinkler irrigation, which spread

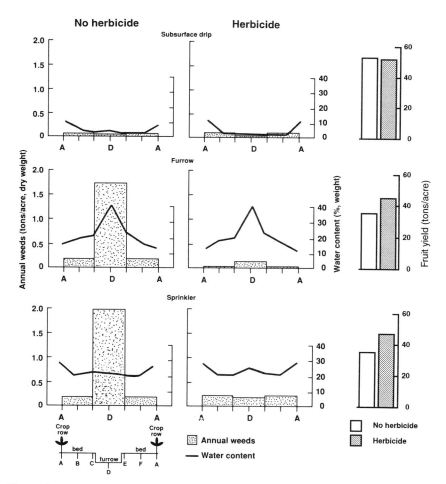

Figure 4   Biomass of annual weeds, soil water content, and tomato (*Lycopersicon esculentum* Miller) fruit yield as affected by herbicide application and three types of irrigation. The crop row (A) and bed furrow separating the raised beds (D) are depicted schematically at the bottom of the figure (adapted with permission from Grattan *et al.*, 1988).

water uniformly over the entire plot; furrow irrigation, which concentrated water between crop rows; and subsurface drip irrigation, which concentrated water directly beneath the crop. In the absence of herbicides, weed growth in the sprinkler and furrow irrigation systems was substantial, and tomato yield was reduced by weed competition. In contrast, when subsurface drip irrigation was used, weed growth was extremely low, even in the absence of herbicides, and crop yield was unaffected by weed competition. It should be noted that the experiment was conducted at a site at which precipitation during the tomato production season is nil or extremely low.

Seed placement may affect the availability of soil moisture and the outcome of crop–weed interactions. In a review of weed management tactics useful for production of agronomic crops in Nebraska, Bender (1994, p. 37) identified planting conditions as optimum when adequate moisture to germinate the crop is present below the soil surface but the surface is sufficiently dry to prevent germination of weeds until the next rain. In this case, seed placement in moister soil provides a competitive advantage to the crop by permitting it to emerge before the weeds.

Manipulating the location of nutrients appears to offer some potential as a means of increasing both crop yield and weed suppression. In a review of the effects of fertilizer placement on crop and weed performance, DiTomaso (1995) noted that banding fertilizers within the crop row of bean (*Phaseolus vulgaris* L.), soybean, peanut (*Arachis hypogaea* L.), wheat, alfalfa (*Medicago sativa* L.), Lincoln bromegrass (*Bromus inermis* Leyss.), littleseed canary grass (*Phalaris minor* Retz.), and rice not only increased crop yield compared to broadcast applications but also reduced weed density. Application of nutrients in a deep band (e.g., 7 cm below seed level) appeared to be more effective for improving crop yield and weed suppression than did application of nutrients in a band on the soil surface.

In a field study conducted in Florida, banded applications of P (5 cm depth) failed to increase lettuce yield compared to broadcast P but did reduce the strength of weed competition measured as the percentage difference between weed-free and weed-infested crop yield (Shrefler *et al.*, 1994). However, lower rates of P were applied in the banded treatment than in the broadcast treatment, thus confounding fertilizer rate with placement effects.

Reinartsen *et al.* (1984) compared the effects of surface broadcasting and deep banding (5 cm depth) N fertilizer in a no-tillage winter wheat production system in eastern Washington. Banding N increased wheat yield but had no effect on wild oat biomass. In the same area of Washington, Cochran *et al.* (1990) compared the effects of broadcasting and banding (5 cm depth) N fertilizer on yield of winter wheat and biomass production of two grass weeds, downy brome and jointed goat grass (*Aegilops cylindrica* Host.). Comparisons were made in three tillage systems: moldboard plowing, shallow rototilling, or no-tillage prior to planting. Wheat grain yields were higher with band-applied N compared to broadcast N in all tillage systems, with or without the presence of grass weeds. Weed biomass did not differ between the two fertilizer application treatments in the moldboard plow and no-tillage systems, but less weed growth occurred with banded N in the rototilled system. Cochran *et al.* (1990) suggested that the lack of weed suppression they observed with banded N in the moldboard plow and no-tillage systems may have resulted from inherently high soil fertility conditions; greater weed suppression would be expected when unfertilized surface soil limited weed growth and deep banded fertilizer relieved nutrient deficits for crops. Alternatively, crop performance (competitive response) may not have been closely linked to weed suppression (competitive effect). Mechanistic studies of nutrient competition are needed to address this issue.

Rasmussen *et al.* (1996) found that banding nitrogen fertilizer 5 cm below the depth at which spring barley was sown decreased weed biomass 55% and increased grain yield 28% compared with broadcast fertilizer application. Supporting the notion that fertility placement effects should be more pronounced in low relative to high fertility soils, the yield increase from fertilizer placement, averaged over 3 years, was 17% on sandy loam compared with 38% on coarse sand. Although all treatments were subjected to spring tine harrowing for weed control, the authors' opinion was that "fertilizer placement favored the crop at the expense of the weeds."

2. Temporal Aspects Because many weed species are capable of earlier and more rapid uptake of nutrients than are associated crops (Alkamper, 1976), some research effort has focused on whether delayed application of nutrients deprives weeds of nutrients early in the growing season and better matches the timing of crop uptake ability. Data indicate that temporal differences in nutrient supply can be important in determining competitive interactions between certain weed and crop species, but that results may not be consistent. In pot experiments with corn, wild mustard [*Sinapis arvensis* L. syn. *Brassica kaber* (DC) L.C. Wheeler var. *pinnatifida* (Stokes) L.C. Wheeler], and common lambsquarters grown in mixture, Alkamper *et al.* (1979) observed that applying half a dose of NPK fertilizer at planting and the remainder at corn ear emergence improved crop growth and reduced weed growth compared with application of the full nutrient dose at planting. In field competition experiments with rice and barnyard grass, Smith and Shaw (1966) reported that application of N fertilizer after barnyard grass had reached the reproductive stage increased rice yields much more than earlier application; effects on the weed were not reported, however. In field experiments with winter wheat infested with downy brome, effects of delayed application of N fertilizer on weed growth and crop yield varied between years in a manner that was suggested to reflect variations in precipitation (Anderson, 1991). In experiments with two competing weed species (*Abutilon theophrasti* Medic. and *Datura stramonium* L.) supplied with nutrients at different times, Benner and Bazzaz (1987) concluded that "the timing of nutrient availability may influence plant growth and competition, but its effects are not readily predictable, and they appear to be less important than the effects of factors such as emergence time" (p. 243).

Crop–weed interactions can be tilted in favor of the crop through higher seeding densities (Blackshaw, 1993b; Carlson and Hill, 1985; Tollenaar *et al.*, 1994a; Wilson *et al.*, 1995) and narrower row widths (Teasdale and Frank, 1983), which often increase early season light capture by the crop (Teasdale and Frank, 1983; Yelverton and Coble, 1991; Forcella *et al.*, 1992; Egli, 1994; Board and Harville, 1994) and which may also increase early season water and nutrient capture (Sojka *et al.*, 1988). Several caveats apply to this approach, however: (i) If preemergence herbicides are not used, early season cultivation or postemergence herbicides may still be needed, in which case row spacing or plant vigor must accommodate this traffic; (ii) density cannot be increased too greatly in crops for which the size of each marketable unit is negatively density dependent; (iii) in nonirrigated areas with high evapotranspiration potential, excessive crop density and early season canopy cover may exhaust limited moisture supplies and reduce yield because of late season moisture deficits; and (iv) in humid areas, dense canopy cover may increase susceptibility to certain pathogens because of changes in microclimate.

3. Physiological Aspects   Most leguminous crop species have access to atmospheric N through bacterial fixation; nonleguminous weeds do not. Because of this physiological difference, legume crops may effectively shade and suppress associated nonleguminous weeds under conditions of low soil N availability, as was observed by Liebman (1989) in studies with barley–pea intercrops grown in competition with white mustard. Conversely, weeds may be more competitive toward legume crops under conditions of high soil N availability. Staniforth (1962) reported that weed growth and resulting soybean yield reductions were greater when N fertilizer was applied to a previous year's corn crop than when it was not applied. Dotzenko *et al.* (1969) observed that weed seed and seedling densities in sugar beet increased with increasing rates of N fertilizer applied to previous crops of barley, corn, or bean. Thus, manipulation of soil N fertility levels may be an important component of weed management in crop rotations containing legumes and other crops. For example, a fall cover crop that immobilized soil N following corn harvest might improve the competitive performance of a subsequent soybean crop.

Species can also differ in their response and sensitivity to forms of mineral nitrogen. Teyker *et al.* (1991) fertilized corn and pigweed with ammonium or nitrate nitrogen. Corn biomass was unaffected by N source; however, use of ammonium and the addition of a nitrification inhibitor reduced pigweed shoot dry weight by 75% compared with the nitrate N source. Thus, the form of nitrogen fertilizer applied may be chosen so as to benefit the crop over certain weeds.

4. Intercropping Crop species that differ spatially, temporally, and physiologically in their patterns of resource use can often capture more light, water, and nutrients when planted in mixtures (intercrops) than when planted as single-species stands (sole crops) (Ofori and Stern, 1987; Willey, 1990; Fukai and Trenbath, 1993; Keating and Carberry, 1993; Morris and Garrity, 1993a,b). Complementarity in resource use between intercrop components can limit competition between them and can lead to higher yields per unit land (Vandermeer, 1989, pp. 68–105; Willey, 1990; Liebman, 1995). Conversely, resource complementarity between intercrop components can also lead to greater resource preemption from weeds and less weed growth compared to sole crops (Liebman and Dyck, 1993b). In a field study integrating ecophysiological and agronomic measurements, Abraham and Singh (1984) observed that a sorghum–fodder cowpea intercrop captured more macronutrients, intercepted more light, suppressed weed growth more effectively, and provided higher crop yield than did a sorghum sole crop.

Because intercrop mixtures are often planted at higher total density than those used for the component species grown as sole crops, it can be difficult to ascertain whether intercrop weed control advantages are results of increased crop diversity or increased crop density. Results of replacement series experiments, in which total crop density is maintained constant for both intercrops and sole crops, indicate that some intercrops can reduce weed growth below levels obtained from sole crops of the components without an increase in total crop density (Fleck *et al.*, 1984; Sharaiha and Gliessman, 1992), whereas in other cases, weed growth in intercrops is intermediate between that obtained from the most weed-suppressive and least weed-suppressive sole crops (Shetty and Rao, 1981; Mohler and Liebman, 1987). Increases in crop density in intercropping systems definitely contribute to improved weed suppression (Liebman and Dyck, 1993b) and can increase total yield of intercrops assessed on a unit land area basis (Willey, 1979; Ofori and Stern, 1987).

Although intercropping practices are widely used in tropical areas, they are employed in limited ways in many mechanized, temperate farming systems. Examples include forage grasses and legumes intercropped with corn, soybean, barley, oat, or wheat; soybean intercropped with wheat; field pea intercropped with barley, oat, or wheat; and grasses and legumes planted as understories in fruit and nut orchards (Liebman, 1995). Expansion of opportunities to use intercropping practices would be enhanced by more attention from agricultural engineers (Vandermeer, 1989, pp. 199–201).

D. Strategies to Exploit Differential Growth Responses and Susceptibilities

Differential responses may exist between crop and weed species to abiotic and biotic environmental factors. Abiotic factors, such as temperature, moisture, fertility, and light conditions, may result in differential growth responses, reflecting differential responses at levels of resource capture, conversion, or allocation. Alternatively, biotic factors, such as pathogens, herbivores, and crop residues, may have species-specific effects in suppressing plant growth. Identification of differential responses may allow field conditions to be manipulated to the benefit of crops and detriment of weeds.

1. Temperature Conditions   The outcome of interspecific competition between plants can be markedly affected by temperature conditions (Pearcy *et al.,* 1981; Flint and Patterson, 1983; Holt, 1988; Radosevich and Roush, 1990; Wall, 1993). Although very little research effort has been directed toward exploiting differential responses to temperature in weed management strategies, results of an experiment conducted by Weaver *et al.* (1988) indicate that more attention to this subject could be worthwhile. After quantifying differences between tomato and four weed species in their germination and emergence responses to temperature, Weaver *et al.* (1988) predicted the sowing temperatures at which tomato would emerge before the weeds and consequently experience less competition from them. In addition to choosing planting times based on soil temperature as a means of improving weed control, soil temperature might be manipulated intentionally for weed management purposes. Significant differences can occur, for example, between tillage systems in early season soil temperature characteristics (Johnson and Lowery, 1985; Cox *et al.,* 1990; Dwyer *et al.,* 1995) that affect crop, and probably, weed emergence.

2. Moisture Conditions   Competitive interactions between crops and weeds may also be affected by moisture conditions. For example, Ogg *et al.* (1994) reported that mayweed chamomile (*Anthemis cotula* L.) was more aggressive in reducing growth of pea under conditions of low moisture availability than under conditions of greater moisture availability because of differential susceptibility to moisture stress between the crop and weed species. Conversely, cocklebur was more detrimental in its effects against soybean in wetter years than in drier years (Mortensen and Coble, 1989). This effect was ascribed to (i) greater reduction in vegetative growth of cocklebur, compared to soybean, under dry conditions and consequently reduced canopy growth and reduced light competition; and (ii) reduced yield potential of soybean under dry conditions. Water management and moisture conditions have a strong impact on the species composition of weeds infesting rice and the outcome of rice–weed competition (Moody and Drost, 1983; Sarkar and Moody, 1983; Seaman, 1983). Because weed species adapted to particular irrigation (flooding) regimes build up rapidly with repeated production of rice under the same set of management conditions, variation in crop establishment techniques and irrigation practices may be necessary to prevent weed proliferation where rice is grown in near-continuous monocultures.

3. Fertility Conditions   Application of nutrients (especially N) may either increase or alleviate competitive suppression of crops by weeds. Supporting Alkamper's (1976) contention that weeds are often more responsive to fertilizer application than are crops, Carlson and Hill (1985) found that application of N to wild oat–spring wheat mixtures increased wild oat growth and decreased wheat yields. Yield reductions of wheat due to interference from Italian ryegrass (*Lolium multiflorum* Lam.) were greater under N fertilized conditions than under unfertilized conditions (Appleby *et al.*, 1976). Conversely, Tollenaar *et al.* (1994b) found that applying N to corn and a mixture of weed species resulted in lower weed biomass and increased corn yields, suggesting greater crop interference against the weeds at higher fertility levels. Wells (1979) reported that application of N fertilizer to wheat in Australian experiments increased wheat yield and had no effect on the strength of weed interference against the crop. A mechanistic, resource-based explanation is needed to reconcile these opposing types of competitive outcomes. As discussed previously, timing of nutrient availability may be a critical factor affecting differential responses between crop and weed species. Results may also depend on crop cultural practices, such as planting date; the identities of the species and genotypes involved and their intrinsic growth responses to N; and differential responses to other environmental factors, including ratios of N to other nutrients (Tilman, 1982, pp. 139–189).

4. Light Conditions   Coffee (*Coffea arabica* L.), cocoa (*Theobroma cacao* L.), and tea (*Thea sinensis* L.) are well adapted to production in the shaded understory of taller tree species (Willey, 1975). Production of these crops in shade may confer weed control advantages. In an experiment conducted in Mexico, Nestel and Altieri (1992) reported that weed biomass in shaded coffee systems was substantially less than that in unshaded coffee or in control plots without coffee or shade trees. In Nicaraguan coffee plantations, shading has been observed to reduce the proliferation of particularly aggressive weed species and to shift composition of the weed community toward species that are less competitive toward the crop (C. P. Staver, personal communication).

5. Pathogens and Herbivores   Differential susceptibility to pathogens and herbivores can reduce weed growth and shift competitive interactions between crop and weed species to favor the crop. Exploitation of this principle serves as the core component of weed biocontrol strategies. A large number of microbe species are potentially available as inundative biocontrol agents for weeds (Charudattan and DeLoach, 1988; Charudattan, 1991). Less attention has been directed toward the use of herbivores as weed biocontrol agents, except in rangeland systems.

In field trials conducted in California, Pantone *et al.* (1989a,b) observed that addition of fiddleneck flower gall nematode [*Anguina amsinckiae* (Steiner and Scott, 1935) Thorne, 1961] to mixtures of wheat and coast fiddleneck (*Amsinckia intermedia* Fischer and Meyer) greatly increased wheat's ability to suppress fiddleneck's flower and seed production. Conversely, addition of the nematode reduced fiddleneck's negative effect on wheat and increased wheat seed yield. The nematode does not attack wheat and is considered a possible biocontrol agent for the weed (Pantone *et al.*, 1985).

Rhizobacteria can be selective in their effects between weeds and crops. Kennedy *et al.* (1991) conducted laboratory and fieldwork to examine the effects of *Pseudomonas fluorescens* strain D7 on the weed downy brome. Downy brome is a major pest in winter wheat regions of the northwestern United States and there is a lack of effective chemical practices with which to control it. Application of the bacterial isolate reduced germination, biomass production, and seed production of downy brome but did not have inhibitory effects on wheat. In plots infested with downy brome, wheat seed yields were increased significantly at two of three sites by application of D7. The suppressive effect of the bacterial isolate has been attributed to a phytotoxin produced by strain D7 (Tranel *et al.*, 1993; Gurusiddaiah *et al.*, 1994). Although much more work needs to be conducted to establish the widescale efficacy of this approach, it merits attention. The ability of soil microbial factors to suppress or convey resistance to crop pathogens has been documented by van Bruggen (1995); it follows that "weed suppressive soils" may exist, a phenomenon that could offer unique opportunities for management. To bring this concept to fruition, however, requires greater understanding of interactions between soil microbes, weeds, crops, and various environmental factors.

Foliar pathogens may also serve as weed biocontrol agents. Paul and Ayres (1987) conducted field experiments with lettuce (*Lactuca sativa* L.), the weed common groundsel (*Senecio vulgaris* L.), and a rust fungus (*Puccinia lagenophorae* Cooke) that attacks the weed. Lettuce fresh weight was significantly reduced by competition from uninfected groundsel at sowing densities from 250 to 65,000 seeds m^{-2} but was not reduced by rust-infected groundsel until the weed sowing density exceeded 4000 seeds m^{-2}. Lettuce yield in plots with infected groundsel was two or three times greater than that in plots in which the weed had not been infected with the pathogen. The strength of competitive suppression of groundsel by lettuce was increased by rust infection. Auld and Morin (1995) noted that mycoherbicides have great potential in tropical regions of high humidity and predictable rainfall. Furthermore, traditional production of fermented foods offers expertise that could be used on-farm or in cottage industries for the production of local supplies of weed control pathogens.

In a field study investigating impacts of insects introduced as biocontrol agents of tansy ragwort (*Senecio jacobaea* L.), both herbivory and plant competition proved important for control of this weed species (McEvoy *et al.,* 1993). At the local scale of small plots used in the experiment, competition from existing meadow vegetation acted in an additive manner with herbivory by ragwort flea beetle (*Longitarsus jacobaeae* Waterhouse) to eliminate all ragwort individuals, except those within the soil seedbank, in a period of months.

Crop management practices may determine whether herbivores attack weeds selectively. In an assessment of weed biocontrol agents in rice cropping systems, Templeton (1983) noted that tadpole shrimp (*Triopus* spp.), which feed on seedlings and disturb their roots by agitation of the soil, can have both deleterious and beneficial effects depending on planting practices. If rice is seeded directly into water, as is the practice in California, tadpole shrimp are considered pests because they feed on both rice and weed seedlings. However, the problem is avoided in transplanted rice because the plants are larger and their roots are adequately covered with soil. Populations of 20–30 tadpole shrimp m^{-2} have significantly reduced weed populations in transplanted rice fields. Field trials suggested that use of tadpole shrimp as weed biocontrol agents could reduce hand labor requirements for weed control in rice by 70–80% (Templeton, 1983).

6. Crop Residue   Crop residues may selectively reduce weed growth. For example, Liebl *et al.* (1992) demonstrated that use of a rye cover crop in conjunction with minimum tillage was a highly effective approach for limiting weed competition in soybean with minimal reliance on herbicides. Weed growth in the rye mulch system was significantly lower than that in a corn stubble system. The weed-suppressive effect of rye was attributed to allelopathy as well as shading, cooling, and physical obstruction effects of mulch on the soil surface. Compared to the corn stubble treatment, rye residue had no direct effect on soybean yield. Similarly, Mohler (1991) reported that the presence of a rye mulch decreased weed biomass and had no detrimental effect on sweet corn yield.

Anderson (1993) reported that growth of jointed goat grass (*A. cylindrica* Host) was reduced 70–85% by incorporated residues of wheat, corn, millet (*Pennisetum glaucum* L.), safflower (*Carthamus tinctorius* L.), and sorghum. Wheat growth was reduced 50–70% by the same five crop residues. Suppression of goat grass by crop residues could be overcome by adding N fertilizer, suggesting that N immobilization was responsible for the observed suppression; recovery of the wheat crop was not investigated. Anderson (1993) suggested that the combination of weed-suppressive, N immobilizing crop residues with band application of N to foster the growth of the crop was a possible management strategy to reduce weed growth while maintaining crop performance.

Soil-incorporated residues of hairy vetch and crimson clover were shown to be more detrimental for emergence and dry matter production (at 4 weeks after planting) of pitted morningglory (*Ipomoea lacunosa* L.) than of corn, an effect attributed to soluble allelopathic compounds (White *et al.*, 1989). In contrast to corn, cotton was suppressed by the legume residues, indicating that the selectivity of residue effects is species specific.

Differential responses between corn and common lambsquarters to crimson clover residue were demonstrated in a field study conducted by Dyck *et al.* (1995). Soil management treatments consisted of (i) crimson clover that was grown for 2 months before incorporation into the soil as a N source and (ii) bare fallow that was maintained for 2 months before being amended with different rates of synthetic N fertilizer. Following the clover and bare fallow regimes, sweet corn was grown alone and in combination with a fixed density of common lambsquarters in each soil management treatment. The resulting patterns of corn and common lambsquarters biomass production are shown in Fig. 5.

In the absence of the weed, crimson clover supplied an estimated 58 kg N ha^{-1} to the corn crop (data not shown) and increased corn biomass 6% compared to the unfertilized bare fallow treatment (Fig. 5A). In contrast, in the presence of the weed, corn biomass in the crimson clover treatment was 35% higher than that in the unfertilized bare fallow treatment and 20% higher than that in bare fallow plots fertilized with 45 or 90 kg N ha^{-1}. The substantial increase in corn biomass observed in the crimson clover treatment reflected a reduction in weed competition, as estimated by the difference in corn biomass between weed-free and weed-infested plots. Common lambsquarters competition reduced corn biomass by 2% in the crimson clover treatment compared to 23, 21, and 22% reductions in the bare fallow treatments amended with 0, 45, and 90 kg N ha^{-1}, respectively (Fig. 5B). Common lambsquarters biomass was least in the crimson clover treatment and increased with increasing N application in the bare fallow treatments.

Results of this study and others (Liebman and Ohno, 1997) suggest that leguminous crop residues can have important effects as selective weed-

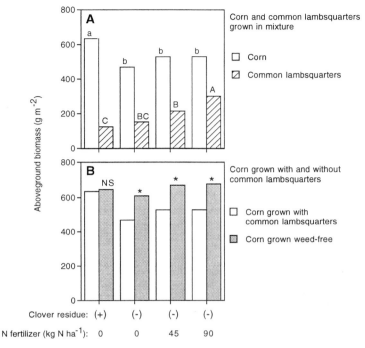

Figure 5 Effects of crimson clover residue and nitrogen (N) fertilizer on biomass of corn and common lambsquarters (*Chenopodium album* L.) grown in mixture (A); bars denoted by common letters of the same case are not significantly different at $P = 0.05$. Comparing corn growth with and without common lambsquarters (B) demonstrates yield loss due to weed competition; NS, not significant; *, significant weed effect ($P = 0.05$). Aboveground crimson clover biomass (4758 kg ha^{-1}), with a calculated N equivalence value of 58 kg N ha^{-1}, was incorporated 1 day before planting corn and common lambsquarters. Established densities were 7.3 corn plants m^{-2} and 81 common lambsquarters plants m^{-2}. Aboveground crop and weed biomass were harvested at 86 days after planting (adapted with permission from *Agric. Ecosyst. Environ.* Dyck, E., Liebman, M., and Erich, M. S., Cropweed interference as influenced by a leguminous or synthetic fertilizer nitrogen source. I. Doublecropping experiments with crimson clover, sweet corn, and lambsquarters. Copyright 1995, Vol. 56, pp. 93–108, Elsevier Science B. V., Amsterdam. The Netherlands).

suppressive agents as well as sources of N fertility. However, the efficacy of legume residues for reducing weed growth and competition may be a function of residue quality. In contrast to the high level of weed suppression obtained from immature, fresh crimson clover (Dyck and Liebman, 1994; Dyck *et al.*, 1995), weaker and less consistent weed suppression was obtained when winter-killed, weathered clover residue was used (Dyck and Liebman, 1995). More research attention is needed to identify the biological, chemical, and physical characteristics of legume residues that may result in selective and effective weed suppression.

E. Strategies Employing Weed-Tolerant and Weed-Suppressive Cultivars

Crop varieties can differ greatly in their abilities to tolerate weeds (i.e., sustain the presence of weeds with little or no yield loss compared to weed-free control treatments) and suppress weeds (i.e., reduce weed emergence, growth, and reproduction). Although few efforts have previously been made to exploit these differences for weed management, there is now increasing interest in choosing and breeding cultivars for use as components of ecologically based weed management strategies (Berkowitz, 1988; Callaway, 1992; Callaway and Forcella, 1993; Jordan, 1993b). One of the crucial steps for making better use of crop genetic resources in weed management is the identification of heritable characteristics that confer greater ability to tolerate or suppress weeds.

Potter and Jones (1977) compared three crop and six weed species and reported that partitioning of dry matter into new leaf area was highly correlated with rapid growth. Thus, the use of crop varieties with high initial leaf area partitioning and expansion rates may be particularly important in weed management strategies (Forcella, 1987; Wortmann, 1993; Callaway and Forcella, 1993). Other characteristics, such as rates of shoot and root extension and ion absorption efficiencies, may also be important (Seibert and Pearce, 1993).

Callaway and Forcella (1993) selected for a soybean genotype with a high rate of leaf area expansion and were successful in developing a line that was superior in suppressing weed growth and providing high yield under both weed-infested and weed-free conditions. Blackshaw (1993a) showed that semidwarf cultivars of winter wheat were more susceptible to yield reduction by downy brome than were taller cultivars; taller cultivars intercepted more light and were better able to suppress growth of downy brome. However, Wicks *et al.* (1986) identified short statured wheat cultivars with superior ability to suppress summer annual weeds. Comparisons of different rice cultivars indicated that weed growth was least and weed-infested crop yield was highest for taller, leafier cultivars (Garrity *et al.,* 1992). Kropff and Spitters (1992) conducted evaluations of the INTERCOM model that indicated morphological characteristics, such as leaf area expansion rate, specific leaf area, and plant height, rather than physiological characteristics such as maximal photosynthetic rate largely determined ability to maintain yield in the presence of weeds. Lindquist and Kropff (1996) used INTERCOM to simulate competition for light between barnyard grass and irrigated rice, evaluating leaf area relative growth rates, i.e., LAI per LAI per growing degree day (RGRL). In these simulations, several items of importance to plant breeders wishing to improve the competitive ability of rice cultivars were noted: (i) Increasing rice RGRL improved tolerance and barnyard grass suppressive ability, (ii) benefits of improved tolerance and suppressive ability varied with barnyard grass density, and (iii) the magnitude of variation in rice RGRL required to improve tolerance and suppressive ability was found to be within the range of values expected to exist in rice genotypes currently available for breeding work.

In addition to the development of new cultivars, weed management efforts could benefit from information on the competitive abilities of currently available cultivars. Herbicide dose, for example, could be reduced for cultivars that are highly competitive with weeds (Christensen, 1994). Alternatively, several management tactics may have to be invoked if an uncompetitive cultivar is planted into a field that contains high weed pressure. In Maine, dry bean farmers typically grow any of four locally adapted varieties—"Jacob's Cattle", "Marafax", "Soldier", or "Yellow Eye"— although others such as "Midnight Black Turtle" could be chosen. Although the locally adapted varieties show a high degree of morphological similarity, they differ in their abilities to tolerate and suppress wild mustard (*S. arvensis* L.), a common weed problem in Maine (Fig.6; E. R. Gallandt, unpublished results). Black Turtle, Yellow Eye, and Jacobs Cattle, for example, showed greater tolerance to wild mustard (Fig. 6, top), as well as greater suppression of wild mustard (Fig. 6; bottom) compared to Marafax. Yield loss was greatest with Soldier (Fig. 6, top). In this instance, although choice of variety will continue to be market driven, the decision should probably affect a grower's weed management plan; e.g., cultivation efforts should be increased in a crop of Marafax or Soldier beans relative to Black Turtle, Yellow Eye, or Jacob's Cattle if fields are known to contain wild mustard.

The recent mechanistic experiments of Kropff and Spitters (1992) and Callaway and Forcella (1993), by identifying specific breeding objectives, could encourage future research on competitive cultivars. Because the yield- and weed-related costs and benefits of using weed-tolerant and weed-suppressive cultivars are not known (Jordan, 1993b), studies that compare conventional and competitive cultivars in various management systems are needed to determine the potential role of competitive cultivars in ecologically based weed management systems.

V. Many Little Hammers

As pointed out by Medd (1987), weeds are influenced by almost every aspect of crop production, including the obvious "direct controls" of herbicide application and cultivation as well as "indirect controls," such as crop genotype, time of sowing, nutrition, and other cultural factors. By understanding and organizing these direct and indirect controls into crop-

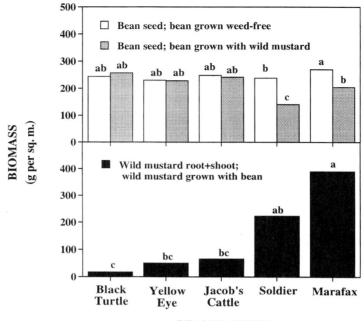

Figure 6   Seed yield of five bean (*Phaseolus vulgaris* L.) varieties grown either weed-free or with wild mustard (*Sinapis arvensis* L.) (top) and total biomass (root plus shoot) of wild mustard grown with the different bean varieties (bottom). Plants were harvested 104 days after planting. Bean varieties were established at 24 plants m^{-2} and were grown with wild mustard at 6 plants per m^2 or under weed-free conditions. Weed-infested bean seed yields were inversely correlated with mustard total biomass ($P < 0.001$) (E. R. Gallandt, unpublished results).

ping strategies that reduce weed problems, environmental and economic objectives of weed management may be more readily met.

There are significant advantages to reliance on a combination of different methods, i.e., the use of "many little hammers", over a single pest control tactic, i.e., "one large hammer". The use of a combination of methods can lead to (i) acceptable control through the additive, synergistic, or cumulative action of tactics that may not be effective when used alone; (ii) reduced risk of crop failure or serious loss by spreading the burden of protection across several methods; and (iii) minimal exposure to any one tactic and consequently reduced rates at which pests adapt and become resistant to the methods (Bottrell and Weil, 1995). The philosophy of using a combination of methods to regulate pest populations is congruent with integrated pest management concepts that were introduced in the early

1970s (Bottrell, 1979). Although the philosophy of using many little hammers to manage weeds does not exclude use of direct controls (the large hammers), it shifts the focus of weed management to the many indirect controls and many possible interactions that can lead to successful management.

At the heart of this approach is the hypothesis that multiple and temporally variable stresses can be imposed on weeds that will result in acceptable crop yield and quality. This approach has been formalized in the "cumulative stress" approach for knapweed (*Centaurea* spp.) control in North America (Müller-Schärer and Schroeder, 1993). The impetus in this instance is likely the lack of economically viable "single hammer" approaches in rangeland ecosystems. The approach could prove useful, however, in all agroecosystems.

Modeling efforts to understand the combined action of plant stresses, although focused on a crop response to pests (e.g., Browde *et al.,* 1994), may help select specific tactics to impose on weeds. Mechanistic models link pest or damage levels to an appropriate physiological or demographic coupling point, e.g., leaf area index, stand density, intercepted light, photosynthesis, assimilated carbon, translocation rate, growth of different organs, and leaf senescense. Simulations in both rice (Pinnschmidt *et al.,* 1995) and potato (Johnson, 1992) showed that the effects of multiple stresses were less than additive and suggested that important pest–pest interactions were occurring. Although certain combinations may evoke a synergistic response, other tactics may be less than additive or antagonistic; many stresses may therefore be required to achieve a cumulative response and attain a desired level of weed suppression.

Because considerable genetic variability may exist within and among weed populations (Barrett, 1988), weeds often have the capacity to evolve resistance to or tolerance of management practices. Although attention has focused largely on variation in herbicide resistance, differences in weed responses to other factors may reflect genetic variation. For example, Tardif and Leroux (1992) observed significant differences among quackgrass genotypes in their shoot and rhizome responses to N application. The evolution of weed genotypes that mimic crop coloration, phenology, seed size, and other characteristics has been noted for a number of weed species (Barrett, 1983). Thus, changes in weed management practices can be expected to be met with dynamic responses in weeds that result in their continued adaptation and persistence. The use of multiple tactics in a concerted manner may be particularly important in limiting the rate of these changes in weed populations and in preventing shifts in the species composition of weed communities toward particularly noxious species. In the study conducted by Covarelli and Tei (1988), for example, using crop rotation (wheat–corn) rather than monoculture (continuous corn) reduced both the total weed density and the dominance of the weed community by the most abundant species (barnyard grass).

As the number of species within a community increases so does the potential for interactions. In a study of weed dynamics in potato cropping systems, Liebman *et al.* (1996) found that significant interspecific interactions occurred among weed species. *Brassica rapa, Sinapis arvensis,* and *Raphanus raphanistrum,* as a group, suppressed the growth of lambsquarters, the other dominant weed in the system; this interaction explained why the authors found either no relationship or a weak relationship between lambsquarters plant density and shoot biomass. Clements *et al.* (1994), in an insightful review, discuss the value of viewing weeds as a community rather than as individual species. Through increased understanding of the relationships between management and community structure and function, weed diversity may be strategically manipulated to suit crop production purposes.

What might a many little hammers weed management strategy look like? A two-hammer approach is common and most often includes the direct controls—herbicides and cultivation (Buhler *et al.,* 1992, 1993; Poston *et al.,* 1992; Mulder and Doll, 1993; Burnside *et al.,* 1994). In general, these studies have shown that herbicide applications and/or rates can be reduced if combined with well-timed cultivation. Less common are studies that include two indirect controls, for example, tactics aimed at processes such as interspecific competition, seed production, dispersal, seed survival, or seedling recruitment.

Combinations of direct controls with indirect tactics that are focused on interspecific competition and seedling mortality have shown that the competitive ability of a crop can assume some of the burden of weed control from a herbicide application. In Denmark, with recent political directives to reduce herbicide use, Christensen (1994) has shown that weeds may be controlled effectively in "competitive" cereal varieties with lower herbicide doses than those required for less competitive varieties. For example, spring rape (*Brassica napus* L.), used as a surrogate weed, was controlled in a competitive spring barley variety with one-third the herbicide required for control in the least competitive variety (Christensen *et al.,* 1990).

Additional bitactical strategies may or may not include herbicides combined with other tactics. Herbicides or plant growth regulators have been employed at low rates to enhance biocontrol programs using weed-feeding insects (Messersmith and Adkins, 1995) and weed pathogens (Scheepens, 1987; Hodgson *et al.,* 1988; Wymore and Watson, 1989). Insect defoliation of weeds and interspecific competition against weeds may also prove to be a useful combination of tactics (Ang *et al.,* 1994).

Examples of three little hammers, although relatively infrequent in the literature, demonstrate some creative management strategies. The common feature of most three-tactic strategies is that they exploit interspecific competition, a pivotal process that can influence both weed seedling mortality and seed production. Teasdale (1995) found that reduced herbicide rates (1/4X) provided weed control and corn yield as good as full herbicide rates (X) if corn was grown with narrow row spacing and the crop population was doubled. Thomsen *et al.* (1996) reported that suppression of yellow starthistle (*Centaurea solstitialis* L.) growth and seed production through grazing and mowing was enhanced by planting subterranean clover (*Trifolium subterraneum* L.) to compete against the weed. Other examples of three-tactic strategies, which have achieved varying degrees of success, include combinations of herbicide banding, cultivation, and intercropping (Samson and Coulman, 1989); crop cultivar, row spacing, and seeding density (Malik *et al.*, 1993); and crop cultivar, tillage intensity, and mulch (Shilling *et al.*, 1995).

The closest one may come to viewing the action of many little hammers is probably demonstrated by studies that examine the effects of crop rotation on crop–weed interactions and weed population dynamics. Rotation increases system diversity, and hence the number of possible ecological interactions and stresses, and is often successful at suppressing weeds (Liebman and Dyck, 1993b; Liebman and Ohno, 1997). Although rotation has been examined in combination with practices such as tillage (Buhler *et al.*, 1994; Blackshaw *et al.*, 1994) or tillage plus herbicides (Young *et al.*, 1994), rotation itself can affect many demographic processes, supporting the notion that many little hammers can be useful for ecologically managing weeds.

VI. Summary and Conclusions

Conventional weed management in the latter half of the 20th century has had a relatively narrow focus on methods to reduce weed seedling recruitment (through the use of preemergence herbicides) and/or weed seedling survival (through the use of postemergence herbicides and cultivation). The many other factors in agroecosystems known to influence weed population dynamics and crop–weed interactions have been largely ignored. The increased incidence of herbicide resistance and greater demands for environmental stewardship signal that a shift from chemical technologies toward ecologically based strategies is now desirable, and that a broader focus for weed management efforts is necessary. Broader, well-integrated weed management strategies that exploit the full range of factors determining weed performance need to include tactics that affect crop–weed competition, weed seed production, seed dispersal, and seed survival. The use of such tactics could (i) lessen the burden placed on management at the seedling stage; (ii) reduce requirements for off-farm inputs, and (iii) retard the adaptation of weeds to the stresses imposed upon them.

Shifting to more integrated approaches that exploit diverse stress and mortality factors will not be easy. In contrast to the simple testing and development of single- or two-tactic management strategies, a major difficulty facing the designers of ecologically based weed management systems is the complexity of interactions involving multiple stresses and mortality factors. As noted by Levins and Vandermeer (1990): "[agroecosystems] might very well exhibit behavior so complicated that even a deep understanding of the way the system is put together will not necessarily ensure that we can predict what will happen under different management strategies" (p. 349).

Comparative cropping systems studies could play an important role in understanding and testing weed management via many little hammers. A large-scale experiment with a historical database can serve as a site for evaluating discoveries made in smaller component studies, which can examine mechanisms of interactions between two or three factors. Perturbation of weed dynamics would demonstrate the effect of a tactic within the context of many interacting factors. This hierarchical approach to research, which includes studies at levels ranging from single factors to the level of the cropping system, is critical to the development of multiple stress weed management systems.

Inspired by the desire to compare conventional with alternative management strategies within an ecosystem context, cropping systems studies are being conducted in the United States (King and Buchanan, 1993; Smolik *et al.*, 1993; Temple *et al.*, 1994; Alford *et al.*, 1996) and Europe (Vereijken, 1989; Besson and Niggli, 1991; Lebbink *et al.*, 1994). Weed scientists have, as yet, played little or no role in evaluating interactions within these systems. We hope that situation will soon change.

Acknowledgments

This chapter is Contribution No. 1947 of the Maine Agricultural and Forest Experiment Station. We thank F. A. Drummond, J.-L. Jannink, N. Jordan, J. L. Lindquist, B. D. Maxwell, D. A. Mortensen, and C. P. Staver for critical reviews of the manuscript and their suggestions for improvements. Errors of omission and commission remain our responsibility.

References

Abernathy, J. R., and Bridges, D. C. (1994). Research priority dynamics in weed science. *Weed Technol.* **8,** 396–399.

Abraham, C. T., and Singh, S. P. (1984). Weed management in sorghum–legume intercropping systems. *J. Agric. Sci. Cambridge* **103**, 103–115.

Akobundu, I. O. (1987). "Weed Science in the Tropics: Principles and Practices." Wiley, Chichester, UK.

Akobundu, I. O. (1991). Weeds in human affairs in sub–Saharan Africa: Implications for sustainable food production. *Weed Technol.* **5**, 680–690.

Aldrich, R. J. (1984). "Weed–Crop Ecology: Principles in Weed Management. Breton, North Scituate, MA.

Aldrich, R. J. (1987). Predicting crop yield reductions from weeds. *Weed Technol.* **1**, 199–206.

Alford, A. R., Corson, S., Drummond, F. A., Gallandt, E. R., Groden, E., Lambert, D. H., Liebman, M., Marra, M. C., McBurnie, J. C., Porter, G. A., and Salas, B. (1996). "The Ecology, Economics, and Management of Potato Cropping Systems: A Report of the First Four Years of the Maine Potato Ecosystem Project." Maine Agricultural and Forest Experiment Station Bulletin No. 843, University of Maine, Orono.

Alkamper, J. (1976). Influences of weed infestation on effect of fertilizer dressings. *Pflanzen.-Nachr.* **29**, 191–235.

Alkamper, J., Pessios, E., and Long, D. V. (1979). Einfluss der Dungung auf die Entwicklung und Nahrstoffaufnahme Verschiedener Unkrauter in Mais. *Proc. Eur. Weed Res. Soc.* **1979**, 181–192.

Anderson, R. L. (1991). Timing of nitrogen application affects downy brome (*Bromus tectorum*) growth in winter wheat. *Weed Technol.* **5**, 582–585.

Anderson, R. L. (1993). Crop residue reduces jointed goatgrass (*Aegilops cylindrica*) seedling growth. *Weed Technol.* **7**, 717–722.

Andow, D. A., and Hidaka, K. (1989). Experimental natural history of sustainable agriculture: Syndromes of production. *Agric. Ecosyst. Environ.* **27**, 447–462.

Ang, B. N., Kok, L. T., Holtzman, G. I., and Wolf, D. D. (1994). Canada thistle (*Cirsium arvense*) response to simulated insect defoliation and plant competition. *Weed Sci.* **42**, 403–410.

Appleby, A. P., Olsen, P. O., and Colbert, D. R. (1976). Winter wheat yield reduction from interference by Italian ryegrass. *Agron. J.* **68**, 463–466.

Ascard, J. (1994). Soil cultivation in darkness reduces weed emergence. *Acta Hort.* **372**, 167–177.

Ascard, J. (1995a). Dose–response models for flame weeding in relation to plant size and density. *Weed Res.* **34**, 377–385.

Ascard, J. (1995b). Effects of flame weeding on weed species at different developmental stages. *Weed Res.* **35**, 397–411.

Aspelin, A. L. (1994). Pesticide industry sales and usage—1992 and 1993 market estimates. U.S. Environmental Protection Agency, Washington, DC.

Auld, B. A., and Morin, L. (1995). Constraints in the development of bioherbicides. *Weed Technol.* **9**, 638–652.

Auld, B. A., and Tisdell, C. A. (1988). Influence of spatial distribution of weeds on crop yield loss. *Plant Prot. Q.* **3**(2), 81.

Ballare, C. L., Scopel, A. L., Ghersa, C. M., and Sanchez, R. A. (1987). The population ecology of *Datura ferox* in soybean crops. A simulation approach incorporating seed dispersal. *Agric. Ecosyst. Environ.* **19**, 177–188.

Barrett, S. C. H. (1983). Crop mimicry in weeds. *Econ. Bot.* **37**(3), 255–282.

Barrett, S. C. H. (1988). Genetics and evolution of agricultural weeds. *In* "Weed Management in Agroecosystems: Ecological Approaches" (M. A. Altieri and M. Liebman, eds.), pp. 57–75. CRC Press, Boca Raton, FL.

Bauer, T. A., Mortensen, D. A., Wicks, G. A., Hayden, T. A., and Martin, A. R. (1991). Environmental variability associated with economic thresholds for soybeans. *Weed Sci.* **39**, 564–569.

Bellinder, R. R., Gummesson, G., and Karlsson, C. (1994). Percentage-driven government mandates for pesticide reduction: The Swedish model. *Weed Technol.* **8**, 350–359.

Bender, J. (1994). "Future Harvest: Pesticide-Free Farming." Univ. of Nebraska Press, Lincoln.

Benner, B. L., and Bazzaz, F. A. (1987). Effects of timing of nutrient addition on competition within and between two annual plant species. *J. Ecol.* **75**, 229–245.

Berkowitz, A. R. (1988). Competition for resources in weed–crop mixtures. *In* "Weed Management in Agroecosystems: Ecological Approaches" (M. A. Altieri and M. Liebman, eds.), pp. 89–120. CRC Press, Boca Raton, FL.

Besson, J. M., and Niggli, U. (1991). DOK-Versuch: Vergleichende langzeit-untersuchungen in den drei anbausystemen biologisch-dynamisch, organisch-biologisch und konventionell. *Schweiz. Landw. Forsch.* **31**, 79–109.

Blackshaw, R. E. (1993a). Downy brome (*Bromus tectorum*) density and relative time of emergence affect interference in winter wheat (*Triticum aestivum*). *Weed Sci.* **41**, 551–556.

Blackshaw, R. E. (1993b). Safflower (*Carthamus tinctorius*) density and row spacing effects on competition with green foxtail (*Setaria viridis*). *Weed Sci.* **41**, 403–408.

Blackshaw, R. E., Larney, F. O., Lindwall, C. W., and Kozub, G. C. (1994). Crop rotation and tillage effects on weed populations on the semi-arid Canadian prairies. *Weed Sci.* **8**, 231–237.

Board, J. E., and Harville, B. G. (1994). A criterion for acceptance of narrow row culture in soybean. *Agron. J.* **86**, 1103–1106.

Bottrell, D. G. (1979). "Integrated Pest Management." Council on Environmental Quality, U.S. Government Printing Office, Washington, DC.

Bottrell, D. G., and Weil, R. R. (1995). Protecting crops and the environment: Striving for durability. *In* "Agriculture and Environment: Bridging Food Production and Environmental Protection in Developing Countries" (A. S. R. Juo and R. D. Freed, eds.), ASA Spec. Publ. No. 60, pp. 63–64. American Society of Agronomy, Madison, WI.

Boydston, R. A., and Hang, A. (1995). Rapeseed (*Brassica napus*) green manure crop suppresses weeds in potato (*Solanum tuberosum*). *Weed Technol.* **9**, 669–675.

Bozsa, R. C., and Oliver, L. R. (1993). Shoot and root interference of common cocklebur (*Xanthium strumarium*) and soybean (*Glycine max*). *Weed Sci.* **41**, 34–37.

Brain, P., and Cousens, R. (1990). The effect of weed distribution on predictions of yield loss. *J. Appl. Ecol.* **27**, 735–742.

Bridges, D. C., and Anderson, R. L. (1992). Crop losses due to weeds in the United States by state. In "Crop Losses Due to Weeds in the United States—1992" (D. C. Bridges, ed.), pp. 1–60. Weed Science Society of America, Champaign, IL.

Browde, J. A., Pedigo, L. P., Owen, M. D. K., and Tylka, G. L. (1994). Soybean yield and pest management as influenced by nematodes, herbicides, and defoliating insects. *Agron. J.* **86**, 601–608.

Brust, G. E., and House, G. J. (1988). Weed seed destruction by arthropods and rodents in low–input soybean agroecosystems. *Am. J. Altern. Agric.* **3**, 19–35.

Buhler, D. D. (1992). Population dynamics and control of annual weeds in corn (*Zea mays*) as influenced by tillage systems. *Weed Sci.* **40**, 241–248.

Buhler, D. D. (1995). Influence of tillage systems on weed population dynamics and management in corn and soybean in the central USA. *Crop Sci.* **35**, 1247–1258.

Buhler, D. D., and Oplinger, E. S. (1990). Influence of tillage systems on annual weed densities and control in solid–seeded soybean (*Glycine max*). *Weed Sci.* **38**, 158–165.

Buhler, D. D., Gunsolus, J. L., and Ralston, D. F. (1992). Integrated weed management techniques to reduce herbicide inputs in soybean. *Agron. J.* **84**, 973–978.

Buhler, D. D., Gunsolus, J. L., and Ralston, D. F. (1993). Common cocklebur (*Xanthium strumarium*) control in soybean (*Glycine max*) with reduced bentazon rates and cultivation. *Weed Sci.* **41**, 447–453.

Buhler, D. D., Stoltenberg, D. E., Becker, R. L., and Gunsolus, J. L. (1994). Perennial weed populations after 14 years of variable tillage and cropping practices. *Weed Sci.* **42**, 205–209.

Burnside, O. C., Ahrens, W. H., Holder, B. J., Wiens, M. J., Johnson, M. M., and Ristau, E. A. (1994). Efficacy and economics of various mechanical plus chemical weed control systems in dry beans (*Phaseolus vulgaris*). *Weed Technol.* **8**, 238–244.

Bussler, B. H., Maxwell, B. D., and Puettmann, K. J. (1995). Using plant volume to quantify interference in corn (*Zea mays*) neighborhoods. *Weed Sci.* **43**, 586–594.

Callaway, M. B. (1992). A compendium of crop varietal tolerance to weeds. *Am. J. Altern. Agric.* **7**, 169–180.

Callaway, M. B., and Forcella, F. (1993). Crop tolerance to weeds. In "Crop Improvement for Sustainable Agriculture" (M. B. Callaway and C. A. Francis, eds.), pp. 100–131. Univ. of Nebraska Press, Lincoln.

Cardina, J. (1995). Biological weed management. In "Handbook of Weed Management Systems" (A. E. Smith, ed.), pp. 279–342. Dekker, New York.

Cardina, J., Sparrow, D. H., and McCoy, E. L. (1995). Analysis of spatial distribution of common lambsquarters (*Chenopodium album*) in no-till soybean (*Glycine max*). *Weed Sci.* **43**, 258–268.

Carlson, H. L., and Hill, J. E. (1985). Wild oat (*Avena fatua*) competition with spring wheat: Effects of nitrogen fertilization. *Weed Sci.* **34**, 29–33.

Chandler, J. M. (1991). Estimated losses of crops to weeds. In "CRC Handbook of Pest Management in Agriculture, Vol. 1" (D. Pimentel, ed.), pp. 53–65. CRC Press, Boca Raton, FL.

Charudattan, R. (1991). The mycoherbicide approach with plant pathogens. In "Microbial Control of Weeds" (D. O. TeBeest, ed.), pp. 24–57. Chapman & Hall, New York.

Charudattan, R., and DeLoach, C. J. (1988). Management of pathogens and insects for weed control in agroecosystems. In "Weed Management in Agroecosystems: Ecological Approaches" (M. A. Altieri and M. Liebman, eds.), pp. 245–264. CRC Press, Boca Raton, FL.

Chikoye, D., and Swanton, C. J. (1995). Evaluation of three empirical models depicting *Ambrosia artemisiifolia* competition in white bean. *Weed Res.* **35**, 421–428.

Chikoye, D., Weise, S. F., and Swanton, C. J. (1995). Influence of common ragweed (*Ambrosia artemisiifolia*) time of emergence and density on white bean (*Phaseolus vulgaris*). *Weed Sci.* **43**, 375–380.

Christensen, S. (1994). Crop weed competition and herbicide performance in cereal species and varieties. *Weed Res.* **34**, 29–36.

Christensen, S., Streibig, J. C., and Haas, H. (1990). Interaction between herbicide activity and weed suppression by spring barley varieties. *Proc. Eur. Weed Res. Soc.* **1990**, 367–374.

Clements, D. R., Weise, S. F., and Swanton, C. J. (1994). Integrated weed management and seed species diversity. *Phytoprotection* **75**, 1–19.

Cochran, V. L., Morrow, L. A., and Schirman, R. D. (1990). The effect of N placement on grass weeds and winter wheat responses in three tillage systems. *Soil Tillage Res.* **18**, 347–355.

Conway, G. R., and Barbier, E. B. (1990). "After the Green Revolution: Sustainable Agriculture for Development." Earthscan, London.

Cousens, R., and Mortimer, M. (1995). "Weed Population Dynamics." Cambridge Univ. Press, Cambridge, UK.

Cousens, R., Brain, P., O'Donovan, J. T., and O'Sullivan, P. A. (1987). The use of biologically realistic equations to describe the effects of weed density and relative time of emergence on crop yield. *Weed Sci.* **35**, 720–725.

Covarelli, G., and Tei, F. (1988). Effet de la rotation culturale sur la flore adventice du mais. In "VIIIeme Colloque International sur la Biologie, l'Ecologie et la Systematique des Mauvaises Herbes, Vol. 2," pp. 477–484. Comite Francais de Lutte Contre Mauvaises Herbes, Paris European Weed Research Society, Leverkusen, Germany.

Cox, W. J., Zobel, R. W., Van Es, H. M., and Otis, D. J. (1990). Tillage effects on some soil physical and corn physiological characteristics. *Agron. J.* **82**, 806–812.

Cruz, R.T., O'Toole, J. C., and Moody, K. (1983). Leaf water potential of weeds and rice (*Oryza sativa*). *Weed Sci.* **31**, 410–414.

Daar, S. (1987). Flame weeding on European farms. *IPM Practitioner* **9**(3), 1–4.

Derksen, D. A., Lafond, G. P., Thomas, A. G., Loeppky, H. A., and Swanton, C. J. (1993). Impact of agronomic practices on weed communities: Tillage systems. *Weed Sci.* **41**, 409–417.

Dinham, B. (1993). "The Pesticide Hazard: A Global Health and Environmental Audit." Zed Books, London.

DiTomaso, J. M. (1995). Approaches for improving crop competitiveness through manipulation of fertilization strategies. *Weed Sci.* **43,** 491–497.

Dotzenko, A. D., Ozkan, M., and Storer, K. R. (1969). Influence of crop sequence, nitrogen fertilizer, and herbicides on weed seed populations in sugar beetfields. *Agron. J.* **61,** 34–37.

Dwyer, L. M., Ma, B. L., Hayhoe, H. N., and Culley, J. L. B. (1995). Tillage effects on soil temperature, shoot dry matter accumulation and corn grain yield. *J. Sust. Agric.* **5,** 85–99.

Dyck, E., and Liebman, M. (1994). Soil fertility management as a factor in weed control: The effect of crimson clover residue, synthetic nitrogen fertilizer, and their interaction on emergence and early growth of lambsquarters and sweet corn. *Plant Soil* **167,** 227–237.

Dyck, E., and Liebman, M. (1995). Crop–weed interference as influenced by a leguminous or synthetic fertilizer nitrogen source. II. Rotation experiments with crimson clover, field corn, and lambsquarters. *Agric. Ecosyst. Environ.* **56,** 109–120.

Dyck, E., Liebman, M., and Erich, M. S. (1995). Crop–weed interference as influenced by a leguminous or synthetic fertilizer nitrogen source. I. Doublecropping experiments with crimson clover, sweet corn, and lambsquarters. *Agric. Ecosyst. Environ.* **56,** 93–108.

Eadie, A. G., Swanton, C. J., Shaw, J. E., and Anderson, G. W. (1992). Banded herbicide applications and cultivation in a modified no-till corn (*Zea mays*) system. *Weed Technol.* **6,** 535–542.

Egley, G. H. (1983). Weed seed and seedling reduction by soil solarization with transparent polyethylene sheets. *Weed Sci.* **31,** 404–409.

Egli, D. B. (1994). Mechanisms responsible for soybean yield response to equidistant planting patterns. *Agron. J.* **86,** 1046–1049.

Elmore, C. L., Roncoroni, J. A., and Giraud, D. D. (1993). Perennial weeds respond to control by soil solarization. *Calif. Agric.* **47**(1), 19–22.

Evans, G. C. (1972). "The Quantitative Analysis of Plant Growth." Univ. of California Press, Berkeley.

Exley, D. M., and Snaydon, R. W. (1992). Effects of nitrogen fertilizer and emergence date on root and shoot competition between wheat and blackgrass. *Weed Res.* **32,** 175–182.

Firbank, L. G., Mortimer, A. M., and Putwain, P. D. (1985). *Bromus sterilis* in winter wheat: A test of a predictive population model. *Asp. Appl. Biol.* **9,** 59–66.

Fleck, N. G., Machado, C. M. N., and De Souza, R. S. (1984). Eficiencia da consorciacao de culturas no controle de plantas daninhas. *Pesq. Agropec. Brasil.* **19,** 591–598.

Flint, E. P., and Patterson, D. T. (1983). Interference and temperature effects on growth in soybean (*Glycine max*) and associated C_3 and C_4 weeds. *Weed Sci.* **31,** 193–199.

Forcella, F. (1987). Tolerance of weed competition associated with high leaf area expansion rate in tall fescue. *Crop Sci.* **27,** 146–147.

Forcella, F., and Lindstrom, M. J. (1988). Movement and germination of weed seeds in ridge-till crop production systems. *Weed Sci.* **36,** 56–59.

Forcella, F., Westgate, M. E., and Warnes, D. D. (1992). Effect of row width on herbicide and cultivation requirements in row crops. *Am. J. Altern. Agric.* **7,** 161–167.

Forcella, F., Eradat–Oskoui, K., and Wagner, S. W. (1993). Application of weed seedbank ecology to low input crop management. *Ecol. Appl.* **3,** 74–83.

Froud-Williams, R. J. (1988). Changes in weed flora with different tillage and agronomic management systems. *In* "Weed Management in Agroecosystems: Ecological Approaches" (M. A. Altieri and M. Liebman, eds.), pp. 213–236. CRC Press, Boca Raton, FL.

Fukai, S., and Trenbath, B. R. (1993). Processes determining intercrop productivity and yields of component crops. *Field Crops Res.* **34,** 247–271.

Garrity, D. P., Movillon, M., and Moody, K. (1992). Differential weed suppression ability in upland rice cultivars. *Agron. J.* **84,** 586–591.

Gassman, P. W. (1993). Pesticide fate research trends within a strict regulatory environment: The case of Germany. *J. Soil Water Conserv.* **48**, 178–187.

Gianessi, L. P., and Puffer, C. (1990). "Herbicide Use in the United States." Resources for the Future, Washington, DC.

Gill, G. S. (1995). Development of herbicide resistance in annual ryegrass (*Lolium rigidum* Gaud.) populations in the cropping belt of Western Australia. *Aust. J. Exp. Agric.* **35**, 67–72.

Gliessman, S. R. (1986). Plant interactions in multiple cropping systems. *In* "Multiple Cropping Systems" (C. A. Francis, ed.), pp. 82–95. Macmillian, New York.

Goldberg, D. E. (1990). Components of resource competition in plant communities. *In* "Perspectives on Plant Competition" (J. B. Grace and D. Tilman, eds.), pp. 27–50. Academic Press, San Diego.

Goldberg, D. E., and Landa, K. (1991). Competitive effect and response: Hierarchies and correlated traits in the early stages of competition. *J. Ecol.* **79**, 1013–1030.

Goolsby, D. A., Coupe, R. C., and Markovchick, D. J. (1991). Distribution of selected herbicides and nitrate in the Mississippi River and its major tributaries, April through June 1991. U.S. Geological Survey, Water Resources Investigations Report 91-4163, Denver, CO.

Goolsby, D. A., Battaglin, W. A., and Thurman, E. M. (1993). Occurrence and transport of agricultural chemicals in the Mississippi River basin, July through August 1993. U.S. Geological Survey Circular 1120-C. U.S. Government Printing Office, Washington, DC.

Graf, B., and Hill, J. E. (1992). Modelling the competition for light and nitrogen between rice and *Echinochloa crus–galli*. *Agric. Syst.* **40**, 345–359.

Graham, P. L., Steiner, J. L., and Wiese, A. F. (1988). Light absorption and competition in mixed sorghum–pigweed communities. *Agron. J.* **80**, 415–418.

Grattan, S. R., Schwankl, L. J., and Lanini, W. T. (1988). Weed control by subsurface drip irrigation. *Calif. Agric.* **42**(3), 22–24.

Gurusiddaiah, S., Gealy, D. R., Kennedy, A. C., and Ogg, A. G. (1994). Isolation and characterization of metabolites from *Pseudomonas fluorescens*-D7 for control of downy brome (*Bromus tectorum*). *Weed Sci.* **42**, 492–501.

Gutfeld, R. (1993). U.S. commits to cut use of pesticides as study finds high levels for children. *The Wall Street Journal*, June 28.

Hall, M. R., Swanton, C. J., and Anderson, G. W. (1992). The critical period of weed control in grain corn (*Zea mays* L.). *Weed Sci.* **40**, 441–447.

Hallberg, G. R. (1989). Pesticide pollution of groundwater in the humid United States. *Agric. Ecosyst. Environ.* **26**, 299–367.

Hamill, A. S., Surgeoner, G. A., and Roberts, W. P. (1994). Herbicide reduction in North America: In Canada, an opportunity of motivation and growth in weed management. *Weed Technol.* **8**, 366–371.

Harvey, S. J., and Forcella, F. (1993). Vernal seedling emergence model for common lambsquarters (*Chenopodium album*). *Weed Sci.* **41**, 309–316.

Hilje, L., Castillo, L. E., Thrupp, L., and Wesseling, I. (1992). "El Uso de los Plaguicidas en Costa Rica." Heliconia, Editorial Universidad Estatal a Distancia, San Jose, Costa Rica.

Hodgson, R. H., Wymore, L. A., Watson, A. K., Snyder, R. H., and Collette, A. (1988). Efficacy of *Colletotrichum coccodes* and thidiazuron for velvetleaf (*Abutilon theophrasti*) control in soybean (*Glycine max*). *Weed Technol.* **2**, 473–480.

Holt, J. S. (1988). Ecological and physiological characteristics of weeds. *In* "Weed Management in Agroecosystems: Ecological Approaches" (M. A. Altieri and M. Liebman, eds.), pp. 7–23. CRC Press, Boca Raton, FL.

Holt, J. S. (1992). History of identification of herbicide resistant weeds. *Weed Technol.* **6**, 615–620.

Holt, J. S., and LeBaron, H. M. (1990). Significance and distribution of herbicide resistance. *Weed Technol.* **4**, 141–149.

Horowitz, M., Regev, Y., and Herzlunger, G. (1983). Solarization for weed control. *Weed Sci.* **31,** 170–179.

House, G. J., and Brust, G. E. (1989). Ecology of low-input, no-tillage agroecosystems. *Agric. Ecosyst. Environ.* **27,** 331–345.

Hunt, R. (1982). "Plant Growth Curves: The Functional Approach to Plant Growth Analysis." University Park Press, Baltimore, MD.

Johnson, G. A., Mortensen, D. A., and Martin, A. R. (1995). A simulation of herbicide use based on weed spatial distribution. *Weed Res.* **35,** 197–205.

Johnson, K. B. (1992). Evaluation of a mechanistic model that describes potato crop losses caused by multiple pests. *Phytopathology* **82,** 363–369.

Johnson, M. D., and Lowery, B. (1985). Effect of three conservation tillage practices on soil temperature and thermal properties. *Soil Sci. Soc. Am. J.* **49,** 1547–1552.

Jordan, N. (1993a). Simulation analysis of weed population dynamics in ridge–tilled fields. *Weed Sci.* **41,** 468–474.

Jordan, N. (1993b). Prospects for weed control through crop interference. *Ecol. Appl.* **3,** 84–91.

Jordan, N., Mortensen, D. A., Prenzlow, D. M., and Curtis-Cox, K. (1995). Simulation analysis of crop rotation effects on weed seedbanks. *Am. J. Bot.* **82,** 390–398.

Keating, B. A., and Carberry, P. S. (1993). Resource capture and use in intercropping: Solar radiation. *Field Crops Res.* **34,** 273–301.

Keeley, P. E. (1987). Interference and interaction of purple and yellow nutsedges (*Cyperus rotundus* and *Cyperus esculentus*) with crops. *Weed Technol.* **1,** 74–81.

Kennedy, A. C., Elliot, L. F., Young, F. L., and Douglas, C. L. (1991). Rhizobacteria suppressive to the weed downy brome. *Soil Sci. Soc. Am. J.* **55,** 722–727.

King, L. D., and Buchanan, M. (1993). Reduced chemical input cropping systems in the southeastern United States. I. Effect of rotations, green manure crops and nitrogen fertilizer on crop yields. *Am. J. Altern. Agric.* **8,** 58–77.

Knezevic, S. Z., Weise, S. F., and Swanton, C. J. (1994). Interference of redroot pigweed (*Amaranthus retroflexus*) in corn (*Zea mays*). *Weed Sci.* **42,** 568–573.

Knezevic, S. Z., Weise, S. F., and Swanton, C. J. (1995). Comparison of empirical models depicting density of *Amaranthus retroflexus* L. and relative leaf area as predictors of yield loss in maize (*Zea mays*). *Weed Res.* **35,** 207–214.

Kremer, R. J. (1993). Management of weed seed banks with microorganisms. *Ecol. Appl.* **3,** 42–52.

Kremer, R. J., and Spencer, N. R. (1989a). Interaction of insects, fungi, and burial on velvetleaf (*Abutilon theophrasti*) seed viability. *Weed Technol.* **3,** 322–328.

Kremer, R. J., and Spencer, N. R. (1989b). Impact of a seed–feeding insect and microorganisms on velvetleaf (*Abutilon theophrasti*) seed viability. *Weed Sci.* **37,** 211–216.

Kropff, M. J. (1993a). Eco-physiological models for crop-weed competition. *In* "Modelling Crop–Weed Interactions" (M. J. Kropff and H. H. van Laar, eds.), pp. 25–32. CAB International, Wallingford, UK.

Kropff, M. J. (1993b). General introduction. In "Modelling Crop–Weed Interactions" (M. J. Kropff and H. H. van Laar, eds.), pp. 1–7. CAB International, Wallingford, UK.

Kropff, M. J., and Lotz, L. A. P. (1992). Optimization of weed management systems: The role of ecological models of interplant competition. *Weed Technol.* **6,** 462–470.

Kropff, M. J., and Lotz, L. A. P. (1993a). Empirical models for crop–weed competition. *In* "Modelling Crop–Weed Interactions" (M. J. Kropff and H. H. van Laar, eds.), pp. 9–24. CAB International, Wallingford, UK.

Kropff, M. J., and Lotz, L. A. P. (1993b). Ecophysiological characterization of the species. *In* "Modelling Crop–Weed Interactions" (M. J. Kropff and H. H. van Laar, eds.), pp. 83–104. CAB International, Wallingford, UK.

Kropff, M. J., and Spitters, C. J. T. (1991). A simple model of crop loss by weed competition from early observations on relative leaf area of the weeds. *Weed Res.* **31,** 97–105.

Kropff, M. J., and Spitters, C. J. T. (1992). An eco-physiological model for interspecific competition, applied to the influence of *Chenopodium album* L. on sugar beet. I. Model description and parameterization. *Weed Res.* **32,** 437–450.

Kropff, M. J., Weaver, S. E., and Smits, M. A. (1992). Use of ecophysiological models for crop–weed interference: Relations amongst weed density, relative time of weed emergence, relative leaf area, and yield loss. *Weed Sci.* **40,** 296–301.

Kropff, M. J., Lotz, L. A. P., and Weaver, S. E. (1993a). Practical applications. *In* "Modelling Crop–Weed Interactions" (M. J. Kropff and H. H. van Laar, eds.), pp. 149–167. CAB International, Wallingford, UK.

Kropff, M. J., van Keulen, N. C., van Laar, H. H., and Schnieders, B. J. (1993b). The impact of environmental and genetic factors. *In* "Modelling Crop–Weed Interactions" (M. J. Kropff and H. H. van Laar, eds.), pp. 137–147. CAB International, Wallingford, UK.

Kropff, M. J., Weaver, S. E., Lotz, L. A. P., Lindquist, J. L., Joenje, W., Schnieders, B. J., van Keulen, N. C., Migo, T. R., and Fajardo, F. F. (1993c). Understanding crop–weed interaction in field situations. *In* "Modelling Crop–Weed Interactions" (M. J. Kropff and H. H. van Laar, eds.), pp. 105–136. CAB International, Wallingford, UK.

Kumar, B., Yaduraju, N. T., Ahuja, K. N., and Prasad, D. (1993). Effect of soil solarization on weeds and nematodes under tropical Indian conditions. *Weed Res.* **33,** 423–429.

Lawson, H. M. (1994). Changes in pesticide usage in the United Kingdom: Policies, results, and long-term implications. *Weed Technol.* **8,** 360–365.

Lebbink, G., van Faassen, H. G., van Ouwerkerk, C., and Brussaard, L. (1994). The Dutch programme on soil ecology of arable farming systems: Farm management monitoring programme and general results. Agric. Ecosyst. Environ. **51,** 7–20.

Leistra, M., and Boesten, J. J. T. I. (1989). Pesticide contamination of groundwater in western Europe. *Agric. Ecosyst. Environ.* **26,** 369–389.

Levins, R., and Vandermeer, J. H. (1990). The agroecosystem embedded in a complex ecological community. *In* "Agroecology" (C. R. Carroll, J. H. Vandermeer, and P. R. Rosset, eds.), pp. 341–362. McGraw–Hill, New York.

Liebl, R., Simmons, F. W., Wax, L. M., and Stoller, E. W. (1992). Effect of rye (*Secale cereale*) mulch on weed control and soil moisture in soybean (*Glycine max*). *Weed Technol.* **6,** 838–846.

Liebman, M. (1989). Effects of nitrogen fertilizer, irrigation, and crop genotype on canopy relations and yields of an intercrop/weed mixture. *Field Crops Res.* **22,** 83–100.

Liebman, M. (1995). Polyculture cropping systems. *In* "Agroecology: The Science of Sustainable Agriculture, 2nd Edition" (M.A. Altieri, ed.), pp. 205–218. Westview Press, Boulder, CO.

Liebman, M., and Dyck, E. (1993a). Weed management: A need to develop ecological approaches. *Ecol. Appl.* **3,** 39–41.

Liebman, M., and Dyck, E. (1993b). Crop rotation and intercropping strategies for weed management. *Ecol. Appl.* **3,** 92–122.

Liebman, M., and Janke, R. R. (1990). Sustainable weed management practices. *In* "Sustainable Agriculture in Temperate Zones" (C. A. Francis, C. B. Flora, and L. D. King, eds.), pp. 111–143. Wiley, New York.

Liebman, M., and Ohno, T. (1997). Crop rotation and legume residue effects on weed emergence and growth: Applications for weed management. *In* "Weed Biology, Soil Management, and Weed Management" (J. L. Hatfield and D. D. Buhler, eds.), in press. Ann Arbor Press, Ann Arbor, MI.

Liebman, M., and Robichaux, R. H. (1990). Competition by barley and pea against mustard: Effects on resource acquisition, photosynthesis, and yield. *Agric. Ecosyst. Environ.* **31,** 155–172.

Liebman, M., Drummond, F. A., Corson, S., and Zhang, J. (1996). Tillage and rotation crop effects on weed dynamics in potato production systems. *Agron. J.* **88,** 18–26.

Lindquist, J. L. and Kropff, M. J. (1996). Applications of an ecophysiological model for irrigated rice (*Oryza sativa*)—Echinochloa competition. *Weed Sci.* **44,** 52–56.

Lindquist, J. L., Mortensen, D. A., Clay, S. A., Schmenk, R., Kells, J. J., Howatt, K., and Westra, P. (1996). Stability of coefficients in the corn yield loss–velvetleaf density relationship across the north central U.S. *Weed Sci.* **44,** 309–313.

Lockeretz, W., Shearer, G., and Kohl, D. H. (1981). Organic farming in the corn belt. *Science* **211,** 540–547.

Lotz, L. A. P., Groeneveld, R. M. W., Habekotte, B., and van Oene, H. (1991). Reduction of growth and reproduction of *Cyperus esculentus* by specific crops. *Weed Res.* **31,** 153–160.

Lotz, L. A. P., Kropff, M. J., Wallinga, J., Bos, H. J., and Groeneveld, R. M. W. (1994). Techniques to estimate relative leaf area and cover of weeds in crops for yield loss prediction. *Weed Res.* **34,** 167–175.

Lund, R. D., and Turpin, F. T. (1977). Carabid damage to weed seeds found in Indiana cornfields. *Econ. Entomol.* **6,** 695–698.

Malik, V. S., Swanton, C. J., and Michaels, T. E. (1993). Interaction of white bean (*Phaseolus vulgaris* L.) cultivars, row spacing, and seeding density with annual weeds. *Weed Sci.* **41,** 62–68.

Maxwell, B. D., and Ghersa, C. (1992). The influence of weed seed dispersion versus the effect of competition on crop yield. *Weed Technol.* **6,** 196–204.

Maxwell, B. D., Wilson, M. V., and Radosevich, S. R. (1988). Population modelling approach for evaluating leafy spurge (*Euphorbia esula*) development and control. *Weed Technol.* **2,** 132–138.

Maxwell, B. D., Zasada, J. C., and Radosevich, S. R. (1993). Simulation of salmonberry and thimbleberry population establishment and growth. *Can. J. For. Res.* **23,** 2194–2203.

McEvoy, P. B., Rudd, N. T., Cox, C. S., and Huso, M. (1993). Disturbance, competition, and herbivory effects on ragwort (*Senecio jacobaea*) populations. *Ecol. Monogr.* **63,** 55–75.

Medd, R. W. (1987). Weed management on arable lands. *In* "Tillage: New Directions in Australian Agriculture" (P. S. Cornish and J. E. Pratley, eds.), pp. 222–259. Inkata Press, Melbourne, Australia.

Messersmith, C. G., and Adkins, S. W. (1995). Integrating weed-feeding insects and herbicides for weed control. *Weed Technol.* **9,** 199–208.

Mohler, C. L. (1991). Effects of tillage and mulch on weed biomass and sweet corn yield. *Weed Technol.* **5,** 545–552.

Mohler, C. L. (1993). A model of the effects of tillage on emergence of weed seedlings. *Ecol. Appl.* **3,** 53–73.

Mohler, C. L., and Liebman, M. (1987). Weed productivity and composition in sole crops and intercrops of barley and field pea. *J. Appl. Ecol.* **24,** 685–699.

Moody, K., and Drost, D. C. (1983). The role of cropping systems on weeds in rice. *In* "Proceedings of the Conference on Weed Control in Rice, 31 August–4 September 1981," pp. 73–86. International Rice Research Institute, Los Banos, Philippines.

Mooney, H. A., and Chiariello, N. R. (1984). The study of plant function: The plant as a balanced system. *In* "Perspectives on Plant Population Biology" (R. Dirzo and J. Sarukhan, eds.), pp. 305–323. Sinauer, Sunderland, MA.

Morris, R. A., and Garrity, D. P. (1993a). Resource capture and utilization in intercropping: Water. *Field Crops Res.* **34,** 303–317.

Morris, R. A., and Garrity, D. P. (1993b). Resource capture and utilization in intercropping: Non-nitrogen nutrients. *Field Crops Res.* **34,** 319–334.

Mortensen, D. A., and Coble, H. D. (1989). The influence of soil water content on common cocklebur (*Xanthium strumarium*) interference in soybeans (*Glycine max*). *Weed Sci.* **37,** 76–83.

Mortensen, D. A., Johnson, G. A., and Young, L. J. (1993). Weed distribution in agricultural fields. *In* "Soil Specific Crop Management" (P. Robert, R. H. Rust, and W. E. Larson, eds.), pp. 113–124. American Society of Agronomy, Madison, WI.

Mortensen, D. A., Dieleman, J. A., and Johnson, G. A. (1997). Weed spatial variation and weed management. *In* "Weed Biology, Soil Management, and Weed Management" (J. L. Hatfield and D. D. Buhler, eds.), in press. Ann Arbor Press, Ann Arbor, MI.

Mortimer, A. M. (1983). On weed demography. *In* "Recent Advances in Weed Research" (W. W. Fletcher, ed.), pp. 3–40. Commonwealth Agricultural Bureaux, Farnham Royal, UK.

Mulder, T. A., and Doll, J. D. (1993). Integrating reduced herbicide use with mechanical weeding in corn (*Zea mays*). *Weed Technol.* **7**, 382–389.

Mulder, T. A., and Doll, J. D. (1994). Reduced input corn weed control: The effects of planting date, early season weed control, and row–crop cultivator selection. *J. Prod. Agric.* **7**, 256–260.

Müller-Schärer, H., and Schroeder, D. (1993). The biological control of *Centaurea* spp. in North America: Do insects solve the problem? *Pesticide Sci.* **37**, 343–353.

National Research Council (1989). "Alternative Agriculture." National Academy Press, Washington, DC.

Nelson, H., and Jones, R. D. (1994). Potential regulatory problems associated with atrazine, cyanazine, and alachlor in surface water source drinking water. *Weed Technol.* **8**, 852–861.

Nestel, D., and Altieri, M. A. (1992). The weed community of Mexican coffee agroecosystems: Effect of management upon plant biomass and species composition. *Acta Ecol.* **13**, 715–726.

Norris, R. F. (1992). Case history for weed competition/population ecology: Barnyardgrass (*Echinochloa crus-galli*) in sugarbeets (*Beta vulgaris*). *Weed Technol.* **6**, 220–227.

Ofori, F., and Stern, W. R. (1987). Cereal–legume intercropping systems. *Adv. Agron.* **41**, 41–90.

Ogg, A. G., Stephens, R. H., and Gealy, D. R. (1994). Interference between mayweed chamomile (*Anthemis cotula*) and pea (*Pisum sativum*) is affected by form of interference and soil water regime. *Weed Sci.* **42**, 579–585.

Oliver, L. R. (1988). Principles of weed threshold research. *Weed Technol.* **2**, 398–403.

Pantone, D. J., Brown, S. M., and Womersley, C. (1985). Biological control of fiddleneck. *Calif. Agric.* **39**, 4–5.

Pantone, D. J., Williams, W. A., and Maggenti, A. R. (1989a). An alternative approach for evaluating the efficacy of potential biocontrol agents of weeds. 1. Inverse linear model. *Weed Sci.* **37**, 771–777.

Pantone, D. J., Williams, W. A., and Maggenti, A. R. (1989b). An alternative approach to evaluating the efficacy of potential biocontrol agents of weeds. 2. Path anaysis. *Weed Sci.* **37**, 778–783.

Paul, N. D., and Ayres, P. G. (1987). Effects of rust infection of *Senecio vulgaris* on competition with lettuce. *Weed Res.* **27**, 431–441.

Pearcy, R. W., Tumosa, N., and Williams, K. (1981). Relationships between growth, photosynthesis, and competitive interactions for a C_3 and a C_4 plant. *Oecologia* **48**, 371–376.

Pearcy, R. W., Ehleringer, J., Mooney, H. A., and Rundel, P. W. (eds.) (1991). "Plant Physiological Ecology: Field Methods and Instrumentation." Chapman & Hall, London.

Pimentel, D., Acquay, H., Biltonen, M., Rice, P., Silva, M., Nelson, J., Lipner, V., Giordano, S., Horowitz, A., and D'Amore, M. (1992). Environmental and economic costs of pesticide use. *Bioscience* **42**, 750–760.

Pinnschmidt, H. O., Batchelor, W. D., and Teng, P. S. (1995). Simulation of multiple species pest damage in rice using CERES-rice. *Agric. Syst.* **48**, 193–222.

Poston, D. H., Murdock, E. C., and Toler, J. E. (1992). Cost-efficient weed control in soybean (*Glycine max*) with cultivation and banded herbicide applications. *Weed Technol.* **6**, 990–995.

Potter, J. R., and Jones, J. W. (1977). Leaf area partitioning as an important factor in growth. *Plant Physiol.* **59**, 10–14.

Putnam, A. R., and DeFrank, J. (1983). Use of phytotoxic plant residues for selective weed control. *Crop Prot.* **2**(2), 173–181.

Putnam, A. R., DeFrank, J., and Barnes, J. P. (1983). Exploitation of allelopathy for weed control in annual and perennial cropping systems. *J. Chem. Ecol.* **9**, 1001–1010.

Radosevich, S. R., and Roush, M. L. (1990). The role of competition in agriculture. *In* "Perspectives on Plant Competition" (J. B. Grace and D. Tilman, eds.), pp. 341–366. Academic Press, San Diego.

Rasmussen, J. (1992). Testing harrows for mechanical control of annual weeds in agricultural crops. *Weed Res.* **32,** 267–274.

Rasmussen, J., and Svenningsen, T. (1995). Selective weed harrowing in cereals. *Biol. Agric. Hort.* **12,** 29–46.

Rasmussen, K., Rasmussen, J., and Petersen, J. (1996). Effects of fertilizer placement on weeds in weed harrowed spring barley. *Acta Agric. Scand.* **46,** 192–196.

Reganold, J. P., Palmer, A. S., Lockhart, J. C., and Macgregor, A. N. (1993). Soil quality and financial performance of biodynamic and conventional farms in New Zealand. *Science* **260,** 344–349.

Regnier, E. E., and Janke, R. R. (1990). Evolving strategies for managing weeds. *In* "Sustainable Agricultural Systems" (C. A. Edwards, R. Lal, P. Madden, R. H. Miller, and G. House, eds.), pp. 174–202. Soil and Water Conservation Society, Ankeny, IA.

Reinertsen, M. R., Cochran, V.L., and Morrow, L. A. (1984). Response of spring wheat to nitrogen fertilizer placement, row spacing, and wild oat herbicides in a no-till system. *Agron. J.* **76,** 753–756.

Repetto, R. (1985). "Paying the Price: Pesticide Subsidies in Developing Countries." World Resources Institute, Washington, DC.

Rydberg, T. (1993). Weed harrowing—Driving speed at different stages of development. *Swed. J. Agric. Res.* **23,** 107–113.

Samson, R. A., and Coulman, B. E. (1989). Herbicide banding, cultivation and interseeding as an integrated weed management system for corn. Proceedings of the 44th Annual Northeastern Corn Improvement Conference, 16–17 February 1989, pp. 33–38.

Sarkar, P. A., and Moody, K. (1983). Effects of stand establishment techniques on weed populations in rice. *In* "Proceedings of the Conference on Weed Control in Rice, 31 August–4 September 1981," pp. 57–71. International Rice Research Institute, Los Banos, Philippines.

Scheepens, P. C. (1987). Joint action of *Cochliobolus lunatus* and atrazine on *Echinochloa crusgalli* (L.) Beauv. *Weed Res.* **27,** 43–47.

Seaman, D. E. (1983). Farmers' weed control technology for water–seeded rice in North America. *In* "Proceedings of the Conference on Weed Control in Rice, 31 August–4 September 1981," pp. 167–176. International Rice Research Institute, Los Banos, Philippines.

Seibert, A. C., and Pearce, R. B. (1993). Growth analysis of weed and crop species with reference to seed weight. *Weed Sci.* **41,** 52–56.

Shainsky, L. J., and Radosevich, S. R. (1992). Mechanisms of competition between Douglas fir and red alder seedlings. *Ecology* **73,** 30–45.

Sharaiha, R., and Gliessman, S. (1992). The effects of crop competition and row arrangement in the intercropping of lettuce, favabean, and pea on weed biomass and diversity and on crop yields. *Biol. Agric. Hort.* **9,** 1–13.

Shetty, S. V. R., and Rao, A. N. (1981). Weed managment studies in sorghum/pigeonpea and pearl millet/groundnut intercrop systems—Some observations. *In* "Proceedings of the International Workshop on Intercropping, Hyderabad, India, 10–13 January 1979," pp. 238–248. International Crops Research Institute for the Semi-Arid Tropics, Patencheru, India.

Shetty, S. V. R., Sivakumar, M. V. K., and Ram, S. A. (1982). Effect of shading on the growth of some common weeds of the semi–arid tropics. *Agron. J.* **74,** 1023–1029.

Shilling, D. G., Brecke, B. J., Hiebsch, C., and MacDonald, G. (1995). Effect of soybean (*Glycine max*) cultivar, tillage, and rye (*Secale cereale*) mulch on sicklepod (*Senna obtusifolia*). *Weed Technol.* **9,** 339–342.

Shrefler, J. W., Dusky, J. A., Shilling, D. G., Brecke, B. J., and Sanchez, C. A. (1994). Effects of phosphorus fertility on competition between lettuce (*Lactuca sativa*) and spiny amaranth (*Amaranthus spinosus*). *Weed Sci.* **42,** 556–560.

Smith, R. J., and Shaw, W. C. (1966). Weeds and their control in rice production. *USDA Agric. Handbook* **292**, 1–64.

Smolik, J. D., Dobbs, T. L., Rickerl, D. H., Wrage, L. J., Buchenau, G. W., and Machacek, T. A. (1993). Agronomic, economic, and ecological relationships in alternative (organic), conventional, and reduced–till farming systems, Bull. No. 718. South Dakota Agricultural Experiment Station, Brookings.

Snaydon, R. W. (1979). A new technique for studying plant interactions. *J. Appl. Ecol.* **16**, 281–286.

Sojka, R. E., Karlen, D. L., and Sadler, E. J. (1988). Planting geometries and the efficient use of water and nutrients. In "Cropping Strategies for Efficient Use of Water and Nitrogen" (W. L. Hargrove, ed.), pp. 43–68. American Society of Agronomy, Madison, WI.

Soni, P., and Ambasht, R. S. (1977). Effect of crop–weed competition on the mineral structure of a wheat crop. *Agroecosystems* **3**, 325–336.

Standifer, L., Wilson, P. H., and Porche-Sorbet, R. (1984). Effect of solarization on soil weed seed populations. *Weed Sci.* **32**, 569–573.

Staniforth, D. W. (1962). Responses of soybean varieties to weed competition. *Agron. J.* **54**, 11–13.

Stapleton, J. J., and DeVay, J. E. (1986). Soil solarization: A non-chemical approach for management of plant pathogens and pests. *Crop Prot.* **5**, 190–198.

Stone, J. F., Eichner, M. L., Kim, C., and Koehler, K. (1988). Relationships between clothing and pesticide poisoning: Symptoms among Iowa farmers. *J. Environ. Health* **50**, 210–215.

Streibig, J. C., Combellack, J. H., Pritchard, G. H., and Richardson, R. G. (1989). Estimation of thresholds for weed control in Australian cereals. *Weed Res.* **29**, 117–126.

Tardif, F. J., and Leroux, G. D. (1992). Response of three quackgrass biotypes to nitrogen fertilization. *Agron. J.* **84**, 366–370.

Teasdale, J. R. (1993). Interaction of light, soil moisture, and temperature with weed suppression by hairy vetch residue. *Weed Sci.* **41**, 46–51.

Teasdale, J. R. (1995). Influence of narrow row/high population corn (*Zea mays*) on weed control and light transmittance. *Weed Technol.* **9**, 113–118.

Teasdale, J. R., and Frank, J. R. (1983). Effect of row spacing on weed competition with snap beans (*Phaseolus vulgaris*). *Weed Sci.* **31**, 81–85.

Teasdale, J. R., and Mohler, C. L. (1993). Light transmittance, soil temperature, and soil moisture under residue of hairy vetch and rye. *Agron. J.* **85**, 673–680.

Teasdale, J. R., Beste, C. E., and Potts, W. E. (1991). Response of weeds to tillage and cover crop residue. *Weed Sci.* **39**, 195–199.

Temple, S. R., Friedman, D. B., Somasco, O., Ferris, H., Scow, K., and Klonsky, K. (1994). An interdisciplinary, experiment station-based participatory comparison of alternative crop management systems for California's Sacramento Valley. *Am. J. Altern. Agric.* **9**, 64–71.

Templeton, G. E. (1983). Integrating biological control of weeds in rice into a weed control program. In "Proceedings of the Conference on Weed Control in Rice, 31 August–4 September 1981," pp. 219–223. International Rice Research Institute, Los Banos, Philippines.

Terpstra, R., and Kouwenhoven, J. K. (1981). Inter-row and intra-row weed control with a hoe-ridger. *J. Agric. Eng. Res.* **26**, 127–134.

Teyker, R. H., Hoelzer, H. D., and Liebl, R. A. (1991). Maize and pigweed response to nitrogen supply and form. *Plant Soil* **135**, 287–292.

Thomsen, C. D., Williams, W. A., Olkowski, W., and Pratt, D. W. (1996). Grazing mowing, and clover plantings control yellow starthistle. *IPM Practitioner* **18**(2), 1–4.

Thurman, E. M., Goolsby, D. A., Meyer, M. T., and Kolpin, D. W. (1991). Herbicides in surface waters of the midwestern United States: The effect of the spring flush. *Environ. Sci. Tech.* **25**, 1794–1796.

Tilman, D. (1982). "Resource Competition and Community Structure." Princeton Univ. Press, Princeton, NJ.

Tollenaar, M., Dibo, A. A., Aguilera, A., Weise, S. F., and Swanton, C. J. (1994a). Effect of crop density on weed interference in maize. *Agron. J.* **86**, 591–595.

Tollenaar, M., Nissanka, S. P., Aguilera, A., Weise, S. F., and Swanton, C. J. (1994b). Effect of weed interference and soil nitrogen on four maize hybrids. *Agron. J.* **86**, 596–601.

Tranel, P. J., Gealy, D. R., and Kennedy, A. C. (1993). Inhibition of downy brome (*Bromus tectorum*) root growth by a phytotoxin from *Pseudomonas flurescens* strain D7. *Weed Technol.* **7**, 134–139.

United States Environmental Protection Agency (USEPA) (1992). "Another Look: National Survey of Pesticides in Drinking Water Wells. Phase 2 Report." USEPA, Washington, DC.

Van Acker, R. C., Swanton, C. J., and Weise, S. F. (1993a). The critical period of weed control in soybean (*Glycine max* [L.] Merr.). *Weed Sci.* **41**, 194–200.

Van Acker, R. C., Weise, S. F., and Swanton, C. J. (1993b). Influence of interference from a mixed weed sprecies stand on soybean (*Glycine max* [L.] Merr.) growth. *Can. J. Plant Sci.* **73**, 1293–1304.

van Bruggen, A. H. C. (1995). Plant disease severity in high-input compared to reduced-input and organic farming systems. *Plant Disease* **79**, 976–984.

Vandermeer, J. (1989). "The Ecology of Intercropping." Cambridge Univ. Press, Cambridge, UK.

Vangessel, M. J., Schweizer, E. E., Lybecker, D. W., and Westra, P. (1995). Compatibility and efficiency of in-row cultivation for weed management in corn (*Zea mays*). *Weed Technol.* **9**, 754–760.

Vengris, J., Colby, W. G., and Drake, M. (1955). Plant nutrient competition between weeds and corn. *Agron. J.* **47**, 213–216.

Vereijken, P. (1989). The DFS farming systems experiment. *In* "Development of Farming Systems. Evaluation of the Five-Year Period 1980–1984" (J. C. Zadoks, ed.), pp. 1–8. Pudoc, Wageningen, The Netherlands.

Vitta, J. I., Satorre, E. H., and Leguizamon, E. E. (1993). Using canopy attributes to evaluate competition between *Sorghum halapense* (L.) Pers. and soybean. *Weed Res.* **33**, 89–97.

Volz, M. G. (1977). Infestations of yellow nutsedge in cropped soil: Effects on soil nitrogen availability to the crop and on associated N transforming bacterial populations. *Agroecosystems* **3**, 313–323.

Walker, R. H. (1995). Preventive weed management. *In* "Handbook of Weed Management Systems" (A. E. Smith, ed.), pp. 35–50. Dekker, New York.

Wall, D. A. (1993). Comparison of green foxtail (*Setaria viridis*) and wild oat (*Avena fatua*) growth, development, and competitivenesss under three temperature regimes. *Weed Sci.* **41**, 369–378.

Warwick, S. I. (1991). Herbicide resistance in weedy plants: Physiology and population biology. *Annu. Rev. Ecol. Syst.* **22**, 95–114.

Weaver, S. E., Smits, N., and Tan, C. S. (1987). Estimating yield losses of tomato (*Lycopersicon esculentum*) caused by nightshade (*Solanum* spp.) interference. *Weed Sci.* **35**, 163–168.

Weaver, S. E., Tan, C. S., and Brain, P. (1988). Effect of temperature and soil moisture on time of emergence of tomatoes and four weed species. *Can. J. Plant Sci.* **68**, 877–886.

Weaver, S. E., Kropff, M. J., and Cousens, R. (1994). A simulation model of competition between winter wheat and *Avena fatua* for light. *Ann. Appl. Biol.* **124**, 315–331.

Weinberg, A. C. (1990). Reducing agricultural pesticide use in Sweden. *J. Soil Water Conserv.* **45**, 610–613.

Wells, G. J. (1979). Annual weed competition in wheat crops: The effect of weed density and applied nitrogen. *Weed Res.* **19**, 185–191.

White, R. H., Worsham, A. D., and Blum, U. (1989). Allelopathic potential of legume debris and aqueous extracts. *Weed Sci.* **37**, 674–679.

Wicks, G. A., Ramsel, R. E., Nordquist, P. T., Schmidt, J. W., and Challaiah (1986). Impact of wheat cultivars on establishment and suppression of summer annual weeds. *Agron. J.* **78**, 59–62.

Wiles, L. J., and Wilkerson, G. C. (1991). Modelling competition for light between soybean and broadleaf weeds. *Agric. Syst.* **35**, 37–51.

Wilkerson, G.G., Jones, J. W., Coble, H. D., and Gunsolus, J. L. (1990). SOYWEED: A simulation model of soybean and common cocklebur growth and competition. *Agron. J.* **82**, 1003–1010.

Willey, R. W. (1975). The use of shade in coffee, cocoa, and tea. *Hort. Abst.* **45**, 791–798.

Willey, R. W. (1979). Intercropping—Its importance and research needs. Part 2. Agronomy and research approaches. *Field Crop Abst.* **32**, 73–84.

Willey, R. W. (1990). Resource use in intercropping systems. *Agric. Water Manage.* **17**, 215–231.

Williams, E. D. (1970). Effects of decreasing light intensity on the growth of *Agropyron repens* in the field. *Weed Res.* **10**, 360–366.

William, R. D., and Warren, G. F. (1975). Competition between purple nutsedge and vegetables. *Weed Sci.* **23**, 317–323.

Wilson, B. J., Wright, K. J., Brain, P., Clements, M., and Stephens, E. (1995). Predicting the competitive effects of weed and crop density on weed biomass, weed production, and crop yield in wheat. *Weed Res.* **35**, 265–278.

Wilson, J. B. (1988a). The effect of initial advantage on the course of plant competition. *Oikos* **51**, 19–24.

Wilson, J. B. (1988b). Shoot competition and root competition. *J. Appl. Ecol.* **25**, 279–296.

Wilson, P. J., and Aebischer, N. J. (1995). The distribution of dicotyledenous arable weeds in relation to distance from the field edge. *J. Appl. Ecol.* **32**, 295–310.

Woebbecke, D. M., Meyer, G. E., Von Bargen, K., and Mortensen, D. A. (1995). Color indices for weed identification under various soil, residue, and lighting conditions. *Trans. Am. Soc. Agric. Eng.* **38**, 259–269.

Woolley, B. L., Michaels, T. E., Hall, M. R., and Swanton, C. J. (1993). The critical period of weed control in white bean (*Phaseolus vulgaris*). *Weed Sci.* **41**, 180–184.

World Health Organization (WHO) (1990). "Public Health Impact of Pesticides Used in Agriculture." WHO, Geneva.

Wortmann, C. S. (1993). Contribution of bean morphological characteristics to weed suppression. *Agron. J.* **85**, 840–843.

Wymore, L. A., and Watson, A. K. (1989). Interaction between a velvetleaf isolate of *Colletotrichum coccodes* and thidiazuron for velvetleaf (*Abutilon theophrasti*) control in the field. *Weed Sci.* **37**, 478–483.

Wyse, D. L. (1992). Future of weed science research. *Weed Technol.* **6**, 162–165.

Wyse, D. L. (1994). New technologies and approaches for weed management in sustainable agricultural systems. *Weed Technol.* **8**, 403–407.

Yelverton, F. H., and Coble, H. D. (1991). Narrow row spacing and canopy formation reduces weed resurgence in soybeans (*Glycine max*). *Weed Technol.* **5**, 169–174.

Young, F. L., Wyse, D. L., and Jones, R. J. (1983). Effect of irrigation on quackgrass (*Agropyron repens*) interference in soybeans (*Glycine max*). *Weed Sci.* **31**, 720–727.

Young, F. L., Ogg, A. G., Jr., Papendick, R. I., Thill, D. C., and Alldredge, J. R. (1994). Tillage and weed management affects winter wheat yield in an integrated pest management system. *Agron. J.* **86**, 147–154.

Zimdahl, R. L. (1980). "Weed–Crop Competition: A Review." International Plant Protection Center, Oregon State University, Corvallis.

Zimdahl, R. L. (1988). The concept and application of the critical weed-free period. *In* "Weed Management in Agroecosystems: Ecological Approaches" (M. A. Altieri and M. Liebman, eds.), pp. 145–155. CRC Press, Boca Raton, FL.

Zimdahl, R. L. (1991). Weed science: A plea for thought. U.S. Department of Agriculture, Cooperative State Research Service. Washington, DC.

III

Ecosystem Processes

10

Nitrogen Use Efficiency in Row-Crop Agriculture: Crop Nitrogen Use and Soil Nitrogen Loss

G. Philip Robertson

I. Introduction

Nitrogen loss from cropping systems has interested agronomists since the early recognition by Liebig and others that most crops are nitrogen limited. The widespread early adoption of crop rotations that include legumes (Oakley, 1925; Francis and Clegg, 1990) provides historical acknowledgment of the importance of nitrogen gains and losses to cropping system success. It is now recognized that even in semiarid regions (Breman and deWit, 1983) nitrogen is usually the principal resource limiting crop production (Fig. 1).

It was not until the post-World War II era, when munitions plants were converted to fertilizer factories and synthetic nitrogen became cheap and readily available, that nitrogen use efficiency became a moot point for most producers in the United States and other economically developed regions. Even today it is easier and usually more cost-effective at the farm scale to apply another 50 kg N ha^{-1} to an already fertilized field than to devise a means to prevent an equivalent 50 kg^{-1} N ha^{-1} loss. In fact, the price of nitrogen fertilizer in constant dollars has declined over past decades to its current average global price of about \$0.10 kg^{-1} N for urea and anhydrous ammonia (Bumb, 1995; Fee, 1995), providing an even further disincentive for nitrogen conservation. This has resulted in an exponential increase in nitrogen fertilizer use over the past 50 years—first in developed countries and now, with about a 20-year lag, in developing regions (Fig. 2). Globally,

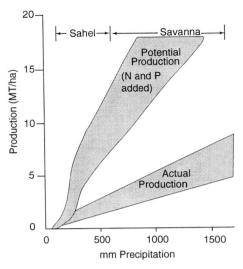

Figure 1 Ranges of actual and potential production in West African pastures at different rates of rainfall. Redrawn from Breman and deWit (1983).

today almost as much fertilizer is fixed from industrial sources as appears to be fixed biologically (Soderland and Svensson, 1976; Vitousek and Matson, 1993).

There is also an off-farm, environmental cost to fertilizer use, however, and the emerging recognition of this cost has renewed interest in finding ways to minimize nitrogen loss from fertilized cropping systems. Moreover, in some developing regions where fertilizer remains unavailable because

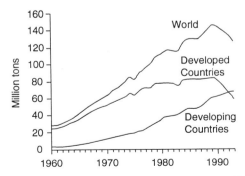

Figure 2 World nitrogen fertilizer consumption since 1960. From FAO (1960–1992) and Bumb (1995).

of economic constraints, and elsewhere where there is interest in reducing farm chemical use (NRC, 1989; Harwood, 1990), the conservation of endogenous nitrogen remains an important management goal. Regardless of the magnitude of fertilizer use, the effective containment of cropping system nitrogen requires a greater nitrogen use efficiency at the ecosystem scale. This is a difficult management goal despite decades of research that have provided substantial insights into process-level details of the nitrogen cycle in many different types of ecosystems.

A. An Overview of the Nitrogen Cycle

The nitrogen cycle is distinguished from many other nutrient cycles by its complexity: Nitrogen in the biosphere exists in a number of different oxidation states, each of which is differentially available to plants, microbes, and other organisms and each of which is differentially reactive with the chemical and physical environment. Moreover, most transformations among nitrogen's different oxidation states are biologically mediated, including changes among ionic (NH_4^+, NO_2^-, and NO_3^-), organic (both particulate and dissolved), and gaseous (N_2, N_2O, NO_x, and NH_3) forms.

Figure 3 shows a generalized nitrogen cycle for a typical row-crop ecosystem. Nitrogen enters the plant in inorganic form: primarily as nitrate (NO_3^-), and perhaps ammonium (NH_4^+; see Chapter 5, this volume) from the soil solution. Leguminous crops can also acquire significant nitrogen

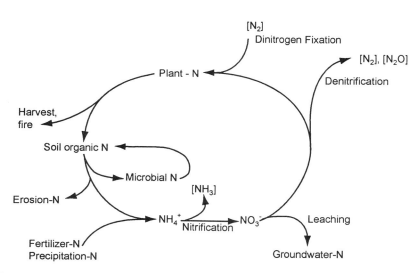

Figure 3 A simplified nitrogen cycle for a typical row-crop ecosystem; only major pathways are shown. Gas-phase nitrogen appears in brackets. For nonleguminous crops, N_2 fixation will be nil.

from the atmosphere when soil solution nitrogen pools are low. At harvest, some proportion of the aboveground plant nitrogen is removed from the system with the crop yield; the remainder, as well as nitrogen contained in roots, joins the crop residue pool as soil organic nitrogen unless a portion is burned.

In erosive environments a portion of soil organic nitrogen can be lost to surface water or to another portion of the field. Most of the residue nitrogen, however, will be oxidized and released to the soil solution as ammonium, although a portion may temporarily remain as microbial biomass nitrogen or in a more recalcitrant organic form. In agricultural soils ammonium is usually quickly nitrified to nitrate, which will be subject to leaching, denitrification, and also available again for plant uptake. All three of these soil nitrogen pools can be supplemented with exogenous inputs: Manure and, more commonly, urea-N can be added to the organic N pool, and inorganic fertilizers (in particular, anhydrous ammonia) can supplement the ammonium and nitrate pools. In most soils urea-N quickly hydrolyzes to the ammonium pool (Ladd and Jackson, 1982), and for both urea and anhydrous ammonia a significant portion of the added nitrogen can be volatilized as NH_3 within a few days of application.

Nitrogen availability represents the net rate at which the pools of inorganic nitrogen are replenished. For most row-crop systems the relevant pool is nitrate (cf. Chapter 5, this volume). Often, the size of the nitrate pool can be used to assess its availability, as is used, for example, in the widely prescribed presidedress nitrogen test. Recognize, however, that nitrogen availability can be high even when the nitrate pool is small if nitrate consumption is high (and vice versa if nitrate consumption is low). Thus, nitrogen mineralization assays (e.g., Hart *et al.*, 1994) are usually the better measure of endogenous nitrogen availability.

B. Nitrogen-Efficient Cropping Systems

Nitrogen-efficient cropping systems are those in which nitrogen availability is high and from which losses occur principally via nitrogen removed from the field in crop yield. Nitrogen efficiency thus requires

1. knowledge of crop nitrogen demands vs systemwide nitrogen inputs and outputs; whether inputs are synthetic or biological, they must be sufficient to replace the nitrogen lost in yield plus that lost via other pathways, principally leaching, denitrification, and, especially in tropical systems, fire;

2. synchrony between soil nitrogen availability and crop nitrogen demand, such that availability—inputs to soil inorganic nitrogen pools—is low except during periods of grand crop growth when available nitrogen must be well matched to crop needs;

3. the placement of available nitrogen in soil such that it coincides with locations of high sink strength, i.e., placement that is coincident with crop

root distributions both at the local row–interrow scale and at a broader field scale that includes underlying patterns of soil heterogeneity; and

4. knowledge of important nitrogen loss pathways—especially leaching and denitrification—sufficient to suggest strategies to minimize their impact.

In the following pages is an overview of the nitrogen use efficiency of modern row-crop ecosystems, the significant features of these systems that either promote or hinder conservative nitrogen use, and emerging management strategies that use this knowledge to improve efficiency. Although most of this discussion is focused on annual row crops typical of the U.S. Midwest, the principles should be applicable in row-crop systems everywhere, temperate and tropical, perennial and annual.

II. Crop Nitrogen Demand

Crop nitrogen needs vary widely, largely as a function of crop species and growing conditions. For most crops nitrogen needs are well known and easily determined from yield data so long as the crop is not overfertilized. Typical rates of nitrogen extraction for fertilized row crops appear in Table I. Amounts of nitrogen removed with crop yields vary widely, largely as a function of crop and yield; in general 100–200 kg N ha^{-1} is removed during harvest, although when the whole aboveground plant is removed—such as for alfalfa and silage corn—extraction values may be 50% higher.

Extraction values represent the minimum levels of nitrogen that must be added back to a cropping system after each harvest if the system is to remain sustainable. Additionally, of course, nitrogen must also be added to cover other nitrogen losses as described later. Failure to completely resupply nitrogen from exogenous sources means that the following equivalent crop will be nitrogen deficient to the extent that it cannot be provided nitrogen that is newly mineralized from native soil organic matter (SOM). In many arable soils nitrogen in native SOM sums to only 3–15 MT N ha^{-1} prior to the onset of cultivation (assuming an A horizon depth of 25 cm with a bulk density of 1.2 g/cm^3 and a nitrogen content of 0.1–0.5 %N). At crop nitrogen extraction rates of >0.1 MT N ha^{-1} year^{-1} (Table I), it is easy to see why continuous cropping—even in the absence of other losses that can be of a similar magnitude—requires exogenous nitrogen and why soil nitrogen pools are typically depleted substantially following only 30–40 years of crop production (Haas *et al.,* 1957; Bauer and Black, 1981; Paustian *et al.,* 1995). Moreover, only a portion of the 3–15 MT N ha^{-1} present prior to cultivation is actually mineralizable—much of it will be in passive or

Table I Typical Nitrogen Extraction Rates for Major Row Crops under High Yield Conditions[a]

Crop	Tissue	Yield (MT/ha)	% N	Total N (kg/ha)
Alfalfa	Above ground	9.0	2.8	252
Maize	Grain	10.0	2.6	260
	Residue	9.0	0.7	63
Soybeans	Grain	2.8	6.3	176
	Residue	5.4	0.9	49
Wheat	Grain	5.4	2.0	108
	Residue	6.0	0.8	48
Rice	Grain	7.9	1.8	142
	Residue	10.0	0.5	50
Cotton	Lint and seed	4.2	2.9	122
	Residue	5.0	1.3	65
Potatoes	Tubers	56.0	0.3	168
	Residue	5.0	1.8	90
Sorghum	Grain	9.0	3.0	270
	Residue	5.0	0.7	35
Sugar beets	Roots	68.0	0.2	136
	Residue	36.0	0.4	144
Sugarcane	Stalks	112.0	0.1	112
	Residue	50.0	0.2	100

[a] From Olson and Kurtz (1982) and Robertson and Rosswall (1986).

resistant soil organic matter fractions that turn over on the order of centuries or millennia (Juma and Paul, 1984; Paustian *et al.*, 1992).

Typically, exogenous nitrogen is supplied as anhydrous ammonia, urea, or some other form of synthetically fixed nitrogen (Jones, 1982; Neeteson, 1995), as organic nitrogen in the form of manure or waste sludge (e.g., Baldock and Musgrave, 1980), and/or as nitrogen biologically fixed either by the crop itself (in the case of a grain or forage legume) or by a preceding cover crop such as the winter annual *Vigna villosa* (e.g., Ebelhar *et al.*, 1984; Harris *et al.*, 1994). Some exogenous nitrogen also arrives in precipitation, though typically values are small (ca. 10 kg N ha^{-1} year^{-1}) relative to crop needs, as are amounts of N_2 fixed by associative and free-living N_2 fixers.

The amount of nitrogen removed in harvested biomass is also a function of the crop's nitrogen use efficiency (NUE). NUE is the efficiency with which carbon is fixed relative to available nitrogen. Operationally it is defined as plant tissue C:N ratio or plant nitrogen concentration: Plants with a higher C:N ratio (or lower N concentration) are more nitrogen efficient (cf. Chapter 2, this volume). In natural populations NUE appears to be a well-characterized response to a limited nitrogen supply. Evidence for this comes mainly from studies of noncultivated ecosystems in which it appears that nitrogen is used more efficiently—i.e., NUE is higher—where it is less available. Inferential evidence includes changes in litterfall C:N ratios along nitrogen availability gradients (e.g., Boerner, 1984; Birk and Vitousek, 1986). Experimental evidence dates from Turner's (1977) work showing increased C:N ratios in the litterfall of Douglas fir fertilized with sugar to immobilize available soil nitrogen and from Shaver and Melillo's (1984) work showing lower C:N ratios in marsh grasses following nitrogen fertilization.

In annual crops NUE shows similar trends: Nitrogen fertilization generally leads to lower whole plant C:N ratios as crop yield begins to fail to respond to added N above some near-saturating level. In Fig. 4, for example, the response of maize to added nitrogen was greatest between 0 and 112 kg ha^{-1} N addition, falling to little or no additional response beyond 224 kg N ha^{-1} (Broadbent and Carlton, 1978). This means that for most crops nitrogen demands increase disproportionately with yield—in this case (Fig. 3) the final 20% yield increase required 50% more fertilizer N.

III. Matching Nitrogen Availability to Demand

A. Synchrony between Nitrogen Supply and Demand

Synchrony between nitrogen release from either SOM or fertilizer and the demand for nitrogen by plants is a critical component that controls

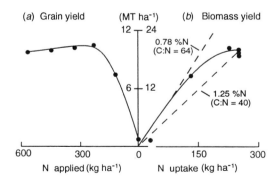

Figure 4 The response of corn (maize) to nitrogen fertilizer. (a) Grain yield response to fertilization rate; (b) total plant biomass as a function of nitrogen uptake. Dashed lines represent the minimum and maximum nitrogen use efficiencies, i.e., the amounts of biomass produced per kilogram N taken up at low and high fertilization rates. Data from Broadbent and Carlton (1978), after Loomis and Conner (1992).

systemwide nitrogen use efficiency and thus can strongly affect nutrient retention. In mixed-species native communities and in many cropped perennial systems there will be few times during the year when soil microbial activity does not coincide with periods of at least some plant uptake. In systems cropped to annual monocultures, on the other hand, this synchrony may be largely absent. Most grain crops, for example, are part of the ecosystem for only 12–16 weeks, and for only a few weeks of this period will biomass be accumulating at a significant rate. During the grand phase of vegetative growth of corn, for example, nitrogen can be taken up at the astonishing rate of 4 kg N ha^{-1} day^{-1}, but uptake is only sustained at this rate for 3 or 4 weeks and falls to nil within the following 2 or 3 weeks (Olson and Kurtz, 1982). This matches rather poorly the much longer periods during which soil temperature and moisture will be sufficient to support microbial activity. During these periods microbes continue to mineralize nitrogen from crop residue and other SOM pools, and this asynchrony can contribute to a large potential for nitrogen loss and a correspondingly low systemwide NUE (Fig. 5).

Asynchrony between nutrient supply and plant demand can in some cases be ameliorated by other attributes of the system. If, for example, soil water flux is low prior to plant growth, then not much nitrogen will be lost by leaching, and if denitrification is equally low then soil nitrate may remain available for crop uptake later. Likewise, if soil carbon remains available prior to plant growth, microbes may reduce nitrogen loss by immobilizing

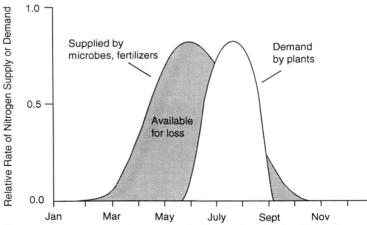

Figure 5 Asynchrony between nitrogen supply (from mineralized soil organic matter and fertilizers) and the demand for nitrogen by plants in a hypothetical north temperate annual cropping system. The shaded area represents the period during which the system is especially vulnerable to nitrogen loss via leaching and denitrification.

springtime nitrogen in microbial biomass, prior to releasing it later in the season as labile carbon stores are depleted.

Asynchrony can also be buffered by serial intercrops that temporarily immobilize nitrogen. In temperate regions winter annuals such as vetch (*Vigna* spp.) or some native weeds may be particularly well suited in this regard because they can begin growth in early fall prior to crop senescence, lay dormant over winter, then accumulate mineralized or newly fixed nitrogen rapidly prior to their killing before crop planting the following spring. If properly managed, the nitrogen accumulated should eventually be available to the crop during its 6- to 8-week growth phase (Harris *et al.*, 1994). Theoretically, as long as moisture conditions are favorable, this cover crop nitrogen could be available to the target crop within the same growing season: Nitrogen mineralization within a 25-cm-deep A horizon (with a bulk density as above) must sum to 1.3 mg g^{-1} day^{-1} N to meet crop nitrogen needs of 4–6 kg ha^{-1} day^{-1} during the target crop's period of grand vegetative growth (Olson and Kurtz, 1982). This rate of N mineralization is well within the range of mineralization rates for most forest and prairie soils of 1–4 mg g^{-1} day^{-1} N under favorable moisture and temperature conditions (e.g., Birch, 1960; Keeney, 1980; Robertson, 1982).

Other mechanisms might also be used to buffer asynchrony in specific systems. For example, many highly weathered tropical soils have a net positive charge (Uehara and Gillman, 1981) and this anion exchange capacity could temporarily retain mineralized nitrogen that might otherwise be lost as leached NO_3^- (Sollins *et al.*, 1988; Matson *et al.*, 1987). Nitrification inhibitors have been developed for a similar purpose—to protect mineralized nitrogen from leaching—and have been used with mixed success (Meisinger *et al.*, 1980). By and large, however, asynchrony in cropping systems has not been addressed in a comprehensive manner, and it deserves serious attention where system wide NUE is an important management goal.

B. Spatial Coincidence

Equal to the importance of ensuring that soil nitrogen availability and uptake are well synchronized is the concept of spatial symmetry—ensuring that N availability and uptake are well matched spatially. Management—particularly the management of row crops—does not necessarily result in a spatial arrangement of plants and resources within a field that are well matched, and a poor match will correspondingly reduce systemwide NUE.

The spatial variability of soil resources occurs at a number of scales, ranging from the sub-mm scale of the rhizosphere to the 10^3 km scale of regional landscapes (Robertson and Gross, 1994). From a local management perspective only two of these scales are relevant—variability associated with row spacing, i.e., row–interrow heterogeneity, and variability associated with field-scale heterogeneity, i.e., historical features of the landscape that can lead to an uneven distribution of soil resources across individual fields.

Row–interrow effects on nutrient availability have been recognized for decades (e.g., Linn and Doran, 1984; Klemedtsson et al., 1987), and a number of management strategies take these differences into consideration to increase both water and nutrient use efficiencies of row crop systems. Such strategies range from drip irrigation in vegetable, citrus, and sugarcane to banding fertilizer and/or herbicides within crop rows. Placing fertilizer within crop rows, rather than broadcasting or injecting it indiscriminately, typically increases fertilizer use efficiency substantially (e.g., Malhi and Nyborg, 1985). Other techniques can also rearrange row–interrow nutrient availability. Ridge tillage, for example, an important soil management technique for many low-input farmers in the U.S. Midwest (NRC, 1989), is a tillage technique whereby spatial asymmetry is minimized by periodically mounding the between-row A_p horizon into semipermanent ridges into which the crop has been planted. This appears to effectively concentrate labile organic matter and soil biotic activity within the rows, achieving the same effect as fertilizer banding.

Spatial heterogeneity at larger scales is also emerging as a management issue. Evidence is accumulating to suggest that soil nitrogen availability is highly variable in natural communities, with variable patches of fertility— e.g., higher soil organic matter levels or higher rates of potential nitrogen mineralization—occurring within fields at scales that can affect individual plants (e.g., Robertson et al., 1988; Hook et al., 1991). One might expect conversion to agriculture to remove this patchiness where it is not related to landscape-level geomorphological features such as slope position, but this does not appear to be the case. Rather, long-term cultivation appears to simply attenuate and enlarge precultivation fertility patches (Robertson et al., 1993), thus underscoring the persistence of biological legacies in even agricultural ecosystems (Franklin and Forman, 1987; Magnuson, 1990).

Whether the result of biological or geomorphologic processes, or—more likely—some combination of the two, field-scale soil variability (e.g., Fig. 6) appears to be a major feature of most cropping systems (Webster, 1985; Trangmar et al., 1985; Aiken et al., 1991; Dessaint et al., 1991; Robertson et al., 1997). This knowledge has led to the development of soil-specific crop management techniques (e.g., Mulla, 1993) that allow crop inputs to be applied at variable rates across individual fields. This approach holds great promise for better matching crop inputs to actual needs, and thus great promise for improving systemwide NUE. Consider, as a simple example, differences in plant productivity across a 100-ha corn field. It is likely— based on what we now know about spatial variability—that across perhaps 25% of the field maximum yields will be achieved with N inputs of 100 kg ha^{-1} N, across another 50% yields will respond to as much as 140 kg ha^{-1}

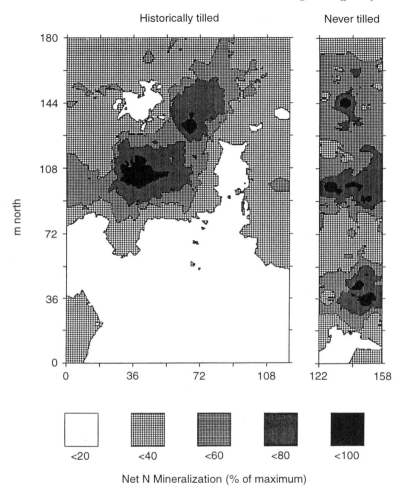

Figure 6 The variability of soil nitrogen availability (nitrogen mineralization potentials) across a 0.5-ha field cropped for decades in southwest Michigan. From Robertson *et al.* (1993).

N, and across the remaining patches of low fertility yields will respond to 180 kg ha^{-1} N. Standard management practices designed to maximize yield would proscribe fertilization for the lowest common denominator, i.e., as much as 180 kg ha^{-1} N across the entire field—despite the fact that only 25% of the field will respond to this level of fertilization. Over the remaining 75% of the field some 40–80 kg ha^{-1} N will be ignored by the crop and thus is highly likely to be lost. An agronomic approach that allows rates of fertilizer application across this field to vary by geographic position from 100 to 180 kg ha^{-1} N will substantially decrease the nitrogen available for total field loss. The direct economic benefit will likely be minor at least for the next decade or two—fertilizer is currently inexpensive and the cost of the technology required to implement site-specific farming practices is still prohibitive for most producers—but the environmental benefit could be substantial.

IV. Major Pathways of Nitrogen Loss from Cropping Systems

Nitrogen not removed in harvest is lost through three major pathways in most row-crop ecosystems: fire, leaching, and denitrification. Other pathways of loss can be important in specific systems or climates but are less important in general because of improved crop management practices. Nitrogen losses from erosion and runoff, for example, can still be high in some cropping systems but have been minimized elsewhere by the adoption of conservation tillage and sound residue management techniques. Protection from fire, leaching, and denitrification, however, appears to be more difficult to implement, though probably more often because of economic or management trade-offs than because of a lack of agronomic knowledge.

A. Fire

Fire has historically contributed to nitrogen losses in traditional bush-fallow rotation systems, and fire remains an important loss vector in many parts of the tropics. Robertson and Rosswall (1986), for example, estimated that in 1978 in West Africa, 5.5×10^9 kg N was lost annually from mainly mid-successional forests cleared for the crop phase of shifting cultivation. Annual residue burning during this phase and from continuous cultivation systems can also remove substantive amounts of nitrogen from the local system. In West Africa, ca. 253×10^6 kg N year^{-1} was volatilized by fire from all annual crops in 1978, a value that is not far from the 870×10^6 kg N estimated to have been removed from these cropping systems in primary yield.

Burning losses from individual crops range widely owing to crop-specific management practices and residue nitrogen contents. Losses from West African sorghum summed to 43×10^6 kg N with 152×10^6 kg N removed in yield, whereas burning losses from cassava were less than 0.1×10^6 vs. 43×10^6 kg N removed in yield. Even in developed economies with fully mechanized agriculture, residue burning remains a conventional agronomic practice in specific situations, usually for pathogen control. For example, in South Australia and Kansas burning is used to control take-all disease in continuous wheat rotations, and in Texas and California it has been common to burn lowland rice residue.

B. Leaching

Nitrogen losses to leaching are important in all cropping systems in which precipitation exceeds actual evapotranspiration (AET), i.e., in all systems with water flow through the solum. In semiarid cropping systems leaching may thus occur only rarely, and this may also be the case in more humid regions during the growing season—in the northern United States, for example, growing season AET is often sufficient to keep most rainwater from percolating through the soil profile and thus sufficient to keep leachable nitrogen within the rooting zone (e.g., Allison, 1973). Outside of the growing season, however, when AET is low relative to precipitation and when precipitation remains moderately high (ca. 75 mm month^{-1}), leaching losses can be high (e.g., Chichester, 1977). This can be especially true in the fall and spring when soil nitrate pools are also moderately high (ca. 5–10 μg N g soil $^{-1}$) from a combination of low plant uptake and the ongoing mineralization of soil organic matter nitrogen while soil temperatures and moisture are sufficient to support microbial activity.

Actual amounts of nitrogen leached from the pool of available nitrate (NO_3^-) are a complex function of the amount of water that moves through the profile and its path and residence time—controlled largely by the soil's pore geometry and aggregate structure. Water that moves quickly through a profile may do so primarily through macropores and may thus pick up very little of the nitrate in smaller pores and inside soil aggregates (e.g., Thomas and Phillips, 1979; Helling and Gish, 1991; Smucker *et al.*, 1995). Slower moving wetting fronts may, on the other hand, spend sufficient time in the profile to equilibrate with micropore and intraaggregate spaces that may contain high concentrations of nitrate from high rates of microbial activity and prior protection from the fast-moving wetting fronts. These slower-moving fronts may thus leach proportionately more nitrogen from the system.

There is a large body of literature regarding nitrogen leaching at the plot, field, and landscape scales, mostly dating from the 1970s when health concerns caught up with widespread fertilizer use (see reviews in Singh and Sekhon, 1979; Burden, 1982; Keeney, 1982; Legg and Meisinger, 1982; Magee, 1982; Hallberg, 1989). In general, reported studies have delineated many of the important interrelationships among the matrix of factors that regulate leaching—climate, soil, crop, agronomic management practices, and landscape position—and have fueled the development of quantitative models for predicting leaching losses from cropped ecosystems in general (e.g., Godwin and Jones, 1992; Tanji and Nour el Din, 1992; Paustian *et al.*, 1992). Many of these models do an excellent job of predicting leaching losses under specific conditions, but it is clear from their performance under other conditions and in other systems that our understanding of leaching losses at the ecosystem level is still incomplete.

C. Denitrification

Denitrification is the conversion of soil nitrate to the nitrogen gases N_2O and N_2 by a diverse array of bacteria that use nitrate as a terminal electron acceptor in the absence of oxygen. The importance of this process as a pathway of N loss in upland cropping systems is evident primarily from mass balance studies, in which denitrification is assumed to equal the difference between known inputs and harvest plus leaching outputs (e.g., Allison, 1953; Frissel, 1977; Legg and Meisinger, 1982; Paustian et al., 1990). Not until the development of a technique for studying in situ denitrification directly—using acetylene to block N_2O reductase (Yoshinari et al., 1977)—was the importance of this pathway fully appreciated. Various studies have since proceeded to confirm the magnitude of losses estimated from mass balance approaches and to identify the mechanisms whereby denitrification can be active even in well-drained soils.

It is now well accepted that denitrification can occur wherever available carbon, nitrate, and low oxygen concentrations co-occur—in aggregates (e.g., Sexstone et al., 1985a), soil organic matter particles (e.g., Parkin, 1987), and in bulk soil following precipitation events. Cultivation can further favor denitrification via its effects on soil bulk density, infiltration rates, and soluble C levels. Field estimates of denitrification rates range from nil to >100 kg N ha^{-1} year^{-1} in different upland cropping systems (von Rheinbaben, 1990; Aulakh et al., 1992; Peoples et al., 1995), a range similar to that of leaching losses. In general, annual rates appear to be on the order of 20–50 kg N ha^{-1} year^{-1}. As a proportion of total N loss, the importance of denitrification relative to leaching losses appears to range from no leaching with moderate denitrification, denitrification and leaching about equal, to moderate leaching with no denitrification (e.g., Jones et al., 1977; Winteringham, 1980).

Part of the difficulty of measuring—and therefore of understanding—denitrification sufficiently well to accurately model the process at a field scale is related to denitrification's high spatial (e.g., Folorunso and Rolston, 1984; Robertson et al., 1988) and temporal (e.g., Sexstone et al., 1985b; Parsons et al., 1991) variability in most ecosystems. Perhaps because of this variability, minimizing denitrification losses has not been a specific target of management techniques to date except in lowland rice production, where denitrification is known to represent a very high proportion of nitrogen loss.

V. Management to Maximize Nitrogen Use in Row-Crop Ecosystems

Strategies to maximize nitrogen use in row-crop ecosystems must necessarily concentrate on nitrogen retention. For some cropping systems—in particular for low-yield or low-fertility systems—our understanding of the processes involved and their major field-level controls is probably sufficient to allow the design of nitrogen-efficient rotations. This is not the case for most high-yield systems, however, in which an enormous amount of nitrogen is made available to the crop over a very short-term period of rapid growth. In these systems, making 100–200 kg N ha^{-1} available over a 6- to 8-week period without saturating the system with excess nitrogen remains a substantial challenge.

The principle barriers to efficient nitrogen use in high-yield cropping systems are both economic and scientific. From a scientific standpoint, we lack basic knowledge of the ecological interactions in soil that regulate the seasonal timing of nitrogen release from organic matter. We also lack knowledge about the field-level controls on important pathways of loss— notably about controls on denitrification and leaching, and in particular on ways in which they interact.

From an agronomic standpoint, in some cases minimizing nitrogen loss will be easy and relatively inexpensive. Devising for tropical regions a means for residue control other than burning can save as much as a third of the nitrogen typically removed in grain yields for many crops. The emerging adoption of new technology that allows crop inputs to be applied differentially across individual fields can help to better match crop demands for nitrogen with the supply of fertilizer N, thereby reducing excess application. On the other hand, devising a cover crop or residue management strategy that can help to synchronize soil nitrogen availability with nitrogen release from soil organic matter may first require the basic ecological knowledge noted previously.

In many regions, however, the implementation of nitrogen-efficient cropping practices may ultimately depend on economic factors. Until the cost of nitrogen fertilizer increases, or until the environmental costs that are now externalized are made more direct, there may be little incentive for growers to adopt nitrogen conservation strategies that require additional expense or the management of additional information.

To summarize, agronomists today have many strategies at their disposal for designing nitrogen-efficient cropping systems, and, with the development of new knowledge about ecological interactions in cropping systems, will have additional strategies in the coming years. The implementation of these strategies will not be cost-free, however, and may require marketplace and societal pressures in order to be enacted.

Acknowledgments

I thank P. L. Ambus, T. T. Bergsma, J. C. Boles, and M. A. Cavigelli for insightful comments on the manuscript and for helpful discussions during its preparation; L. E. Jackson, R. R. Harwood, and an anonymous reviewer also made many helpful comments on an earlier draft. Funding for this contribution was provided by the NSF LTER Program and the Michigan Agricultural Experiment Station.

References

Aiken, R. M., Jawson, M. D., Grahammer, K., and Polymenopoulos, A. D. (1991). Positional, spatially correlated, and random components of variability in carbon dioxide efflux. *J. Environ. Qual.* **20,** 301–308.

Allison, F. E. (1955). The enigma of soil nitrogen balance sheets. *Adv. Agron.* **7,** 213–250.

Allison, F. E. (1973). "Soil Organic Matter and Its Role in Crop production." Elsevier, New York.

Angus, J. F. (1995). Modeling nitrogen fertilization requirements for crops and pastures. *In* "Nitrogen Fertilization in the Environment" (P. E. Bacon, ed.), pp. 109–128. Dekker, New York.

Aulakh, M. S., Doran, J. W., and Mosier, A. R. (1992). Soil denitrification—significance, measurement, and effects of management. *Adv. Soil Sci.* **18,** 1–57.

Baldock, J. O., and Musgrave, R. B. (1980). Manure and mineral fertilizer effects in continuous and rotational crop sequences in central New York. *Agron. J.* **72,** 511–518.

Bauer, A., and Black, A. L. (1981). Soil carbon, nitrogen and bulk density comparisons in two cropland tillage systems after 25 years and in virgin grassland. *Soil Sci. Soc. Am. J.* **45,** 1166–1170.

Birch, H. F. (1960). Nitrification in soils after different periods of dryness. *Plant Soil* **12,** 81–96.

Birk, E. M., and Vitousek, P. M. (1986). Nitrogen availability, nitrogen cycling, and nitrogen use efficiency in loblolly pine stands on the Savannah River Plant, South Carolina. *Ecology* **67,** 69–79.

Boerner, E. J. (1984). Foliar nutrient dynamics and nutrient use efficiency of four deciduous tree species in relation to site fertility. *J. Appl. Ecol.* **21,** 1029–1040.

Breman, H., and deWit, C. T. (1983). Rangeland productivity and exploitation in the Sahel. *Science* **221,** 1341–1344.

Broadbent, F. E., and Carlton A. B. (1978). Field trials with isotopically labelled nitrogen fertilizer. *In* "Nitrogen in the Environment" (D. R. Nielsen and J. G. MacDonald, eds.), pp. 1–41. Academic Press, New York.

Bumb, B. L. (1995). World nitrogen supply and demand: An overview. *In* "Nitrogen Fertilization in the Environment" (P. E. Bacon, ed.), pp. 1–41. Dekker, New York.

Burden, R. J. (1982). Nitrate contamination of New Zealand aquifers: A review. *New Zealand J. Soil Sci.* **25,** 205–220.

Chichester, F. W. (1977). Effects of increased fertilizer rates on nitrogen content of runoff and percolate from monolith lysimeters. *J. Environ. Qual.* **6,** 211–216.

Dessaint, F., Chadoeuf, R., and Barralis, G. (1991). Spatial pattern analysis of weed seeds in the cultivated soil seed bank. *J. Appl. Ecol.* **28,** 703–720.

Ebelhar, S. A., Frye, W. W., and Blevins, R. L. (1984). Nitrogen from legume cover crops for no-tillage corn. *Agron. J.* **76,** 51–55.

Fee, R. (1995). Nitrogen prices ready to rise. *Successful Farming* **98,** 11.

Folorunso, O. A., and Rolston, D. E. (1984). Spatial variability of field-measured denitrification gas fluxes. *Soil Sci. Soc. Am. J.* **48,** 1214–1219.

Food and Agriculture Organization (FAO) (1945–1992). "Fertilizer Yearbooks" (annual issues). FAO, Rome.

Francis, C. A., and Clegg, M. D. (1990). Crop rotations in sustainable agricultural systems. *In* "Sustainable Agricultural Systems" (C. A. Edwards, R. Lal, P. Madden, R. H. Miller, and G. House, eds.), pp. 107–122. Soil and Water Conservation Society, Ankeny, IA.

Franklin, J. F., and Forman, R. T. T. (1987). Creating landscape patterns by forest cutting: Ecological consequences and principles. *Landscape Ecol.* **1,** 5–18.

Frissel, M. J. (1977). Cycling of mineral nutrients in agricultural ecosystems. *Agro-Ecosystems* **4,** 1–354.

Godwin, D. C., and Jones, C. A. (1992). Nitrogen dynamics in soil plant systems. *In* "Modeling Plant and Soil Systems" (J. Hanks, and J. T. Ritchie, eds.), pp. 287–322. American Society of Agronomy, Madison, WI.

Haas, H. J., Evans, C. E., and Miles, E. F. (1957). Nitrogen and carbon changes in Great Plains soils as influenced by cropping and soil treatments. USDA Technical Bull. No. 1164.

Hallberg, G. R. (1989). Nitrate in ground water in the United States. *In* "Nitrogen Management and Ground Water Protection" (R. F. Follett, ed.). Elsevier, New York.

Harris, G. H., Hesterman, O. B., Paul, E. A., Peters, S. E., and Janke, R. R. (1994). Fate of legume and fertilizer ^{15}N in a long-term cropping systems experiment. *Agron. J.* **86,** 910–915.

Hart, S. C., Stark, J. M., Davidson, E. A., and Firestone, M. K. (1994). Nitrogen mineralization, immobilization, and nitrification. *In* "Methods of Soil Analysis: Part 2 Microbiological and Biochemical Properties" (J. M. Bigham and S. H. Mickelson, eds.), pp. 985–1018. Soil Science Society of America, Madison, WI.

Harwood, R. R. (1990). A history of sustainable agriculture. *In* "Sustainable Agricultural Systems" (C. A. Edwards, R. Lal, P. Madden, R. H. Miller, and G. House, eds.), pp. 3–19. Soil and Water Conservation Society, Ankeny, IA.

Helling, C. S., and Gish, T. J. (1991). Physical and chemical processes affecting preferential flow. *In* "Preferential Flow Proceedings of the National Symposium, Dec. 1991, Chicago" (T. J. Gish and A. Shirmohammadi, eds.), pp. 77–86.

Hook, P. B., Burke, I. C., and Lauenroth, W. K. (1991). Heterogeneity of soil and plant N and C associated with individual plants and openings in North American shortgrass steppe. *Plant Soil* **138,** 247–256.

Jones, M. B., Delwiche, C. C., and Williams, W. A. (1977). Uptake and losses of ^{15}N applied to annual grass and clover in lysimeters. *Agron. J.* **69,** 1019–1023.

Jones, U. S. (1982). "Fertilizers and Soil Fertility." Prentice Hall, New York.

Juma, M., and Paul, E. A. (1984). Kinetic analysis of soil nitrogen mineralization in soil. *Soil Sci. Soc. Am. J.* **48,** 753–758.

Keeney, D. R. (1980). Prediction of soil nitrogen availability in forest ecosystems: A literature review. *For. Sci.* **26,** 159–171.

Keeney, D. R. (1982). Nitrogen management for maximum efficiency and minimum pollution. *In* "Nitrogen in Agricultural Soils" (F. J. Stevenson, ed.), pp. 605–650. American Society of Agronomy, Madison, WI.

Klemedtsson, L., Berg, M., Clarholm, J., Schnurer, J., and Roswall, T. (1987). Microbial nitrogen transformations in the root environment of barley. *Soil Biol. Biochem.* **19,** 551–558.

Ladd, J. N., and Jackson, R. B. (1982). Biochemistry of ammonification. *In* "Nitrogen in Agricultural Soils" (F. J. Stevenson, ed.), pp. 173–228. American Society of Agronomy, Madison, WI.

Legg, J. O., and Meisinger, J. J. (1982). Soil nitrogen budgets. *In* "Nitrogen in Agricultural Soils" (F. J. Stevenson, ed.), pp. 503–566. American Society of Agronomy, Madison, WI.

Linn, D. M., and Doran, J. W. (1984). Effect of water-filled pore space on CO_2 and N_2O production in tilled and non-tilled soils. *Soil Sci. Soc. Am. J.* **48,** 1267–1272.

Loomis, R. S., and Connor, D. J. (1992). "Crop Ecology: Productivity and Management in Agricultural Systems." Cambridge Univ. Press, Cambridge, UK.

Magee, P. N. (1982). Nitrogen as a potential health hazard. *Philos. Trans. R. Soc. London. B* **296,** 241–248.

Magnuson, J. J. (1990). Long-term ecological research and the invisible present. *BioScience* **40,** 495–500.

Malhi, S. S., and Nyborg, M. (1985). Methods of placement for increasing the efficiency of nitrogen fertilizers applied in the fall. *Agron. J.* **77,** 27–32.

Matson, P. A., Vitousek, P. M., Ewel, J. J., Mazzarino M. J., and Robertson, G. P. (1987). Nitrogen cycling following land clearing in a premontane wet tropical forest near Turrialba, Costa Rica. *Ecology* **68,** 491–502.

Meisinger, J. J., and Randall, G. W. (1995). Estimating nitrogen budgets for soil-crop systems. *In* "Managing Nitrogen for Groundwater Quality" (R. F. Follett, D. R. Keeney, and R. M. Cruse, eds.), pp. 85–124. Soil Science Society of America, Madison, WI.

Meisinger, J. J., Randall, G. W., and Vitosh, M. L. (eds.) (1980). "Nitrification Inhibitors—Potentials and Limitations." American Society of Agronomy, Madison, WI.

Mulla, D. J. (1993). Mapping and managing spatial patterns in soil fertility and crop yield. *In* "Soil Specific Crop Management" (P. C. Robert, R. H. Rust, and W. E. Larson, eds.), pp. 15–26. American Society of Agronomy, Madison, WI.

National Research Council. (NRC) (1989). "Alternative Agriculture." National Academy Press, Washington, DC.

Neeteson, J. J. (1995). Nitrogen management for intensively grown arable crops and field vegetables. *In* "Nitrogen Fertilization and the Environment" (P. E. Bacon, ed.), pp. 295–325. Dekker, New York.

Oakley, R. A. (1925). The economics of increased legume production. *J. Am. Soc. Agron.* **17,** 373–389.

Olson, R. A., and Kurtz, L. T. (1982). Crop nitrogen requirements, utilization and fertilization. *In* "Nitrogen in Agricultural Soils" (F. J. Stephenson, ed.), pp. 567–604. American Society of Agronomy, Madison, WI.

Parkin, T. B. (1987). Soil microsites as a source of denitrification variability. *Soil Sci. Soc. Am. J.* **51,** 1194–1199.

Parsons, L. L., Murray, R. E., and Smith, M. S. (1991). Soil denitrification dynamics: Spatial and temporal variations of enzyme activity, populations, and nitrogen gas loss. *Soil Sci. Soc. Am. J.* **55,** 90–95.

Paustian, K., Andren, O., Clarholm, M., Hansson, A. C., Johansson, G., Lagelof, J., Lindberg, T., Pettersson, R., and Sohlenius, B. (1990). Carbon and nitrogen budgets of four agroecosystems with annual and perennial crops, with and without N fertilization. *J. Appl. Ecol.* **27,** 60–84.

Paustian, K., Parton, W. J., and Persson, J. (1992). Modeling soil organic matter in organic-amended and nitrogen-fertilized long-term plots. *Soil Sci. Soc. Am. J.* **56,** 476–488.

Paustian, K., Robertson, G. P., and Elliott, E. T. (1995). Management impacts on carbon storage and gas fluxes (CO_2, CH_4) in mid-latitude cropland and grassland ecosystems. *In* "Soil Management and Greenhouse Effect" (R. Lal, J. Kimble, E. Levine, and B. A. Stewart, eds.), pp. 69–84. CRC Press, Boca Raton, FL.

Peoples, M. B., Mosier, A. R., and Freney, J. R. (1995). Minimizing gaseous losses of nitrogen. *In* "Nitrogen Fertilization in the Environment" (P. E. Bacon, ed.), pp. 565–602. Dekker, New York.

Robertson, G. P. (1982). Nitrification in forested ecosystems. *Philos. Trans. R. Soc. London* **296,** 445–457.

Robertson, G. P., and Gross, K. L. (1994). Assessing the heterogeneity of belowground resources: Quantifying pattern and scale. *In* "Plant Exploitation of Environmental Heterogeneity" (M. M. Caldwell and R. W. Pearcy, eds.), pp. 237–253. Academic Press, New York.

Robertson, G. P., and Rosswall, T. (1986). Nitrogen in West Africa: The regional cycle. *Ecol. Monogr.* **56**, 43–72.

Robertson, G. P., Huston, M. A., Evans, F., and Tiedje, J. M. (1988). Spatial variability in a successional plant community: Patterns of nitrogen availability. *Ecology* **69**, 1517–1524.

Robertson, G. P., Crum, J. R., and Ellis, B. G. (1993). The spatial variability of soil resources following long-term disturbance. *Oecologia* **96**, 451–456.

Robertson, G. P., Klingensmith, K. M., Klug, M. J., Paul, E. A., Crum, J. C., and Ellis, B. G. (1997). Soil resources, microbial activity, and plant productivity across an agricultural ecosystem. *Ecol. Appl.* **7**, 158–170.

Sexstone, A. J., Revsbech, N. P., Parkin, T. P., and Tiedje, J. M. (1985a). Direct measurement of oxygen profiles and denitrification rates in soil aggregates. *Soil Sci. Soc. Am. J.* **49**, 645–651.

Sexstone, A. J., Parkin, T. P., and Tiedje, J. M. (1985b). Temporal response of soil denitrification rates to rainfall irrigation. *Soil Sci. Soc. Am. J.* **49**, 99–103.

Shaver, G. R., and Melillo, J. M. (1984). Nutrient budgets of marsh plants: Efficiency concepts and relation to availability. *Ecology* **65**, 1491–1510.

Singh, B., and Sekhon, G. S. (1979). Nitrate pollution of groundwater from farm use of nitrogen fertilizers: A review. *Agric. Environ.* **4**, 207–224.

Soderlund, R., and Svensson, B. H. (1976). The global nitrogen cycle. *In* "Nitrogen, Phosphorus and Sulphur—Global Cycles" (B. H. Svensson and R. Soderlund, eds.), Ecological bull., pp. 23–74. Stockholm, Sweden.

Smucker, A. J. M., Richner, W., and Snow, V. O. (1995). Bypass flow via root-induced macropores (RIMS) in subirrigated agriculture. *In* "Clean Water, Clean Environment for the 21st Century Conf. Proc. Vol. III: Practices, Systems, and Adoption, Kansas, City, KS," pp. 255–258.

Sollins, P., Robertson, G. P., and Uehara, G. (1988). Nutrient mobility in variable- and permanent-charge soils. *Biogeochemistry* **6**, 181–199.

Tanji, K. K., and Nour el Din, M. (1992). Nitrogen solute transport. *In* "Modeling Plant and Soil Systems" (J. Hanks and J. T. Ritchie, eds.), pp. 341–364. American Society of Agronomy, Madison, WI.

Thomas, G. W., and Phillips, R. E. (1979). Consequences of water movement in macropores. *J. Environ. Qual.* **8**, 149–152.

Trangmar, B., Yost, R., and Uehara, G. (1985). Application of geostatistics to spatial studies of soil properties. *In* "Advances in Agronomy" (N. C. Brady, ed.), pp. 45–94. Academic Press, New York.

Turner, J. (1977). Effect of nitrogen availability on nitrogen cycling in a douglas-fir stand. *For. Sci.* **23**, 307–316.

Uehara, G., and Gillman, G. (1981). "The Mineraology, Chemistry and Physics of Tropical Soils with Variable Charge Clays." Westview Press, Boulder, CO.

Vitousek, P. M., and Matson, P. A. (1993). Agriculture, the global nitrogen cycle and trace gas flux. *In* "The Biogeochemistry of Global Change: Radiatively Active Trace Gases" (R. Oremland, ed.), pp. 193–208. Chapman & Hall, New York.

von Rheinbaben, W. (1990). Nitrogen losses from agricultural soils through denitrification—A critical evaluation. *Z. Pflanzenernahr. Bodenk* **153**, 157–166.

Webster, R. (1985). Quantitative spatial analysis of soil in the field. *In* "Advances in Soil Science" (B. A. Stewart, ed.), pp. 1–70. Springer-Verlag, New York.

Winteringham, F. P. W. (1980). Nitrogen balance and related studies: A global review. *In* "Soil Nitrogen as Fertilizer or Pollutant." IAEA, Vienna, Austria.

Yoshinari, T., Hynes, R., and Knowles, R. (1977). Acetylene inhibition of nitrous oxide reduction and measurement of denitrification and nitrogen fixation in soil. *Soil Biol. Biochem.* **9**, 177–183.

11

Soil Microbial Communities and Carbon Flow in Agroecosystems

Kate M. Scow

I. Introduction

The importance of the soil carbon (C) cycle is often overlooked in traditional agricultural studies because the primary focus is on the crop, which is not subject to carbon limitations, and on those nutrients such as nitrogen (N) that do limit productivity. The decomposition portion of the carbon cycle, however, governs many agronomic processes that occur below ground and manifest themselves above ground. Microorganisms control the decomposition of C and their activity regulates nutrient cycling in soils. Even though the consequences of their activities can be quite obvious, the presence of most microorganisms is usually taken for granted in cropping systems. In studies of ecosystems, microorganisms are generally considered not as organisms, in an autoecological sense, but as disembodied rates and pools of nutrients. A better mechanistic understanding of many ecosystem processes could be obtained from knowledge of microbial physiology and metabolism, of the factors controlling microbial populations and activities, and of the spatial and temporal distribution of microorganisms.

Steady-state levels of organic carbon in soil are governed by both the input of C, primarily as plant products, and the output of C mediated by the activity of decomposers (Fig. 1). A large fraction of photosynthesized carbon flows through the belowground ecosystem (Heal and MacLean, 1975; Coleman *et al.*, 1976) with most of it passing through microorganisms. Carbon cycling in agroecosystems has a significant impact at the global

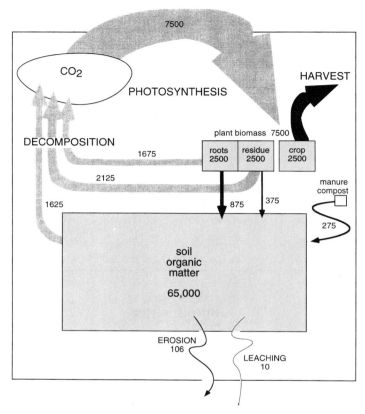

Figure 1   Carbon cycling in an agroecosystem (adapted from Nature and Properties of Soils, 11/E by Brady/Weil, © 1996. Reprinted by permission of Prentice-Hall, Inc., Upper Saddle River, NJ).

scale because agriculture occupies approximately 11% of the land surface area of the earth (Paustian *et al.,* 1995).

This review is concerned with carbon flow through soils in agroecosystems and with the relationships between microbial communities and the biogeochemical cycling of carbon. Primarily, the decomposition portion of the carbon cycle is considered. Emphasis is on microbial populations and with only minor consideration of community interactions and the soil food web. Carbon flow is considered first at the individual microbial cell level, then at the scale of the population and community, and finally in the soil environment, at the pore and aggregate scale. Implications of carbon cycling for the sustainability of agroecosystems are discussed as is the potential for managing soil biology and carbon flow.

In an ecological context, the following concepts will be developed. Microbial autoecology and community ecology affect C flows at the system level. Microbial ecophysiology is an important factor for C flow in regard to production and assimilation and given starvation of soils. Community structure can influence C dynamics. Spatial and temporal factors affect C availability to microorganisms with ramifications for C flow. All this, in turn, affects agricultural productivity and N and C budgets of environmental importance, e.g., global greenhouse gases.

II. Carbon Flow at the Cellular Level

Because many important transformations of carbon are governed by microbial activities, it is important to understand such processes as they occur at the level of the individual microbial cell. Although there is tremendous diversity in the metabolic pathways of microorganisms, certain generalizations can be made for the aerobic, heterotrophic organisms that dominate soil communities in upland agricultural systems. Figure 2 is a simplified depiction of the fate of an organic C compound during its metabolism by

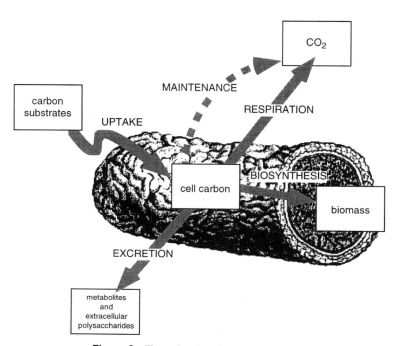

Figure 2   Flow of carbon in a bacterial cell.

a bacterium. This substrate not only provides a source of energy, resulting in the formation of carbon dioxide (respiration), but also a source of carbon, resulting in production of new cells (assimilation). Other potential fates of C are the formation of extracellular polymers and intermediate metabolites that ultimately become associated with soil organic matter. Fixation of carbon dioxide by autotrophic bacteria, such as lithotrophs and phototrophs, and by many organisms via carboxylation reactions also occurs. However, these processes are not considered further because they have little impact on C flow at the ecosystem scale. The following sections provides an overview of each of the major processes and discusses the implications, for agricultural soils, of how microorganisms use carbon.

A. Respiration

The use by microorganisms of organic C compounds as electron donors usually produces carbon dioxide, regardless of whether the terminal electron acceptor is oxygen (aerobic conditions) or nitrate (oxygen limited). Conversion from organic to inorganic forms of molecules is termed mineralization. Most organisms capable of C respiration possess the tricarboxylic acid cycle (TCA), at least in part, during which carbon dioxide is released in two reaction steps in the cycle (Brock and Madigan, 1991). Carbon dioxide is also released in the decarboxylation of pyruvate to form acetyl CoA before entry into the TCA cycle. Additionally, decarboxylation occurs in gluconeogenesis, the pentose phosphate pathway, and other reactions; however, the majority of carbon dioxide produced by growing cells is associated with the TCA cycle.

As conditions become more reduced, such as in fine-textured, wet, and/ or carbon-amended soils, then less carbon dioxide is produced. During fermentation, far less carbon dioxide is generated than through respiration and the majority of the carbon is excreted as organic compounds (organic acids and alcohols). Predominance of such reducing conditions is usually temporary in agricultural soils, however, and fermentation products, still rich in energy, are readily mineralized as soon as sufficient oxygen is present. Under highly reduced conditions, there is virtually no production of carbon dioxide; instead, processes carried out by methanogenic bacteria result in the generation of methane from the consumption of carbon dioxide as a terminal electron acceptor. With the exception of flooded agroecosystems such as rice, however, such highly reduced conditions are uncommon in agricultural soils.

B. Assimilation and the Yield Coefficient

A consequence of the growth of any organism is the incorporation of carbon from the environment into cellular tissue. Thus, the other major function of the TCA cycle is the formation of key intermediates for biosynthesis. Intermediates from the TCA cycle provide the C skeletons for the production of saccharides, amino acids, nucleic acids, and fatty acids. These monomers are then assembled into the polymers that constitute the majority of the dry weight of a microbial cell. Important groups of polymers include the polysaccharides, proteins, DNA, RNA, and lipids. In general, the monomers, which are often associated with cell cytoplasm, are readily degradable in the soil environment. The polymers, many of which are associated with the cell wall and membranes, are more slow to decompose. Among the polymers are highly resistant structures, for example, lignin and tannin, often associated with cell walls and pigments.

Major differences are evident between microorganisms and animals in regard to the flow of carbon through the individual. Assimilation efficiency, which is the proportion of ingested carbon that is assimilated into tissues rather than excreted, is helpful for describing how animals relate to the C cycle because it estimates how much carbon passes untransformed through animal biomass. Assimilation efficiency is less useful for describing microorganisms because microbial uptake and assimilation are often equivalent (e.g., efficiency is 100%) (Heal and MacLean, 1975). Large, complex organic molecules are digested by enzymes outside the cell wall and converted into small compounds that are easy for microorganisms to take up. Also, whereas respiration accounts for a small portion of the carbon consumed by animals and plants, it is a major fate of the carbon consumed by microorganisms. Thus, the production efficiency, which is the proportion of a food source that is used to make biomass rather than respired as carbon dioxide (Heal and MacLean, 1975), is useful for analyzing the impact of microorganisms on the C cycle because of the information it provides about the fate of carbon. Production efficiency is synonymous with the concept of yield, a term commonly used in microbiology.

There are numerous approaches for calculating yield, including calculations based on mass or percentage C of substrate consumed, number of electrons transferred, or amount of ATP formed in a reaction (Payne, 1970; McCarty, 1971). In this article, yield is defined as gram cells per gram carbon source, which is directly equivalent to P/A.

The calculated yield is presumed to be constant in pure cultures of bacteria when organisms are cultivated in standardized liquid medium under constant conditions (Payne, 1970). The value of the yield coefficient is dependent on the organism but, more important on the particular substrate and electron acceptor used. Yields of many bacteria in pure culture are approximately 60% (Payne, 1970) when growing on simple organic compounds such as sugars and amino acids. From these earlier studies, the widespread use of a yield factor of 50–60% has been adopted as a rule of thumb in many microbiological studies. However, growth on methane or long-chained hydrocarbons results in lower yields, on the order of 25%

(Payne, 1970). Yields of aerobic growth on a variety of substrates ranged from 12 to 60% depending on the compound and bacterium involved (McCarty, 1971, 1988). Yields of fungi are usually assumed higher than those of bacteria (Alexander, 1977); however, limited data are available even for common soil fungi.

Yield varies as a function of various factors. Microorganisms using electron acceptors other than oxygen (e.g., sulfate or carbon dioxide) or obtaining energy from fermentation have yields that are quite low, in the range from 4 to 20% (McCarty, 1971, 1988). Exceptions are organisms using nitrate as an electron acceptor (denitrifiers) that have yields equivalent to organisms using oxygen (McCarty, 1971, 1988). Yields for lithoautotrophs, organisms that obtain their electrons from inorganic molecules, are dependent on the reduction potential of the redox pair that serves as the electron donor.

Consistency in cultural and environmental conditions is not an attribute of nature and this makes generalizations about yields more difficult for soil than for pure cultures. Yields of approximately 60% have been observed in mixed cultures of soil and sewage bacteria on pure chemicals as well as on more complex substrates (Payne, 1970). However, in soil studies Jenkinson and Powlson (1976b) and Shields *et al.* (1974) reported yield values on the order of 40–60% on simple sugars; Heal and MacLean (1975) reported yields ranging from 35 to 60% for mixed populations growing on relatively simple substrates; and Sparling *et al.* (1981) found values ranging from 8 to 84%.

In mixed cultures of microorganisms, one may expect variation in yield for the same substrate under identical growth conditions because of variations in the structure of the microbial community involved in metabolism (Gaudy and Gaudy, 1980). Increasing the carbon to nitrogen ratio of the substrate can increase the apparent cell yield (Gaudy and Gaudy, 1980). When nutrients are limiting, the yield may be half of the value measured under nutrient-sufficient conditions (Chesbro *et al.*, 1979). Yield can be reduced by accumulation of polymers, dissipation of heat by ATPase mechanisms, the presence of inhibitors, or unbalanced ionic concentrations (McCarty, 1971). These stresses, many of which are commonly encountered by soil populations, probably increase the amount of energy used for maintenance rather than growth, thus causing a decrease in yield.

C. Maintenance

A portion of the C respired as CO_2 is associated with expenditure of energy to maintain existing microbial cells. The importance of maintenance is obvious in organisms such as animals that may spend a large portion of their life span not actively growing. Because of the rapid growth rates of microorganisms, and the false assumption that microbial populations must grow to be active, maintenance is usually overlooked in microorganisms. Maintenance costs may be trivial for organisms growing in rich laboratory media and in short-term microbiological studies; however, as will be discussed below, these costs may consume a large portion of an organism's substrate under the carbon-starved conditions common to soils.

Maintenance C is the C used for purposes other than production of new cell material (Pirt, 1965). Because consumption of C for maintenance results in generation of energy—for osmotic regulation, maintenance of pH gradients, and motility—its ultimate fate is carbon dioxide. Carbon for maintenance may come from a substrate being metabolized or, under starvation conditions, from C stored in reserved cell material (see below). Maintenance has been defined in numerous ways, resulting in some debate; various approaches are well described by Smith *et al.* (1986). Most considerations of C cycling in soil use the original formulation of Marr *et al.* (1963) wherein the equation representing the growth rate is modified by addition of a maintenance cofficient that is multiplied by population:

$$(dS/dt)\ Y = (dx/dt) + ax,$$

where S is substrate concentration, Y is yield, x is the population size, and a is the maintenance rate per cell. Evolution of carbon dioxide is commonly used to measure C flux, in which case the change in the amount of product (e.g., carbon dioxide), corrected for maintenance (Pirt, 1975; Smith *et al.*, 1986), is

$$dP/dt = Y_{(p/x)}\ (dx/dt) + ax,$$

where P is product concentration and $Y_{(p/x)}$ is the yield of product per unit biomass formed. As mentioned previously, discrepancies in the yield term, often seen when comparing different substrate concentrations, may be due to a higher portion of the substrate C being diverted for maintenance energy that would be manifested as a lower apparent yield (Pirt, 1965, 1975, 1982).

Consideration of maintenance is hardly a trivial issue in the study of C flow in agricultural soils. Soil populations are usually starved for carbon (Sparling *et al,* 1981; Behara and Wagner, 1974; Anderson and Domsch, 1978; Jenkinson and Powlson, 1976a). Attempts to calculate generation times or turnover times of the soil microbial biomass have shown that annual C inputs to soil are less than the amount required to support the measured levels of microbial biomass and their hypothetical maintenance requirements, let alone to support increases in the biomass (Smith and Paul, 1990).

One of the problems with the analyses described previously is uncertainty regarding the magnitude of the maintenance coefficient (a as defined previously). Maintenance coefficients determined for laboratory cultures of bacteria were too high for soil populations, probably because laboratory cultures had not adapted to long-term exposures to nutrient deprivation (Anderson and Domsch, 1985). *Arthrobacter globiformas,* a soil bacterium, was found to decrease its maintenance coefficient substantially when starved for C (Chapman and Gray, 1981). Maintenance coefficients for soil microbial populations, assuming no net growth, were calculated based on data describing substrate consumption and steady-state biomass levels. The values ranged from 0.001 to 0.003 per hectare (Babiuk and Paul, 1970; Shields *et al.,* 1973; Behera and Wagner, 1974; Paul and Voroney, 1980), approximately an order of magnitude lower than values measured for laboratory cultures. Low maintenance demands of soil microorganisms are presumed to help organisms endure the oligotrophic conditions of soil (Dawes, 1985).

Smith and Paul (1989) and Smith (1989) estimated yearly requirements for microbial maintenance in soil based on estimated C inputs and biomass values in various ecosystems. Even assuming the lowest maintenance value reported for soil microorganisms (0.001 per hectare), maintenance demands still far exceeded annual C inputs for almost every system. Smith *et al.* (1986) proposed a model that includes an active and dormant microbial biomass, cryptic growth, and recycling of dead biomass. In a sensitivity analysis for individual parameters in his model, Smith (1989) found that calculated maintenance rates were most sensitive to total biomass and yield and insensitive to the ratio of the active to dormant biomass and the death rate. When the model was used to calculate maintenance for three ecosystems, and these values were compared to C inputs, inputs were found to exceed maintenance requirements such that microbial biomass could turn over up to eight times annually in a wheat cropping system (Smith, 1989). The calculated maintenance values were approximately 25–50% of the 0.001 per hectare value commonly used for maintenance (Smith, 1989). Although the estimates of maintenance improved in being more compatible with measured data, Smith (1989) concluded the estimates were still flawed by lack of consideration of temporal fluctuations in both C inputs and microbial physiology, both of which were represented by average annual values in the model.

D. Other Metabolic Products

Organisms not limited by carbon can undergo uncoupled growth, due to lack of other required nutrients, resulting in energy-spilling reactions that result in the synthesis of storage polymers and excretion of extracellular polymeric substances (Dawes, 1985). Storage compounds in microorganisms include carbohydrates (polyglucans, and glycogen), lipids (poly-β-hydroxybutyrate and polyalkanoates), polyphosphates, and cyanophycin/phycocyanin (Dawes, 1985). Poly-β-hydroxybutyrate may also serve as an electron sink for excess reducing power when cells become oxygen limited and activity of the electron transport chain ceases (Dawes, 1985). Accumulation of storage compounds appears to enhance survival of microbial isolates under starvation conditions. Other common products of microbial synthesis are alkyl C compounds such as polymethylene (Kinchesh *et al.*, 1995; Baldock *et al.*, 1990).

Many of the extracellular polymeric substances are polysaccharides containing common sugars (e.g., glucose, galactose, and mannose), uronic acids, pyruvate and N-acetylated amino sugars (e.g., glucosamine and galactosamine), as well as smaller amounts of more unusual sugars (Christensen, 1989). Most are hydrophilic and have zero or net negative charge. Both bacteria and fungi produce these substances. Examples of specific types of polysaccharides include cellulose, dextran, alginate, xanthan, scleroglucan, and gellan (Kenne and Lindberg, 1983). At high C input rates, especially when nitrogen or other nutrients are limiting, a considerable amount of substrate C may be converted to extracellular metabolites (Sparling *et al.*, 1981).

E. Implications of the Metabolic Fate of C for Soil Processes

What occurs at the level of the microbial cell has enormous impacts at the scale of the agroecosystem. Figure 3 summarizes potential interactions between the products of microbial metabolism and the soil. The most striking result of heterotrophic microbial metabolism is the loss of carbon from soil. It is common that from 40 to 60% of the C taken up by microorganisms is immediately released as carbon dioxide during metabolism. Higher microbial yields would therefore be desirable if the objective is to accumulate C in soil.

Assimilation has implications beyond simply the retention of C in microbial tissue. Not all compounds created by microorganisms are rapidly degraded. Cell wall components, being polymers, are less biodegradable than cytoplasm constituents. Some fungal cell walls, in particular, are highly resistant to breakdown because they are composed of polysaccharide–melanin complexes or of heteropolysaccharides (Martin and Haider, 1979; Alexander, 1977). For example, in soil, radiolabeled cell wall material from the fungus *Aspergillis* degraded more slowly than cell walls of the bacterium *Azotobacter* (Mayaudon and Simonart, 1963). Labeled melanins extracted from two fungal species were mineralized less than 10% in 3 months (Martin and Haider, 1979), which is far less than amounts mineralized from plant residues in the same time. Fungi and actinomycetes can directly produce humic-like material very similar to humus extracted from soil (Martin and Haider, 1971; Huntjens, 1972).

Also important to C flow is the physical location of microbial-produced substances and microorganisms in the soil matrix. Using electron microscopy, Foster (1981) found that soil extracellular polymers are most often

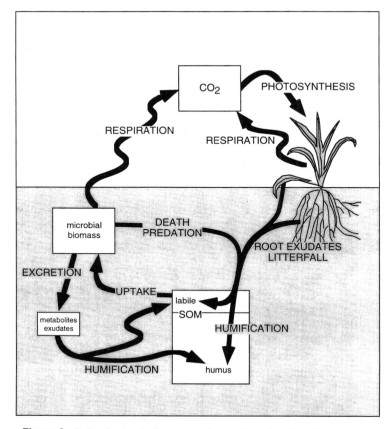

Figure 3   Role of microbial processes in the flow of carbon through soil.

associated with the clay fraction. Polysaccharides are protected from decomposition through their interaction with clay, metal ions, and tannins (Tisdall and Oades, 1982). Intact bacterial colonies and fungal hyphae in soil are commonly coated with fine clay, C of microbial origin is often extracted in the clay fraction of soil, and microbial–mineral associations are important in the formation of soil aggregates (Tisdall and Oades, 1982). The common association of microbial products (such as alkyl C compounds) with clay increases their resistance to mineralization (Guggenberger *et al.*, 1995; Kinchesh *et al.*, 1995). Interactions between microorganisms and the soil matrix are discussed in greater detail in a later section.

III. Carbon Flow: Populations and Communities

Although among microorganisms there are some similarities in how individual cells utilize carbon, C flow in soil is also affected by the size of microbial populations and the composition of their communities. Although the entire microbial biomass is usually treated as a single entity, major functional or taxonomic groups can sometimes be distinguished. Few studies concern interactions among functional groups, let alone among individual species, within the microbial community. Potential interactions have strong impacts, at least theoretically, on C flow and include (i) the exchange of transformed carbon compounds, often resulting in more efficient utilization of the potential energy present in organic material; (ii) the exchange of other chemical compounds, such as electron acceptors or nutrients; (iii) competition for C and other nutrients; and (iv) predation and parasitism, which lead to the introduction of C into the food web. The following sections will briefly consider C flow as it is related to the microbial biomass, to differences between bacteria and fungi, and to the soil food web.

A. Microbial Biomass

Given the enormous diversity of microorganisms in soil, it is not practical or even feasible to consider differences in C flow among species. Both the structure (e.g., community diversity and species distribution) and function (e.g., process rates) are important in characterizing ecosystems, but it has long been assumed in ecology that the latter becomes more important as the size of organism decreases (Odum, 1971). There have been serious methodological limitations in our ability to characterize microbial communities, although recently there have been major advances (Ritz *et al.*, 1994). New methods include extracting microbial cellular material, such as phospholipid fatty acids or DNA, directly from soil and using this information for estimating overall diversity or identifying taxonomic groups (Tunlid and White, 1992; Zelles *et al.*, 1992; Torsvik *et al.*, 1994). In studies of C flow, it is common to ignore species diversity and, at least implicitly, assume the total microbial biomass functions as a supraorganism, as represented as a single compartment in many C flow models (e.g., Jenkinson and Rayner, 1977). Similarly, flows between pools of C, largely controlled by microorganisms, are represented by single rate constants that reflect the collective activity of many species.

Substantial research over the past two decades has focused on the microbial biomass in soil (Smith *et al.*, 1993; Lynch and Panting, 1982; Paul and Voroney, 1980; Smith and Paul, 1989). Numerous methods have been developed to measure the soil microbial biomass, including the chloroform fumigation incubation method, the chloroform fumigation extraction method, substrate-induced respiration, and adenosine triphosphate analysis (Horwath and Paul, 1994). Microbial biomass varies significantly among soil types and vegetation systems. Values for biomass range from 30 to 2780 μg C per gram of soil (Bolton *et al.*, 1985). The C contained in microbial biomass ranges from 0.5 to 5% of the total organic C in the soil (Jenkinson and Ladd, 1981; Marumoto *et al.*, 1982). Being one of the most labile pools of soil organic matter, microbial biomass is an important reservoir of plant nutrients (Marumoto *et al.*, 1982; Jenkinson and Ladd, 1981; Smith and Paul, 1986), as will be discussed in a later section. Because process rates are strongly dependent on the size of microbial populations, quantification of total microbial populations is important in estimating the rates at C turnover. However, microbial biomass is too gross a measure to be useful for most specific processes and better progress can be made in quantifying specific groups using most probable number methods (Woomer, 1994), immunofluorescent direct counts (Bottomley, 1994), or biomass estimation in conjunction with specific inhibitors (Anderson and Domsch, 1973).

Another important issue in linking microbial populations to carbon flow is distinguishing between the dormant and active microbial biomass (Smith *et al.*, 1993). Smith et al. (1986, 1993) distinguish between the active portion of the microbial population, which grows, and a sustaining population, which has only maintenance demands. This distinction is important for estimating turnover times of microbial biomass, where the turnover times of the total biomass range from 2 to 6 years and that of the active biomass from 0.3 to 1 year (Smith *et al.*, 1986)

Microbial biomass is sensitive to perturbation by agronomic practices or other factors (Smith and Paul, 1986; Doran, 1987). The size of the soil microbial biomass is related to the mass of C and other nutrients available from plant residues and root exudates (Fraser *et al.*, 1988; Adams and McLaughlin, 1981) and is strongly influenced by rainfall or irrigation (Campbell and Biederbeck, 1976). Microbial biomass is also positively correlated to an estimate of the organic nitrogen available to crops in no-tillage surface soil (Doran, 1987). The magnitude of microbial biomass depends on environmental conditions (e.g., rainfall) as well as degree of disturbance or age of an ecosystem (Insam *et al.*, 1989). It has been proposed that the ratio of microbial biomass carbon to total organic carbon (C_{mic}/C_{org}) reflects the "stability" of the organic matter content of an agricultural ecosystem (Anderson and Domsch, 1986, 1989); this hypothesis remains to be tested with long-term studies.

B. Composition of Microbial Communities: Bacteria and Fungi

Differences in microbial community composition may directly affect carbon flow. For example, the low mineralization potential of an occluded, labile organic matter pool in a no-till soil was attributed to its origin as fungal cellular material (Cambardella and Elliot, 1994). Many assumptions about the microbiology of C dynamics in soil are biased by the fact that bacterial processes are better understood than fungal processes. Soil fungi, however, often constitute a greater portion of the total soil biomass than do bacteria. Forest litter and soils, for example, are largely dominated by fungi (Cromack and Caldwell, 1992). Shields et al. (1973) and Nannipieri et al. (1978) found, using direct counting techniques, that bacteria comprised only 15–20% of the biomass in agricultural and grassland soils. Chemical inhibitors that selectively interfere with 70 S- or 80 S-type ribosomes are often used to separate eukaryotes from prokaryotes. Using such inhibitors, Anderson and Domsch (1973, 1980) also estimated that fungi contributed 60–90% of the total short-term respiration in 14 arable soils. The ratio of prokaryotic to eukaryotic biomass was 1 : 4 in two agricultural soils and 1 : 9 in a forest soil (Anderson and Domsch, 1978). However, not all agricultural systems are dominated by fungi. Bacteria are more abundant than fungi in arable reclaimed marine sediments in The Netherlands (Moore and de Ruiter, 1991) and in Colorado soils cultivated with spring wheat (Holland and Coleman, 1987). Abundances of particular predatory groups, e.g., bacterial versus fungal feeding nematodes, may reflect differences in the proportions of bacteria and fungi (Moore and de Ruiter, 1991).

Composition and management of plant residues are important determinants of the relative importance of fungi and bacteria. Among cellulolytic microorganisms, fungi tend to be selected for by plant residues with high C/N ratios, whereas bacteria tend to be selected by lower C/N ratio material (Broder and Wagner, 1988). Fungi had a greater influence on decomposition in no-till systems in which surface residues select for organisms that can withstand low water potentials and obtain nutrients from the underlying soil profile (Beare *et al.*, 1992). Bacteria were more important in tilled soils where residues were well mixed into the soil profile.

There are also a number of important physiological and ecological differences between bacteria and fungi. Anderson and Domsch (1980) found large differences in the elemental composition of 24 species of common soil bacteria and fungi. Carbon to nitrogen ratios averaged around 5.5 for the bacteria and 8.3 for the fungi and average C contents (mg per dry weight of cells) were 430 for the bacteria and 373 for the fungi. Fungi conserve more energy than do bacteria and can recycle nutrients internally by translocating cytoplasm. Within the fungal biomass, levels of N vary considerably. Paustian and Schnürer (1987) found levels of N to be 1 or 2% (dry weight) in cell walls and 10% in the cytoplasm of fungi. Fungi often produce proportionally more cell wall than cytoplasmic material when starved for N and thus can extend into new regions of the soil without requiring balanced growth conditions. Filamentous fungi can grow more effectively than bacteria under conditions of low moisture and low pH

(Beare *et al.*, 1992; Holland and Coleman, 1987). Because bacteria are smaller than fungi, they can occupy smaller pores and thus potentially have greater access to material contained within these pores. Although neither group of organism is very mobile in soil, the filamentous growth structure of a fungus permits it to access C in one location, e.g., on the surface, and nutrients in another location, e.g., in the soil profile (Holland and Coleman, 1987).

Degree of disturbance associated with cultivation is inversely related to fungal biomass (Hunt *et al.*, 1989). Fungi are also more disrupted than are bacteria by the tillage practices commonly used in agriculture (Johnson and Pfleger, 1992). Many conclusions about the importance or lack of importance of fungi in agricultural soils are based on laboratory studies using very disturbed soils. Therefore, the importance of fungi in agricultural soils may be underestimated (Moorhead and Reynolds, 1992).

The bacterial species composition of soil communities is also important to C cycling, although methodological limitations have previously hindered studies that consider diversity and community structure. Cycling of nutrients can be dependent on close relationships among various trophic groups. For example, C flow in anaerobic systems is dependent on close spatial and temporal interactions and exchange of C sources and electron donors among different functional groups of bacteria (Brock and Madigan, 1991). These groups include fermenters, hydrogen-producing acetogenic bacteria, and sulfate-reducing bacteria and methanogens. Similarly, an aerobic community in close proximity to an anaerobic community will metabolize and assimilate reduced substances (e.g., methane) and thus conserve C within the system. Whether greater species diversity within functional groups can influence C dynamics is not yet known.

Recent developments in the application of phospholipid fatty acid (PLFA) and molecular analysis to whole communities are rapidly improving our capabilities to consider microbial diversity and community structure in soils (Ritz *et al.*, 1994). Both approaches utilize microbial cellular material, directly extracted from soil, to provide information about the composition and metabolic status of soil communities. Phospholipids are integral components of cell membranes and are rapidly metabolized when a cell dies in soil; therefore, they provide an accurate measurement of living biomass (White *et al.*, 1979). Principle types of fatty acids are defined on the basis of chain length, degree of unsaturation, and the presence of substituents (e.g., methyls, hydroxyls, and cylopropane rings). Whereas bacterial fatty acids usually range from C-12 to C-20 and include odd-chain, methyl-branched, and cyclopropane fatty acids, fungal fatty acids are typically longer in length and contain saturated even-chain and polyenoic fatty acids (Tunlid and White, 1992). The entire PLFA profile can be used as a "fingerprint" of the whole soil community. Also, specific signature lipids are used to discriminate among different microbial taxonomic groups.

Microbial community structure in agricultural ecosystems has been characterized based on phospholipids (Bossio and Scow, 1997; Zelles *et al.*, 1992, 1995). PLFA profiles were measured in soil microbial communities from organic (amended with cover crops and manure), low-input (cover crops and mineral fertilizer), and conventional (mineral fertilizer) farming systems of the Sustainable Agriculture Farming System Project in California (Temple *et al.*, 1994; Gunapala and Scow, 1997). Sustained increases in microbial biomass and activity, resulting presumably from high organic matter inputs, had been previously observed in the organic and low-input tomato soils (Gunapala and Scow, 1997). PLFA profiles of the communities in organic and conventional soils were significantly different throughout the growing season based on multivariate statistical tests. The low-input profiles fell intermediate between the organic and conventional soils on most sample dates. The organic and low-input soils had higher ratios of fungal to bacterial signature PLFAs than did the conventional soil. Diversity indices, using PLFAs as "species," were the same in all farming systems and across all dates. Changes in PLFA profiles over the season were larger than changes associated with farming system practices, except immediately following cover crop or fertilizer amendment. Spatial variation in PLFA profiles across field blocks was relatively insignificant. Comparison of the PLFA profile from the SAFS tomato soils with soils cultivated with rice showed that the differences between the two soil types were far greater than any differences within soil type associated with farming system and seasonal variations (Bossio and Scow, 1997). PLFA analysis thus proved to be a promising approach for relating microbial community composition to agricultural practices.

C. Soil Food Web

Microorganisms are members of complex soil food webs that also include numerous species of protozoa, arthropods, oligochaetes, annelids, molluscs, and other organisms (Killham, 1994). These micro- and macrofaunal species are important as predators of bacteria and fungi, as well as processors and mixers of newly added organic material in soil. Food webs have been described by (i) "connectedness" web models that emphasize the diets and trophic relationships among groups of organisms (common in studies of communities); (ii) energy flow models that estimate the fluxes of material (e.g., C as the carrier for energy) through different groups (common in studies of ecosystems); and (iii) functional web models that address biological interactions that determine the size and activity of functional groups (often integrate aspects of i and ii) (Paine, 1980; Moore and de Ruiter, 1991). Energy flow models tend to be biased toward, and most sensitive to, organisms with high biomass and turnover rates (e.g., microorganisms). Higher trophic levels may seem insignificant in such models if one considers the small portion of material, energy or elemental, that flows through their biomass. The regulation and stimulation of microbial populations by the higher trophic levels, however, can significantly affect C flow (Moore and de Ruiter, 1991; de Ruiter *et al.*, 1995). For example, the removal of mites from plots in the Chihuahuan desert resulted in larger populations of bacterial-feeding nematodes, reduction in bacterial populations, and reduction in rates of decomposition of desert shrub litter (Santos *et al.*, 1981). The presence of nematodes can increase the fraction of C that is respired by microorganisms (Ingham *et al.*, 1985). Through their predation on bacteria, protozoa can substantially increase the mineralization and plant uptake of nitrogen in soil (Clarholm, 1994; Kuikman *et al.*, 1990).

IV. Carbon Flow at the Soil Level

The majority of C enters the soil in the form of complex organic matter containing highly reduced, polymeric substances. During decomposition, energy is obtained from oxidation of C–H bonds in the organic material. When oxygen becomes limiting, many inorganic compounds are reduced during their use by bacteria as electron acceptors. When decomposition slows down, oxygen again becomes available, and the reduced inorganic compounds are reoxidized because they provide electron donors and reducing equivalents for chemolithotrophic microorganisms. These oxidations and reductions are the reactions underpinning the biogeochemical cycles of C, as well as nitrogen, sulfur, and other elements, that are essential to the functioning of agroecosystems.

A. Inputs of C to Soil

Plants are the major sources of C to soil in agroecosystems. Crop-associated C enters in the form of litter, stubble, roots, and root exudates. Organic amendments, such as cover crops, compost, and manures, are other important sources of C. Although organic pesticides are a potential source of C, they contribute a very small amount of C, usually on the order of $< 0.001\%$ of the total organic C in soil.

Distinguishing between types of organic inputs, even within a plant, is important because of differences in the time and frequency of their addition to soil, as well as in their chemical composition. Although quantification of the aboveground biomass is relatively simple, estimating total C inputs from plants is complicated by uncertainties about the true root biomass and about losses due to plant respiration, grazing, and death. Shoot to root ratios (not accounting for losses from roots) for temperate, herbaceous plants, including cultivated crops, ranged between 1.2 and 1.5 (Buyanovsky and Wagner, 1986). Significantly higher ratios, ranging from 2 to 5, have also been reported for cultivated crops (Beauchamp and Voroney, 1993). Perennial crops rarely enter the reproductive phase because of frequent cutting, and thus allocate more of their total productivity below round than do annual crops. The shoot to root ratio may vary with fertilization (Hansson *et al.*, 1990) and growth stage (Keith *et al.*, 1986).

Amounts of C associated with crop residues incorporated into the soil depend on the particular crop and vary widely (Smith *et al.*, 1993). For example, the estimated C inputs to soil with postharvest residue of wheat and soybean were 343–372 g/m^2, whereas inputs following corn harvest were almost three times higher (Buyanovsky and Wagner, 1986).

Root exudates, including sloughing of root cells, can account for 10–33% of the net plant photosynthate (Beauchamp and Voroney, 1993). Bowen and Rovira (1991) estimated that total exudation and grazing loss by soil predators may account for as much as 150% of C in the root biomass at harvest.

The temporal and spatial distribution of inputs can vary substantially. Root exudates enter the soil throughout the growing season, whereas the residue from crops, as stubble and root mass, enters the soil in a pulse following harvest. Spatial distribution of the root biomass depends on the crop; approximately 76, 64, and 79% of the total root biomass of wheat, soybean, and corn, respectively, was in the top 10 cm of soil at harvest (Buyanovsky and Wagner, 1986). Incorporation of cover crops and manures occurs in a single pulse, in spring, before planting. Distribution of these inputs is in the top 10–15 cm of soil, if tilled, or on the soil surface in no-till systems (Beare *et al.,* 1992).

B. Pools of Carbon within Soil

Soil organic matter (SOM) has been the subject of numerous recent reviews and books (Stevenson, 1994; Coleman *et al.*, 1989; Chen and Avimelech, 1986; MacCarthy *et al.*, 1990; Wilson, 1991) and will only briefly be considered here. SOM consists of a broad spectrum of chemical classes, including (but not limited to) amino acids, lignin, polysaccharides, proteins, cutins, chitins, melanins, suberins, and paraffinic macromolecules, as well as anthropogenic chemicals. A major organic fraction, associated with mineral surfaces, is collectively known as humic material, which consists of amorphous macromolecules derived from the preceding list of biogenic precursors (Hayes, 1991). There is no consensus on the exact structure of humic material in soil, though a number of hypotheses have been proposed (Stevenson, 1994; Schnitzer and Khan, 1978; Hayes, 1985).

The pools of organic matter in soil have been characterized by numerous approaches, including chemical fractionation, physical fractionation, and biological methods (Stevenson, 1994; Stevenson and Elliot, 1989). Although the pools defined are often not equivalent across methods, the use of multiple approaches has increased our understanding of C dynamics in soil.

Organic matter fractions obtained by chemical methods are operationally defined and may be artifacts or products of harsh extraction methods (Stevenson and Elliot, 1989). Chemically fractionated SOM is usually divided into nonhumic substances, such as carbohydrates, lipids, and proteins, and humic substances, such as humic acid, fulvic acid, and humin (Stevenson, 1994). The humic substances are highly complex and not well characterized. Humin may consist of tightly bound and condensed humic material, fungal melanins, and paraffinic molecules (Stevenson and Elliot, 1989). It has proven difficult to relate amounts and structures of humic substances to their function or behavior in soil.

New approaches in analytical chemistry have improved our ability to analyze chemical fractions extracted from soil. Nuclear magnetic resonance (NMR) spectroscopic techniques are now used to study humic substances (Malcolm and McCarthy, 1991). Analytical pyrolysis (py) is used to study soil organic matter via gas chromatography-mass spectrometry (pyGC-MS) and py-direct mass spectrometry (Haider and Schulten, 1985; Schnitzer and Schulten, 1992). pyGC-MS can be used to analyze whole soils and does not require extraction of humic substances.

Emerging methods for characterizing soil C pools are aimed at maintaining the interrelationships between organic and mineral components. Combinations of these methods have been more fruitful than chemical approaches alone in investigations of C flow in soils. Figure 4 shows a flowchart of the physical fractionation approach (Stevenson and Elliot, 1989). Physical fractionation techniques include separation by sieving (usually particles and aggregates >50 μm), sedimentation (particle sizes >2 μm), and densitometry. Studies using different fractionation methods are often difficult to compare because of differences in soil preparation, size fractions, and reagents used in density separation.

Various methods for sieving and sedimentation of soil are described by Stevenson and Elliot (1989) and Christensen (1992). These approaches separate out soil fractions by diameter or size, shape, and density. Following separation, the organic matter associated with each fraction may be quantified as total C, analyzed for specific chemical constituents, incubated to determine the bioavailability of its C, or carbon dated (Stevenson and Elliot, 1989; Anderson and Paul, 1984). Studies of the C associated with particles and aggregates are discussed in a later section. Care must be taken in making generalizations about C flow because excessive sonication can fragment the soil and generate new material within size fractions. In addition, the nonideality of soil particles may violate Stoke's law, generating uncertainty

SOIL

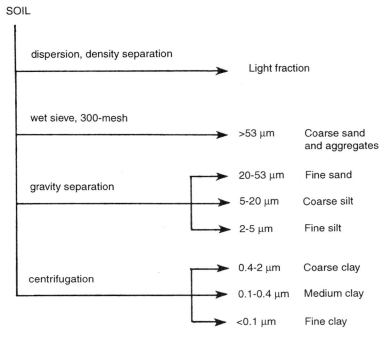

Figure 4   Flow-chart of the physical fractionation approach (Stevenson and Elliot, 1989, with permission of University of Hawaii Press).

about the distribution of particles across the size fractions (Cambardella and Elliot, 1994).

Densitometry methods are used to separate lighter from heavier fractions of organic matter. The approach can be applied to whole soils as well as to specific particle or aggregate groups. Organic matter lighter than a density of 1.6 (Wander *et al.*, 1994), 1.85 (Cambardella and Elliot, 1992), or 2.0 g/cm^3 (Sollins *et al.*, 1984) is considered to be the light fraction (LF) or, in other terminologies, particulate organic matter. The LF is relatively mineral free and consists of partially decomposed plant material, fine roots, and microbial biomass with a rapid turnover time. It can account for almost half of the total soil C in grassland and forest soils (Cambardella and Elliot, 1992), is a source readily mineralizable C and N (Janzen *et al.*, 1992), and declines rapidly under cultivation. The heavy fraction (HF), on the other hand, is organic matter adsorbed onto mineral surfaces and sequestered within organomineral aggregates (Sollins *et al.*, 1984). The HF is less sensitive to disturbance and chemically more resistant than the LF.

Another pool that is considered by some to represent a portion of organic matter involved in short-term denitrification and mineralization is soluble carbon (Burford and Bremner, 1975; DeLuca and Keeney, 1994; Sikora and McCoy, 1990; McGill *et al.,* 1986), where both total soluble organic C and specific components, such as anthrone-reactive C, have been measured. Attempts to correlate levels of soluble C to microbial process rates have been both successful (Burford and Bremner, 1975) and unsuccessful (Cook and Allan, 1992). Soluble C may measure humic materials that are not readily degradable (Nelson *et al.,* 1994) and solubility limits in the aqueous phase may exclude material that would otherwise be degradable. Measurements that include nonsoluble C, such as the LF, are probably more inclusive of the organic C pool that is readily degradable.

Biologically defined pools of available organic C are usually based on the amount of carbon dioxide released during short-term respiration assays (Davidson *et al.,* 1987; Burford and Bremner, 1975) or long-term incubations (Paul *et al.,* 1995). Such methods use indigenous microbial populations or add a defined microbial inoculum to perform the respiration. Using a respiration assay, Paul *et al.* (1995) found available C in corn–soybean plots to increase 70% in 4 years in plots that had not been cultivated and been allowed to revert to old field status.

In most models of C flow, organic matter is represented by C pools of different lability and nutrient availability that are, for the most part, defined by their kinetics (Van Veen *et al.,* 1984) as will be discussed below. Figure 5 illustrates an example of such an approach. Different researchers have assigned different names to the various pools; however, most distinguish between microbial biomass, plant residues (sometimes with two or three pools), chemically protected organic matter, and physically protected organic matter. To date, attempts to relate the pools of C measurable in the lab, with the exception of microbial biomass, to the pools described in models have been unsuccessful (Christensen, 1992). However, continued development and application of particularly physical fractionation methods in the study of C flow may help bridge this gap (Cambardella and Elliot, 1994).

C. Factors Influencing Carbon Flow

Physical and chemical factors exert strong influences on C flow in soil (Parr and Papendick, 1978). The relative importance of different factors varies as a function of the spatial and temporal scale of concern.

1. Substrate Properties and Nutrients   The impact of chemical structure on decomposition rates is very important on small timescales (Oades, 1988), as attested by large differences on the scale of weeks in decomposition of sugars and amino acids in comparison to cellulose and lignin (Alexander, 1977; Killham, 1994; Paul and Clark, 1989). Chemical composition, however, is generally less important to C flow in the long-term than is the

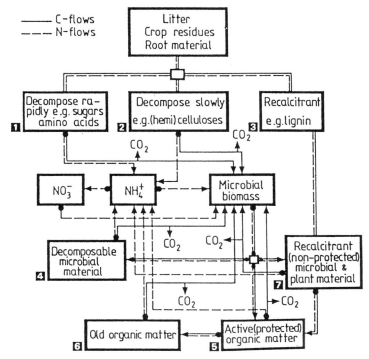

Figure 5 A conceptual model of C and N flow from plant litter through soil (reproduced from *Plant Soil* **76**, 1984, pp. 257–274. Modelling C and N turnover through the microbial biomass in soil. J. A. Van Veen, J. N. Ladd, and M. J. Frissel. © 1984 Kluwer Academic Publishers, with kind permission from Kluwer Academic Publishers).

interaction of residues with the soil matrix and the development of resistant fractions (Van Veen and Paul, 1981).

Lignin content and C/N ratio of residues may be important and both parameters are frequently used to estimate the short-term decomposition of plant residues. For example, the amount of carbon dioxide evolved from plant residues was found to be a function of the square root of percentage carbohydrate divided by the C/N ratio multiplied by the percentage lignin (Herman *et al.*, 1977). Others have not found the C/N ratio to be a good predictor of decomposition rate (Smith *et al.*, 1993).

Complicating the applicability of rate information on individual plant components is that the components never exist individually in plant tissue but are physically and chemically associated in a complex mixture. Cellulose decomposition is far slower in its natural state, which is complexed with lignin, than in purified form (Alexander, 1977). Another important set of associations affecting plant components are their interactions with the soil matrix. For example, polysaccharides and amino acids are strongly associated with clay, and protein interactions with soil polyphenols in "tanning" reactions lead to the formation of irreversible covalent bonds. Interactions of chemicals with the soil matrix usually reduce their rates of decomposition.

2. Climatic Factors   Temperature and moisture are the most important climatic variables determining C flow in soils. Climate can impact soil organic matter levels by influencing plant productivity as well as microbial activity. This is evident by looking at global scale differences in the accumulation of C in soils. As long as there is sufficient net primary productivity, the greatest amount of organic matter accumulation occurs in cold climates or in soils that are saturated a large portion of the year, e.g., poorly drained soils or marshes. In a large number of Australian soils, C content of soils increased with increasing precipitation and decreased with increasing temperature (Oades, 1988). Little organic matter accumulates in arid soils, primarily due to low NPP. Irrigation may increase NPP but any increases in organic inputs may be offset by enhanced rates of decomposition.

Temperature and moisture are also important because they regulate microbial populations directly. In cold climates, there are often higher levels of microbial biomass in summer (Lynch and Panting, 1982; McGill *et al.*, 1986; Patra *et al.*, 1990; Van Gestel *et al.*, 1992). However, DeLuca and Keeney (1994) found the highest levels in February after freeze-thaw cycles due to metabolism by surviving organisms of the C contained in tissues of organisms killed by the cycle. Low moisture results in reduced diffusion of soluble substrates and reduced mobility of microorganisms that lessens their ability to acquire substrates and thus to grow (Grant and Rochette, 1994). At the other extreme, very high moisture levels decrease microbial activity primarily by decreasing oxygen diffusion. Campbell and Biederbeck (1976) found peaks in bacterial numbers after rainfall events in summer fallow and continuous cropped wheat plots. Rapid, short-term increases in microbial biomass followed irrigation and tillage in both cover cropped and bare fallow soils in lettuce and broccoli production systems (Wyland *et al.*, 1995, 1996).

Common approaches to account for the effect of temperature are to use the Q_{10} relationship or the Arhhenius relationship (Paul and Clark, 1989; Alexander, 1977). Dimensionless scaling factors are used to model the effects of moisture (e.g., Molina *et al.*, 1983). Another approach that integrates the effects of different environmental variables is that of Nicolardot and Molina (1994), who use a fitting parameter (a "reduction factor") that takes into account the effect of climatic and soil conditions on decomposition rates.

3. Edaphic Factors and Physical Protection   It is readily apparent that soil is not a well-mixed system and that many potentially degradable substrates are not available to microorganisms. This problem is well recognized in the study of biodegradation of pollutants in soil because it often severely limits the efficacy of bioremediation (Scow, 1993; Scow and Johnson, 1997). The concept of physical protection of organic matter came, in part, from the observation of Martel and Paul (1974) that a resistant fraction of organic residue, as measured by acid hydrolysis, did not decline substantially over time (Paul *et al.*, 1995). Reduced availability of organic matter results from adsorption to clay, isolation in micropores, and physical protection within stable macroaggregates (Beare *et al.*, 1994; Oades, 1988). A consequence of the physical protection of organics is that the complex matrices produced by residual C compounds influences the spatial distribution and fate of microorganisms. Pore size may control access of predatory species and hence small pores may provide refuges for bacteria in soil (Elliot *et al.*, 1980; Elliot and Coleman, 1988). Recent advances in physical fractionation methods have generated new information on C flow and, in particular, increased our understanding of the importance of physical protection in conserving C in soil (Christensen, 1992).

The primary building blocks of soil are mineral particles, e.g., sand, silt, and clay (ranging from <0.2 to 100 μm diameter), and aggregates (20 to >2000 μm diameters). These two categories of soil components overlap and, in some size ranges, may be methodologically indistinguishable. Tisdall and Oades' (1982) model for soil aggregate formation (Fig. 6) describes particles <0.2 μm diameter (stage I) as discrete aggregates of particles (domains) held together by organics or oxyhydroxides. Binding mechanisms involving cation bridges between clay particles, and negatively charged organic molecules such as polyuronides and polycarboxylic residues (bacterial products), lead to formation of aggregates of 2–20 μm (stage II). The stage II aggregates are organized into larger units of 20–250 μm (stage III) by some of the same mechanisms involved in stabilization of the stage II materials. Aggregates >250 μm (stage IV or macroaggregates), the most frail of the structural units, are held together by plant roots and fungal hyphae.

4. Association of Organic C with Different Soil Fractions   Macroaggregates are more transient than microaggregates because their organic-binding agents, roots and hyphae, are more rapidly degradable than the older humified material making up some of the mineral–organic complexes binding the microaggregates. Recently added organic matter is associated with macroaggregates or, when separated by densiometric methods, with the light fraction. Larger aggregates and sand-sized particles contain organic matter that is younger, less aromatic, more aliphatic, and/or more polar than material associated with smaller aggregates. Macroaggregates are far more susceptible to disruption by cultivation and environmental perturbation (e.g., wet–

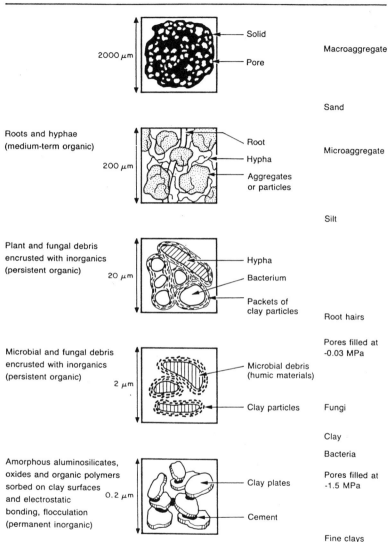

BINDING AGENT   SIZE   SOIL COMPONENT

Figure 6   A conceptual model of soil aggregate formation (reproduced from Tisdall and Oades, 1982, with permission of Blackwell Science Ltd.).

dry cycles) than are microaggregates (Tisdall and Oades, 1982). For example, organic matter lost from cultivated grasslands is largely from macroaggregates (Elliott, 1986). Interestingly, the decay of macroaggregate structure has a shape similar to the exponential curve characteristic for the decay of plant residues (Oades, 1988).

The organic matter associated with microaggregates (fine clay-sized aggregates <0.02 μm diameter) has a lower C : N ratio than does silt-associated material (Tiessen and Stewart, 1983). The silt-size aggregate fraction is rich in humic materials and resistant compounds of microbial origin. Organic matter in silt-size (2–5 μm) and coarse clay-size (0.2–2 μm) classes is more stable than that in other size fractions (Catroux and Schnitzer,1987); is present at relatively high concentrations than in coarser- and finer-sized particles (Tisdall and Oades, 1982; Turchenek and Oades, 1979; Oades, 1988); and is composed of highly stable, aromatic materials of intermediate C : N ratios including humic acids (Tiessen and Stewart, 1983; Turchenek and Oades, 1979). Although humic acids are associated with both clay- and silt-size fractions, they are more aromatic, less readily hydrolyzable, and more stable in particle size fractions larger than clay size (Anderson *et al.*, 1981; Catroux and Schnitzer, 1987). Mineralizable C and N are lower in silt-sized than in clay particles. Bioavailability of C was lower in silt than in clay, light fraction, or whole soil samples at different depths (Nelson *et al.*, 1994). Baldock et al. (1990) found, using solid-state CP/MAS ^{13}C NMR, significant amounts of alkyl, oxygenated alkyl (O-alkyl), acetal, and carboxyl C in the clay fraction of soil after incubation of soil with glucose, which was attributed to metabolism by microorganisms.

The estimated radiocarbon ages were 385, 900, 905, and 965 years, respectively, for organic matter associated with coarse silt, fine silt, coarse clay, and fine clay in cultivated soil from Saschatchewan (Anderson, 1995), supporting the idea of an increase in age of organic matter as one moves from coarse silt to coarse clay. Bonde et al. (1992), using natural abundance of ^{13}C and isotopic dilution, found that an oxisol originally forested with C3 vegetation, then cropped for 50 years with C4 crops, had C turnover times of 59, 6, and 4 years, respectively, in the clay-, silt-, and sand-size fractions of soil. They also found that even within each size class, there was an initially rapid loss of C in the first 12 years after initiation of tillage, followed by a slower rate of loss over the subsequent 38 years. In both studies, the fine clay-size fraction had the oldest organic matter, which is not necessarily compatible with the idea of clay-sized particles playing a role in intermediate turnover of C.

Baldock *et al.* (1992), using ^{13}C NMR, found that as plant materials decompose, the released C undergoes changes in its distribution among soil particle-size fractions with time. At first, carbohydrate and protein structures are degraded, as seen by a decrease in the O-alkyl carbon ^{13}C NMR signal (C/N of 40). Subsequently, there is loss in aromatic structure such as in lignin (C/N of 12). Finally, there is a loss of the most recalcitrant alkyl carbon structures (C/N of 8). Corresponding to these chemical changes was a decrease in particle size with which the organic matter is primarily associated from >20 μm to 2–20 μm to <2 μm in diameter. The role of chemical structure in the association of C with different soil components needs further investigation.

The importance of clay in C flow is manifested in whole soils as well as aggregates. Soils with high clay content are more likely to preserve microbial biomass, provide an environment for close interactions between microorganisms and their metabolites, and may support higher yield coefficients for growth on glucose and their metabolites (Van Veen *et al.*, 1985, 1987; Sorensen, 1983). Using radiolabeled carbon, Oades (1988) and Van Veen et al. (1985) found a greater retention of biomass and carbon in high-clay (42%) than low-clay (12-18%) soils. Sorensen (1975, 1981) also found more accumulation of C, and lower respiration, in high-clay soils and attributed this to sorption of amino acids to clays. Ladd (1989) found a direct correlation between texture and the portion of added C that went to soil organic matter rather than biomass. Clay affects microbial activity indirectly by sequestering what would otherwise be readily decomposable organic matter (Adu and Oades, 1978; Gregorich *et al.*, 1989). Fine-textured soils were less sensitive with respect to loss of C to tillage (Paustian *et al.*, 1995), illustrating the importance of clay in preserving C, even in the presence of physical disruption.

Studies of agroecosystems that compared till and no-till farming systems have greatly increased our understanding of the phenomenon of physical protection. Beare *et al.* (1994) identified three pools of C, including unprotected, protected, and resistant C, in cultivated soil. Unprotected C pools associated with aggregates were 21–65% higher in no-till than in conventional-till surface soils. Macroaggregate protected pools of carbon were more labile than unprotected pools, in part because degradation of the more labile fraction of the exposed C pool had already occurred. Also, rates of mineralization from protected and unprotected pools of C were higher in till than in no-till surface soils. Macroaggregate-protected organic matter accounted for 19% of the mineralizable C in the top layer of no-till but only 10% of the till system. Crushing of macroaggregates increased mineralization in the no-till but not the tilled surface soils (Beare *et al.*, 1994).

D. Microscale Spatial Distribution of Microorganisms

The spatial distribution of microbial communities at the pore and aggregate scale of soil is largely unknown. Little of the total surface area of soil is occupied by microorganisms (Hissett and Gray, 1976; Foster, 1988). Microbial biomass was found to be concentrated in particle size fractions of 2–50 μm (Monrozier *et al.*, 1991). Pore size distribution, which is determined by the arrangement of soil particles, has also been related to the spatial distribution of microbial populations. Using electron microscopy, Kilbertus (1980) found that in three soils the majority of bacteria occupied pores with diameters ≤ 2 μm. Given that a large fraction of those pores containing water at field capacity are smaller than 2 μm in diameter (e.g., approximately 38% in a silt loam) (Papendick and Campbell, 1981), a significant portion of the soil solution and thus soluble C may be inaccessible to microorganisms (Alexander and Scow, 1989).

Most of the microbial metabolic activity responding to added glucose is associated with the surface or outer layers of aggregates (Priesack and Kisser-Priesack, 1993). The interiors of larger soil aggregates are often anaerobic, even in well-aerated soils (Myrold and Tiedje, 1985), so aerobic microorganisms are unlikely to be active deep within large aggregates. Kilbertus (1980) found that only 6% of the inner volume of soil aggregates was colonized by microorganisms. Under aerobic conditions, carbon dioxide production and microbial biomass C were found to be greatest in microaggregates (<0.25 mm), and biomass declined with increasing aggregate diameters up to 20 mm (Seech and Beauchamp, 1988). Concentration of biomass in the outer layer is attributed either to diffusional limitation of oxygen or their carbon source into the aggregate or to rapid consumption of these resources by the outer layer of microorganisms before diffusion of the substrate inside the aggregate.

Microorganisms play important roles in aggregate formation and their activities can provide feedback in determining their own distribution. Bacterial colonies and fungal hyphae in soil are often found coated with particles of fine clay (Tisdall and Oades, 1982). This association can lead to formation of small aggregates on the order of 2–20 μm diameter, the silt-size fraction. When cells die they leave behind fibrous polysaccharides in the interior of the colony (Foster, 1988).

V. Quantitative Descriptions of Carbon Flow in Soil

Numerous models have been developed to describe C flow in soils. Short-term models often look at transient conditions and put more emphasis on plant residue decomposition. Long-term models, often steady-state models, emphasize the residence time of soil organic matter pools. Both types of models are potentially useful for predicting outcomes (e.g., soil organic matter levels) for multiple combinations of crops, soil types, management practices, and climatic conditions. Experimentally, it would be impossible to test all these combinations.

The simplest models describing C flow in soil can be represented by single equations. Thus, an approach often used to estimate C levels in different ecosystems is to assume a single pool of C decomposing under constant conditions via first-order kinetics, where

$$dC/dt = -kC,$$

where C is soil carbon content, t is time, and k is a first-order decay constant. If the system is at equilibrium, then from this relationship can be derived the turnover time, which is the soil organic C content divided by the annual input of C to the soil.

A slight increase in the complexity of this simple model is found in multiexponential models with a first-order decay expression for each important compartment (Woodruff, 1949). The two-compartment or double-exponential equation, which considers a readily degradable and slowly available pool of C, is one such approach (Jenkinson, 1977). Many models of this type focus on the decomposition of plant material and do not consider soil organic matter pools.

These approaches do not consider the biology of microorganisms and the decomposition rate is assumed to be an intrinsic property of the C pool. More complex models are based on the same assumption but differ in that the carbon is divided into fractions of differing degradability and climatic and edaphic factors are considered. Models emphasizing long-term C dynamics, and that assume soil contains multiple pools of C, are those of Jenkinson and Rayner (1977) and Parton et al. (1988). The models differ in how they define the C pools.

Jenkinson and Rayner's (1977) five-compartment model of C flow was developed to describe organic matter dynamics in an agricultural soil at the Rothamsted Experimental Station in Great Britain. The compartments include decomposable and resistant plant material, microbial biomass, and physically stabilized and chemically stabilized organic matter. Data for the pool sizes and fluxes were obtained using radiocarbon dating, ^{14}C derived from bomb testing, data from incubation of labeled plant residues, and direct measurement of microbial biomass. The model could predict the impact of manure additions on soil organic matter levels as well as microbial biomass levels in soils receiving mineral fertilizer or no external inputs (Jenkinson, 1990). The same model was used to compare turnover of microbial biomass in continuous wheat plots at Rothamsted, wheat fallow plots in Canada, and sugar cane plots in Brazil (Paul and Voroney, 1984). Turnover times of microbial biomass were 10 times lower in Brazil and 4 times higher in Canada, relative to the turnover time in the Rothamsted soils. These differences reflected variations in organic inputs, climate, and plant uptake.

The CENTURY model (Parton *et al.*, 1988) simulates dynamics of C, N, P, and S cycling in cultivated and noncultivated grassland soils. The C pools include structural and metabolic C derived from plant residues, as well as active, slow, and passive pools of C making up the soil organic matter. The model, which has been modified to include crop rotations and manure inputs, has been used to analyze how agricultural management practices affect greenhouse gases and soil organic matter contents (Voroney and Angers, 1995).

Models that incorporate aspects of microbial physiology, but still do not consider species or group differences, have focused on the relationships between the microbial populations, C flow, and nutrient fluxes (Parnas, 1975; Bosatta and Berendse, 1984; Van de Werf and Verstraete, 1987). These models emphasizing microbial phenomena usually have simpler descriptions of plant residue and organic matter pools than do the models described previously. Smith *et al.*'s (1986, 1993) model distinguishes between an active and total microbial biomass, between C used for growth versus maintenance, and accounts for decline in microbial biomass due to death. Relationships between the size of the microbial biomass and nitrogen availability could be described using the model (Smith, *et al.*, 1993). Van Veen *et al.* (1984) describe a C flow model that considers C availability, differences in quality and availability of C, biomass turnover, and death (Fig. 5). Plant residues are divided into three pools and organic matter into five pools varying in their rates of decomposition and turnover. Microbial biomass can decay and consume C for maintenance needs. In addition to a microbial biomass pool, there is a metabolite pool that is directly linked to development of a resistant fraction of organic matter.

The usefulness of the models described previously is limited by difficulties in obtaining independent measurements of many of the C and microbial pool sizes and rate constants. Additionally, most models assume steady-state conditions and thus do not capture the short-term, but sometimes major, impacts of transient events such as tillage, irrigation, and other management practices. Finally, important spatial issues, such as the distribution of microorganisms and their substrates, are not considered.

Several mechanistic models have been developed to describe the impact of heterogeneity in the distributions of microorganisms and their substrates on rates of metabolism. Leffelaar (1993) measured and modeled microbial respiration in soil aggregates under different moisture conditions. Hysteresis in the soil water potential curve strongly affected the moisture distribution in the aggregate and was suggested to result in long-term maintenance of anaerobic conditions within the aggregate. Scow and Hutson (1992) and Scow and Alexander (1992) showed that the presence of clay aggregates reduced the turnover rate and kinetics of metabolism of organic compounds in model systems because the aggregates sequestered the C from bacterial populations. Priesack and Kisser-Priesack (1993) also developed and tested mathematical models of microbial metabolism in the presence of aggregates. They found aggregate interiors to contain low levels of microbial biomass because those organisms in the outer layer consumed any available carbon before it could diffuse inside the aggregate. Most of these models have been tested in simple experimental systems and are too complex, as formulated, to include directly in larger models of C flow.

VI. Comparison of Agricultural to Other Ecosystems

Agroecosystems are the most intensively managed, and thus often the most disturbed, of the earth's ecosystems. Carbon flow in agroecosystems, in comparison to natural or less managed ecosystems, differs in several important ways. First, although C inputs may be similar in terms of total amount, substrate quality differs. Carbon inputs in agricultural systems have a greater portion that is readily decomposable than do inputs in forests, deserts, and savannah systems (Greenland, 1995). Second, shoot to root ratios are often higher in plants in row-crop than in native ecosystems (Schimel, 1986). Finally, carbon storage, in the form of labile organic pools or total carbon, is generally reduced in cultivated soils relative to their noncultivated counterparts (Davidson and Ackerman, 1993). Due to intensive management practices, physical disturbance and destruction of soil structure are obviously much greater in agroecosystems than in native systems.

In a comparison of C dynamics in winter wheat and tallgrass prairie, Buyanovsky *et al.* (1987) found that annual production of dry matter was greater in the cultivated than in the native system. Combined loss of carbon dioxide from the decay of above- and belowground litter was twice as high in the cultivated system. Total litter accumulation was 5 times greater in the native than in the cultivated system even though the cultivated system had a higher plant productivity. This was reflected in soil organic matter contents in the top 35 cm that were 1.5 to 2 times higher in the native than in the cultivated system. Estimated first-order decay constants were relatively similar for soil organic matter in the two systems. However, decay constants were considerably higher for both below- and aboveground plant biomass. The higher decay constants were attributed to greater physical disturbance, the absence of an irrigated summer crop reducing water limitations on microbial activity, and increased surface area for plant litter in the cultivated compared to native system. More of the net primary productivity was partitioned into production of SOM (including microbial, labile, and resistant fractions) in the native than in the cultivated system. The difference in timing of the inputs is also important. In the cultivated system, more of the C entered the system during periods when rates of microbial activity were at their highest potentials with respect to favorable moisture and temperature.

Comparison of C and N dynamics in paired grassland and wheat–fallow plots on three soil types revealed higher rates of C mineralization and lower N-immobilization rates in cultivated than in native systems (Schimel *et al.,* 1985; Schimel, 1986). Microbial biomass C was 1.6–2.4 times higher in the grassland than in the cultivated soils, with differences among soils increasing with increased clay content. Microbial populations appeared to be more strongly limited by N than C in the grassland, based on respiration and N-immobilization data, whereas the reverse appeared to be true in the cultivated system. In studies using physical fractionation of soil, Elliot (1986) and Gupta and Germida (1988) found higher C and N mineralization potential in undisturbed aggregates from grassland than in the same soils under cultivation. Also, the macroaggregates had greater mineralization potentials than did the microaggregates. Cambardella and Elliott (1992) found 5-20% of the organic C to be contained in particulate organic matter in agricultural soils, whereas more than 30% was associated with this fraction in grasslands.

Many of the differences between cultivated and native systems are manifested in the top layer of soil. The estimated ^{14}C age of the C in the Ap horizon of a cultivated soil in Saskatchewan was 1100 $\pm$ 145 years, in comparison to an age of 385 $\pm$ 110 for the same soil that had not been cultivated (Anderson, 1995). The carbon in the B horizons of the cultivated and native soils, on the other hand, had equivalent ages. This indicates that the more resistant fractions dominate the soil organic matter in the cultivated soil.

Comparisons of native and cultivated systems have led to interest in designing agroecosystems that capture some of the nutrient-conserving properties of native systems while maintaining high levels of productivity (Schimel, 1986). Thus, as discussed in the following section, the use of reduced tillage and increased inputs of C are seen as important variables for increasing the sustainability of agroecosystems.

VII. Impacts of Soil Carbon–Microbial Interactions on Agroecosystems

A. Biomass as a Pool of N: Short-Term Impact of C Dynamics

Microbial biomass is an important pool of available nitrogen, in part because of its low C/N ratio and in part because of its lability. Whereas 1–4% of the organic C in soil is contained in microbial biomass (e.g., Paul and Clark, 1989), approximately 2–6% of the organic N is in microbial tissue (Brookes *et al.,* 1985). Fluxes of N through the microbial biomass can be significant on an ecosystem scale; the amount of N passing through microorganisms was estimated to be more than double that going through plants, the latter of which constitute a far greater biomass (Heal and MacLean, 1975). An estimated 1–4% of the soil organic N is mineralized annually and potentially available for crop uptake (Anderson and Domsch, 1980). This amount of N is similar in magnitude, and partly equivalent, to the amount of N contained in microbial biomass (Anderson and Domsch, 1980). Because at least some portion of the microbial community will turn over every year, the contribution of nutrients from the microbial biomass is usually greater than what is contained in the biomass at any one time. One estimate of the amount of turnover of mineral N derived from microbial biomass is 23 g N m^2 per year in a Swedish agronomic system (Andren, 1990).

The concept that there is a critical balance between the mineralization and immobilization of N by soil microorganisms is historically important in soil fertility research and practice. The balance between immobilization and mineralization is governed by the metabolic status of the decomposer organisms and the nutrient composition of their substrates. The C/N ratio is often used to estimate whether a residue will support net N mineralization and thus provide adequate fertility for the crop (Smith *et al.*, 1993; Alexander, 1977). The ratio does not indicate the availability of either the N or the C pool to microorganisms (Parr and Papendick, 1978). This concept is simplistic and needs reevaluation because there is increasing recognition of the importance of microbial biomass as a direct source of nitrogen to crops (Smith *et al.*, 1986; Robertson *et al.*, 1988). The negative impact of a high C/N ratio may be short-lived and harmful only if plant demand for N is high during periods of substantial immobilization. Immobilization can be of benefit at the ecosystem level, however, because it retains N in the soil profile. In prairie grassland, very little inorganic N accumulates because it is rapidly immobilized following mineralization (DeLuca and Keeney, 1994; Schimel, 1986). This reduces N leaching or denitrification losses from the soil system.

For a long time, there has been an interest in the impact of climatic extremes on soil organic carbon (Birch, 1959; Birch and Friend, 1961). During cycles of wetting and drying or freezing and thawing, new organic carbon is made available in soil and is often rapidly mineralized (Soulides and Allison, 1961; Campbell *et al.*, 1970; Shields *et al.*, 1974). Part of the carbon originates directly from organic matter but a large portion of it derives from the death of microorganisms. Quantities of plant nutrients released after fumigation or drying of soils are similar to quantities available in freshly killed biomass (Marumoto, 1984; Anderson and Domsch, 1980). Both mineralizable C and potentially mineralizable N were closely related to the amounts of microbial biomass initially present and to their changes over time following disturbance (Nicolardot, 1988). The physiological status of microbial populations may influence the amount of C released, with the most susceptible organisms being those in the growth phase (Soulides and Allison, 1961). There are also likely to be impacts of spatial distribution of the microorganisms on their vulnerability to disturbance. Microorganisms are less vulnerable to the effects of drying in small rather than large pores due to the higher matric potential associated with smaller pores. Also, organisms in small pores are less susceptible to fumigants because of slow diffusion of vapor into small pores, especially inside aggregates.

It should be noted that another potential consequence of freeze–thaw cycles is leaching of soluble organic C from the soil profile (Wang and Bettany, 1993). However, moderate C leaching may also work to conserve C within an agroecosystem. Carbon moving into the lower soil profile layers, where there is low microbial activity and high clay contents, is subject to reduced mineralization.

An interesting consequence of these early observations was the development of methods to estimate the size of the microbial biomass (Jenkinson and Powlson, 1976a,b). When it was realized that part of the flush of carbon dioxide following soil disturbance was due to metabolism of dead microbial tissue, efforts were made to develop quantitative relationships between C liberated from purposeful disturbances (including wetting–drying and chemical treatment) and the size of the microbial biomass. Fumigation with an organic solvent, such as chloroform, was found to be more selective than physical methods of C associated with microorganisms and less likely to release abiotic C. From these investigations the fumigation–incubation method for determination of microbial biomass was developed (Jenkinson and Powlson, 1976a,b; Horwath and Paul, 1984)

B. Maintaining and Increasing Soil Carbon Pools: Long–Term Impact of Carbon Dynamics

There have been substantial losses of C from agricultural soils, with predictions of 30–60 pg C globally in the past 100 years (Greenland, 1995). There is an estimated 20–40% loss in the initial soil C inventory after cultivation of native soils (Davidson and Ackerman, 1993; Mann, 1989). Most of the losses occur very soon after cultivation begins (Davidson and Ackerman, 1993; Stewart, 1995). Loss of organic C follows an exponential decay curve with far greater losses in the initial part of the curve (Jenny, 1933; Stewart, 1995). After the first few decades of cultivation, soils reach an apparent new steady state with respect to C (Stewart, 1995). Thus, most losses of C from U.S. agronomic soils occurred decades ago, whereas substantial losses are now occurring from recently cultivated soils in the tropics.

On a landscape scale, there is interest in increasing organic carbon to improve soil structure and reduce erosion. Because there is an inverse relationship between the amount of C conserved and the amount of C released from soil, global-scale concerns about greenhouse gases (carbon dioxide, CH_4, and N_2O), for which soil can be both a source and a sink, are directly related to concerns about soil organic matter management. Interest in global climate change has generated a large volume of literature, much of which is also relevant to the management of C in agroecosystems (Lal *et al.*, 1995a,b).

Increasing SOM storage involves increasing C inputs and/or decreasing rates of C decomposition (Paustian *et al.*, 1995). Management practices to achieve these objectives are discussed below from the perspective of the role that microbial communities play in these efforts.

There is a direct relationship between the concentration of soil C and the annual additions of C to soil (Paustian *et al.*, 1995). Increasing crop productivity by increasing inorganic fertilizers can sometimes result in an increase in organic C, particularly in soils in which yields are far below their maximum attainable yields (Stewart, 1995). Fertilization and improved management has doubled or tripled the amount of crop residue produced over the past few decades and thus presumably increased soil organic C (Allison, 1973). In the long-term study at Rothamsted, however, increased crop productivity from inorganic sources of fertilizer did not result in higher organic matter contents after almost a century of management. The plots receiving inorganic fertilizer had levels of soil C only slightly above levels in plots that were neither fertilized nor manured and considerably below levels in plots that received manure (Jenkinson *et al.*, 1992). Addition of straw, in conjunction with mineral fertilizer, however, was estimated to double the C input compared to plots receiving only mineral fertilizer (Greenland, 1995). It is possible that fertilizer costs, as well as fossil fuel emissions of carbon dioxide during field operations, may decrease potential gains provided by increased productivity (Schlesinger, 1995).

Use of fallow may lead to reduction in soil organic matter, in part because of the absence of organic inputs during the fallow period. Fallow is used in semiarid climates, such as the Great Plains and eastern Washington, to conserve water, preserve inorganic N for following crops, and to reduce soilborne diseases. Microorganisms consume soil C because there is little C available to them from plant residues and microbial activity may be stimulated by field operations during fallow. Cihacek and Ulmer (1995) and Janzen (1987) found that reducing the frequency of fallow, and thus increasing the time when a crop was present, reduced annual losses of carbon dioxide from wheat crop–fallow cropping systems.

Rates of decomposition of organic residues can be potentially controlled by management practices. Unfortunately, the same factors that can reduce decomposition, e.g., decreasing soil moisture, decreasing oxygen levels, and decreasing soil temperature by crop residue manipulation, can also interfere with crop growth.

Physical protection of organic material clearly is important in conserving C in soils. Many practices used in agriculture, particularly tillage, destroy the macroaggregate structure that is important in protecting C from microbial decomposition. Thorough mixing of plant residues into the soil, previously thought necessary for decomposition and subsequent planting of the crop, also leads to greater loss of C from soils. Tillage and cultivation have been implicated for the substantially higher amounts of C mineralized in agroecosystems (80% of inputs) compared to native prairie (<60%) on the same soil type (Buyanovsky and Wagner, 1995). Stubble mulching buries only 15% of crop residue per operation and thus even with three or four operations, only 50% of residue will be buried (Stewart, 1995). Increases in conservation tillage from 27 to 57% would reduce losses of C from soil by almost half (Kern and Johnson, 1991). Using reduced tillage to achieve higher soil C is not considered to be feasible by all. Donigian *et al.* (1995) argues that increased conservation tillage practices alone will not significantly increase soil pool sizes. Combining C inputs with reduced tillage, however, may enhance C retention. Paustian *et al.* (1995) predicted substantial increases in soil C pools in Michigan if a winter cover crop was used and no-till was employed. Consumption by soils of greenhouse gases is also influenced by tillage. Methane uptake was seven or eight times lower in tilled agriculture than in non-tilled grassland soils, and agricultural soils managed using organic practices were intermediate in their uptake of methane (Robertson, 1993).

Decomposition can also be retarded if a greater portion of the organic inputs consist of high lignin-containing residues (e.g., peat or sawdust) (Paustian *et al.*, 1995; Herman *et al.*, 1977). Breeding of crops to contain higher levels of lignin is possible (Paustian *et al.*, 1995).

Another, less commonly considered, mechanism for decreasing C losses from soil is to decrease the ratio of C respired to that assimilated by microorganisms or, in other words, to increase the yield coefficient. The portion of C that goes into biomass and eventually into organic matter has been called the "humus yield" (Paustian *et al.*, 1995). Unfortunately, aerobic microorganisms are very efficient and therefore accumulation of organic carbon is slow (Schlesinger, 1995). Adoption of no-till can lead to a greater proportion of fungi making up the soil community and, because they generally have higher yields, no-till management may lead to greater retention of C in the soil (Paustian *et al.*, 1995). This has yet to be tested as a management tool. Fertilization with nitrogen may decrease respiration by increasing the amount of C assimilated, possibly in reserve material, when conditions are carbon limited (Knapp *et al.*, 1983). Adding N fertilizer may increase C storage in soil if the carbon sources provided are relatively recalcitrant, e.g., high lignin. This would enhance production of two major

402   *Kate M. Scow*

constituents of soil humus: lignin degradation products and proteinaceous compounds (Paustian *et al.,* 1995).

In conclusion, attempts to preserve existing organic matter in agricultural soils do not seem to be as successful as adding new organic matter to soil. Nyborg *et al.* (1995) found that increases in organic C in Canadian soils planted with barley was a linear function of total residue C added in the first decade. If soil organic C levels are already high, then additions may not result in a net increase in soil C (Campbell *et al.,* 1991), as found in a mollisol in Canada. Schlesinger (1990) argues that the loss of organic matter from arable land greatly exceeds the rate of formation of organic matter elsewhere and the net effect is a gain in atmospheric carbon dioxide. Also, any annual increases in soil C constitute a small fraction of the annual release of C from fossil fuels (Schlesinger, 1990).

VIII. Conclusion

The decomposition portion of the carbon cycle directly impacts many important components of agroecosystems, including the availability of nutrients to crops, soil structure, water retention and infiltration, and particularly soil organic matter content. The activities of specific microbial populations, as well as interactions among members of soil communities, determines not only how much C is retained or lost from soil but also the chemical forms and micro- and macroscale locations of C remaining in soil. The immense diversity of soil microorganisms makes it impossible to even begin to survey the broad range of metabolic pathways, reaction rates, and life strategies involved in the cycling of C in agroecosystems. The challenge is to utilize what information exists, identify and fill the major gaps in knowledge, and then develop generalizations essential for translating basic knowledge into practical applications. In particular, we need to identify where significant improvements can be made by incorporating more knowledge about soil biology into the simple models used as management tools to predict responses of soil C to agronomic practices and environmental perturbations.

Acknowledgments

I thank Jenn Macalady, Niklaus Grunwald, Louise Jackson, and two anonymous reviewers for many suggested improvements to the manuscript. Sandra Uesugi is appreciated for her tireless work on the reference list. Jenn Macalady created Figs. 1–3. This work was supported by the Kearney Foundation of Soil Science; U.S. EPA (No. R819658) Center for Ecological Health Research at UC Davis; the USDA National Research Initiative program; and the USDA Cooperative State Research Service Sustainable Agriculture Research and Education Program

(SAREP). Although the information in this document has been funded in part by the United States Environmental Protection Agency, it may not necessarily reflect the views of the agency and no official endorsement should be inferred.

References

Adams, T. M., and McLaughlin, R. J. (1981). The effects of agronomy on the carbon and nitrogen contained in the soil biomass. *J. Agric. Sci. Cambridge* **97**, 319–327.

Adu, J. K., and Oades, J. M. (1978). Physical factors influencing decomposition of organic materials in soil aggregates. *Soil Biol. Biochem.* **10**, 109–115.

Alexander, M. (1977). "Introduction to Soil Microbiology." Wiley, New York.

Alexander, M., and Scow, K. M. (1989). Kinetics of biodegradation in soil. *In* "Reactions and Movement of Organic Chemicals in Soil," SSSA Spec. Publ., 22, pp. 243–269. Soil Science Society of America, Madison, WI.

Allison, F. E. (1973). "Soil Organic Matter and Its Role in Crop Production." Elsevier, Amsterdam.

Anderson, D. W. (1995). Decomposition of organic matter and carbon emissions from soils. *In* "Soils and Global Change" (R. Lal, J. Kimble, E. Levine, and B. A. Stewart, eds.), Advances in Soil Science, pp. 165–175. Lewis, Boca Raton, FL.

Anderson, D. W., and Paul, E. A. (1984). Organo–mineral complexes and their study by radiocarbon dating. *Soil Sci. Soc. Am. J.* **48**, 298–301.

Anderson, D. W., Saggar, S., Bettany, J. R., and Stewart, J. W. B. (1981). Particle size fractionation and their use in studies of soil organic matter: I. The nature and distribution of forms of carbon, nitrogen and sulfur. *Soil Sci. Soc. Am. J.* **45**, 767–772.

Anderson, J. P., and Domsch, K. H. (1980). Quantities of plant nutrients in the microbial biomass of selected soils. *Soil Sci.* **130**, 211–216.

Anderson, J. P., and Domsch, K. H. (1985). Determination of ecophysiological maintenance carbon requirements of soil microorganisms in a dormant state. *Biol. Fertil. Soils* **1**, 81–89.

Anderson, J. P. E., and Domsch, K. H. (1973). Quantification of bacterial and fungal contributions to soil respiration. *Arch. Mikrobiol.* **93**, 113–127.

Anderson, J. P. E., and Domsch, K. H. (1978). A physiological method for the quantitative measurement of microbial biomass in soils. *Soil Biol. Biochem.* **10**, 215–221.

Anderson, T. H., and Domsch, K. H. (1986). Carbon link between microbial biomass and soil organic matter. *In* "Perspectives in Microbial Ecology" (F. Meguar, and M. Gantar, eds.), Proc. 4th Int. Symp. Microb. Ecol., pp. 467–471. Slovene Society for Microbiology, Ljubljana, Yugoslavia.

Anderson, T. H., and Domsch, K. H. (1989). Ratios of microbial biomass carbon to total organic carbon in arable soils. *Soil Biol. Biochem.* **21**, 471–479.

Andren, O. (1990). Organic carbon and nitrogen flows. *In* "Ecology of Arable Land: Organisms, Carbon and Nitrogen Cycling," *Ecol. Bull.*, Vol. 40, pp. 85–126. Munksgaard Inter., Copenhagen.

Babiuk, L. A., and Paul, E. A. (1970). The use of fluorescein isothiocyanate in the determination of the bacterial biomass of grassland soils. *Can. J. Microbiol.* **16**, 57–62.

Baldock, J. A., Oades, J. M., Vassalo, A. M., and Wilson, M. A. (1990). Solid state CP/MAS 13C NMR analysis of particle size and density fractions of a soil incubated with uniformly labelled 13C-glucose. *Aust. J. Soil Res.* **28**, 193–212.

Baldock, J. A., Oades, J. M., Waters, A. G., Peng, X., Vassallo, A. M., and Wilson, M. A. (1992). Aspects of the chemical structure of soil organic materials as revealed by solid-state ^{13}C NMR spectroscopy. *Biogeochemistry* **16**, 1–42.

Beare, M. H., Parmalee, R. W., Hendrix, P. F., Cheng, W., Coleman, D. C., and Crossley, D. A., Jr. (1992). Microbial and faunal interactions and effects on litter N and decomposition in agroecosystems. *Ecol. Monogr.* **62**, 569–591.

Beare, M. H., Cabrera, M. L., Hendrix, P. F., and Coleman, D. C. (1994). Aggregate-protected and unprotected organic matter pools in conventional- and no-tillage soils. *Soil Sci. Soc. Am. J.* **58**, 787–795.

Beauchamp, E. G., and Voroney, R. P. (1993). Crop carbon contribution to the soil with different cropping and livestock systems. *J. Soil Water Conserv.* **49**, 205–209.

Behara, B., and Wagner, G. H. (1974). Microbial growth rate in glucose-amended soil. *Proc. Soil Sci. Soc. Am.* **38**, 591–594.

Birch, H. F. (1959). Further observations on humus decomposition and nitrification. *Plant Soil* **11**, 162–286.

Birch, H. F., and Friend, M. T. (1961). Resistance of humus to decomposition. *Nature (London)* **191**, 731–732.

Bolton, H., Jr., Elliot, S. F., Papendick, R. I., and Bezdicek, D. F. (1985). Soil microbial biomass and selected soil enzyme activities: Effects of fertilization and cropping practices. *Soil Biol. Biochem.* **17**, 297–302.

Bonde, T. A., Christensen, B. T., and Cerri, C. C. (1992). Dynamics of soil organic matter as reflected by natural 13C abundance in particle size fractions of forested and cultivated oxisols. *Soil Biol. Biochem.* **24**, 275–277.

Bosatta, E., and Berendse, F. (1984). Energy or nutrient regulatin of decomposition: Implications for the mineralization–immobilization response to perturbations. *Soil Biol. Biochem.* **16**, 63–67.

Bossio, D. A., and Scow, K. M. (1997). Impact of carbon and flooding on PLFA profiles and substrate utilization patterns of soil microbial communities. *Microb. Ecol.* (in press).

Bossio, D. A., Scow, K. M., Gunapala, N., and Graham, K. J. (1997). Determinants of soil microbial communities: Effects of carbon availability, agricultural management, time, and soil type on phospholipid fatty acid profiles. Submitted for publication.

Bottomley, P. (1994). Light microscopic methods for studying soil microorganisms. *In* "Methods of Soil Analysis, Part 2. Microbiological and Biochemical Properties" (R. W. Weaver, S. Angle, P. Bottomley, D. Bezdicek, S. Smith, A. Tabatabai, and A. Wollum, eds.), SSSA Book Ser. No. 5, pp. 81–106. Soil Science Society of America, Madison, WI.

Bowen, G. D., and Rovira, A. D. (1991). The rhizosphere—The hidden half of the hidden half. *In* "Plant Roots: The Hidden Half" (Y. Waisel, A. Eshel, and U. Kafkafi, eds.). Dekker, New York.

Brady, N. C., and Weil, R. R. (1996). "The Nature and Properties of Soils," 11th Ed. Prentice Hall, New York.

Brock, T. D., and Madigan, M. T. (1991). "Biology of Microorganisms," 6th Ed. Prentice Hall, New York.

Broder, M. W., and Wagner, G. H. (1988). Microbial colonization and decomposition of corn, wheat and soybean residues. *Soil Sci. Soc. Am. J.* **52**, 112–117.

Brookes, P. C., Landman, A., Pruden, G., and Jenkinson, D. S. (1985). Chloroform fumigation and the release of soil nitrogen: A rapid direct extraction method to measure microbial biomass nitrogen in soil. *Soil Biol. Biochem.* **17**, 837–842.

Bunnel, F. L., Tait, D. E. N., and Flanagan, P. W. (1977). Microbial respiration and substrate weight loss-II. A model of the influences of chemical composition. *Soil Biol. Biochem.* **9**, 41–47.

Burford, J. R., and Bremner, J. M. (1975). Relationship between the denitrification capacities of soils and total, water-soluble and readilty decomposable organic matter. *Soil Biol. Biochem.* **7**, 387–394.

Buyanovsky, G. A., and Wagner, G. H. (1986). Post-harvest residue input to cropland. *Plant Soil.* **93**, 57–65.

Buyanovsky, G. A., and Wagner, G. H. (1995). Soil respiration and carbon dynamics in parallel native and cultivated ecosystems. *In* "Soils and Global Change" (R. Lal, J. Kimble, E. Levine, and B. A. Stewart, eds.), Advances in Soil Science, pp. 209–217. Lewis, Boca Raton, FL.

Buyanovsky, G. A., Kucera, C. L., and Wagner, G. H. (1987). Comparative analysis of carbon dynamics in native and cultivated ecosystems. *Ecology* **68**, 2023–2031.

Campbell, C. A., and Biederbeck, V. O. (1976). Soil bacterial changes as affected by growing season weather conditions. A field and laboratory study. *Can. J. Soil Sci.* **56**, 293–310.

Cambardella, C. A., and Elliot, E. T. (1992). Carbon and nitrogen dynamics of soil organic matter fractions from cultivated grassland soils. *Soil Sci. Soc. Am. J.* **58**, 123–130.

Cambardella, C. A., and Elliot, E. T. (1994). Carbon and nitrogen dynamics of soil organic matter fractions from cultivated grassland soils. *Soil Sci. Am. J.* **58**, 123–130.

Campbell, C. A., Biederbeck, V. O., and Warder, F. G. (1970). Simulated early spring thaw conditions injurious to soil microflora. *Can. J. Soil Sci.* **50**, 157–259.

Campbell, C. A., Bowren, K. E., Schnitzer, M., Zentner, R. P., and Townley-Smith, L. (1991). Effect of crop rotations and fertilization on soil organic matter and some biological properties of a thick Black Chernozem. *Can. J. Soil Sci.* **71**, 377–387.

Catroux, G., and Schnitzer, M. (1987). Chemical, spectroscopic and biological characteristics of the organic matter in particle size fraction separated from an Aquoll. *Soil Sci. Soc. Am. J.* **51**, 31–40.

Chapman, S. J., and Gray, T. R. G. (1981). Endogenous metabolism and macromolecular composition of *Arthrobacter globiformis*. *Soil Biol. Biochem.* **13**, 11–18.

Chen, Y., and Avimelech, Y. (eds.) (1986). "The Role of Organic Matter in Modern Agriculture." Nijhoff, Dordrecht.

Chesbro, W. R., Evans, T., and Eifert, R. (1979). Very slow growth of Escherichia coli. *J. Bacteriol.* **139**, 625–638.

Christensen, B. E. (1989). The role of extracellular polysaccharides in biofilms. *J. Biotechnol.* **10**, 181–202.

Christensen, B. T. (1992). Physical fractionation of soil and organic matter in primary particle size and density separates. *Adv. Soil Sci.* **20**, 1–90.

Christensen, B. T., and Sorensen, L. H. (1985). The distribution of native and labelled carbon between soil particle size fractions isolated from long-term incubation experiments. *J. Soil Sci.* **36**, 219–229.

Cihacek, L. J., and Ulmer, M. G. (1995). Estimated soil organic carbon losses from long-term crop-fallow in the northern Great Plains of the USA. *In* "Soil Management and Greenhouse Effect" (R. Lal, J. Kimble, E. Levine, and B. A. Stewart, eds.), Advances in Soil Science, pp. 85–92. Lewis, Boca Raton, FL.

Clarholm, M. (1994). The microbial loop in soil. *In* "Beyond the Biomass" (K. Ritz, J. Dighton, and K. E. Giller, eds.), pp. 221–230. Wiley, Chichester, UK.

Coleman, D. C., Andrews, R., Ellis, J. E., and Singh, J. S. (1976). Energy flow and partition in selected managed and natural ecosystems. *Agro-ecosystems* **3**, 45–54.

Coleman, D. C., Oades, J. M., and Uehara, G. (eds.) (1989). "Dynamics of Soil Organic Matter in Tropical Ecosystems." Univ. of Hawaii Press, Honolulu.

Cook, B. D., and Allan, D. L. (1992). Dissolved organic carbon in old field soils: Total amounts as a measure of available resources for soil mineralization. *Soil Biol. Biochem.* **24**, 585–594.

Cromack, K., Jr., and Caldwell, B. A. (1992). The role of fungi in litter decomposition and nutrient cycling. *In* "The Fungal Community. Its Organization and Role in the Ecosystem" (G. C. Carroll, and D. T. Wicklow, eds.), 2nd Ed., pp. 653–668. Dekker, New York.

Davidson, E. A., and Ackerman, I. L. (1993). Changes in soil carbon inventories following cultivation of previously untilled soils. *Biogeochemistry* **20**, 161–193.

Davidson, E. A., Galloway, L. F., and Strand, M. K. (1987). Assessing available carbon: Comparison of techniques across selected forest soils. *Comm. Soil Sci. Plant Anal.* **18**, 45–64.

Dawes, E. A. (1985). Starvation, survival and energy reserves. *In* "Bacteria in their Natural Environments" (M. Fletcher, and G. D. Floodgate, eds.), pp. 43–79. Academic Press, London.

DeLuca, T. H., and Keeney, D. R. (1994). Soluble carbon and nitrogen pools of prairie and cultivated soils: Seasonal variation. *Soil Sci. Soc. Am. J.* **58**, 835–840.

de Ruiter, P. C., Neutel, A.-N., and Moore, J. C. (1995). Energetics, patterns of interaction strengths, and stability in real ecosystems. *Science* **269**, 1257–1260.

Donigian, A. S., Jr., Patwardhan, A. S., Jackson, R. B., IV, Barnwell, T.O., Jr., Weinrich, K. B., and Rowell, A. L. (1995). Modeling the impacts of agricultural management practices on soil carbon in the central U.S. *In* "Soil Management and Greenhouse Effect" (R. Lal, J. Kimble, E. Levine, and B. A. Stewart, eds.), Advances in Soil Science, pp. 121–135. Lewis, Boca Raton, FL.

Doran, J. W. (1987). Microbial biomass and mineralization nitrogen distributions in no-tillage and plowed soils. *Biol. Fertil. Soils.* **5**, 68–75.

Elliott, E. T. (1986). Aggregate structure and carbon, nitrogen, and phosphorus in native and cultivated soils. *Soil Sci. Soc. Am. J.* **50**, 627–633.

Elliott, E. T., and Coleman, D. C. (1988). Let the soil work for us. *Ecol. Bull.* **39**, 23–32.

Elliot, E. T., Anderson, R. V., Coleman, D. C., and Cole, C. V. (1980). Habitable pore space and microbial trophic interactions. *Oikos* **35**, 327–335.

Foster, R. C. (1981). Polysaccharides in soil fabrics. *Science* **214**, 665–667.

Foster, R. C. (1988). Microenvironments of soil microorganisms. *Biol. Fertil. Soils* **6**, 189–203.

Fraser, D. G., Doran, J. W., Sahs, W. W., and Lesoing, G. W. (1988). Soil microbial populations and activities under conventional and organic management. *J. Environ. Qual.* **17**, 585–590.

Gaudy, A. F., Jr., and Gaudy, E. T. (1980). "Microbiology for Environmental Scientists and Engineers." McGraw-Hill, New York.

Grant, R. F., and Rochette, P. (1994). Soil microbial respiration at different water potentials and temperatures: Theory and mathematical modeling. *Soil Sci. Soc. Am. J.* **58**, 1681–1690.

Greenland, D. J. (1995). Land use and soil carbon in different agroecological zones. *In* "Soil Management and Greenhouse Effect" (R. Lal, J. Kimble, E. Levine, and B. A. Stewart, eds.), Advances in Soil Science, pp. 9–24. Lewis, Boca Raton, FL.

Gregorich, E. G., Kachanoski, R. G., and Voroney, R. P. (1989). Carbon mineralization in soil size fractions after various amounts of aggregate disruption. *J. Soil Sci.* **40**, 649–659.

Guggenberger, G., Zech, W., Haumaier, L., and Christensen, B. T. (1995). Land-use effects on the composition of organic matter in particle-size separates of soils: II. CPMAS and solution ^{13}C NMR analysis. *Eur. J. Soil Sci.* **46**, 147–158.

Gunapala, N., and Scow, K. M. (1997). Dynamics of soil microbial biomass and activity in conventional and organic farming systems. *Soil Biol. Biochem.*, in press.

Gupta, V. V. S. R., and Germida, J. J. (1988). Distribution of microbial biomass and its activity in different soil aggregate size classes as affected by cultivation. *Soil Biol. Biochem.* **20**, 777–786.

Haider, K., and Schulten, H-R. (1985). Pyrolysis field ionization mass spectrometry of lignins, soil humic compounds, and whole soil. *J. Anal. Appl. Pyrolysis* **8**, 317–331.

Hansson, A.-C., Andrén, O., Boström, S., Boström,U., Clarholm, M., Lagerlöf, J., Lindberg, T., Paustian, K., Pettersson, R., and Sohlenius, B. (1990). Structure of the agroecosystem. *Ecol. Bull.* **40**, 41–83.

Hayes, M. H. B. (1985). Extraction of humic substances from soil. *In* "Humic Substances in Soil, Sediment, and Water: Geochemistry, Isolation, and Characterization" (G. A. Aiken, D. M. McKnight, and R. L. Wershaw, eds.). Wiley, New York.

Hayes, M. H. B. (1991). Concepts of the origins, composition, and structures of humic substances. *In* "Advances in Soil Organic Matter Research: The Impact on Agriculture and the Environment" (W. S. Wilson, ed.), pp. 3–22. Royal Society of Chemistry, Cambridge, UK.

Heal, O. W., and MacLean, S. F., Jr. (1975). Comparative productivity in ecosystems—Secondary productivity. *In* "Unifying Concepts in Ecology" (W. H. Van Dobben, and R. H. Lowe-McConnell, eds.), pp. 89–108. Junk, The Hague.

Herman, W. A., McGill, W. B., and Dormaar, J. F. (1977). Effects of initial chemical composition on decomposition of roots of three grass species. *Can. J. Soil Sci.* **57**, 205–215.

Hissett, R., and Gray, T. R. G. (1976). Microsites and time changes in soil microbe ecology. *In* "The Role of Terrestrial and Aquatic Organisms in Decomposition Processes" (J. M. Anderson and A. MacFadyen, eds.), pp. 23–39. Blackwell, Oxford, UK.

Holland, E. A., and Coleman, D. C. (1987). Litter placement effects on microbial and organic matter dynamics in an agroecosystem. *Ecology* **68**, 425–433.

Horwath, W. R., and Paul, E. A. (1994). Microbial biomass. *In* "Methods of Soil Analysis, Part 2. Microbiological and Biochemical Properties" (R. W. Weaver, S. Angle, P. Bottomley, D. Bezdicek, S. Smith, A. Tabatabai, and A. Wollum, eds.), SSSA Book Ser. No. 5, pp. 753–773. Soil Science Society of America, Madison, WI.

Hu, S., Coleman, D. C., Hendrix, P. F., and Beare, M. H. (1995). Biotic manipulation effects on soil carbohydrates and microbial biomass in a cultivated soil. *Soil Biol. Biochem.* **27**, 1127–1135.

Hunt, H. W., Coleman, D. C., Ingham, E. R., Ingham, R. E., Elliot, E. T., Moore, J. C., Rose, S. L., Reid, C. P. P., and Morley, C. R. (1987). The detrital food web in a shortgrass prairie. *Biol. Fertil. Soils* **3**, 57–68.

Hunt, H. W., Elliott, E. T., and Walter, D. E. (1989). Inferring trophic transfers from pulse-dynamics in detrital food webs. *Plant Soil* **115**, 247–259.

Huntjens, J. L. M. (1972). Amino acid composition of humic acid-like polymers produced by streptomyces and of humic acids from pasture and arable land. *Soil Biol. Biochem.* **4**, 330–345.

Ingham, R. E., Trofymow, J. A., Ingham, E. R., and Coleman, D. C. (1985). Interactions of bacteria, fungi, and their nematode grazers: Effects on nutrient cycling and plant growth. *Ecol. Monogr.* **55**, 119–140.

Insam, H., Parkinson, D., and Domsch, K. H. (1989). Influence of macroclimate on soil microbial biomass. *Soil Biol. Biochem.* **21**, 211–221.

Janzen, H. H. (1987). Soil organic matter characteristics after long-term cropping to various spring wheat rotatations. *Can. J. Soil Sci.* **67**, 845–856.

Janzen, H. H., Campbell, C. A., Brandt, S. A., Lafond, G. P., and Townley Smith, L. (1992). Light-fraction organic matter in soils from long-term crop rotations. *Soil Sci. Soc. Am. J.* **56**, 1799–1806.

Jenkinson, D. S. (1977). Studies on the decomposition of plant material in soil. V. The effects of plant cover and soil type on the loss of carbon from ^{14}C labelled ryegrass decomposing under field conditions. *J. Soil Sci.* **28**, 424–434.

Jenkinson, D. S., and Ladd, J. N. (1981). Microbial biomass in soil: Measurement and turnover. *In* "Soil Biochemistry," Vol. 5, pp. 415–471. Dekker, New York.

Jenkinson, D. S., and Powlson, D. S. (1976a). The effects of biocidal treatments on metabolism in soil—I. Fumigation with chloroform. *Soil Biol. Biochem.* **8**, 167–177.

Jenkinson, D. S., and Powlson, D. S. (1976b). The effects of biocidal treatments on metabolism in soil—V. A method for measuring soil biomass. *Soil Biol. Biochem.* **8**, 209–213.

Jenkinson, D. S., and Rayner, J. H. (1977). The turnover of soil organic matter in some of the Rothamsted classical experiments. *Soil Sci.* **123**, 298–305.

Jenkinson, D. S., Harkness, D. D., Vance, E. D., Adams, D. E., and Harrison, A. F. (1992). Calculating net primary production and annual input of organic matter to soil from the amount and radiocarbon content of soil organic matter. *Soil Biol. Biochem.* **24**, 295–308.

Jenny, J. (1933). Soil fertility losses under Missouri conditions. *MO. Agric. Exp. Stn. Bull.* **324**. As cited in Stewart, 1995.

Johnson, N. C., and Pfleger, F. L. (1992). Vesicular–arbuscular mycorrhizae and cultural stress. *In* "Mycorrhizae in Sustainable Agriculture" (G. J. Bethlenfalvay, and R. G. Linderman, eds.), ASA Spec. Publ. No. 54, pp. 71–99. American Society of Agronomy, Madison, WI.

Keith, H., Oades, J. M., and Martin, J. K. (1986). The input of carbon to soil from wheat plants. *Soil Biol. Biochem.* **18,** 445–449.

Kenne, L., and Lindberg, B. (1983). Bacterial polysaccharides. *In* "The Polysaccharides" (G. O. Aspinall, ed.), Vol. 2, pp. 287–363. Academic Press, London.

Kern, J. S., and Johnson, M. G. (1991). The impact of conservation tillage use on soil and atmospheric carbon in the contiguous United States. US EPA Report No. EPA//600/3-91/056. United States Environmental Protection Agency, Washington, DC.

Kilbertus, G. (1980). Etudé des microhabitats contenus dans les aggrégats du sol. Leur relation avec la biomass bacteriénne et la taille des procaryotes présents. *Rev. Ecol. Biol. Sol.* **17,** 543–557.

Killham, K. (1994). "Soil Ecology." Cambridge Univ. Press, Cambridge, UK.

Kinchesh, P., Powlson, D. S., and Randall, E. W. (1995). ^{13}C NMR studies of organic matter in whole soils: II. A case study of some Rothamsted soils. *Eur. J. Soil Sci.* **46,** 139–146.

Knapp, E. B., Elliott, L. F., and Campbell, G. S. (1983). Microbial respiration and growth during the decomposition of wheat straw. *Soil Biol. Biochem.* **15,** 319–323.

Kuikman, P. J., Jansen, A. G., van Veen, J. A., and Zehnder, A. J. B. (1990). Protozoan predation and the turnover of soil organic carbon and nitrogen in the presence of plants. *Biol. Fertil. Soils* **10,** 22–28.

Ladd, J. N. (1989). The role of the soil microflora in the degradation of organic matter. *In* "Recent Advances in Microbial Ecology" (T. Hattori, Y. Ishida, Y. Muruyama, R. Y. Morita, and A. Uchida, eds.), pp. 168–174. Japan Scientific Societies Press, Tokyo.

Lal, R., Kimble, J., Levine, E., and Stewart, B.A. (eds.) (1995a). "Soils and Global Change," Advances in Soil Science. Lewis, Boca Raton, FL.

Lal, R., Kimble, J., Levine, E., and Stewart, B. A. (eds.) (1995b). "Soil Management and Greenhouse Effect," Advances in Soil Science. Lewis, Boca Raton, FL.

Leffelaar, P. A. (1993). Water movement, oxygen supply and biological processes on the aggregate scale. *Geoderma* **57,** 143–165.

Lynch, J. M., and Panting, L. M. (1982). Effects of season, cultivation and nitrogen fertilizer on the size of the soil microbial biomass. *J. Sci. Food Agric.* **33,** 249–252.

MacCarthy, P., Clapp, C. E., Malcolm, R. L., and Bloom, P. R. (eds.) (1990). "Humic Substances in Soil and Crop Sciences: Selected Readings." Soil Science Society of America, Madison, WI.

Malcolm, R. L., and McCarthy, P. (1991). The individuality of humic substances in diverse environments. *In* "Advances in Soil Organic Matter Research: The Impact on Agriculture and the Environment." (W. S. Wilson, ed.), pp. 23–34. Royal Society of Chemistry, Cambridge, UK.

Mann, L. K. (1989). Changes in soil carbon storage after cultivation. *Soil Sci.* **142,** 279–288.

Marr, A. G., Nilson, E. H., and Clark, D. J. (1963). The maintenance requirement of *Escherichia coli. Ann. N.Y. Acad. Sci.* **102,** 536–548.

Martel, Y. A., and Paul, E. A. (1974). The use of radiocarbon dating of organic matter in the study of soil genesis. *Soil Sci. Soc. Am. Proc.* **38,** 501–506.

Martin, J. P., and Haider, K. (1971). Microbial activity in relation to soil humus formation. *Soil Sci.* **111,** 54–63.

Martin, J. P., and Haider, K. (1979). Biodegradation of 14C-labelled model and cornstalk lignins, phenols, model phenolase humic polymers and fungal melanins as influenced by a readily available carbon source and soil. *Appl. Environ. Microbiol.* **38,** 283–289.

Marumoto, T. (1984). Mineralization of C and N from microbial biomass in paddy soil. *Plant Soil* **76,** 165–173.

Marumoto, T.,. Anderson, J. P. E., and Domsch, K. H. (1982). Mineralization of nutrients from soil microbial biomass. *Soil Biol. Biochem.* **14,** 469–475.

Mayaudon, J., and Simonart, P. (1963). Humification des microorganismsm marques par 14C dans le sol. *Ann. Inst. Pasteur* **105,** 257–266.

McCarty, P. L. (1971). Energetics and bacterial growth. *In* "Organic Compounds in Aquatic Environments" (S. D. Faust, and J. V. Hunter, eds.), pp. 495–531. Dekker, New York.

McCarty, P. L. (1988). Bioengineering issues related to in situ remediation of contaminated soils and groundwater. *Basic Life Sci.* **45**, 143–163.

McGill, W. B., Hunt, H. W., Woodmansee, R. G., and Reuss, J. O. (1981). PHOENIX, a model of the dynamics of carbon and nitrogen in grassland soils. *Ecol Bull. NFR* **33**, 49–115.

McGill, W. B., Cannon, K. R., Robertson, J. A., and Cook, F. D. (1986). Dynamics of soil microbial biomass and water-soluble organic C in Breton L after 50 years of cropping two rotations. *Can. J. Soil Sci.* **66**, 1–19.

Molina, J. A. E., Clapp, C. E., Shaffer, M. J., Chichester, F. W., and Larson, W. E. (1983). NCSOIL, a model of nitrogen and carbon transformations in soil: Description, calibration and behavior. *Soil Sci. Soc. Am. J.* **47**, 85–91.

Monrozier, L. J., Ladd, J. N., Fitzpatrick, R. W., and Foster, R. C. (1991). Components and microbial biomass content of size fractions in soils of contrasting aggregation. *Geoderma* **50**, 37–62.

Moore, J., and de Ruiter, P. C. (1991). Temporal and spatial heterogeneity of trophic interactions within below-ground food webs. *Agric. Ecosyst. Environ.* **34**, 371–397.

Moorhead, D. L., and Reynolds, J. F. (1992). Modeling the contributions of decomposer fungi in nurient cycling. *In* "The Fungal Community. Its Organization and Role in the Ecosystem" (G. C. Carroll, and D. T. Wicklow, eds.), 2nd Ed., pp. 691–714. Dekker, New York.

Myrold, D. D., and Tiedje, J. M. (1985). Diffusional constraints on denitrification in soil. *Soil Sci. Soc. Am. J.* **49**, 651–657.

Nannipieri, P., Johnson, R. L., and Paul, E. A. (1978). Criteria for measurement of microbial growth and activity in soil. *Soil Biol. Biochem.* **10**, 223–229.

Nelson, P. N., Dictor, M.-C., and Soulas, G. (1994). Availability of organic carbon in soluble and particle-size fractions from a soil profile. *Soil Biol. Biochem.* **26**, 1549–1555.

Nicolardot, B. (1988). Evolution du niveau de biomasse microbienne du sol au cours d'une incubation de longue durée: relations avec la minéralisation du carbone et de l'azote organique. *Rev. Écol. Biol. Sol.* **25**, 287–304.

Nicolardot, B., and Molina, J. A. E. (1994). C and N fluxes between pools of soil organic matter: Model calibration with long-term field experimental data. *Soil Biol. Biochem.* **26**, 245–251.

Nicolardot, B., Molina, J. A. E., and Allard, M. R. (1994). C and N fluxes between pools of soil organic matter: Model calibration with long-term incubation data. *Soil Biol. Biochem.* **26**, 235–243.

Nyborg, M., Solberg, E. D., Malhi, S. S., and Izaurralde, R. C. (1995). Fertilizer N, crop residue, and tillage alter soil C and N content in a decade. *In* "Soil Management and Greenhouse Effect" (R. Lal, J. Kimble, E. Levine, and B. A. Stewart, eds.), Advances in Soil Science, pp. 93–99. Lewis, Boca Raton, FL.

Oades, J. M. (1988). The retention of organic matter in soils. *Biogeochemistry* **5**, 35–70.

Odum, E. P. (1971). "Fundamentals of Ecology," 3rd Ed. Saunders, Philadelphia.

Paine, R. T. (1980). Food webs: Linkage interaction strengths and community infrastructure. *J. Anim. Ecol.* **49**, 667–685.

Papendick, R. I., and Campbell, G. S. (1981). Theory and measurement of water potential, *In* "Water Potential Relations in Soil Microbiology" (J. F. Parr, ed.), SSSA Spec. Publ. No. 9, pp. 1–11. American Society of Agronomy, Madison, WI.

Parnas, H. (1975). Model for decomposition of organic material by microorganisms. *Soil Biol. Biochem.* **7**, 161–169.

Parr, J. F., and Papendick, R. I. (1978). Factors affecting the decomposition of crop residues by microorganisms. *In* "Crop Residue Management Systems" (W. R. Oschwald, ed.), ASA Spec. Publ. No. 31, pp. 101–129. American Society of Agronomy, Madison, WI.

Parton, W. J., Schimel, D. S., Cole, C. V., and Ojima, D. S. (1987). Analysis of factors controlling soil organic matter levels in Great Plains grasslands. *Soil Sci. Am. J.* **51**, 1173–1179.

Parton, W. J., Stewart, J. W. B., and Cole, C. V. (1988). Dynamics of C,N,P, and S in grassland soils: A model. *Biogeochemistry* **5**, 109–131.

Patra, D. D., Brookes, P. C., Colemen, K., and Jenkinson, D. S. (1990). Seasonal changes in soil microbial biomass in an arable and a grassland soil which have been under uniform management for many years. *Soil Biol. Biochem.* **22**, 739–742.

Paul, E. A., and Clark, F. E. (1989). "Soil Microbiology and Biochemistry." Academic Press, New York.

Paul, E. A., and Voroney, R. P. (1980). Nutrient and energy flow through soil microbial biomass. *In* "Contemporary Microbial Ecology" (D. C. Ellwood, J. N. Hedger, M. L. Latham, L. M. Lynch, and J. H. Slater, eds.), pp. 215–237 Academic Press, London.

Paul, E. A., and Voroney, R. P. (1984). Field interpretation of microbial biomass and activity measurements. *In* "Current Perspectives in Microbial Ecology" (M. J. Klug, and C. A. Reddy, eds.), pp. 509–514. American Society of Microbiology, Washington, DC.

Paul, E. A., Horwath, W. R., Harris, D., Follett, R., Leavitt, S. W., Kimball, B. A., and Pregitzer, K. (1995). Establishing the pool sizes and fluxes in CO_2 emissions from soil organic matter turnover. In "Soils and Global Change" (R. Lal, J. Kimble, E. Levine, and B. A. Stewart, eds.), Advances in Soil Science, pp. 297–305. Lewis, Boca Raton, FL.

Paustian, K., and Schnürer, J. (1987). Fungal growth responses to carbon and nitrogen limitation: A theoretical model. *Soil Biol. Biochem.* **19**, 613–620.

Paustian, K., Robertson, G. P., and Elliot, E. T. (1995). Management impacts on carbon storage and gas fluxes (CO2, CH4) in mid-latitude cropland. *In* "Soil Management and Greenhouse Effect" (R. Lal, J. Kimble, E. Levine, and B. A. Stewart, eds.), Advances in Soil Science, pp. 69–83. Lewis, Boca Raton, FL.

Payne, W. J. (1970). Energy yields and growth of heterotrophs. *Annu. Rev. Microbiol.* **24**, 17–52.

Pirt, S. J. (1965). The maintenance energy of bacteria in growing cultures. *Proc. R. Soc. London Ser. B* **163**, 224–231.

Pirt, S. J. (1975). "Principles of Microbe and Cell Cultivation." Blackwell, Oxford, UK.

Pirt, S. J. (1982). Maintenance energy: A general model for energy-limited and energy-sufficient growth. *Arch. Microbiol.* **133**, 300–302.

Preisack, E., and Kisser-Preisack, G. M. (1993). Modeling diffusion and microbial uptake of 13C-glucose in soil aggregates. *Geoderma* **56**, 561–573.

Ritz, K., Dighton, J., and Giller, K. E. (1994). "Beyond the Biomass" Wiley, Chichester, UK.

Robertson, G. P. (1993). Fluxes of nitrous oxide and other nitrogen trace gases from intensively managed landscapes: A global perspective. *In* "Agricultural Ecosystem Effects on Trace Gases and Global Climate Change," ASA Spec. Publ. No. 55, pp. 95–108. ASA-CSA-SSSA, Madison, WI.

Robertson, K., Schnurer, J., Clarholm, M., Bonde, J. A., and Rosswall, T. (1988). Microbial biomass in relation to carbon and nitrogen mineralization during laboratory incubations. *Soil Biol. Biochem.* **20**, 281–286.

Santos, P. F., Phillips, J., and Whitford, W. G. (1981). The role of mites and nematodes in early stages of buried litter decomposition in a desert. *Ecology* **62**, 664–669.

Schimel, D. S. (1986). Carbon and nitrogen turnover in adjacent grassland and cropland ecosystems. *Biogeochemistry* **2**, 345–357.

Schimel, D. S., Coleman, D. C., and Horton, K. H. (1985). Microbial carbon and nitrogen transformations and soil organic matter dynamics in paired rangeland and cropland catenas. *Geoderma* **36**, 201–214.

Schlesinger, W. H. (1990). Evidence from chronosequence studies for a low carbon-storage potential of soils. *Nature* **348**, 232–234.

Schlesinger, W. H. (1995). An overview of the carbon cycle. *In* "Soils and Global Change" (R. Lal, J. Kimble, E. Levine, and B. A. Stewart, eds.), Advances in Soil Science, pp. 9–25. Lewis, Boca Raton, FL.

Schnitzer, M., and Schulten, H. R. (1992). The analysis of soil organic matter by pyrolysis-field ionization mass spectrometry. *Soil Sci. Soc. Am. J.* **56,** 1811–1817.

Schnitzer, M., and Khan, S. U. (1978). "Soil Organic Matter." Elsevier, New York.

Scow, K.M. (1993). Effect of sorption–desorption and diffusion processes on the kinetics of biodegradation of organic chemicals in soil. *In* "Sorption and Degradation of Pesticides and Organic Chemicals in Soil" (D. Linn, ed.), SSSA Spec. Publ. No. 32, pp. 73–114. American Society of Agronomy, Madison, WI.

Scow, K. M., and Alexander, M. (1992). Effect of diffusion and sorption on the kinetics of biodegradation: Experimental results with synthetic aggregates. *Soil Sci. Soc. Amer. J.* **56,** 128–134.

Scow, K. M., and Hutson, J. (1992). Effect of diffusion and sorption on the kinetics of biodegradation: Theoretical considerations. *Soil Sci. Soc. Am. J.* **56,** 119–127.

Scow, K. M., and Johnson, C. R. (1997). Effect of sorption on biodegradation of soil pollutants. *Adv. Agon.* **58,** 1–56.

Scow, K. M., Somasco, O., Gunapala, N., Lau, S., Venette, R., Ferris, H., Miller, R., and Shennan, C. (1994). Changes in soil fertility and biology during the transition from conventional to low-input and organic farming systems. *Calif. Agric.* **48,** 20–26.

Seech, A. G., and Beauchamp, E. G. (1988). Denitrification in soil aggregates of different sizes. *Soil Sci. Soc. Am. J.* **52,** 1616–1621.

Shields, J. A., Paul, E. A., Lowe, W. E., and Parkinson, D. (1973). Turnover of microbial tissue in soil under field conditions. *Soil Biol. Biochem.* **5,** 753–764.

Shields, J. A., Paul, E. A., and Lowe, W. E. (1974). Factors influencing the stability of labelled microbial materials in soils. *Soil Biol. Biochem.* **6,** 31–37.

Sikora, L. J., and McCoy, J. L. (1990). Attempts to determine available carbon in soil. *Biol. Fertil. Soils* **9,** 19–24.

Smith, J. L. (1989). Sensitivity analysis of critical parameters in microbial maintenance-energy models. *Biol. Fertil. Soils* **8,** 7–12.

Smith, J. L., and Paul, E. A. (1986). The role of soil type and vegetation on microbial biomass and activity. IVth Proceedings of the International Society of Microbial Ecology, pp.460–466.

Smith, J. L., and Paul, E. A. (1989). Significance of soil microbial biomass estimates in soil. *In* "Soil Biochemistry" (J. M. Bollag, and G. Stotzky, eds.), Vol. 6, pp. 357–396. Dekker, New York.

Smith, J. L., McNeal, B. L., Cheng, H. H., and Campbell, G. S. (1986). Calculation of microbial maintenance rates and net nitrogen mineralization in soil at steady-state. *Soil Sci. Soc. Am. J.* **50,** 332–338.

Smith, J. L., Papendick, R. I., Bezdicek, D. F., and Lynch, J. M. (1993). Soil organic matter dynamics and crop residue management. *In* "Soil Microbial Ecology" (F. B. Metting, Jr., ed.), pp. 65–94. Dekker, New York.

Sollins, P., Spycher, G., and Glassman, C. A. (1984). Net nitrogen mineralization from light- and heavy-fraction forest soil organic matter. *Soil Biol. Biochem.* **16,** 31–37.

Sorensen, L. H. (1975). The influence of clay on the rate of decay of amino acid metabolites synthesized in soils during decomposition of cellulose. *Soil Biol. Biochem.* **7,** 171–177.

Sorensen, L. H. (1981). Carbon–nitrogen relationships during the humification of cellulose in soils containing different amounts of clay. *Soil Biol. Biochem.* **7,** 171–177.

Sorensen, L. H. (1983). The influence of stress treatments on the microbial biomass and the rate of decomposition of humified matter in soils containing different amounts of clay. *Plant Soil* **75,** 107–119.

Soulides, D. A., and Allison, F. E. (1961). Effect of drying and freezing soils on carbon dioxide production, available mineral nutrients, aggregation, and bacterial population. *Soil Sci.* **91,** 291–298.

Sparling, G. P., Ord, B. G., and Vaughan, D. (1981). Microbial biomass and activity in soils amended with glucose. *Soil Biol. Biochem.* **13,** 99–104.

Stevenson, F. J. (1994). "Humus Chemistry. Genesis, Composition, Reactions," 2nd Ed. Wiley, New York.

Stevenson, F. J., and Elliot, E. T. (1989). Methodologies for assessing quantity and quality of SOM. In "Dynamics of Soil Organic Matter in Tropical Ecosystems" (D. C. Coleman, J. M. Oades, and G. Uehara, eds.), pp. 173–199. Univ. of Hawaii Press, Honolulu.

Stewart, B. A. (1995). Soil management in semiarid regions. In "Soil Management and Greenhouse Effect" (R. Lal, J. Kimble, E. Levine, and B. A. Stewart, eds.), Advances in Soil Science, pp. 251–258. Lewis, Boca Raton, FL.

Temple, S. R., Friedman, D. B., Somasco, O., Ferris, H., Scow, K., and Klonsky, K. (1994). An interdisciplinary, experiment station-based participatory comparison of alternative crop management systems for California's Sacramento Valley. Am. J. Altern. Agric. 9, 64–71.

Tiessen, H. and Stewart, J. W. B. (1983). Particle-size fractions and their use in studies of soil organic matter: II. Cultivation effects on organic matter composition in size fractions. Soil Sci. Soc. Am. J. 47, 509–514.

Tisdall, J. M., and Oades, J. M. (1982). Organic matter and water stable aggregates in soils. J. Soil Sci. 33, 141–161.

Torsvik, V., Goksøyr, J., Daae, F., Sørheim, R., Michalsen, J., and Salte, K. (1994). Use of DNA analysis to determine the diversity of microbial communities. In "Beyond the Biomass" (K. Ritz, J. Dighton, and K. E. Giller, eds.), pp. 39–48. Wiley, Chichester, UK.

Tunlid, A., and White, D. C. (1992). Biochemical analysis of biomass, community structure, nutritional status, and metabolic activity of microbial communities in soil. In "Soil Biochemistry" (G. Stotzky, and J. M. Bollag, eds.), Vol. 7, pp. 229–262. Dekker, New York.

Turchenek, L. W., and Oades, J. M. (1979). Fractionation of organo–mineral complexes by sedimentation and density techniques. Geoderma 21, 311–343.

Van de Werf, H., and Verstraete, W. (1987). Estimation of active soil microbial biomass by mathematical analysis of respiration curves: Development and verification of the model. Soil Biol. Biochem. 19, 253–260.

Van Gestel, M., Ladd, J.N., and Amato, M. (1991). Carbon and nitrogen mineralization from two soils of contrasting texture and microaggregate stability: Influence of sequetial fumigation, drying and storage. Soil Biol. Biochem. 23, 313–322.

Van Gestel, M., Ladd, J. N., and Amato, M. (1992). Microbial biomass responses to seasonal change and imposed drying regimes at increasing depths of undisturbed top soil profiles. Soil Biol. Biochem. 24, 103–111.

Van Veen, J. A., and Paul, E. A. (1981). Organic carbon dynamics in grassland soils. I. Background information and computer simulation. Can. J. Soil Sci. 61, 185–201.

Van Veen, J. A., Ladd, J. N., and Frissel, M. J. (1984). Modelling C and N turnover through the microbial biomass in soil. Plant Soil 76, 257–274.

Van Veen, J. A., Ladd, J. N., and Amato, M. (1985). Turnover of carbon and nitrogen through the microbial biomass in a sandy loam and a clay soil incubated with [^{14}C(U)] glucose and [(NH$_4$)$_2$] SO$_4$ under different moisture regimes. Soil Biol. Biochem. 17, 747–756.

Van Veen, J. A., Ladd, J. N., Martin, J. K., and Amato, M. (1987). Turnover of carbon, nitrogen and phosphorus through the microbial biomass in soils incubated with ^{14}C C-, ^{15}N- and ^{32}P-labelled bacterial cells. Soil Biol. Biochem. 19, 559–565.

Voroney, R. P., and Angers, D. A. (1995). Analysis of the short-term effects of management on soil organic matter using the CENTURY model. In "Soil Management and Greenhouse Effect" (R. Lal, J. Kimbal, E. Levine, and B. A. Stewart, eds.), Advances in Soil Science. Lewis, Boca Raton, FL.

Wander, M. M., Traina, S. J., Stinner, B. R., and Peters, S. E. (1994). Organic and conventional management effects on biologically active soil organic matter pools. Soil Sci. Am. J. 58, 1130–1139.

Wang, F. L., and Bettany, J. R. (1993). Influence of freeze-thaw and flooding on the loss of soluble organic carbon and carbon dioxide from soil. J. Environ. Qual. 22, 709–714.

White, D. C., Davis, W. M., Nickels, J. S., King, J. D., and Bobbie, R. J. (1979). Determination of sedimentary microbial biomass by extractable lipid phosphate. *Oecologia* **40**, 51–62.

Williams, B. L., Cheshire, M. V., and Sparling, G. P. (1987). Distribution of ^{14}C between particle size fractions and carbohydrates separated from a peat incubated with ^{14}C-glycine. *J. Soil Sci.* **38**, 659–666.

Wilson, W. S. (ed.) (1991). "Advances in Soil Organic Matter Research: The Impact on Agriculture and the Environment." Redwood Press, Wiltshire, UK.

Woodruff, C. M. (1949). Estimating the nitrogen delivery of soil from the organic matter determination as reflected by Sanborn field. *Soil Sci. Soc. Am. Proc.* **14**, 208–212.

Woomer, P. L. (1994). Most probable number counts. *In* "Methods of Soil Analysis, Part 2. Microbiological and Biochemical Properties" (R. W. Weaver, S. Angle, P. Bottomley, D. Bezdicek, S. Smith, A. Tabatabai, and A. Wollum, eds.), SSSA Book Series No. 5, pp. 59–80. Soil Science Society of America, Madison, WI.

Wyland, L. J., Jackson, L. E., and Schulbach, K. F. (1995). Soil–plant nitrogen dynamics following incorporation of a mature rye cover crop in a lettuce production system. *J. Agric. Sci. Cambridge* **124**, 17–25.

Wyland, L. J., Jackson, L. E., Chaney, W. E., Klonsky, K., Koike, S. T., and Kimple, B. (1996). Winter cover crops in a vegetable cropping system: Impacts on nitrate leaching, soil water, crop yield, pests and management costs. *Agric. Ecosyst. Environ.* **59**, 1–17.

Zelles, L., Bai, Q. Y., Beck, T., and Beese, F. (1992). Signature fatty acids in phospholipids and lipopolysaccharides as indicators of microbial biomass and community structure in agricultural soils. *Soil Biol. Biochem.* **24**, 317–323.

Zelles, L., Rackwitz, R., Bai, Q. Y., Beck, T., and Beese, F. (1995). Discrimination of microbial diversity by fatty acid profiles of phospholipids and lipopolysaccharides in differently culti- vated soils. *Plant Soil* **170**, 115–122.

12

Effects of Global Change on Agricultural Land Use: Scaling Up from Physiological Processes to Ecosystem Dynamics

Rik Leemans

I. Introduction

Human activities have been changing the surface of the earth since the earliest settlements and the appearance of primitive agriculture. Initially, these changes had only minor and/or local effects but surely were not globally significant. The industrial progress during the past centuries, the increasing use of fossils fuels as the major energy source, and an expansion of intensive agriculture, along with increasing productivities, allowed for a rapid sustained growth of the human population. During the past few centuries, however, with the development of better tools, management practices, crops with higher yields, and the expansion of modern agriculture, the influence of humans on the earth's surface became more pronounced. Currently, humans use approximately 3.2% of the global net primary production (NPP; Vitousek *et al.*, 1986) of ecosystems for food and fodder. However, total NPP used directly (e.g., food, fuelwood, and fiber), indirectly (e.g., land clearing), or lost as a consequence of human activities is much higher (ca. 38.8% of terrestrial NPP; Vitousek *et al.*, 1986). Furthermore 11 and 25% of the natural land cover have been converted to cropland and pastures, respectively, whereas only 6% is protected in its natural state (Morris, 1995). Thus, humans and their activities are nowadays among the major agents shaping the earth's surface (Turner *et al.*, 1990a).

Human-induced changes have not been without consequences. The burning of fossil fuels and conversion of forests and other ecosystems (e.g., wetlands) to pastures and arable lands have increased the atmospheric concentrations of CO_2 and other greenhouse gases (GHGs; Houghton *et al.*, 1996). This increase in GHG concentrations alters the radiative balance of the atmosphere by absorbing and reemitting some of the infrared radiation emitted from the earth's surface. The result is that less heat is "lost" to space than in the absence of GHGs and, consequently, the earth's surface is becoming warmer. This phenomenon was already described a century ago by Arrhenius (1896). Although the magnitude, regional patterns, and timing of such climatic changes are not yet completely understood, the consequences can be pronounced. Although some regions and/or socioeconomic sectors may benefit, climate change generally leads to further stress on natural ecosystems and society (Watson *et al.*, 1996).

Increases in atmospheric GHG concentrations and climate change are followed by rises in sea level, changes in hydrology, and impacts on ecosystems and society. The complex processes leading to emissions of GHGs (influencing atmospheric chemistry and climate), the impacts of these changes and feedbacks, and interactions between them, are collectively known as "global environmental change." Global environmental change incorporates both the systemic earth system components, such as atmospheric chemistry and the climate system (which can only be comprehensively described at the global level), and many other heterogeneous regional and local components. These smaller-scale components become globally significant through their cumulative aspects (Turner *et al.*, 1990b). Initially, global-change research placed the strongest emphasis on the systemic properties. For example, the rapid increase of atmospheric CO_2 resulting from anthropogenic activities has stimulated a great deal of interest in the global carbon cycle (Melillo *et al.*, 1996). Global environmental change thus links a diversity of human activities to the changing atmospheric composition, climate, and land cover. These linkages become apparent in the fluxes of energy, water, and substances between the different components (atmosphere, biosphere, and oceans).

Concerns on global environmental change have led to important international multidisciplinary research programs [e.g., International Geosphere Biosphere Programme (IGBP), 1994; International Human Dimension Programme (IUDP), 1994], assessments of current scientific understanding [e.g., Intergovernmental Panel on Climate Change (IPCC); Houghton *et al.*, 1996; Watson *et al.*, 1996; Bruce *et al.*, 1996], and international treaties, such as the Framework Convention on Climate Change (FCCC), the Biodiversity Convention and Agenda 21, working toward sustainable development. The development of plausible future scenarios for global environmental change is a special requirement for the adequate implementation of such international conventions. These scenarios should link different emission pathways through biospheric and atmospheric processes to impacts on agrosystems, ecosystems, and society so as to objectively allow the evaluation of different response and adaptation policies. In these scenarios all the critical components of the earth's system must be considered and should be based on state-of-the-art scientific and/or technical understanding. IPCC was one of the first to develop such scenarios (e.g., Leggett *et al.*, 1992) and the methodology to do so has been improved considerably since these first attempts (e.g., Wigley *et al.*, 1996; Alcamo *et al.*, 1996).

Scenario studies, for example, show that global CO_2 emissions will increase continuously over the next century if no adequate emission-reduction programs are implemented worldwide. Such a trend is consistent with observations. CO_2 concentrations increased from approximately 280 ppmv in 1870 to 356 ppmv in 1992. After a short drop in CO_2 growth rates during the early 1990s, the annual growth rate is again at the earlier level of 1.6 ppmv and is increasing. Thus, important decisions need to be made now about future tolerable levels of atmospheric GHG concentrations as well as the strategies that will permit us to achieve these. Land use, agriculture, and physiological and ecological processes play an important role in this respect.

In this chapter, I will introduce integrated assessment models (IAMs) for the earth's system (Fig. 1). IAMs are models that address many critical aspects of environmental change by integrating knowledge and understanding from different disciplines. The aim of these models is thus to simulate

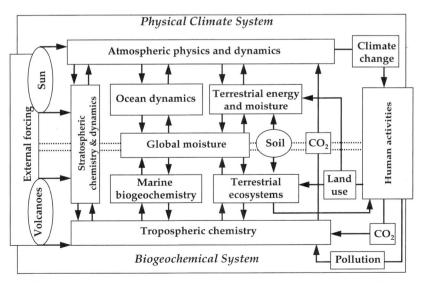

Figure 1   Flow diagram of the earth's system as defined by natural scientists.

the most important processes, interactions, and feedbacks in the earth's system, allowing evaluation of the efficiency of different global-change policies. I will highlight the current limitations of such disciplinary understanding and models and propose ways to enhance their development. Most current models address especially energy use-related emissions, the C cycle, atmospheric processes, and climate change. They do not strongly emphasize impacts and the feedbacks within the earth's system. For example, increasing yields through enhanced plant growth by the direct effects of CO_2 concentrations (Melillo *et al.*, 1996; see Appendix) could reduce total area necessary for crops while maintaining at least current levels of productivity. Such excess land could be reforested, thereby sequestering carbon (e.g., Dixon *et al.*, 1993). Other mitigation options aim at offsetting fossil fuel use and emphasize the use of biofuels and biomass. Their cultivation, however, requires land. Land, land use, and land cover are thus important inclusions in integrated assessments.

Agricultural and ecological models capable of simulating potential crop distribution and yields and vegetation patterns can be used to determine the impacts of global environmental change and, when linked to socioeconomic models, define effects on the agricultural sector and future land-cover patterns. Linking such models with C-cycle and climate-change models allows an evaluation of many different feedback processes. Results from stand-alone and integrated applications of such models will be presented. The advantages of fully integrated earth system models simulating the complex societal and natural dynamics of the earth's system will then be highlighted with examples from the Integrated Model to Assess the Greenhouse Effect (IMAGE 2) model (Alcamo, 1994). This model, among the most developed IAMs, has been used for both the recent IPCC assessments and global environmental change scenario development (e.g., Alcamo *et al.*, 1996). Its applications illustrate important issues and feature the added value of integrated frameworks.

II. Global Modeling of Ecosystem and Crop Patterns

A. Global Ecosystem and Crop Models

Natural ecosystem and agricultural crop models can be classified similarly. An important group of these models consider the productivity of ecosystems and crops and have been in development since the 1960s (e.g., de Wit, 1965). There is a large diversity of such productivity models and many are well adapted to local circumstances, species, and conditions. Another group focuses more on the distribution of ecosystems and crops. These biogeographical models, which have generally been linked to climate and soil patterns, often form an important component of land evaluation systems (e.g., FAO, 1978). In recent years these models have been coupled. They are used to assess some of the effects of global environmental change.

The biogeographical ecosystem models are often derived from climate classification schemes. Generally, these models are relatively simple and can be implemented based on readily available soil and climate data, for example, the classification of Köppen (1936; Table I) or the life zone classification of Holdridge (1947). These models regress broad-scale vegetation patterns or biomes to indices of temperature, moisture, and seasonality. Currently, more process-based biome models that incorporate ecophysiological constraints on these distributions have been developed (e.g., Prentice *et al.*, 1992). These classifications were among the first to be used to

Table I   The Köppen Climate Classification Scheme; Each Class Delimited by Specific Temperature, Moisture, and Seasonality Indices[a]

Major climatic zones	Climatic types	Climatic specifiers
A Tropical rainy climates	**f** No drought periods: Evergeen tropical rainforest	
	w Drought period: Tropical deciduous forests and savannas	
	m Pronounced drought period: Monsoonal deciduous forests	
B Arid climates	**S** Semiarid climates such as grasslands, dry savannas, and low shrubs	
	W Desert climates with sparse vegetation cover	
		h Dry hot
		k Dry and cold
C Temperate rainy climates	**f** No drought periods: Deciduous and evergreen forests	
	s Summer droughts: Oak and eucalyptus forests	
		w Winter droughts
		a Hot summer
		b Warm summer
		c Cool, short summer
D Boreal climates	**w** Winter drought: Deciduous coniferous forests	
	f No drought periods: Deciduous and evergreen coniferous forests	
		a Hot summer
		b Warm summer
		c Cool, short summer
		d Very cold winters
E Snow climates	**T** Climates with a short growing season: Treeless tundra	
	F Climates with no growing season: Ice	

[a] From Köppen (1936). The bold characters specify climate types (e.g., Af, BWh, and Dfc).

determine the impacts of climate change on ecosystems (Emanuel *et al.*, 1985; Guetter and Kutzbach, 1990). Such simulations show large shifts in ecosystem patterns; they also emphasize that impacts of climate change should not be underestimated.

The use of biogeographical models for crop distributions also has a long history. These models' illustration of the climatological constraints on crops is seen, for example, in growth hindrance by temperatures too high and too low. Different climate indices, such as the length of the growing season or effective temperature sums, are applied to regress against observed crop distribution. Such simple regressions can then be applied to a changed climate. Shifts in distribution, similar to those of the biogeographical models for ecosystems, were simulated with such approaches (e.g., Blasing and Solomon, 1984; Rosenzweig, 1985; Parry *et al.*, 1988).

These simple biogeographical models have many disadvantages. First, they are only applicable to stable equilibrium conditions. Climate variability and the transient responses to climate change are not considered. Second, changes in crop physiology and ecosystem productivity, life cycles, and other properties are not adequately included: for example, warming tends to accelerate plant growth, reducing the required growth period. If growth is accelerated during the period in which the grain is filling out, the quality of the yield may decline. These models give only a limited assessment of climatic impacts. Finally, adaptive human behavior and land use, especially important for agriculture, are left out. Although these models illustrate that crop distributions are sensitive to climate change, this conclusion cannot be directly extrapolated to the societal agricultural sector. This sector is probably not very vulnerable because its current management practices create many options for adaptive capabilities. Only under extreme, mostly arid, conditions with marginal agriculture can the impacts of climate change on agriculture become a significant limiting factor here (see discussion below).

Some of the disadvantages of simple biogeographical models are reduced by linking such models to other types. Solomon (1986), for example, linked a simple biogeographic model to a forest succession model to assess changes in species composition and succession. He explicitly included forest dynamics and concluded that significant time lags could occur in the response of forests to climate change. Since then, such studies have been repeated with gradual improvements in detail and resolution. Solomon *et al.* (1993) defined a "climatic envelope" for current agricultural regions and applied it to future conditions. Their assessment showed that agriculture could strongly expand into northern regions, thereby reducing the terrestrial global C storage through conversion of forests to agricultural land. This would increase atmospheric CO_2 concentrations. Linked models are strongly based on parameterizations (i.e., simpler descriptions, often derived empirically from complex models) of the relevant processes (Smith *et al.*, 1994). These studies show that climate-change impacts are significant and important for altering the emissions of GHGs.

The biogeographic models have made one aspect very clear: Environmental constraints largely define limits on crop growth and ecosystem productivity. However, to fully comprehend the impact of crop and ecosystem response to environmental change, productivity models are required. The simplest of these models are the so-called box models for the C cycle. They consist of a number of ecosystems and their characteristic C density, which is defined as the total amount of stored C per unit area. The approach adopted by Smith *et al.* (1992) is a variety of this type. They defined a specific C density for each ecosystem type and calculated its extent with a biogeographical model. Their simulation showed that in a warmer climate more C could be stored globally in the vegetation than under the current climate.

The approach has been elaborated on by Goudriaan and Ketner (1984), who did not use the C density but characterized ecosystem-specific NPPs. These NPPs were subsequently influenced by the atmospheric CO_2 concentration, mimicking the CO_2 fertilization effect (see Appendix). The resulting NPP was then allocated to different components in each ecosystem (Fig. 2). Each compartment has its own specific turnover time for C. This allows for an adequate determination of the global terrestrial C fluxes. Recent models are, in principle, extensions of this approach (e.g., Klein

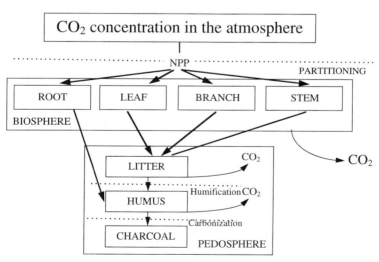

Figure 2 Schematic diagram of a simple global C cycle.

Goldewijk *et al.,* 1994). The most advanced ones also link the cycling of C to that of N, water, and energy; they are geographically explicit and the NPP characterization is replaced by a more process-based parameterization of photosynthesis and respiration (e.g., Melillo *et al.,* 1993). The problems with or challenges in scaling processes between cell, leaf, individual, and ecosystem levels have received considerable attention (Ehleringer and Field, 1993).

Process-based models are the most valid approach to date to assess the impacts of global environmental change comprehensively. The only kind of environmental change that has not been adequately included in such models is land-use change (Turner *et al.,* 1995). This is still approached by applying a simple land-use mask. Such masks depict the current agricultural areas with its rapid C turnover rates and low C storage. Interactions with land use and responses to environmental change are neglected. Furthermore, despite recent progress, the different models simulate different response of natural ecosystems (Melillo *et al.,* 1995). This is due to different parameterizations of plant and ecosystem responses to higher CO_2 levels and climate change in each model. It suggests that the experimental and observational evidence of ecosystem response to CO_2 and climate is not adequate to understand the actual responses over larger time and spatial scales. We therefore still have limited ability to project the future state of the terrestrial environment.

There are similarities between the development of C-cycle models and crop productivity models. The early crop productivity models were simple implementations of local photosynthetic and respiration rates. Yield was then defined as a specific part of biomass production, the harvestable fraction. An example of such an approach is used by the agroecological zone approach of FAO (1978). The influence of increased CO_2 concentrations and climate can be directly calculated. This approach is a central component of the future FAO projections of crop distributions and productivity (e.g., Alexandratos, 1995). Unfortunately, Alexandratos (1995) did not use scenarios for environmental change. He just considered developments under current conditions. The linkage of such a modeling system with a (global) geographic information system allows the establishment of the productivity patterns of different crop species and is capable of determining the impacts of climate change. Leemans and Solomon (1993), in using this approach, concluded that crop responses to climate change differed regionally. In certain regions productivity increased, whereas in other regions it decreased. Their study showed that geographically explicit models were required to fully simulate the envisioned changes. Global assessment tended to average out regional differences and underestimated the overall impacts. This is especially true for agriculture, in which productivity effects are larger than the distributional effects. This response is less obvious in natural ecosystems.

The most advanced crop models have now been included in these approaches (e.g., Easterling *et al.,* 1992a). These models simulate the full phenological development of a crop (often at the plant or leaf level) and consider the specific environmental requirements for each developmental phase. Water, light, and nutrient limitations are explicitly included. They integrate many different factors over the growing period. These models are data demanding and often calibrated for specific crop varieties and localities. However, they do give a comprehensive insight into the impacts of environmental change and different management practices on crop growth and yield. A well-known example is the CERES model family for wheat, maize, and rice (Rosenzweig and Parry, 1994). Similar to the ecosystem models, recent comparisons of wheat productivity models for increased CO_2 concentrations have shown that the outcome between models still varies considerably (S. van de Geijn, personal communication). With some exceptions, most models still fail to consider the interactions of CO_2 effects and climate change (e.g., Reilly *et al.,* 1996). Furthermore, they consider only the first-order effects, such as CO_2 fertilization and climate change, and ignore indirect effects, such as the socioeconomic responses of the agricultural sector.

B. Assessment of the Impacts of Global Change on Agriculture

The short review of crop and ecosystem models illustrates that many assessments of global environmental change impacts are limited, partly because not all biological, physiological, or ecological processes are considered, but more important because of ignoring of cultural and socioeconomic aspects. During the past 5 years several studies have attempted to include the last two aspects. Several national studies have been conducted, but I will limit my review to three trendsetting studies. The following studies highlight the progress: The impact of climatic variations on agriculture study by Parry *et al.* (1988); Assessing agricultural consequences of climate change for the Missouri–Iowa–Nebraska–Kansas region (MINK) study by Crosson and Rosenberg (1994); and The potential impact of climate change on world food supply, an assessment by Rosenzweig and Parry (1994). A more extensive literature review on these types of studies has been given by Reilly *et al.* (1996) as part of IPCC's second assessment report. To introduce these studies, some of the approaches to derive climate scenarios are discussed. This is important to elucidate the underlying assumptions of these studies.

1. Assumptions of Climate-Change Scenarios   Reliable environmental change scenarios are needed for the assessment of agricultural and ecological impacts. In earlier efforts, only climate change or the direct effects of CO_2 were implemented but were seldom combined together. Recently, the climate scenarios have become more linked with the changes in concentrations of CO_2 and other GHGs. There are several ways to create such scenarios (Carter *et al.*, 1994; Table II). The simplest approach is the arbitrary or systematic prescription of a specific climatic change attainable by varying the temperature and precipitation (or other) model input. This synthetic approach is a simple sensitivity analysis that highlights the vulnerability of the system modeled. The major disadvantage of this approach is that the scenarios do not represent a realistic future climate. Although popular in simple and rapid assessments, the approach is no longer used in more thorough assessments. Recent developments in this approach are to link such changes to weather generators. Such algorithms describe the current variability of local and regional climate and stochastically produce climatic series that resemble realistic climatic patterns. They can be used for future climate as well, but the limitations of synthetic approaches remain to a large extent.

The next approach is to use an analog climate, either from a past warmer or cooler period or from a region with a climate assumed similar to the future climate. For example, the warmer period in the Holocene (the so-called climatic optimum) has often been proposed as a good analog for early GHG-induced climate change. Unfortunately, the position/inclination of the earth relative to the sun was different, which generates climatic conditions largely different from a future greenhouse climate (Mitchell, 1990). The other approach is to use single warm or dry years from the historical record. In principle, this is a valid approach to test the vulnerability of ecosystems and/or agriculture, but single years may not show long-term impacts. The disadvantage of the regional analog is that nonclimatic factors, such as soil types, land uses, and socioeconomic parameters, differ. These factors could influence more strongly local crop and ecosystem patterns than climatic factors. Conclusions from analog studies can therefore be misleading.

The most generally used approach is to derive scenarios from three-dimensional Atmospheric Global Circulation Models (AGCMs). These full three-dimensional climate models simulate global climatic patterns on a grid between 2 and 9° longitude and/or latitude and several vertical layers thick by simultaneously solving the energy-balance equations of each grid cell. The temperature exchange with the oceans is prescribed and there are no systemic interactions between the oceans and atmosphere. Generally, such equilibrium AGCM simulations perform well for current climate conditions. Seasonal temperature patterns and pressure fields are similar to observed climatologies. However, the characteristic patterns of precipitation, critical for agriculture and ecosystems, only resemble reality at the

Table II Different Approaches for Creating Climate Scenarios

Approach	Advantage	Disadvantage	Example
Synthetic climate change by varying T and P	Easy to assess vulnerability	No realistic climate change	Holten and Carey (1992), Boer and de Groot (1990)
Weather generators	Contains the climatic variability of local and regional climate	Simplifying assumptions on the climate system lead to deviations; probably no realistic scenario under future conditions	Jones et al. (1994)
Historic analogs	Realistic climate patterns	Probably no realistic future patterns	Mitchell (1990)
Regional analogs	Realistic climate and land-use patterns	Different socioeconomic patterns	Parry et al. (1988)
Equilibrium simulation with global climate models (AGCMs)	Realistic global climate-change patterns	Rough simulated patterns of climate change; only double CO_2 conditions tested; limited regional applications	Leemans (1992)
Equilibrium simulations with global climate models combined with nested regional models	Realistic global and regional climate-change patterns	Only double CO_2 conditions tested; statistical downscaling methods often used that are less appropriate for changed climate	Giorgi et al. (1994)
Transient simulation with coupled global climate and ocean models	Realistic development of global climate-change patterns; unambiguously coupled to GHG concentrations	Computer demanding: only few scenarios possible; rough simulated patterns of climate change	Manabe et al. (1991)
Simulations with Integrated Assessment models	Transient GHG concentrations and climate change considered; feedbacks between components of the climate system considered	Relatively simple models used; difficult to implement because of lack of understanding of linkages and interactions in the climate system	Alcamo et al. (1996), Hulme et al. (1994)

highest grid resolutions of 2 or 3°. These resolutions are very computationally demanding. AGCMs are perturbed by changed GHG concentrations. Double CO_2 runs generally assume that the additional forcing resembles that of a doubling of CO_2 concentrations (often 560 or 600 ppmv) without considering the contribution of other GHGs. The simulated climate change (control run for current climate minus doubled CO_2 run) is then used for impact studies.

Generally, such AGCM-derived climate-change scenarios are overlaid on a higher resolution database with observed climate to obtain plausible future climatologies (for review, see Leemans, 1992). To obtain climate-change patterns for different levels of CO_2 and other GHGs, the doubled-CO_2 climate change is normalized using its global mean temperature change (the so-called climate sensitivity) and the patterns are then scaled for the desired level of GHGs (Viner *et al.*, 1995). This method represents one of the most reliable and flexible approaches for obtaining "realistic" future climatologies with consistency between GHG emission scenarios, climatic change, and impacts. However, the resolution of these AGCM-based climatologies is coarse. Several approaches for linking AGCM with regional climate models and/or regional time series have been developed to improve spatial and temporal resolutions. Unfortunately, the systematic errors in the AGCM simulations have not been reduced.

Runs of AGCMs coupled with ocean circulation models have been performed. Here, the ocean and atmosphere interact with each other continuously. These coupled models are disturbed through time with increasing levels of greenhouse gases and simulate the transient response of the climate system (biosphere, oceans, and atmosphere). These coupled models demand huge computing resources and only a few scenarios are available (Houghton *et al.*, 1996). Generally, they show a lower (or slower) climatic change than the equilibrium AGCMs. This is mainly due to the inertia of the oceans. These models can also include the simultaneous effect of sulphur aerosols, which has a regional cooling effect. The use of these complex models has led to projected future warming being less than that in the earlier simulations. Some impact assessments have already used transient scenarios but due to the amount of data to process and limitation in resolution, these transient scenarios have been used only in a quasiequilibrium mode (e.g., Prentice *et al.*, 1993). Development of truly transient scenarios will significantly advance the science.

Another approach has been developed: the IAMs. These models include simplified formulations for important components of the climate system and simulate the GHG emissions from human activities, fluxes in the earth system, atmospheric chemistry, circulation and radiative forcing (with one- or two-dimensional models), and several impact modules, of which sea level rise, ecosystem distribution, and damage costs are the most widespread.

The large advantage of these models is that they simulate the emissions from land-use change and energy use, both derived from regional socioeconomic, demographic, and technological developments. Important aspects of the C cycle and atmospheric chemistry are considered in defining the atmospheric concentrations of GHGs. Major feedbacks between the biosphere, ocean, and atmosphere are explicitly simulated. Besides, the impacts can alter such outcomes as land use or terrestrial C sequestration. This approach therefore gives an integrated view of the transient response of the climate/earth system. Relatively simple models in this respect are IMAGE 1 (Rotmans *et al.*, 1990), ESCAPE (Hulme *et al.*, 1994), or GCAM (Edmonds *et al.*, 1994). A more advanced model is IMAGE 2 (Alcamo, *et al.*, 1996). Often, the simpler one- or two-dimensional climate modules in these IAMs are improved by linking them to AGCM scenarios to get more realistic global climate-change patterns using the scaling approach described previously. The result of simulations with IAMs is a time-dependent emission and concentration path combined with consistent climate change. This is especially suited for impact assessments of ecosystems and agriculture.

Many diverse approaches for generating climate scenarios clearly exist. Although often hybrid approaches are used, no common, generally accepted approach has emerged to date. This is probably one of the reasons for the inconclusiveness of the impact section of the second assessment report of IPCC (Watson *et al.*, 1996). Its main conclusion, that "climate change adds a significant stress to many systems," seems trivial. However, the diversity of scenarios has led to very high heterogeneity in the results of impact studies and a seemingly very high uncertainty about impacts of climate change. I hope that, with commonly accepted and widely available scenarios of coupled transient GCMs and IAMs, this uncertainty will in the near future be reduced, with clearer regional impact patterns emerging.

2. The Impact of Climatic Variations on Agriculture Study This study by Parry *et al.* (1988) was initiated to assess the effects of climatic variability on agriculture. The aim was to determine the impacts not only on crop yields but also on the indirect effects on the agricultural sector. The study was conducted for several temperate and arid regions. In the temperate regions, temperature was the major constraint, whereas in the latter regions at low latitudes, moisture availability determined productivity. Simple biogeographic models were used to define the shift in forest vegetation and agricultural crops. Several approaches were used to link the models to climate scenarios. Only results from the GISS [This AGCM was develped by the Goddard Institute for Space Studies (GISS) in New York] AGCM were available and used. Visual interpretation of AGCM climate scenarios resulted in the definition of analog regions. For example, Finland's future climate was expected to be similar to that of present-day southern Sweden and Denmark. The productivity levels of these analog regions were then attributed to the original regions. Direct effects of CO_2 concentrations were not included. The resulting levels of yield and forest productivity were used to determine the higher-order effects in the agricultural sector, such as income and production levels. Different levels of adjustments to climatic change were assumed to have occurred. The lowest was the continuation of management and technology levels like those in 1980. The highest levels included changes in crop variety, soil management, and agricultural policy.

The study was the first to emphasize that climate change varies at different locations and that the vulnerability of agricultural regions is strongly dependent on current baseline conditions. For example, yields in marginal agricultural areas in Finland and Canada could disproportionately improve when compared with other regions. The major finding was that crops in the northern hemisphere would experience a longer growing season with higher temperatures than at present, generally enhancing growth. In some regions, moisture could become limiting because of increased evapotranspiration. The impacts of climate change were found to be nonlinear between different levels of temperature change. These large changes affected regional production costs and influenced crop composition. Regionally, there were large differences in the farmer's response to the impacts of climate change. For example, wheat production in Saskatchewan, Canada, would fall by 18% despite increasing productivity, whereas rice production in Japan would increase by 5%. These results were mainly caused by changes in farm-level incomes. The potential adaptation possibilities of farmers partly offset the effects of climate change.

This study was very important in defining the approach to integrated impact assessments. Despite all the limitations of the approach, such as use of regional analogs, this assessment already highlighted the need for improved climate scenarios with higher temporal and spatial resolution and for a better integration of different models to simulate the effects of changes in CO_2 concentration, climate, adaptation on crops, forests, ecosystems, and society.

3. Assessing Consequences of Climate Change for the MINK Region   The MINK study (Easterling *et al.*, 1992b) set another benchmark in agricultural assessments of climate change. The region was selected because it is highly dependent on both forest and agricultural resources for its economy. The advancement lies in the more process-based models used to link first- and higher-order effects. Both the effect of climate change and changes in CO_2 levels were included. These effects constitute the first-order effects. The higher-order effects, such as changes in prices of agricultural products, are a consequence of the first-order effects. The weather record of the warmer and drier Dust Bowl Era (1931–1940) was used as an analog for a future climate. This allowed a good calibration and validation of the models because all relevant data were available. Easterling *et al.* (1992b) then developed different adaptation scenarios with the appropriate assumptions for "dumb" (= no adaptation) and "smart" farmers (= adequate adaptation). The smart farmers changed first crop management, then selected better adapted varieties and finally altered their mix of crops. One of the results of the study was that water use did not increase under climate change. This was mainly due to changes in water use efficiency (WUE) under higher CO_2 levels. The conclusion of this study was that by the time the climate change has materialized, negative impacts will be offset by adaptation by farmers. This would be especially true if the agricultural research community directed its research toward mitigating the effects of climate change.

The original MINK study was expanded into a broader framework in which other resource sectors, such as cattle, forestry, energy, and water, were also included (Rosenberg *et al.*, 1994). Here, equilibrium AGCM scenarios were used to determine the impacts on crops and forest, water, and energy resources, with advanced models for the first-order impacts [e.g., EPIC for crop and crop management (Easterling *et al.*, 1994); FORENA for forest production (Bowes *et al.*, 1994)]. The selection criteria for their models was that they should track the transient responses to atmospheric and climate change and have the capability of being linked with changing management practices and higher-order models. An additional advantage of Easterling *et al.*'s (1992b) approach was that not only they did define response levels but also were capable of defining the levels of land-use-related CO_2, methane, and nitrous oxide emissions. This is an important advancement, especially when mitigation potential has to be evaluated. The MINK study is thus a very good example of a sectorial integrated assessment.

The simulations for the analog climate of the dustbowl showed a decline in yields for all crops. These declines increased the costs of producing the various crops and therefore reduced farm income. Including CO_2 fertilization and WUE effects into the calculations partly offset these declines. Only for wheat (the most responsive crop to CO_2 fertilization) was a slight yield increase simulated. The most vulnerable crop for the regions was corn. Also, on-farm adjustments (smart farmers) reduced the decline, but not as strongly as the direct effects of CO_2. The cattle is dependent on feedgrains and hay. The yield of those crops declined and should have repercussions on the region's production of meat and the meat packaging industry. In the long term this could mean that meat production will become less dominant in the region and this has large consequences for the manufacturing sector.

The main finding of the study was that MINK's competitive ability was reduced under climate change. The agricultural sector was not the most important one in the MINK region, but other sectors, such as the meat packaging industry, are dependent on this sector. The region currently produces approximately 30% of the U.S. corn, wheat, soybeans, and meat production. Many of the crops are grown on irrigated land and are strongly dependent on water availability. Increasing water scarcity through fewer rains and higher evapotranspiration could additionally increase production costs. By growing less corn and more wheat, this emerging competitive disadvantage could be limited. The study clearly showed the interconnectiveness of economic sector and activities. For example, higher costs for producing feedgrains lead to lesser availability of feedgrains for cattle farmers. This reduces meat production and has a negative effect on the meat packaging sector. Such a chain can only be reversed if cheap feedgrains can be imported from neighboring states. Changes in one sector thus clearly influenced other sectors. The simulations showed that the higher-order effects were difficult to predict but that they should not be neglected.

4. Potential Impact of Climate Change on World Food Supply   Rosenzweig and Parry (1994) have used a series of models to determine the possibility of changes in the global agricultural sector under climate change. They created a network of local and regional crop researchers from 18 countries who have estimated the potential changes in national grain crop yields using compatible crop models from the CERES family and standardized climate-change scenarios. These scenarios were derived from three different equilibrium AGCM results, the GISS, GFDL [This AGCM was developed at the General Fluid Dynamics Laboratory (GFDL) at Princeton University], and UKMO [This AGCM was developed at the United Kingdom Meteorological Office (UKMO) in Bracknell, UK] models, and were complemented with prescribed scenarios for CO_2 concentrations. Rosenzweig and Parry (1994) assumed that the CO_2 concentration would have reached CO_2-doubling conditions (555 ppmv) in 2060. This is consistent with the IPCC-IS92a scenario (Alcamo *et al.*, 1995). An additional scenario was developed in which no changes in climate or CO_2 concentrations were assumed. This was the reference scenario to which the others were compared. To assess the possibilities and impacts of adaptation, two different levels were assumed: Adaptation level 1 includes changes in crop variety, planting date (less than 1 month), and irrigation level. Adaptation level 2 includes, in addition, changes in crop type, planting date (more than 1 month), and extension of irrigation (Rosenzweig and Parry, 1994).

When direct CO_2 effects were not included in these scenarios, there was a large decrease in global grain yields (Table III). The decrease was most pronounced in the UKMO climate scenarios and less pronounced in the

Table III Percentage Change in Cereal Production under Different AGCM Equilibrium Scenarios for Doubled CO_2 Climate[a]

Region	GISS	GFDL	UKMO
World total			
Climate effects only	−10.9	−12.1	−19.6
Plus physiological effects of CO_2	−1.2	−2.8	−7.6
Plus adaptation level 1	0.0	−1.6	−5.2
Plus adaptation level 2	1.1	−0.1	−2.4
Developed countries			
Climate effects only	−3.9	−10.1	−23.9
Plus physiological effects of CO_2	11.3	5.2	−3.6
Plus adaptation level 1	14.2	7.9	3.8
Plus adaptation level 2	11.0	3.0	1.8
Developing countries			
Climate effects only	−16.2	−13.7	−16.3
Plus physiological effects of CO_2	−11.0	−9.2	−10.9
Plus adaptation level 1	−11.2	−9.2	−12.5
Plus adaptation level 2	−6.6	−5.6	−5.8

[a] Adaptation level 1 includes changes in crop variety, planting date (less than 1 month), and irrigation level. Adaptation level 2 additionally includes changes in crop type, planting date (more than 1 month), and extension of irrigation (adapted from Rosenzweig and Parry, 1994).

GISS scenarios. Regional changes became apparent with the direct effects of CO_2 included. In general, grain yields in the northern hemisphere increased and they decreased in lesser developed regions. The increase was most pronounced for the GISS scenario and less pronounced for the UKMO. The latter scenario has the highest temperatures and the impacts of higher temperatures seem to be important in decreasing yield. The two different farming adaptation levels have a large impact on response to climate change. Negative impacts in several temperate regions, such as China and Europe, were reversed by increasing the degree of farmers' adaptation of new management practices. Only for adaptation level 2 was it possible to maintain current grain yields (Table III).

Rosenzweig and Parry (1994) then linked the results of these crop simulations to a global food trade model, the Basic Linked System (BLS: Fischer *et al.*, 1998), which is also used for the FAO future projections (Alexandratos, 1995). This model simulates the complex dynamic interactions of producers and consumers, interacting through global markets. Technological improvements are also included in BLS. For the determination of changes in demand, population and GDP growth were assumed. These were consistent with the IPCC-IS92a assumptions and thus compatible with the time path of assumed CO_2 concentrations. Additional scenarios were created with BSL: different levels of trade liberalization, population, and economic growth.

The results change when linked to the BLS. This analysis showed that the increase in grain yields for the developed regions increased the disparities between developed and developing countries. Despite farmer adaptation, this increased cereal prices and reduced food security in the developing world. Even at a high adaptation level the agricultural sector could not prevent these negative effects. The scenarios in the absence of climate change and with full trade liberalization and low population growth showed that global food security actually increased. However, this could be counteracted by low economic growth. Even with climate change, the scenarios with trade liberalization and low population growth show better results than the other scenarios. However, the discrepancies between the developed and lesser developed world remained.

This study represents an important benchmark in assessing the impacts of climate change. It highlights the importance of socioeconomic factors not related to climate change. There are, however, several limitations to this study. Working at the national level typically dampens the effect of local and regional responses. This is especially important for large developed countries in which the environments are heterogeneous and many marginal areas exist, possibly leading to overestimation of the positive effects in developed countries. Furthermore, the models used are all temperate crop models. Although they are calibrated to local circumstances, their applicability to changed tropical conditions could be limited, increasing the negative effect. The disparity between developed and lesser developed regions could be partly a model artifact.

5. Conclusions on Impact Assessments   Although very useful to frame important issues, the impact assessments presented all have their limitations. They address only one or several aspects of global environmental change and consider limited interactions with processes or other sectors. In general, the impacts on current conditions are presented as an increased risk factor for failure. Because of the dynamics of adaptation, progress in agricultural development, and changing land uses, comparison with current conditions can be misleading in that the current conditions are fixed as a status quo. Although several studies discuss the adaptive capacity of farmers in changing types of crop and management practices, there are no possibilities in most models for changing the geographic land-use and land-cover patterns in these assessments. This is especially important for regions with increasing populations and/or changing dietary patterns. Changes in land use are important links to issues such as C storage in different types of landscape, which largely determine the buildup of CO_2 in the atmosphere.

III. Feedbacks in the Earth's System

The influence of the C cycle and land-use and land-cover change on atmospheric CO_2 concentrations is clearly important. The rates of basic physiological and ecosystem processes are influenced by temperature, moisture, and nutrient availability. The latter is mostly a soil property but is altered by decomposition processes and management, such as adding fertilizer. Climate change will be different according to region and alters photosynthetic, respiration, and soil decomposition rates accordingly. Furthermore, enhanced CO_2 concentrations lead to CO_2 fertilization and improved WUE (Eamus, 1992; cf. Appendix). Here, plant type (C_3, C_4, or CAM; annual or perennial; herb, shrub, or tree) and the ecosystem structure highly determine the response (Körner, 1993). Systemic responses of the terrestrial biosphere are thus difficult to determine. In reality, the responses are highly nonlinear and ecosystem specific. Land use further diversifies responses.

An example of a C-cycle model illustrating regional responses of changes in the C cycle is presented. The primary objective was to create a terrestrial C-cycle model that was responsive to changes in environmental factors as well as land cover. The latter change included both the natural response to climate change (e.g., growth responses and shifts in vegetation zones) and land-use influences (e.g., deforestation, forestation, forestry, and agriculture). The starting point of our model was the Goudriaan and Ketner (1984) C-cycle model (see Klein Goldewijk *et al.*, 1994). Its basic model structure was used and implemented for each grid cell of a 0.5° longitude and latitude grid. This enabled us to determine the impacts of local climate, terrain, soil, and land-cover characteristics for C cycling. The total flux between the terrestrial C cycle and the atmosphere represents the logical outcome of these local fluxes.

The model was primarily driven by NPP. NPP was partitioned over different plant parts (e.g., leaf, branches, trunks, and roots), each with a specific longevity (Fig. 2). With time, C will enter to the C pool of the soil and subsequently be decomposed. Decomposition resulted in a C flux to the atmosphere. The Net Ecosystems Productivity (NEP) was thus a function of NPP and the soil decomposition rate. Both are strongly influenced by complex environmental factors. Each land-cover type (cf. ecosystem) had a characteristic NPP, which was adjusted for local climatic and soil conditions, and global atmospheric conditions. The C fertilization effects and changes in WUE were also included. Many experiments have shown that the impacts of these effects are important but that the actual outcome depends on plant type, temperature, altitude, nutrients, and moisture availability. Based on an extensive literature review (see van Minnen *et al.*, 1996a), we have implemented simple response functions for these factors and simulated local and ecosystem-specific responses (Fig. 3). We used a similar approach for the influence of temperature and moisture availability on NPP and soil decomposition (Fig. 3).

The model was capable of simulating the consequences of the transitional C fluxes under land-cover changes. One of the obvious changes that occurred was the expansion and intensification of agricultural land use. The obvious consequences for the terrestrial C cycle were considered (e.g., deforestation). The flexibility of the model allowed evaluation of the regional consequences of land-use change. The model was capable of simulating different pathways of land-use change. Furthermore, global vegetation patterns shifted under a changed climate, and this led to a new value for global C storage. The potential land-cover patterns for the C cycle were based on calculations derived from a biogeographical model (Prentice *et al.*, 1992), an agroecological zone model (Leemans and van den Born, 1994), and a land-use change model (Zuidema *et al.*, 1994). All these models

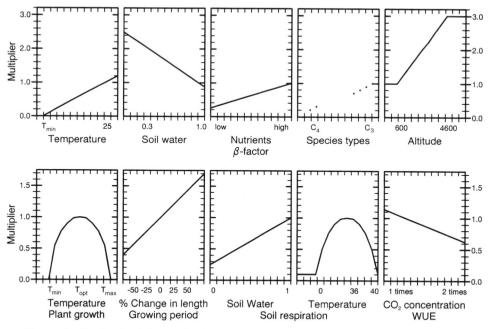

Figure 3 The feedback factors for the C cycle influencing CO_2 fertilization, WUE, plant growth, and soil respiration. [See Klein Goldewijk *et al.* (1994) for derivation of these relationships.]

are components of the IMAGE 2 model (see discussion below), used coarse plant and crop types, and characterized their distributions along climatic and soil gradients. Specific responses for land-cover transitions were defined. This dynamic approach allowed for a semicontinuous estimation of C storage under environmental change, including changes in climate, land use, and atmospheric composition.

van Minnen *et al.* (1996b) have quantified the importance of different feedback processes for the Baseline-A scenario for year 2050 (Table IV) This scenario was the IMAGE 2 implementation of the IS92a scenario (Leggett *et al.*, 1992) with intermediate assumptions for population and economic growth. Globally, the combination of all feedback processes resulted in lower atmospheric CO_2 concentrations. Biospheric uptake of CO_2 was enhanced with higher CO_2 concentrations and in a warmer climate. However, this enhancement became less pronounced toward the end of next century. Further increases in temperature are expected to change the functioning of the terrestrial biosphere and shift it from a sink toward a source.

The most significant process responsible for this result was the increase of plant growth due to higher temperatures at mid- to high latitudes. CO_2 fertilization was less important globally but had a dominant influence in the tropics, where climate change was less pronounced. Climate change also had a large impact on soil respiration in mid- to high latitudes. These processes accelerated the buildup of atmospheric CO_2 but did not offset the gains by increased plant growth in these regions. This strong regional component in the response of the terrestrial biosphere clearly invalidates many of the earlier globally aggregated C-cycle models, which definitely

Table IV   Percentage Change of NEP (pg year^{-1}) between 1990 and 2050 for Forest Areas Based on the IMAGE 2 Simulation for the Baseline Scenario[a]

Feedback processes	Latitudinal belts		
	Low	Middle	High
1. All feedbacks (baseline)	189.3	198.7	226.6
2. Climatic effects on plant growth	144.7	185.1	211.8
3. CO_2 fertilization	222.5	138.3	103.3
4. Climatic effects on soil respiration	231.4	96.5	19.2
5. Feedbacks 3 and 4	238.9	116.7	36.2
6. Feedbacks 2 and 3	147.3	191.2	214.6
7. Feedbacks 2 and 4	179.4	188.8	217.4
8. No feedbacks	211.2	126.8	92.3

[a] From van Minnen *et al.* (1996b).

portrayed some future trends but never the actual response of the terrestrial biosphere because of the complex nonlinear responses of local ecosystems to environmental change. A more rigid implementation of the basic eco-physiological processes is required.

IV. Integrated Modeling of Global Environmental Change

The previous discussions on impacts and the C cycle highlighted the need for strong integrated assessments and the development of models to facilitate this. The development of such models has only just begun (Weyant *et al.*, 1996). To illustrate the capacity of these models, I will demonstrate some applications of the IMAGE 2 model (Alcamo, 1994). This model has the most advanced developments with respect to terrestrial C cycling and its interactions with the atmosphere (see above), land-use change, and agricultural impacts. The only analog model is the regional Asian Integrated Model (AIM) for South and East Asia by Morita *et al.* (1994). There are other integrated assessment models (e.g., Nordhaus, 1992; Wigley *et al.*, 1996), but these more strongly emphasize the causes and/or economics of climate change rather than the effects of land-use and land-cover change and the impacts.

A. Structure of IMAGE 2

Earlier versions of IMAGE proposed a global-average integrated structure for climate change issues by combining (i) an energy model for greenhouse gas emissions, (ii) a global C-cycle model, and (iii) highly parameterized mathematical expressions for atmospheric chemistry, global radiative forcing, temperature change, and sea level rise (Rotmans *et al.*, 1990). The global-average calculations of IMAGE 1.0 were useful for evaluating policies at both the Dutch national and the international levels (e.g., Houghton *et al.*, 1990). IMAGE 2 not only covers the entire globe but also performs many calculations on a high-resolution global grid ($0.5^0 \times 0.5^0$ latitude–longitude); this spatial resolution increases model testability against measurements, allows an improved representation of feedbacks, and provides more detailed information for climate impact analysis. Moreover, the submodels of IMAGE 2 are, in general, more process oriented and contain fewer global parameterizations than previous models, which enhances the scientific credibility of calculations. These developments also add greatly to the computational and data-handling tasks of the model.

The scientific goals of the IMAGE 2 model were to provide insight into the relative importance of different linkages, interactions, and feedbacks in the society–biosphere–climate system and to estimate the most important sources of uncertainty in such a linked system. The policy-related goals of the model are (i) to link important scientific and policy aspects of global environmental change in a geographically explicit manner in order to assist decision making, (ii) to provide a dynamic and long-term (50–100 years) perspective about the consequences of environmental change, (iii) to provide insight into the cross-linkages in the system and the side effects of various policy measures, and (iv) to provide a quantitative basis for analyzing the costs and benefits of various measures (including preventative and adaptive) to address environmental change. These objectives have steered the design and development of the model. The model consists of three fully linked subsystems of models (Fig. 4): (i) the Energy–Industry System (EIS); (ii) the Terrestrial Environment System (TES), and (iii) the Atmosphere–Ocean System (AOS).

EIS computes the emissions of greenhouse gases from world regions as a function of energy consumption and industrial production (de Vries *et al.*, 1994). The EIS models are designed especially for investigating the effectiveness of improved energy efficiency and technological development on future emissions in each region and can be used to assess the consequences of different policies and socioeconomic trends on future emissions. AOS computes the buildup of greenhouse gases in the atmosphere and the resulting zonal-average temperature and precipitation patterns.

TES simulates the changes in global land cover on a grid-scale based on climatic, soil, demographic, and economic factors. The roles of land cover and other factors are then taken into account to compute the flux of CO_2 and other greenhouse gases from the biosphere to the atmosphere. The structure of TES is summarized as follows (Fig. 4): The demand for agricultural and forest products (food, fodder, fiber, and traditional and modern biomass) is linked to the regional availability of agricultural and forest resources. Available land resources are calculated on the spatial grid, characterizing local climate, terrain, soil, and topography. If current resources are inadequate to satisfying demand, land use expands into natural vegetation, converting it into land-cover classes such as agricultural land, pastures, or regrowth forests. This process results in deforestation and increased GHG fluxes toward the atmosphere. Because TES also simultaneously calculates intensification of agricultural productivity on the basis of technological and socioeconomic assumptions, agricultural land can either expand or contract. Abandoned agricultural land converts into the an early successional phase of the potential natural vegetation, often with increasing C densities through time.

There are several direct linkages with other components of IMAGE 2 (Fig. 4). AOS integrates GHG emissions from both TES and EIS and computes the final concentrations, accounting for uptake by the oceans and atmospheric chemistry. It further computes the resulting radiative forcing and latitudinal climate change. Here, the impacts of changed land-cover

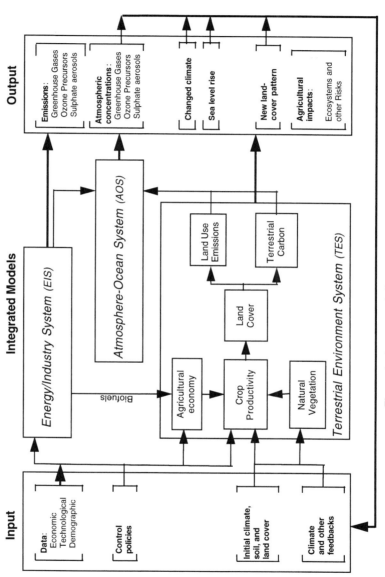

Input

Data:
Economic
Technological
Demographic

Control policies

Initial climate, soil, and land cover

Climate and other feedbacks

Integrated Models

Energy/Industry System (EIS)

Biofuels

Atmosphere-Ocean System (AOS)

Terrestrial Environment System (TES)

Agricultural economy

Crop Productivity

Natural Vegetation

Land Cover

Land Use Emissions

Terrestrial Carbon

Output

Emissions:
Greenhouse Gases
Ozone Precursors
Sulphate aerosols

Atmospheric concentrations:
Greenhouse Gases
Ozone Precursors
Sulphate aerosols

Changed climate

Sea level rise

New land-cover pattern

Agricultural impacts:
Ecosystems and other Risks

Figure 4 A conceptual diagram of the IMAGE 2.

characteristics, such as albedo, are taken into account. The final concentration of GHGs (especially CO_2) and climate change are used again as input to the different models that determine the potential of terrestrial vegetation.

In order to provide a long-term perspective on the consequences of climate change, the model's time horizon extends to the Year 2100. The time steps of different submodels vary, depending on their mathematical and computational requirements, typically from 1 day to 1 year. The structure and underlying assumptions and data sets are described by Alcamo (1994).

Another goal of the model is to provide as much information as possible on the global grid. This is because nearly all potential impacts of climate change (e.g., impacts on ecosystems, agriculture, and coastal flooding) are strongly spatially variable. Moreover, land-use-related greenhouse gas emissions (e.g., nitrous oxide from soils or methane from agricultural activities) greatly depend on "local" environmental conditions and human activities (Turner *et al.*, 1995). In addition, climate feedbacks, such as the effect of temperature on soil respiration or the effect of changing CO_2 levels on plant productivity, also vary substantially from location to location. There are two additional reasons for computing grid-scale information. First, policy makers are interested in regional/national policies to address climate change. Indeed, most climate policies are location specific (e.g., sequestering carbon in forest plantations or reducing nitrous oxide by modifying agricultural practices). Second, grid-scale information makes model calculations more testable against observations compared to more aggregated models.

Nevertheless, it is currently impossible to provide grid-scale calculations for all components of climate change. In particular, this is unfeasible for economic calculations because of the difficulty in specifying economic/demographic factors (e.g., trade relationships, technological development, and similar data) on a subcountry or grid scale for the entire world over the long time horizon of the model. As an intermediate step, economic calculations are performed for 13 world regions (Canada, United States, Latin America, OECD Europe, Eastern Europe, Africa, CIS, Middle East, India + South Asia, China + centrally planned Asian countries, East Asia, Oceania, and Japan), which follows common practice in global economic studies. The criteria for grouping countries together in a particular region are mainly economic similarity and geographic position.

Because the IMAGE 2 model is based on large global data sets and poorly understood global processes, it is unavoidable that many parameters will be ill-defined with large degrees of freedom or uncertainty. With this in mind, our basic approach is to propose submodels that have a comparable level of process detail. We will also adjust a limited number of parameters with the greatest uncertainty to obtain model calculations in reasonable agreement with 1970–1990 data. We selected the period 1970–1990 because of data availability, although we intend to test the model against data from a longer historical period. The individual submodels were tested and their parameters were adjusted, then linked and tested within the three subsystems of models and finally the fully linked model. Results of these validations, given in Alcamo (1994), are beyond the scope of this chapter. Of course, this procedure does not ensure that adjusted parameters and other inputs will be correct for scenario analysis under changed economic and environmental conditions; nevertheless, it does indicate the adequacy of the model in explaining global changes that occurred during the 1970–1990 period, such as the increase in energy-related emissions, estimated changes in deforestation rate and terrestrial carbon fluxes, and the buildup of various greenhouse gases in the atmosphere.

B. Scenario Simulations with IMAGE 2

It is impossible to evaluate policies for climate protection without "no action," "business-as-usual," or benchmark scenarios. IPPC has developed a series of such scenarios (Leggett *et al.*, 1996; Alcamo *et al.*, 1995). I have selected the baseline-A scenario with intermediate assumptions for population growth and economic growth and activities (Alcamo *et al.*, 1996). This scenario leads to an increasing level of GHG emissions and concentrations. Equivalent CO_2 concentrations (equivalent CO_2 concentration is defined as the concentration of CO_2 that would cause a similar amount of radiative forcing or temperature change as the mixture of CO_2 and other greenhouse gases) have doubled by 2060, similar to the assumptions of Rosenzweig and Parry (1994). In this scenario the emissions from human energy use dominate. The population growth here, combined with changing dietary patterns (more meat), led to land-use change. Especially in Africa, India, and China, all potential agricultural land resources were used and this influenced food availability. In most other regions, food productivity increased and less land was required so that abandonment of agricultural land occurred. However, these land-use change patterns were very sensitive to the assumptions of the meat components of the regional diets. Of all land uses, meat production increased the most over the simulated period. This is partly driven by the economic growth (and consequently increasing wealth) in all regions. Such dietary and land-use changes are not considered in the Rosenzweig and Parry (1994) study and could further increase the discrepancies between developed and developing regions.

Several alternative scenarios were developed from this scenario. All of these assume different policy measures and their consequences. Two stabilization scenarios were defined. The FCCC's objective is to stabilize atmospheric GHG concentrations at nondangerous impact levels. In the scenarios, we wanted to obtain such stabilization by 2150 at levels of 450 ppmv (Stab 450 All) or 550 ppmv CO_2 (Stab 550 All), respectively. This was obtained by selectively reducing the energy-related emissions of EIS. Land-use emissions remained the same as in baseline-A. The next two scenarios are actually emission-reduction protocols as proposed during the Climate Negotiations. FCCC distinguishes Annex 1 countries as the developed countries currently emitting most of the global GHGs. The others are the less developed countries that, under FCCC, are allowed to increase their GHG emissions. Here, I have assumed a scenario in which only the Annex 1 countries stabilize their emissions of all GHGs in 2000 and then reduce them subsequently at a rate of 1% annually (St2000 Glob An1 -1%). The other scenarios made similar assumptions, but for all countries (St2000 Glob All -1%).

The last scenario used is the LESS Biomass Intensive scenario (LESS-BI) developed by Ishitani *et al.* (1996) for the IPCC second assessment report. Their aim was to illustrate the potential for Low CO_2-Emissions Energy Supply Systems (LESS). The LESS-BI was one of their scenarios that relied heavily on renewables for energy generation. The implementation of LESS-BI in IMAGE 2 simulates the specific energy requirements and mix of energy sources, including modern biomass, in each region by defining both the locally produced and the imported/exported portions. This implementation mimics the original LESS-BI energy sources mix. An additional assumption in LESS-BI was that total energy use was only half of what it is in the other scenarios. This was obtained by increased energy efficiency employing best available technologies worldwide. The energy mix in LESS-BI was less dependent on fossil fuels. This was obtained through a high dependence on renewable resources, such as solar, nuclear, and modern biomass. The last required land to cultivate such biomass. The original LESS assessment estimated the additional need of 550 Mha biomass plantations, assuming highly productive lands. Our implementation of LESS-BI (Leemans *et al.*, 1996) required 797 Mha, in part because less productive lands were used. Such expansion of agricultural land will influence deforestation patterns and have significance for environmental issues, such as biodiversity.

In Figure 5 some of the results of these IMAGE scenarios are presented. These results are globally aggregated. In reality, information from the model on all local and regional characteristics of crops, vegetation, and climate change patterns, a wealth of other socioeconomic data on food supply, and energy use are also available. Here, I summarize only some major points. All scenarios showed significant changes in land use. Agricultural land increased most in the LESS-BI scenario and this led to significantly more deforestation. After 2030, all scenarios simulated a decrease in the extent of agricultural land. This was caused by the stabilization of demand combined with increased productivity of agricultural land (both

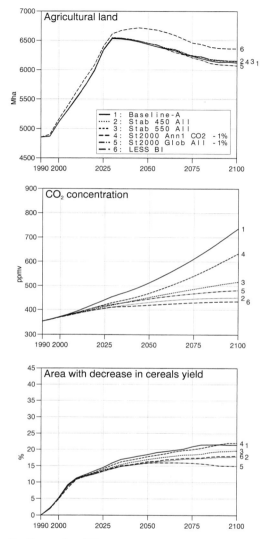

Figure 5 Output for different scenarios simulated with IMAGE 2.

were technology, management, and physiological effects). The scenarios with the higher final GHG concentrations required less agricultural land. This was due to increased productivity of crops as a result of changed climate and CO_2 fertilization. The importance of the plant physiological feedback processes was also clear from the NEP plot. There were large differences between scenarios that could be explained in part by changes

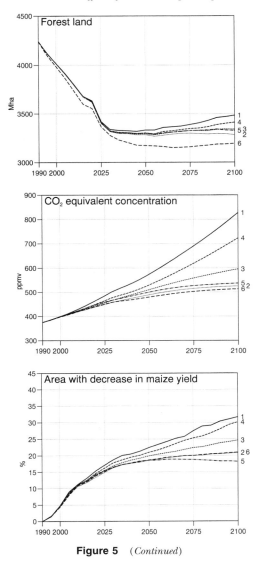

Figure 5   (*Continued*)

in land use (the maximum agricultural land area and NEP around 2025) and the influence of CO_2 fertilization, WUE, climate impacts on plant growth, decomposition, and vegetation shifts.

All scenarios simulated very distinctive patterns of GHG emissions and concentrations. The CO_2-equivalent emissions (all GHGs combined) were generally 10% higher than those of CO_2 alone. This difference was impor-

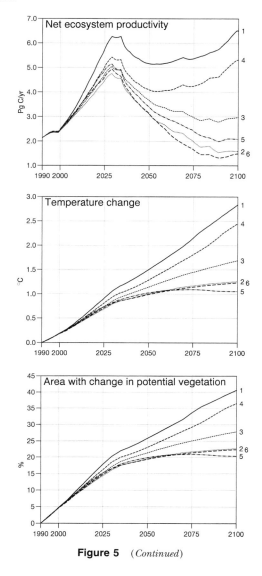

Figure 5   (*Continued*)

tant for examining the direct CO_2 effects and climate change separately. The different concentrations in the scenarios led to different temperature changes. The baseline-A led to an increase in temperature of 2.5°C at 2.5 times the current CO_2 level by 2100. The more stringent scenarios limited the temperature change to 1°C by 2100. Initially, all scenarios followed a similar pattern and the difference in efficiency of these policy measures became apparent only after 2025. This also implies the impacts were similar in these early stages. Only after 2025 did the different scenarios strongly vary. There was one clear difference among the scenarios. In three scenarios, the LESS-BI, St2000 Glob ALL -1% and Stab 450 ALL, the impact levels stabilized, whereas for the other scenarios they continuously increased. The impact levels for temperate cereals (C3 species) were generally lower than those for maize (a C4 species). This was caused by differences in CO_2 fertilization for these species, which offsets the negative impacts of climate change. C3 plants are responsive to higher CO_2 levels, whereas C4 plants are not. This finding is similar to those in the earlier studies. Also, shifts in potential vegetation were clearly different among the scenarios.

These scenario studies show that this type of integrated model is capable of simulating many components of the earth's system comprehensively. Policy options to mitigate the buildup of GHGs can be tested and their consequences for other possible options evaluated. Optimal response strategies can be developed in this way. The LESS-BI scenario is a good example of this. IMAGE 2 also has the capacity to address other environmental issues, such as food security and biodiversity. Impacts on biodiversity are, for example, a clear function of changes in potential vegetation and land use. Within IGBP, IMAGE 2 is currently used to develop scenarios for biodiversity change (Steffen *et al.*, 1996).

V. Concluding Remarks

In this chapter, the different approaches to impact assessments of climate change on agricultural and ecological systems have been discussed. Such impacts should not be treated in isolation from other components of the climate–biosphere–societal system. Many different models have been developed, environmental data sets have been compiled, and experiments have been conducted to increase our understanding of the local physiological and ecological processes. This understanding is relatively high, especially if we treat processes individually. However, the interaction between processes and factors influencing them has been neglected until recently. The importance of agricultural and ecological research on such interactions and under realistic conditions cannot be stressed enough. Furthermore, the linkages with socioeconomic research and understanding are needed to assess the actual responses to climate change. In this field of research we are only starting to enhance our knowledge (Turner *et al.*, 1995).

Integrated approaches and IAMs can highlight gaps or incompatibilities between disciplines and/or scientific understanding. These methods have not only relevance as a policy tool: Scientifically speaking, they are objective tools to emphasize the systemic and highly nonlinear linkages between the different components of the earth's system. The research activities of the major international programs (SCOPE, WCRP, IGBP, and IHDP) will advance the development of these approaches. Linkage with remotely sensed data, for example, can assist in obtaining the geographic coverage needed. Data dissemination centers, such as CIESIN and CDIAC, and connections through the Internet provide a necessary link to standardizing some of the input data sets. Model comparisons, such as VEMAP by Melillo *et al.* (1995), will further highlight discrepancies in our understanding and guide both modeling and experimental research. Scientifically speaking, this is an exciting era, in which large-scale global environmental change is pushing the advancement of both basic and applied research on smaller scales. It is necessary to integrate the knowledge from cell to leaf to plant to plot to ecosystem, and on to region and the world. The relatively small-scale physiological and ecological processes in the biosphere play an important role in this respect.

Despite the complexity of the earth's system, major achievements in combining disciplinary expertise from many different natural and physical sciences have been rapid during the past few years. However, the largest challenge is still to come. These scientific achievements need to be integrated with the expertise of the social, cultural, and economic sciences. Particularly important is the development of adequate policies for global environmental change in which sustainable use of natural resources to accommodate an increasing human population is allowed for; this demands a better understanding of the interactions between the environment and human behavior. Scientists' major challenge is probably to guide policy makers and leaders to appropriate ways for reducing the threats of global environmental change.

VI. Summary

The small-scale physiological processes on plant and ecosystem levels and the environmental constraints of these processes determine the distribution and yields of crops and ecosystems. Plants, fields, and ecosystems define the fluxes of C, N, water, and energy between the biosphere and atmosphere and thus form an important component of the earth's system. Interaction and feedback between atmospheric changes and the biosphere determine the consequences of global environmental change. This chapter discussed different modeling frameworks to assess such consequences at different levels. It is stressed that most disciplinary models, i.e., those that only simulate a component without considering changes in or influences from other components, can lead to misleading results, especially when they are linked to environmental change scenarios of limited quality. Integrated assessment models were described in this chapter. These models link societal developments with the biosphere and other components of the earth's system. Only comprehensive models such as these can bridge the gaps between (i) use of our best scientific understanding of small-scale physiological and ecological processes, of local, regional, and global scales, and of the interactions between different components of the earth's system; and (ii) meeting the requirements of policy advisors and those of policy makers for developing consistent policy measures and strategies to address global environmental change.

VII. Appendix

Direct Effects of Increased CO_2 Concentrations on Plant Growth

Plants fix CO_2 during photosynthesis. This process takes place in all green plant cells. Three very different photosynthesis pathways are distinguished: C_3, C_4, and CAM. The pathways differ in their first carbon organic acid of the photosynthesis process (a three-carbon compound, a four-carbon compound, and crassulacean acid, respectively). These pathways have evolved under specific environmental conditions and can be recognized by different leaf morphologies and physiologies.

C_3 plants are globally the most common. They include all important tree and most crop species, such as wheat, rice, and potato. Their complete photosynthesis occurs in single cells immediately under the epidermis of a leaf and the photosynthesis reaction is reversible and therefore less effective (see Chapter 2, this volume). This results in higher photosynthesis rates at higher ambient CO_2 concentrations. This is the so-called CO_2 fertilization effect. Annual plants tend to be more responsive to CO_2 than perennial plants and herbaceous more than woody plants. The responses are species specific.

C_4 plants tend to grow in warmer, more water-limited regions and include many tropical grasses and the agriculturally important species corn, sugarcane, and sorghum. C_4 plants initially fix CO_2 in specialized cells near the epidermis of the leaf. This fixed C is transported as a C_4 compound to inner cells where the final photosynthesis occurs. This mechanism concentrates internal C levels and this results in an efficient photosynthesis. This pathway is much less responsive to changes in ambient CO_2 concentration. It has been suggested that this pathway evolved in an era with low CO_2 concentrations. The CAM pathway involves plants, such as cacti, adapted to extreme droughts. It is a variant of the C_4 pathway. The difference, however, is that the CO_2 is not only transported internally but also stored as crassulacean acid. In these plants CO_2 uptake occurs at night to reduce water losses during the day.

An important effect of changing CO_2 levels is also changes in WUE. Ambient CO_2 diffuses through stomata into the leaves. Simultaneously, water is lost to the atmosphere. The ratio between C gain and water lost is defined as WUE. Under higher CO_2 levels, stomata close and less water is lost, while photosynthesis continues at normal rates. This increases WUE and allows plants to grow under more arid conditions. Increased WUE occurs in all plant types. This process is probably one of the main causes of long-term adaptation of CO_2 fertilization in C_3 plants: Plants tend to return to "normal" photosynthesis rates when grown under high CO_2 levels. Detailed discussions of these processes can be found in Melillo *et al.* (1996) and Kirschbaum *et al.* (1996).

Acknowledgments

Funding by the Netherlands Ministry of Housing, Spatial Planning and Environment under Contract MAP410 has made this contribution possible. The research on IMAGE 2 was facilitated by funds provided by the National Research Programme on Global Air Pollution and Climate Change and contributes to IGBP-GCTE core research. The author has also received numerous suggestions for this paper from other experts during the development of the Second Assessment Report of IPCC (SAR). Without their comprehensive compilations this chapter would have suffered from many limitations. Specific comments by B. de Vries, G. J. van den Born, R. de Wijs, and an anonymous reviewer on earlier drafts were highly appreciated.

References

Alcamo, J. (ed.) (1994). "IMAGE 2.0: Integrated Modeling of Global Climate Change." Kluwer, Dordrecht.

Alcamo, J., Bouwman, A., Edmonds, J., Grübler, A., Morita, T., and Sugandhy, A. (1995). An evaluation of the IPCC IS92 emission scenarios. *In* "Climate Change 1994: Radiative Forcing of Climate Change and an Evaluation of the IPCC IS92 Emission Scenarios" (J. T. Houghton, L. G. Meira Filho, J. Bruce, H. Lee, B. A. Callander, E. Haites, N. Harris, and K. Maskell, eds.), pp. 247–304. Cambridge Univ. Press, Cambridge, UK.

Alcamo, J., Kreilemans, G. J. J., Bollen, J. C., van den Born, G. J., Krol, M. S., Toet, A. M. C., and de Vries, H. J. M. (1996). Baseline scenarios of global environmental change. *Glob. Environ. Change* **6,** 255–259.

Alexandratos, N. (ed.) (1995). "Agriculture: Towards 2010: An FAO Study." Wiley, Chichester, UK.

Arrhenius, S. (1896). On the influence of the carbonic acid in the air upon the temperature of the ground. *Philos. Mag. Ser. 5* **41,** 237–276.

Blasing, T. J., and Solomon, A. M. (1984). Response of the North American Corn Belt to climatic warming. *Prog. Biomet.* **3,** 311–321.

Boer, M. M., and de Groot, R. S. (eds.) (1990). "Landscape–Ecological Impacts of Climatic Change." IOS Press, Amsterdam.

Bowes, M. D., Sedjo, R. A., Darmstadter, J., Katz, L. A., and Lemon, K. M. (1994). Paper 3. Impacts and responses to climate change in forests of the MINK region. *Climatic Change* **24**, 63–82.

Bruce, J. P., Lee, H., and Haiters, E. F. (eds.) (1996). "Climate Change 1995. Economic and Social Dimensions of Climate Change." Cambridge Univ. Press, Cambridge, UK.

Carter, T. R., Parry, M. L., Harasawa, H., and Nishioka, S. (1994). "IPCC Technical Guidelines for Assessing Impacts of Climate Change." IPCC Spec. Rep. No. CGER-1015-'94, December 1994. Intergovernmental Panel on Climate Change, WMO, and UNEP, Geneva.

Crosson, P. R., and Rosenberg, N. J. (1994). An overview of the MINK study. *Climatic Change* **24**, 159–173.

de Vries, B., van den Wijngaard, R., Kreileman, G. J. J., Olivier, J. A., and Toet, S. (1994). A model for calculating regional energy use and emissions for evaluating global climate scenarios. *Water Air Soil Pollut.* **76**, 79–131.

de Wit, C.T. (1965). "Photosynthesis of Leaf Canopies." Agricultural Research Rep. No. 663. Centre for Agricultural Publicaton and Documentation, Wageningen.

Dixon, R. K., Winjum, J. K., and Schroeder, P. E. (1993). Conservation and sequestration of carbon: The potential of forest and agroforest management practices. *Global Environ. Change* **1993**, 159–173.

Eamus, D. (1992). The interaction of rising CO_2 and temperature with water use efficiency. *Plant Cell Environ.* **14**, 843–852.

Easterling, W. E., McKenny, M. S., Rosenberg, N. J., and Lemon, K. M. (1992a). Simulation of crop responses to climate change: Effects with present technology and no adjustments (the 'dumb farmer' scenario). *Agric. For. Meteorol.* **59**, 53-73.

Easterling, W. E., Rosenberg, N. J., McKenny, M. S., and Jones, C. A. (1992b). An introduction to the methodology, the region of study, and a historical analog of climate change. *Agric. For. Meteorol.* **59**, 3–15.

Easterling, W. E., III, Crosson, P. R., Rosenberg, N. J., McKenny, M. S., Katz, L. A., and Lemon, K. M. (1994). Agricultural impacts of and responses to climate change in the Missouri–Iowa–Nebraska–Kansas (MINK) region. *Climatic Change* **24**, 23–61.

Edmonds, J., Wise, M., and MacCracken, C. (1994). "Advanced Energy Technologies and Climate Change an Analysis Using the Global Change Assessment Model (GCAM)." Draft report, April 1994. Pacific Northwest Laboratory, Washington DC.

Ehleringer, J. R., and Field, C. B. (eds.) (1993). "Scaling Physiological Processes: Leaf to Globe." Academic Press, San Diego.

Emanuel, W. R., Shugart, H. H., and Stevenson, M. P. (1985). Climatic change and the broad-scale distribution of terrestrial ecosystems complexes. *Climatic Change* **7**, 29–43.

Fischer, G., Froberg, K., Keyzer, M. A., and Pahrik, K. S. (1988). "Linked National Models: A Tool for International Food Food Policy Analysis." Kluwer, Dordrecht.

Food and Agriculture Organization (FAO) (1978). "Report on the Agro-Ecological Zones Project. Volume 1. Methodology and Results for Africa." FAO, Rome.

Giorgi, F., Brodeur, C. S., and Bates, G. T. (1994). Regional climate change scenarios over the United States produced with a nested regional climate model. *J. Climate* **7**, 375–399.

Goudriaan, J., and Ketner, P. (1984). A simulation study for the global carbon cycle, including man's impact on the biosphere. *Climatic Change* **6**, 167–192.

Guetter, P. J., and Kutzbach, J. E. (1990). A modified Köppen classification applied to model simulations of glacial and interglacial climates. *Climatic Change* **16**, 193–215.

Holdridge, L. R. (1947). Determination of world plant formations from simple climatic data. *Science* **105**, 367–368.

Holten, J. I., and Carey, P. D. (1992). "Responses of Natural Terrestrial Ecosystems to Climate Change in Norway," NINA Forskningsrapporter No. 29, January 1992. Norsk Institutt for Naturforskning, Trondheim, Norway.

Houghton, J. T., Jenkins, G. J., and Ephraums, J. J. (eds.) (1990). "Climate Change: The IPCC Scientific Assessment." Cambridge Univ. Press, Cambridge, UK.

Houghton, J. T., Meira Filho, L. G., Callander, B. A., Harris, N., Kattenberg, A., and Maskell, K. (eds.) (1996). "Climate Change 1995. The Science of Climate Change." Cambridge Univ. Press, Cambridge, UK.

Hulme, M., Raper, S. C. B., and Wigley, T.M.L. (1994). An integrated framework to address climate change (ESCAPE) and further developments of the global and regional climate modules (MAGICC). *In* "Integrative Assessment of Mitigation, Impacts and Adaptation to Climate Change" (N. Nakícenovíc, W. D. Nordhaus, R. Richels, and F.L. Toth, eds.), pp. 289–308. IIASA, Laxenburg.

International Geosphere Biosphere Programme (IGBP) (1994). "IGBP in Action: Work Plan 1994–1998," Report, February 1994. IGBP, Stockholm.

International Human Dimensions of Global Environmental Change Programme (IHDP) (1994). "Human Dimensions of Global Environmental Change Programme. Workplan 1994–1995," Occasional Paper No. 6, September 1994. ISSC-HDP, Geneva.

Ishitani, H., Johansson, T. B., Al-Khouli, S., Audus, H., Bertel, E., Bravo, E., Edmonds, J. A., Frandsen, S., Hall, D., Heinloth, K., Jefferson, M., de Laquil, P., III, Moreira, J. R., Nakiceno-vic, N., Ogawa, Y., Pachauri, R., Riedacker, A., Rogner, H.-H., Saviharju, K., Sørensen, B., Stevens, G., Turkenburg, W. C., Williams, R. H., Zhou, F., Friedleifsson, I. B., Inaba, A., Rayner, S., and Robertson, J. S. (1996). Energy supply mitigation options. *In* "Climate Change 1995. Impacts, Adaptations and Mitigation of Climate Change: Scientific–Technical Analysis" (R. T. Watson, M. C. Zinyowera, and R. H. Moss, eds.), pp. 587–647. Cambridge Univ. Press, Cambridge, UK.

Jones, E. A., Reed, D. D., and Desanker, P. V. (1994). Ecological implications of projected climate change scenarios in forest ecosystems of central North America. *Agric. For. Meteorol.* **72,** 31–46.

Kirschbaum, M. U. F., Bullock, P., Evans, J. R., Goulding, K., Jarvis, P. G., Noble, I. R., Rounsevell, M., and Sharkey, T. D. (1996). Ecophysiological, ecological, and soil processes in terrestrial ecosystems: A primer on general concepts and relationships. *In* "Climate Change 1995. Impacts, Adaptations and Mitigation of Climate Change: Scientific–Technical Analysis" (R. T. Watson, M. C. Zinyowera, and R. H. Moss, eds.), pp. 57–74. Cambridge Univ. Press, Cambridge, UK.

Klein Goldewijk, K., van Minnen, J. G., Kreileman, G. J. J., Vloedbeld, M., and Leemans, R. (1994). Simulating the carbon flux between the terrestrial environment and the atmosphere. *Water Air Soil Pollut.* **76,** 199–230.

Köppen, W. (1936). Das geographische System der Klimate. *In* "Handbuch der Klimatologie" (W. Köppen and R. Geiger, eds.), pp. 1–46. Gebrüder Borntraeger, Berlin.

Körner, C. (1993). CO_2 fertilization: The great uncertainty in future vegetation development. *In* "Vegetation Dynamics and Global Change" (A. M. Solomon and H. H. Shugart, eds.), pp. 53–70. Chapman & Hall, New York.

Leemans, R. (1992). Modelling ecological and agricultural impacts of global change on a global scale. *J. Sci. Ind. Res.* **51,** 709–724.

Leemans, R., and Solomon, A. M. (1993). The potential response and redistribution of crops under a doubled CO_2 climate. *Climratic Res.* **3,** 79–96.

Leemans, R., and van den Born, G. J. (1994). Determining the potential global distribution of natural vegetation, crops and agricultural productivity. *Water Air Soil Pollut.* **76,** 133–161.

Leemans, R., van Amstel, A., Battjes, C. C., Kreilemans, G. J. J., and Toet, A. M. C. (1996). The land-cover and carbon cycle consequences of large-scale utilization of biomass as an energy source. *Glob. Environ. Change* **96,** 335–358.

Leggett, J., Pepper, W. J., and Swart, R. J. (1992). Emissions scenarios for the IPCC: An update. *In* "Climate Change 1992. The Supplementary Report to the IPCC Scientific Assessment"

(J. T. Houghton, B. A. Callander, and S. K. Varney, eds.), pp. 71–95. Cambridge Univ. Press, Cambridge, UK.

Manabe, S., Stoufer, R. J., Spelman, M. J., and Bryan, K. (1991). Transient responses of a coupled ocean-atmosphere model to gradual changes of atmospheric CO_2. Part 1: Annual mean response. *J. Climate* **4**, 785–818.

Melillo, J. M., McGuire, A. D., Kicklighter, D. W., Moore, B., III, Vorosmarty, C. J., and Schloss, A. L. (1993). Global climate change and terrestrial net primary production. *Nature* **363**, 234–239.

Melillo, J. M., Borchers, J., Chaney, J., Fisher, H., Fox, S., Haxeltine, A., Janetos, A., Kicklighter, D. W., Kittel, T. G. F., Mcguire, A. D., Mckeown, R., Neilson, R., Nemani, R., Ojima, D. S., Painter, T., Pan, Y., Parton, W. J., Pierce, L., Pitelka, L., Prentice, C., Rizzo, B., Rosenbloom, N. A., Running, S., Schimel, D. S., Sitch, S., Smith, T., and Woodward, I. (1995). Vegetation ecosystem modeling and analysis project: Comparing biogeography and biogeochemistry models in a continental-scale study of terrestrial ecosystem responses to climate change and CO_2 doubling. *Glob. Biogeochem. Cycle* **9**, 407–437.

Melillo, J. M., Prentice, I. C., Farquhar, G. D., Schuze, E.-D., Sala, O.E., Bartlein, P. J., Bazzaz, F. A., Bradshaw, R. H. W., Clark, J. S., Claussen, M. C., Collatz, G. J., Coughenhour, M. B., Field, C. B., Foley, J. A., Friend, A. D., Huntley, B., Körner, C. H., Kurz, W., Lloyd, J., Leemans, R., Martin, P. H., McGuire, A. D., McNaughton, K. G., Neilson, R. P., Oechel, W. C., Overpeck, J. T., Parton, W. A., Pitelka, L. F., Rind, D., Running, S. W., Schimel, D. S., Smith, T. M., Webb, T., III, and Whitlock, C. (1996). Terrestrial biotic responses to environmental change and feedbacks to climate. *In* "Climate Change 1995. The Science of Climate Change" (J. T. Houghton, L. G. M. Filho, B. A. Callander, N. Harris, A. Kattenberg, and K. Maskell, eds.), pp. 445–481. Cambridge Univ. Press, Cambridge, UK.

Mitchell, J. F. B. (1990). Greenhouse warming: Is the mid-Holocene a good analogue? *J. Climate* **3**, 1177–1192.

Morita, T., Matsuoka, Y., Kainuma, M., Kai, K., Harasawa, H., and Dong-Kun, L. (1994). "Asian-Pacific Integrated Model to Assess Policy Options for Stabilizing Global Climate," AIM Rep. 1.0, July 1994. National Institute for Environmental Studies, Tsukuba, Japan.

Morris, D. W. (1995). Earth's peeling veneer of life. *Nature* **373**, 25.

Nordhaus, W. D. (1992). An optimal transition path for controlling greenhouse gases. *Science* **258**, 1315–1319.

Parry, M. L., Carter, T. R., and Konijn, N. T. (eds.) (1988). "The Impact of Climatic Variations on Agriculture. Volume 1: Assessments in Cool Temperate and Cold Regions." Kluwer, Dordrecht.

Prentice, I. C., Cramer, W., Harrison, S. P., Leemans, R., Monserud, R. A., and Solomon, A. M. (1992). A global biome model based on plant physiology and dominance, soil properties and climate. *J. Biogeogr.* **19**, 117–134.

Prentice, I. C., Sykes, M. T., and Cramer, W. (1993). A simulation model for the transient effects of climate change on forest landscapes. *Ecol. Model.* **65**, 51–70.

Reilly, J., Baethgen, W., Chege, F. E., van de geijn, S., Erda, L., Iglesias, A., Kenny, G., Patterson., D., Rogasik, J., Rötter, R., Sombroek, W., Westbrook, J., Bachelet, D., Brklacich, M., Dämmgen, U., Howden, M., Joyce, R. J. V., Lingren, P. D., Schimmelpfennig, D., Singh, U., Sirotenko, O., and Wheaten, E. (1996). Agriculture in a changing climate: Impacts and adaptation. *In* "Climate Change 1995. Impacts, Adaptations and Mitigation of Climate Change: Scientific–Technical Analysis" (R. T. Watson, M. C. Zinyowera, and R. H. Moss, eds.), pp. 427–467. Cambridge Univ. Press, Cambridge, UK.

Rosenberg, N. J., Crosson, P. R., Frederick, K. D., Easterling, W. E., III, McKenny, M. S., Bowes, M. D., Sedjo, R. A., Darmstadter, J., Katz, L. A., and Lemon, K. M. (1994). Paper 1. The MINK methodology: Background and baseline. *Climatic Change* **24**, 7–22.

Rosenzweig, C. (1985). Potential CO_2-induced climatic effects on North American wheat-producing regions. *Climatic Change* **7**, 367–389.

Rosenzweig, C., and Parry, M. L. (1994). Potential impact of climate change on world food supply. *Nature* **367**, 133–138.

Rotmans, J., de Boois, H., and Swart, R. J. (1990). An integrated model for the assessment of the greenhouse effect: The Dutch approach. *Climatic Change* **16**, 331–356.

Smith, T. M., Leemans, R., and Shugart, H.H. (1992). Sensitivity of terrestrial carbon storage to CO_2 induced climate change: Comparison of four scenarios based on general circulation models. *Climatic Change* **21**, 367–384.

Smith, T. M., Leemans, R., and Shugart, H. H. (eds.) (1994). "The Application of Patch Models of Vegetation Dynamics to Global Change Issues." Kluwer, Dordrecht.

Solomon, A. M. (1986). Transient responses of forests to CO_2-induced climate change: Simulation modeling in eastern North America. *Oecologia (Berlin)* **68**, 567–579.

Solomon, A. M., Prentice, I. C., Leemans, R., and Cramer, W.P. (1993). The interaction of climate and land use in future terrestrial carbon storage and release. *Water Air Soil Pollut.* **70**, 595–614.

Steffen, W. L., Chapin, F. S., and Sala, O. E. (1996). Global change and ecological complexity: An international research agenda. *Tr. Ecol. Evol.* **11**, 186.

Turner, B. L., II, Clark, W. C., Kates, R. W., Richards, J. F., Mathews, J. T., and Meyer, W.B. (eds.) (1990a). "The Earth as Transformed by Human Action: Global and Regional Changes in the Biosphere over the Past 300 Years." Cambridge Univ. Press, Cambridge, UK.

Turner, B. L., II, Kasperson, R. E., Meyer, W. B., Dow, K. M., Golding, D., Kasperson, J. X., Mitchell, R. C., and Ratick, S.J. (1990b). Two types of global environmental change: Definitional and spatial-scale issues in their human dimensions. *Global Environ. Change* **1**, 14–22.

Turner, B. L., Skole, D. L., Sanderson, S., Fischer, G., Fresco, L., and Leemans, R. (1995). "Land Use and Land-cover Change: Science/Research Plan," IGBP Rep., November 1995. IGBP, Stockholm.

van Minnen, J. G., Klein Goldewijk, C. G. M., Leemans, R., and Kreileman, G. J. J. (1996a). "Documentation of a Geographically Explicit Dynamic Carbon Cycle Model," RIVM Rep., January 1996. Bilthoven.

van Minnen, J. G., Klein Goldewijk, K., and Leemans, R. (1996b). The importance of feedback processes and vegetation transition in the terrestrial carbon cycle. *J. Biogeogr.* **22**, 805–814.

Viner, D., Hulme, M., and Raper, S.C.B. (1995). Climate change scenarios for the assessments of the climate change on regional ecosystems. *J. Therm. Biol.* **20**, 175–190.

Vitousek, P. M., Ehrlich, P. R., Ehrlich, A. H., and Matson, P.A. (1986). Human appropriation of the products of photosynthesis. *BioScience* **36**, 368–373.

Watson, R. T., Zinyowera, M. C., and Moss, R. H. (eds.) (1996). "Climate Change 1995. Impacts, Adaptations and Mitigation of Climate Change: Scientific–Technical Analysis." Cambridge Univ. Press, Cambridge, UK.

Weyant, J., Davidson, O., Dowlabathi, H., Edmonds, J., Grubb, M., Parson, E. A., Richels, R., Rotmans, J., Shukla, P. R., Tol, R. S. J., Cline, W., and Fankhauser, S. (1996). Integrated assessment of climate change: An overview and comparison of approaches and results. *In* "Climate Change 1995. Economic and Social Dimensions of Climate Change" (J. P. Bruce, H. Lee and E. F. Haites, eds.), pp. 367–396. Cambridge Univ. Press, Cambridge, UK.

Wigley, T. M. L., Richels, R., and Edmonds, J. A. (1996). Economic and environmental choices in the stabilization of atmospheric CO_2 concentrations. *Nature* **379**, 240–243.

Zuidema, G., van den Born, G. J., Alcamo, J., and Kreileman, G. J. J. (1994). Simulating changes in global land cover as affected by economic and climatic factors. *Water Air Soil Pollut.* **76**, 163–198.

Index

Physiological Ecology
A Series of Monographs, Texts, and Treatises

M. L. CODY (Ed.). Habitat Selection in Birds, 1985

R. J. HAYNES, K. C. CAMERON, K. M. GOH, and R. R. SHER-
LOCK (Eds.). Mineral Nitrogen in the Plant–Soil System, 1986

T. T. KOZLOWSKI, P. J. KRAMER, and S. G. PALLARDY. The
Physiological Ecology of Woody Plants, 1991

H. A. MOONEY, W. E. WINNER, and E. J. PELL (Eds.). Response
of Plants to Multiple Stresses, 1991

F. S. CHAPIN III, R. L. JEFFERIES, J. F. REYNOLDS, G. R.
SHAVER, and J. SVOBODA (Eds.). Arctic Ecosystems in a
Changing Climate: An Ecophysiological Perspective, 1991

T. D. SHARKEY, E. A. HOLLAND, and H. A. MOONEY (Eds.).
Trace Gas Emissions by Plants, 1991

U. SEELIGER (Ed.). Coastal Plant Communities of Latin
America, 1992

JAMES R. EHLERINGER and CHRISTOPHER B. FIELD (Eds.).
Scaling Physiological Processes: Leaf to Globe, 1993

JAMES R. EHLERINGER, ANTHONY E. HALL, and GRAHAM D.
FARQUHAR (Eds.). Stable Isotopes and Plant Carbon–Water
Relations, 1993

E.-D. SCHULZE (Ed.). Flux Control in Biological Systems, 1993

MARTYN M. CALDWELL and ROBERT W. PEARCY (Eds.). Exploitation of Environmental Heterogeneity by Plants: Ecophysiological Processes Above- and Belowground, 1994

WILLIAM K. SMITH and THOMAS M. HINCKLEY (Eds.). Resource Physiology of Conifers: Acquisition, Allocation, and Utilization, 1995

WILLIAM K. SMITH and THOMAS M. HINCKLEY (Eds.). Ecophysiology of Coniferous Forests, 1995

MARGARET D. LOWMAN and NALINI M. NADKARNI (Eds.).
Forest Canopies, 1995

BARBARA L. GARTNER (Ed.). Plant Stems: Physiology and Functional Morphology, 1995

GEORGE W. KOCH and HAROLD A. MOONEY (Eds.). Carbon
Dioxide and Terrestrial Ecosystems, 1996

CHRISTIAN KÖRNER and FAKHRI A. BAZZAZ (Eds.). Carbon
Dioxide, Populations, and Communities, 1996

THEODORE T. KOZLOWSKI and STEPHEN G. PALLARDY.
Growth Control in Woody Plants, 1997

J. J. LANDSBERG and S. T. GOWER. Applications of Physiological Ecology to Forest Management, 1997

FAKHRI A. BAZZAZ and JOHN GRACE (Eds.). Plant Resource Allocation, 1997

LOUISE E. JACKSON (Ed.). Ecology in Agriculture, 1997